MATHEMATICAL STATISTICS

PRENTICE-HALL INTERNATIONAL, INC., *London*
PRENTICE-HALL OF AUSTRALIA, PTY. LTD., *Sydney*
PRENTICE-HALL OF CANADA, LTD., *Toronto*
PRENTICE-HALL OF INDIA PRIVATE LTD., *New Delhi*
PRENTICE-HALL OF JAPAN, INC., *Tokyo*

MATHEMATICAL STATISTICS

Second Edition

JOHN E. FREUND

Professor of Mathematics
Arizona State University

PRENTICE-HALL, INC., Englewood Cliffs, New Jersey

Current printing (last digit):

10 9

13-562223-9

Library of Congress Catalog Card Number 71-149435

PRINTED IN THE UNITED STATES OF AMERICA

PREFACE

Like the first edition, this book is still designed for an introductory two-semester or three-quarter course in mathematical statistics with the prerequisite of a standard undergraduate course in calculus. Most of the differences between this edition and the first one reflect the extensive changes which have taken place in recent years in statistical thinking and in the teaching of statistics. Primarily, there is an entirely new chapter on Decision Theory, Chapter 8, which serves to lead into the more traditional material on estimation and hypothesis testing, Chapters 9 and 10, in a modern unified way.

Although the emphasis of the book has remained on the mathematics of statistics, it is hoped that the inclusion of more applied exercises and the separation of theoretical exercises and applied exercises will make it easier to adapt this text to individual preferences concerning a "sound balance" between theory and applications. There have been no changes in notation, and it should be pointed out that boldface type is again used for random variables in order to distinguish between functions and the values which they assume.

The author would like to express his appreciation for the many constructive comments which he received from his colleagues on the first edition; practically all of them have been incorporated into this revision. In particular, the author is indebted to Dr. A. Romano, whose critical comments have contributed greatly to this new edition.

Finally, the author would like to express his appreciation and indebtedness to the McGraw-Hill Book Company for their permission to reproduce in Tables I and II material from their *Handbook of Probability and Statistics with Tables;* to the Literary Executor of the late Sir Ronald Fisher, F.R.S., Cambridge, and to Oliver and Boyd, Ltd., Edinburgh, for their permission to reproduce the material in Table IV from their book *Statistical Methods for Research Workers;* and to Professor E. S. Pearson and the *Biometrika* trustees for their permission to reproduce the material in Tables V and VI.

Scottsdale, Arizona JOHN E. FREUND

CONTENTS

3

PROBABILITY FUNCTIONS, 68

4

PROBABILITY DENSITIES, 101

5

MATHEMATICAL EXPECTATION, 141

9

ESTIMATION, 254

10

HYPOTHESIS TESTING: THEORY, 289

11

HYPOTHESIS TESTING: APPLICATIONS, 314

MATHEMATICAL STATISTICS

1

INTRODUCTION

1.1 HISTORICAL BACKGROUND

In recent years, the growth of statistics has made itself felt in almost every phase of human activity. Statistics no longer consists merely of the collection of data and their presentation in charts and tables—it is now considered to encompass not only the science of basing inferences on observed data, but the entire problem of making decisions in the face of uncertainty. Needless to say, perhaps, this covers enormous ground since uncertainties are met when we flip a coin, when a doctor experiments with a new drug, when an actuary determines life insurance premiums, when a quality control engineer inspects manufactured products, when a teacher rates the abilities of different students, when an economist forecasts business cycles, when a newspaper predicts an election, and so forth. It would be presumptious to say that statistics, in its present state of development, can handle *all* situations involving uncertainties, but new techniques are constantly being developed and modern statistics can, at least, provide the framework for looking at these situations in a logical and systematic fashion. In other words, probability and statistics provide the *models* that are needed to study situations involving uncertainties just as calculus provides the *models* that are needed to describe, say, the concepts of Newtonian physics.

Historically speaking, the origin of probability theory dates back to the seventeenth century. It seems that the Chevalier de Méré, claimed to have been an ardent gambler, was baffled by some questions concerning a game of chance—specifically, he wanted to know how to divide the stakes of two players who fail to complete a game in which the winner has to win three matches out of five. He consulted the French mathematician Blaise Pascal (1623–1662), who in turn wrote about this matter to Pierre Fermat (1601–1665); it is this correspondence which is generally considered the origin of modern probability theory.

The eighteenth century saw a rapid growth of the mathematics of probability as it applies to games of chance, but it was not until the work of Karl Gauss (1777–1855) and Pierre Laplace (1749–1827) that this theory found applications in other fields. Noting that the theory developed for "heads or tails," "red or black," etc., in games of chance applied also to situations where the outcomes are "boy or girl," "life or death," "pass or fail," etc., scholars began to apply probability theory to actuarial mathematics and some aspects of the social sciences. Later, probability and statistics were introduced into physics by L. Boltzmann, J. Gibbs, and J. Maxwell, and in this century they have found applications in all phases of human endeavor which in some way involve an element of uncertainty or risk. The names which are connected most prominently with the growth of mathematical statistics in the first half of this century are those of R. A. Fisher, J. Neyman, E. S. Pearson, and A. Wald. More recently, the work of R. Schlaifer, L. J. Savage, and others, has given impetus to statistical theories based essentially on methods which date back to the eighteenth century English clergyman Thomas Bayes.

Since probability and statistics can be taught at various levels of mathematical refinement, let us point out briefly that the mathematical background expected of the reader is a basic course in differential and integral calculus, including an introduction to partial differentiation, multiple integrals, and infinite series. So far as the arrangement of the material is concerned, Chapters 2 through 7 are devoted primarily to the mathematical concepts and techniques which are required for the study of the statistical methods treated in Chapters 8 through 14. Chapter 2 provides a formal introduction to probability, while the other chapters of the first half of the book deal with what might be called *basic distribution theory*. Chapters 8 through 14 contain an introduction to the most widely used methods of statistics, with emphasis on their theoretical foundation; some of the modern concepts of *Bayesian statistics* are introduced in Chapters 8 and 9. The remainder of this chapter is devoted to certain mathematical preliminaries, with which the reader may well be familiar (at least in part) from high school algebra and even earlier studies of mathematics.

1.2 MATHEMATICAL PRELIMINARY: SETS

Statisticians refer to any process of observation or measurement as an *experiment*. In this sense, an experiment may consist of the simple process of checking whether a switch is turned on or off, it may consist of counting the imperfections in a piece of cloth, . . . , and it may consist of the very complicated process of determining the mass of an electron. The result which one obtains from an experiment (a simple "yes" or "no" answer, an instrument reading, the end product of complicated calculations, . . .) is called an *outcome*, and it is generally desirable to represent the various possible outcomes of experiments by means of points. This has the advantage that we can discuss the outcomes of experiments mathematically, without having to go through lengthy verbalizations.

For instance, if we consider someone who shoots at a target, the outcomes can be represented by the two points of Figure 1.1, where the point

FIGURE 1.1 Outcomes for one shot.

labeled 0 represents a miss (0 hits) and the point labeled 1 represents a hit (1 hit). Similarly, the outcomes of one roll of a die can be represented by the six points of Figure 1.2. To consider a slightly more complicated

FIGURE 1.2 Points rolled with one die.

example, the outcomes of an experiment which consists of two shots at a target can be represented by the three points of Figure 1.3 if we are interested only in the total number of hits, or by the four points of Figure 1.4 if we are interested also in which shot was a hit and which shot was a miss. In Figure 1.4, the numbers 0 and 1 again denote a miss and a hit, and the two coordinates represent the respective shots. Note that in Figure 1.4 the event of getting one hit and one miss is represented by the *set* of two points inside the dotted line.

FIGURE 1.3 Outcomes for two shots.

Similarly, in an experiment in which we roll two dice (say, one red and one green), the various outcomes can be represented by eleven points labeled 2, 3, 4, . . . , 12 provided we are interested only in the total rolled with the pair of dice, or by the thirty-six points of Figure 1.5, which tell us also what happened to each die. Note that in Figure 1.5 the event of rolling a 7 is represented by the set of six points inside the dotted line, and the event of rolling 2, 3, or 12 is represented by the *set* of four points circled in that diagram.

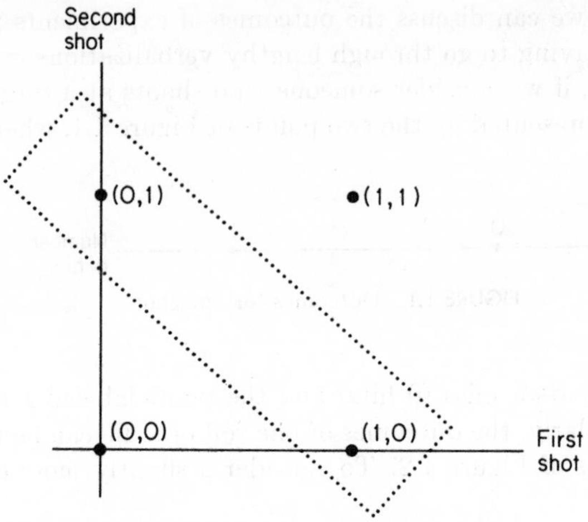

FIGURE 1.4 Outcomes for two shots.

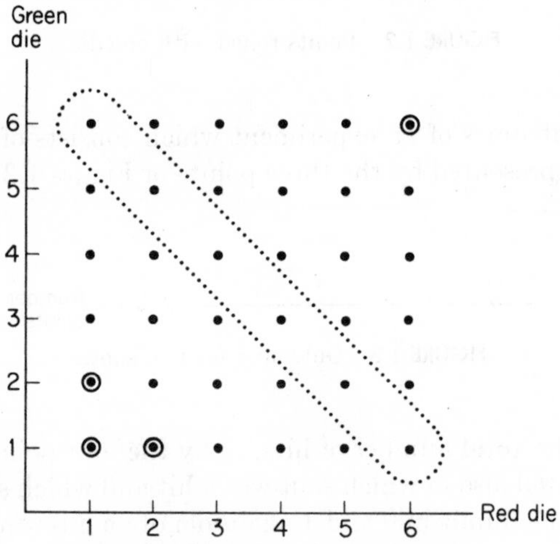

FIGURE 1.5 Outcomes for two dice.

Since points and sets of points thus play an important role in describing the results of experiments, some of the basic concepts of set theory will be introduced in the remainder of this section. However, even though the proofs of many theorems in probability and statistics require some knowledge of the *Algebra of Sets* which we shall discuss in Section 1.2.4, it will be seen that many of them can also be justified quite readily by means of simple diagrams. Thus, some knowledge of the algebra of sets (or *Boolean algebra* as it is also called) is desirable for the study of probability and statistics, but it is not absolutely essential.

Throughout this discussion we have used the term "set" without giving it a formal definition; this was done with the tacit understanding that a set is a collection, group, or class of points or other kinds of objects. Indeed, mathematicians usually leave the term "set" undefined, subject to the qualification that sets are "mathematical objects" which must obey the postulates set down in Section 1.2.4.

The objects that belong to a set, points in all of the preceding examples, are usually referred to as its *members* or its *elements*, and sets are often specified by actually listing (in braces) the individual members, their names, or their descriptions. Thus, the set which consists of the different outcomes of a roll of a die may be indicated as $\{1, 2, 3, 4, 5, 6\}$; similarly, $\{(1, 0), (0, 1)\}$ is the set of outcomes inside the dotted line of Figure 1.4, namely, the set of outcomes representing one hit and one miss. It should also be noted that when the members of a set are thus listed, their order does not matter; $\{0, 1, 2\}$ and $\{2, 0, 1\}$, for instance, represent the *same* set consisting of the possible number of hits in two shots at a target.

Instead of listing the elements, which is often impracticable or even impossible, sets can also be specified by giving a rule according to which one can decide whether any given object does or does not belong to a set. Thus, we can speak of the set of *all college students* without having to list them all, by specifying that a person belongs to this set if and only if he (or she) is enrolled in an institution of higher learning. Similarly, we can speak of the set of *odd positive integers* by specifying that its members must be numbers of the form $2k + 1$, where k is a positive integer or zero.

1.2.1 Discrete Sample Spaces

A set whose elements represent all possible outcomes of an experiment is called a *sample space* for the experiment and it will be denoted by the letter S. [In other applications of set theory, this kind of set (namely, a set which consists of all elements relevant to a given situation) is referred to as a "universal set" or as a "universe of discourse," and it is generally denoted by the letter U or I.]

The fact that the sample space which represents a given experiment need not be unique is illustrated in Figures 1.3 and 1.4. Each of these sample spaces represents the outcomes of an experiment consisting of two shots at a target, the difference being in what we mean by "outcome." *Generally speaking, it is desirable to use sample spaces whose elements cannot be "subdivided" into more primitive or more elementary kinds of outcomes; that is, an element of a sample space should not represent two or more outcomes which are distinguishable in some fashion.* This was the case in Figure 1.4 but not in Figure 1.3; it was also the case in Figure 1.5 but would not have been if we had used eleven points to represent the total rolled with the two dice (as was suggested on page 3). By following this rule one can avoid many of the difficulties which hampered the early development of the theory of probability.

Sample spaces are usually classified according to the *number of elements* which they contain and also according to the *dimension* of the geometrical configuration in which the points are displayed. Thus, the sample space of Figure 1.3 consists of three points arranged in a one-dimensional configuration, while the sample space of Figure 1.4 consists of four points arranged in a two-dimensional configuration. If a sample

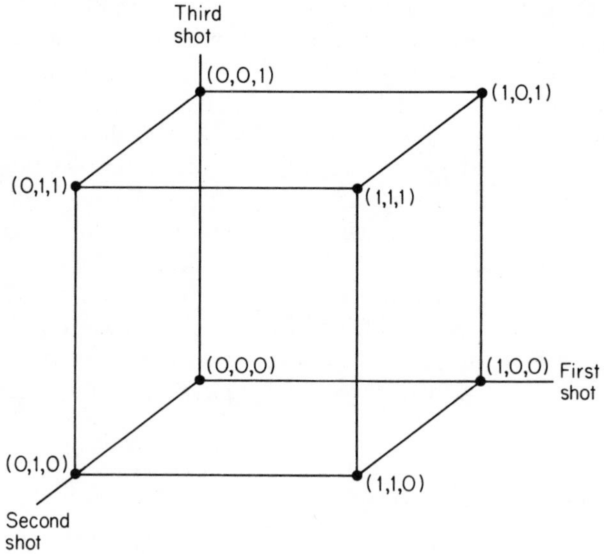

FIGURE 1.6 Outcomes for three shots.

space contains a finite number of elements or a countable infinity, namely, as many as there are positive integers, it is said to be *discrete*. All of the sample spaces which we have discussed so far have been discrete, and so is the three-dimensional sample space of Figure 1.6, whose eight elements (points) represent the possible outcomes of an experiment consisting of three shots at a target. (The three coordinates represent the successive shots, and 0 and 1 again stand for miss and hit.) If a coin is flipped until *head* appears for the first time, this could happen on the first flip, the second flip, the third flip, the fourth flip, . . ., and there are infinitely many possibilities. However, there are only as many possibilities as there are positive integers, and the sample space for this experiment is also discrete.

If the elements (points) of a sample space constitute a continuum, say, all the points on a line, all the points on a line segment, or all the points in a plane, the sample space is said to be *continuous*. Continuous sample spaces arise whenever the outcomes of experiments are measurements of physical properties such as temperature, speed, pressure, length, . . ., which are measured on continuous scales. The remainder of this chapter as well as Chapters 2 and 3 will be devoted to problems involving sample spaces that are discrete; the continuous case will be introduced in Chapter 4.

1.2.2 Subsets and Events

Set A is called a *subset* of set B *if and only if each element of A is also an element of B*. Thus, in Figure 1.5 the six points inside the dotted line form a subset of the sample space, and so do the four points circled in that diagram. According to this definition, *each set is a subset of itself;* furthermore, *the empty set $\varnothing$, the set which has no elements, is considered to be a subset of every set.* Thus, the set of all eight points of Figure 1.6 is a subset of the sample space (for three shots), and the set of all points representing, say, six hits in three shots, obviously an empty set, is also a subset of this sample space in a rather trivial way.

Using the terminology introduced in this and the preceding section, let us now state formally that *all events with which we are concerned in probability are subsets of appropriate sample spaces.* In other words, "event" is the non-technical term and "subset of a sample space" or merely "subset" is the corresponding mathematical counterpart. For instance, in Figure 1.6 the subset which consists only of the point (0, 0, 0) represents the event of missing the target in each of three shots; the subset which consists of the points (1, 0, 0), (0, 1, 0), and (0, 0, 1) represents the event of hitting the target once and missing it twice in three shots; the subset $\{(0, 1, 1), (1, 0, 1), (1, 1, 0)\}$ represents the event of hitting the target twice and missing it once in three shots; and the subset $\{(1, 1, 1)\}$ represents the event of hitting the target in each of three shots. Of course, the entire sample space represents the (certain) event of getting either 0, 1, 2, or 3 hits in three shots at the target.

Similarly, if there are six applicants for a teaching position and R, S, T, U, V, and W stand for the selection of Mr. Reed, Mrs. Stone, Mr. Taylor, Mr. Upton, Miss Vaughn, or Mrs. Watson, then $\{R, T, U\}$ represents the event that the job will go to one of the three men, $\{S, W\}$ represents the event that the job will go to one of the two married women, and $\{V\}$ represents the event that the job will go to the single woman, Miss Vaughn.

(Although it is true that the elements of the sample space are represented by letters and not by points in this example, we could easily identify the different selections with six points, say, those of Figure 1.2.)

1.2.3 Operations on Sets

There are various ways in which sets can be used to form new sets. For instance, in the preceding example, if the selection of a man constitutes the set M, then the selection of one of the women constitutes the set which we call the *complement* of M and write as M'. Similarly, if L represents the event that we get 0 or 1 hits in three shots at a target, then L', the complement of L, represents the event that we get 2 or 3 hits. In general, *if set A belongs to the sample space S, then A' called the complement of A, is the set which is composed of all the elements of S that are not in A.* To give another illustration, suppose that among the three male job applicants Mr. Reed and Mr. Upton are married and Mr. Taylor is single. Then, if we let N represent the selection of one of the single applicants, we have $N = \{T, V\}$ and $N' = \{R, S, U, W\}$. As we have used the equal sign in the last sentence, $N = \{T, V\}$ is meant to imply that N and $\{T, V\}$ are two symbols for one and the same set. More generally, we say that *two sets A and B are equal, and we write $A = B$, if and only if every element of A is also an element of B and vice versa*, namely, if the "membership" of the two sets is the same.

New sets can also be formed by combining the elements of two sets in some fashion. For instance, *the intersection of two sets A and B, denoted $A \cap B$ (which also reads "A cap B" or simply "A and B"), is the set which consists of all the elements belonging to both A and B.* With reference to our example, we can thus write $M \cap N = \{T\}$ for the selection of a single man, $M \cap N' = \{R, U\}$ for the selection of a married man, $M' \cap N = \{V\}$ for the selection of a single woman, and $M' \cap N' = \{S, W\}$ for the selection of a married woman. Also, if K represents the event that we hit the target in each of three shots, while L still represents the event that we get 0 or 1 hits, then $K \cap L = \varnothing$; that is, K and L have no elements in common.

If two sets have no elements in common, we say that they are *disjoint*, and that the corresponding events are *mutually exclusive*. More generally, any number of sets $A_1, A_2, A_3, \ldots$, are said to be *disjoint* (and the events which they represent are said to be *mutually exclusive*) if and only if they have pairwise no elements in common. Thus, with reference to the example of the preceding paragraph, the sets $M \cap N$, $M \cap N'$, $M' \cap N$, and $M' \cap N'$ are disjoint and the selection of a single man, a married man, a single woman, and a married woman are mutually exclusive events. Another example of mutually exclusive events are those representing, respectively, 0, 1, 2, and 3 hits in three shots at a target; on the other hand, hitting the target *at least once* in three shots and hitting it *at most twice* are not mutually exclusive events.

When we form the intersection of two sets we do not literally "combine" their elements; instead, we take only those which are contained in both. If we actually want to combine the elements of two sets A and B

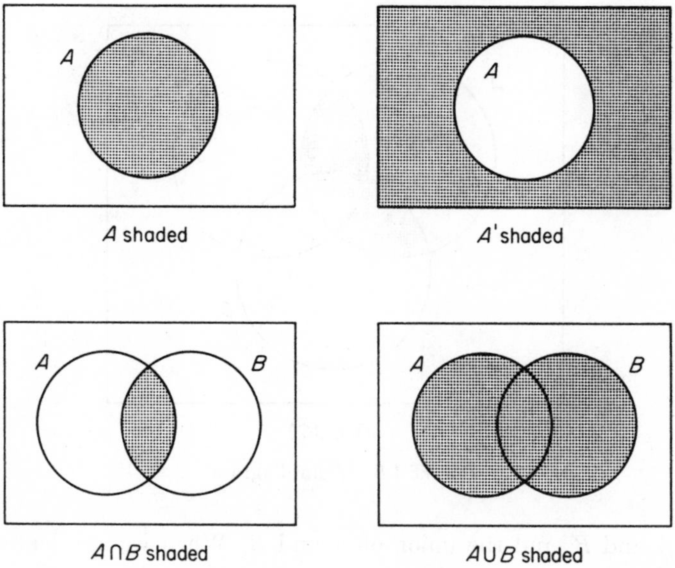

FIGURE 1.7 Venn diagrams.

(that is, take them all together), we form what is called their *union*, denoted $A \cup B$. With reference to the six job applicants, $M \cup N = \{R, T, U, V\}$ represents the event that the person who gets the job is either a man or single, while $M' \cup N' = \{R, S, U, V, W\}$ represents the event that the person who gets the job is either a woman or married. Note that $M \cup N$ includes Mr. Taylor, the man who is single, and that $M' \cup N'$ includes Mrs. Stone and Mrs. Watson, the two women who are married. *Formally, the union of two sets A and B, denoted $A \cup B$, is the set which consists of all elements belonging either to A or to B, including those which belong to both.* It is customary to read $A \cup B$ as "the union of A and B," "A cup B," or simply as "A or B," where the "or" is what logicians call the "inclusive or" because $A \cup B$ also includes the elements which are common to A and B. To give one more example, if the sets K and L are defined as on page 8, then $K \cup L$ represents the event that we will get 0, 1, or 3 hits in three shots at the given target.

Sample spaces, subsets, and events, are often depicted by means of *Venn diagrams* like those of Figures 1.7 and 1.8.* In each case, the sample space S is represented by a rectangle, while subsets (or events) are represented by regions within the rectangles, usually by circles or parts of circles. Thus, the shaded regions of the four Venn diagrams of Figure 1.7 represent, respectively, the set A, the complement of A, the intersec-

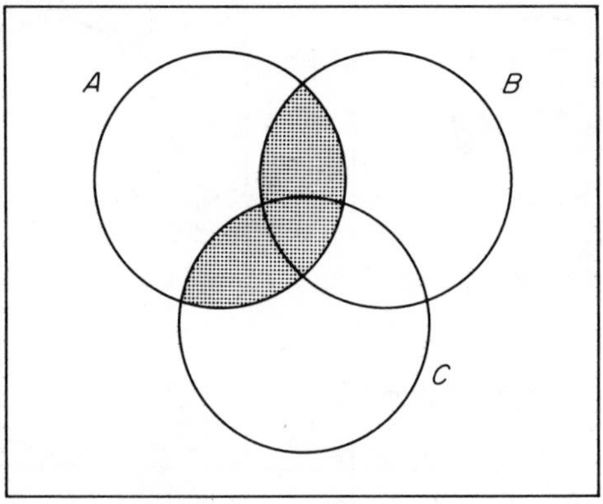

$A \cap (B \cup C)$

FIGURE 1.8 Venn diagram.

tion of A and B, and the union of A and B. When we are dealing with three sets (subsets or events) it is customary to draw the respective circles as in Figure 1.8. Note that the shaded region of this diagram represents the set $A \cap (B \cup C)$—it contains everything in A that also belongs to either B or C.

1.2.4 The Algebra of Sets

As we have indicated on page 5, sets are "mathematical objects" which behave according to certain postulates. These postulates are listed below, and it should be understood that the sets A, B, and C to which we refer are all subsets of some sample space S:

* These diagrams are named after the nineteenth century English logician John Venn.

POSTULATE 1 (*closure laws*):

For each pair of sets A and B there exists a unique set $A \cup B$ *and a unique set* $A \cap B$ *in the sample space S.*

POSTULATE 2 (*commutative laws*):

$A \cup B = B \cup A$ *and* $A \cap B = B \cap A$

POSTULATE 3 (*associative laws*):

$(A \cup B) \cup C = A \cup (B \cup C)$

and

$(A \cap B) \cap C = A \cap (B \cap C)$

POSTULATE 4 (*distributive laws*):

$A \cap (B \cup C) = (A \cap B) \cup (A \cap C)$

and

$A \cup (B \cap C) = (A \cup B) \cap (A \cup C)$

POSTULATE 5 (*identity laws*):

$A \cap S = A$ *for each set A, and there exists a unique set* $\varnothing$ *such that* $A \cup \varnothing = A$ *for each set A.*

POSTULATE 6 (*complementation law*):

For each set A there exists a unique set A′ such that $A \cap A' = \varnothing$ *and* $A \cup A' = S$.

Note that the fifth postulate defines, in fact, what we mean by the empty set $\varnothing$, and that the sixth postulate defines what we mean by the complement A' of the set A.

It is easy to verify that all these postulates hold if we interpret sets intuitively as collections of points or other kinds of objects. This can be done either by checking whether the sets given on both sides of each equation contain the same elements, or by verifying that they are represented by identical regions of Venn diagrams. For instance, to verify the second distributive law, we have only to observe that the region ruled *one way or the other or both ways* in Figure 1.9, representing $A \cup (B \cap C)$, is identical with the region ruled *both ways* in Figure 1.10, representing $(A \cup B) \cap (A \cup C)$. In Figure 1.9, set A is ruled one way, $B \cap C$ is ruled the other way, and their union, $A \cup (B \cap C)$, is given by the region ruled one way or the other or both ways. In Figure 1.10, $A \cup B$ is ruled

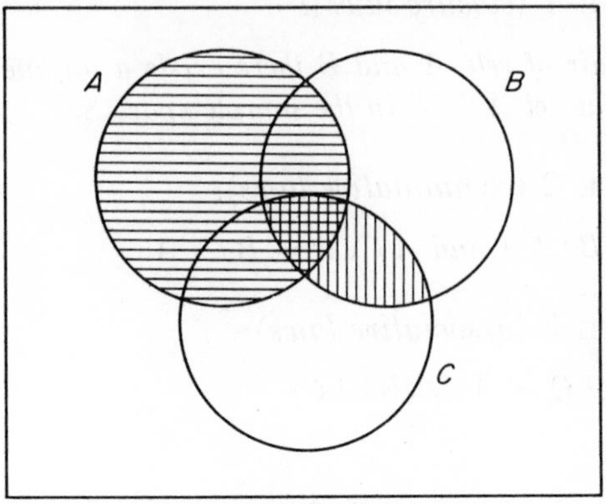

$A \cup (B \cap C)$

FIGURE 1.9 Venn diagram.

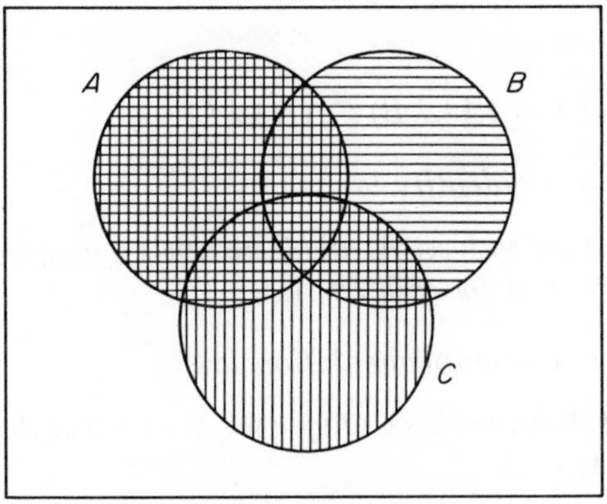

$(A \cup B) \cap (A \cup C)$

FIGURE 1.10 Venn diagram.

one way, $A \cup C$ is ruled the other way, and their intersection $(A \cup B) \cap (A \cup C)$ is given by the region ruled both ways.

By using the basic postulates, we can give algebraic proofs of many further rules (theorems) about sets; for instance, to prove that $A \cap A = A$ for each set A, we can argue that

	Justification
$A = A \cap S$	Postulate 5
$= A \cap (A \cup A')$	Postulate 6
$= (A \cap A) \cup (A \cap A')$	Postulate 4
$= (A \cap A) \cup \varnothing$	Postulate 6
$= A \cap A$	Postulate 5

It is difficult to say *why* we chose this particular sequence of steps; proofs like this are usually tricky and there is no set rule which always tells us how to proceed. Of course, A and $A \cap A$ are represented by the same region of a Venn diagram and, hence, they are equal, but this argument constitutes a *verification* rather than a formal proof.

The following are some other theorems about sets which can be derived from the basic postulates:

$$A \cup A = A \text{ for each set } A$$

$$S' = \varnothing \text{ and } \varnothing' = S$$

$$(A')' = A \text{ for each set } A$$

$$A \cup S = S \text{ and } A \cap \varnothing = \varnothing \text{ for each set } A$$

The first of these is similar to the rule which we have just proved, the next two tell us that $\varnothing$ is the complement of S and vice versa, the one after that tells us that the *complement of the complement* of any set is the set, itself, and the last two are very similar in nature to Postulate 5. Several other important theorems about sets will be given in the exercises that follow. (So far as the algebraic proofs of these theorems are concerned, note that it is always permissible to justify steps by citing not only the postulates but also theorems which have already been proved.)

1.2.5 Set Functions

Another important concept in the theory of sets is that of a *set function*, namely, a function which assigns numbers to the different subsets of a set. A simple example of such a function is the one which assigns to each subset A of a set the number of elements in A, written $N(A)$. To give an example, let us consider the Venn diagram of Figure 1.11, which shows how 80 college students are classified according to whether or not they are freshmen (F or F') and according to whether or not they live at home (H or H'). Symbolically, we can write the given information as $N(F \cap H) = 22$, $N(F \cap H') = 28$, $N(F' \cap H) = 14$, and $N(F' \cap H') = 16$, and by using these numbers we can easily determine what number must be assigned to any other subset (of the set described by the Venn diagram). For instance, to obtain $N(F)$, the number of freshmen in this group, we have only to add the number of freshmen who live at home to the number of freshmen who do not live at home, and we get

$$N(F) = N(F \cap H) + N(F \cap H') = 22 + 28 = 50$$

Similarly, the number of students who live at home is

$$N(H) = N(F \cap H) + N(F' \cap H) = 22 + 14 = 36$$

and it will be left to the reader to verify that $N(F') = 30$ and $N(H') = 44$.

The set function used in this example has the following important properties: (1) *the numbers which it assigns to the various subsets are all positive or zero,* and (2) *if two subsets have no elements in common, the number assigned to their union equals the sum of the numbers assigned to the individual sets.* As we shall see in Exercises 13, 14, and 15 below, knowledge of these properties will come in handy in certain applications, but what is even more important, these properties are characteristic of all so-called *measure functions,* that is, functions which assign weights to objects, areas to geometrical regions, volumes to solids, and so on. We have mentioned all this because in Chapter 2 we shall define probability as a special kind of *measure* (degree of certainty) which we assign to events.

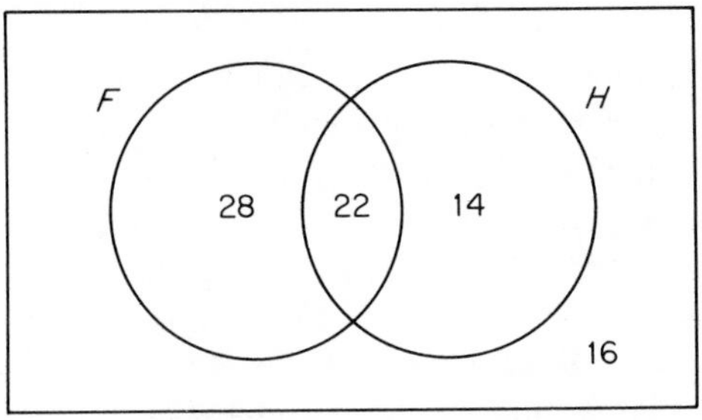

FIGURE 1.11 Venn diagram.

THEORETICAL EXERCISES

1. Using Venn diagrams verify the first distributive law of Postulate 4, namely, $A \cap (B \cup C) = (A \cap B) \cup (A \cap C)$.
2. Using Venn diagrams verify the two *de Morgan* laws
   (a) $(A \cap B)' = A' \cup B'$;
   (b) $(A \cup B)' = A' \cap B'$.
3. Using Venn diagrams verify that for any two sets A and B
   (a) $A \cup (A \cap B) = A$;
   (b) $A \cap (A \cup B) = A$;
   (c) $(A \cap B) \cup (A \cap B') = A$;
   (d) $A \cup B = (A \cap B) \cup (A \cap B') \cup (A' \cap B)$.

4. Indicate what postulate justifies each step of the following proof that $A \cup A = A$ for any set A:

$$\begin{aligned} A &= A \cup \varnothing \\ &= A \cup (A \cap A') \\ &= (A \cup A) \cap (A \cup A') \\ &= (A \cup A) \cap S \\ &= A \cup A \end{aligned}$$

5. Indicate what postulate justifies each step of the following proof that $A \cup S = S$ for any set A:

$$\begin{aligned} A \cup S &= (A \cup S) \cap S \\ &= (A \cup S) \cap (A \cup A') \\ &= A \cup (S \cap A') \\ &= A \cup (A' \cap S) \\ &= A \cup A' \\ &= S \end{aligned}$$

6. Use the postulates of the algebra of sets to prove that $A \cap \varnothing = \varnothing$ for any set A.
7. Prove that $A \cap (B_1 \cup B_2 \cup B_3) = (A \cap B_1) \cup (A \cap B_2) \cup (A \cap B_3)$ for any sets A, B_1, B_2, and B_3.
8. Prove that if $A \cap B = A$, then $A \cap B' = \varnothing$. What relationship between the sets A and B is expressed by either of these two equations?

APPLIED EXERCISES

9. A resort hotel has two station wagons, which it uses to shuttle its guests to and from the airport. If the larger of the two station wagons can carry 5 passengers and the smaller can carry 4 passengers, the point (0, 3) represents the fact that at a given moment the larger station wagon is empty while the smaller one has 3 passengers, the point (4, 2) represents the fact that at the given moment the larger station wagon has 4 passengers while the smaller one has 2 passengers, . . ., draw a figure showing the 30 points of the corresponding sample space. Also, if E stands for the event that at least one of the station wagons is empty, F stands for the event that together they carry 2, 4, or 6 passengers, and G stands for the event that each carries the same number of passengers, list the points of the sample space which belong to each of the following sets: (a) E; (b) F; (c) G;

(d) $E \cup F$; (e) $E \cap F$; (f) $F \cup G$; (g) $E \cup F'$; (h) $E \cap G'$; and (i) $F' \cap E'$.

10. An electronics firm plans to build a research laboratory in Southern California, and its management has to decide between sites in Los Angeles, San Diego, Long Beach, Pasadena, Santa Barbara, Anaheim, Santa Monica, and Westwood. If A represents the event that they will choose a site in San Diego or Santa Barbara, B represents the event that they will choose a site in San Diego or Long Beach, C represents the event that they will choose a site in Santa Barbara or Anaheim, and D represents the event that they will choose a site in Los Angeles or Santa Barbara, list the elements (site selections) of each of the following sets: (a) A'; (b) D'; (c) $C \cap D$; (d) $B \cap C$; (e) $B \cup C$; (f) $A \cup B$; (g) $C \cup D$; (h) $(B \cup C)'$; (i) $B' \cap C'$.

11. Among the eight cars which a dealer has in his showroom, Car 1 is new, has air-conditioning, power steering, and bucket seats, Car 2 is one year old, has air-conditioning, but neither power steering nor bucket seats, Car 3 is two years old, has air-conditioning and power steering, but no bucket seats, Car 4 is three years old, has air-conditioning, but neither power steering nor bucket seats, Car 5 is new, has no air-conditioning, no power steering, and no bucket seats, Car 6 is one year old, has power steering, but neither air-conditioning nor bucket seats, Car 7 is two years old, has no air-conditioning, no power steering, and no bucket seats, and Car 8 is three years old, has no air-conditioning, but power steering as well as bucket seats. If a customer buys one of these cars, and the event that he chooses a new car, for example, is represented by the set {Car 1, Car 5}, indicate similarly the sets which represent the events that
(a) he chooses a car without air-conditioning;
(b) he chooses a car without power steering;
(c) he chooses a car with bucket seats;
(d) he chooses a car that is either two or three years old.
Also state in words what kind of car he will choose, if his choice is given by
(e) the complement of the set of part (a);
(f) the union of the sets of parts (b) and (c);
(g) the intersection of the sets of parts (c) and (d);
(h) the intersection of the sets of parts (f) and (g).

12. If Mr. Jones buys one of the houses advertised for sale in a Detroit newspaper (on a given day), R is the event that the house has two or more baths, S is the event that it has three bedrooms, T is the event that it costs at least \$35,000, and U is the event that it is new, describe (in words) the elements of each of the following sets: (a) R';

(b) S'; (c) T'; (d) U'; (e) $R \cap S$; (f) $R \cap T$; (g) $S' \cap T$; (h) $T \cup U$; (i) $T' \cup U$; (j) $R \cup S$; (k) $R \cup T$; and (l) $T \cap U$.

13. In a group of 200 college students 133 are enrolled in a course in physics, 117 are enrolled in a course in mathematics, and 95 are enrolled in both. How many of these students are not enrolled in either course? (*Hint:* Draw a suitable Venn diagram and fill in the numbers associated with the various regions.

14. A market research organization claims that among 500 housewives interviewed, 310 regularly buy Product X, 265 regularly buy Product Y, 90 regularly buy both, and 65 buy neither on a regular basis. Using a Venn diagram and entering figures corresponding to the number of housewives belonging to the various regions, check whether the results of this survey should be questioned.

15. Among 120 visitors to Disneyland, 74 stayed for at least three hours, 86 spent at least $8.00, 64 went on the Matterhorn ride, 60 stayed for at least three hours and spent at least $8.00, 52 stayed for at least three hours and went on the Matterhorn ride, 54 spent at least $8.00 and went on the Matterhorn ride, and 48 stayed for at least three hours, spent at least $8.00, and went on the Matterhorn ride. Drawing a Venn diagram with three circles and filling in the numbers associated with the various regions, find
(a) how many of the visitors to Disneyland stayed for at least three hours, spent at least $8.00, but did not go on the Matterhorn ride;
(b) how many of the visitors to Disneyland went on the Matterhorn ride, but stayed less than three hours and spent less than $8.00;
(c) how many of the visitors to Disneyland stayed less than three hours, spent at least $8.00, but did not go on the Matterhorn ride.

1.3 MATHEMATICAL PRELIMINARY: COMBINATORIAL METHODS

Before we can study *what is probable* in a given situation we must know *what is possible*, and in the study of *what is possible*, there are essentially two kinds of problems: First there is the problem of describing or *listing all possible outcomes*, and then there is the problem of *determining the number of possible outcomes* (without actually constructing a complete list). The second kind of problem is especially important because there are many situations in which we really do not need a complete list, or in which a complete listing would be prohibitive. Thus, we shall study the second kind of problem in Sections 1.3.2 through 1.3.4, and touch upon the first kind of problem briefly in Section 1.3.1.

1.3.1 Tree Diagrams

Although the problem of listing all possible outcomes may seem straightforward and easy (at least, so long as their number is finite), this is not always the case. Suppose, for instance, that two used-car salesmen make a small bet each week as to who will be the first to make three sales. If we let X denote a sale made by the first salesman and Y a sale made by the second, $XXYX$ is one sequence which leads to the first salesman winning the bet, and so are the sequences $YXXX$ and $XYXYX$. On the other hand, YYY, $YXXYY$, and $XYYY$ are sequences which lead to the second salesman winning the bet. Continuing this way very carefully, we may be able to list all possible sequences of X's and Y's which lead to one of the two sales men winning the bet, and come up with the correct answer that there are altogether 20 possibilities.

To handle problems like this systematically, it is helpful to refer to a *tree diagram* like that of Figure 1.12. This diagram shows that for the first letter there are two possibilities (two branches) corresponding to which salesman makes the first sale; for the second letter there are two branches emanating from each of the two branches, and, similarly, for the third letter there are two branches emanating from each of the four branches. After that things begin to change: for the fourth letter there are two branches emanating from each of the six middle branches, but the first and last branches terminate for they represent XXX and YYY, namely, situations in which one of the salesmen wins. For the fifth letter there are two branches emanating from 6 of the remaining 12 branches while the others terminate, and it can thus be seen that there are altogether 20 different paths along the "branches" of the tree diagram of Figure 1.12. In other words, there are 20 possible outcomes to this "experiment." It can also be seen from the tree diagram that in *two* of the cases the issue was decided after 3 sales, in *six* of the cases the issue was decided after 4 sales, and that in *twelve* of the cases the issue was not decided until the fifth sale. Note that this argument applies also to a play-off in sports where a team must win 3 games out of 5; if a team must win 4 games out of 7, the total number of possible outcomes (arrangements of wins and losses) is 70, as the reader will be asked to show in Exercise 2 on page 29.

To consider another example in which a tree diagram can be of some aid, suppose that someone wants to go by bus, by train, or by plane on a week's vacation to one of the five East North Central states. What we would like to know is *the number of different ways in which this can be done.* Looking at the tree diagram of Figure 1.13, it is apparent that the selection can be made in 15 different ways corresponding to the 15 distinct paths along the branches of the tree. Starting at the top, the first path

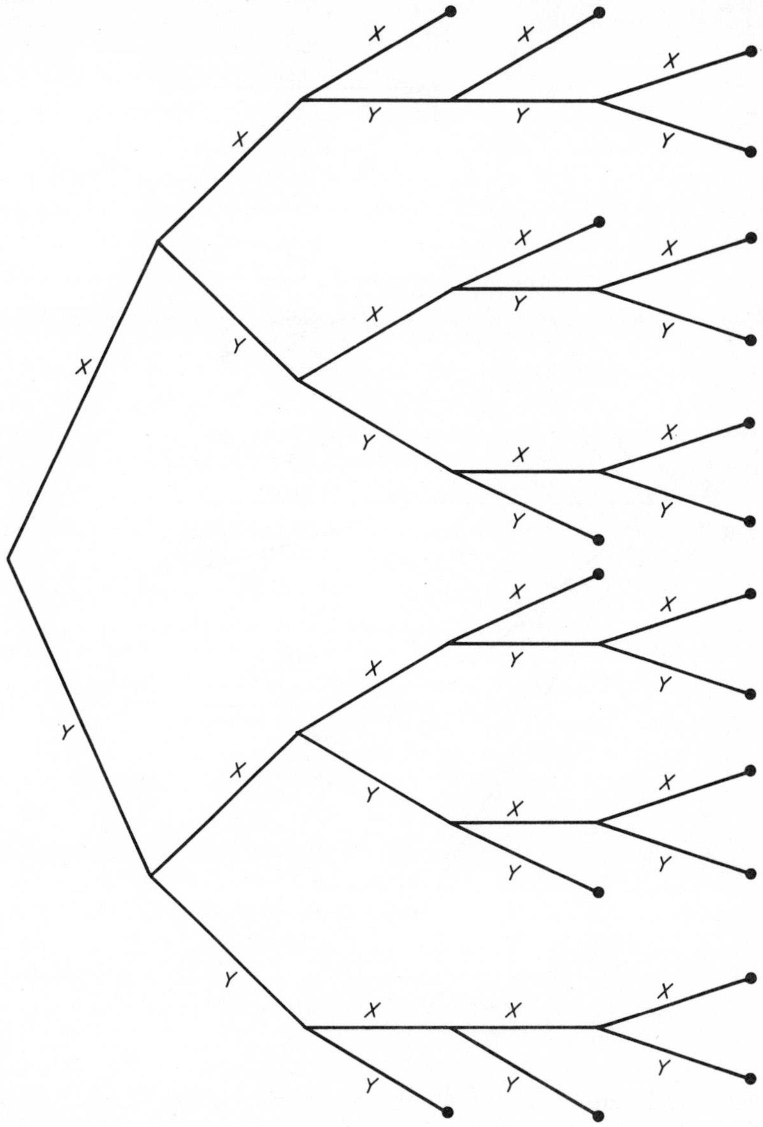

FIGURE 1.12 Tree diagram.

represents a trip to Ohio by bus, the second represents a trip to Ohio by train, . . . , the sixth represents a trip to Indiana by plane, . . . , and the last represents a trip to Wisconsin by plane.

Note that the answer we obtained in this example is the *product* of 5 and 3, namely, the *product* of the number of ways in which we can choose

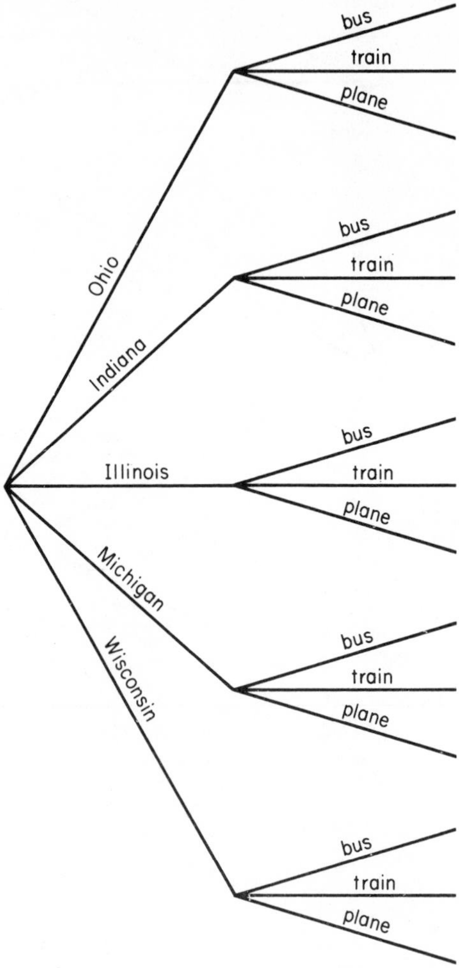

FIGURE 1.13 Tree diagram.

one of the five states and the number of ways in which we can choose one of the means of transportation. In fact, our example illustrates the following general rule:

THEOREM 1.1

If sets A_1 and A_2 have, respectively, n_1 and n_2 elements, there are $n_1 \cdot n_2$ different ways in which one can first select an element of A_1 and then an element of A_2.

To justify this theorem, we have only to construct an array showing all possible pairs or visualize a "tree" with n_1 branches emanating from the starting point and then n_2 branches emanating from the end points of each of the first n_1. (Actually, this theorem holds also when the order in which the selections are to be made is unspecified, but it is desirable to specify the order to make the theorem directly applicable to *repeated selections from one set.*) To give another example, if there are 12 girls and 9 boys in a choir, there are $12 \cdot 9 = 108$ ways in which they can give an award to one of the girls and an award to one of the boys.

Visualizing appropriate tree diagrams, one can easily generalize Theorem 1.1 so that it applies to selections from more than two sets. In that case we have:

THEOREM 1.2

If sets $A_1, A_2, \ldots, A_k$ have, respectively, $n_1, n_2, \ldots, n_k$ elements, there are $n_1 \cdot n_2 \cdots n_k$ different ways in which one can first select an element of A_1, then an element of $A_2, \ldots,$ and finally an element of A_k.

Thus, if a restaurant offers the choice of 6 different salads, 14 different main dishes, and 7 different desserts, there are $6 \cdot 14 \cdot 7 = 588$ different ways in which one can choose a salad, a main dish, and a dessert. Also, if a true-false test consists of 20 questions, there are

$$\underbrace{2 \cdot 2 \cdot 2 \cdot 2 \cdots 2 \cdot 2}_{\text{20 factors}} = 2^{20} = 1{,}048{,}576$$

different ways in which one can mark the test, and only one of these corresponds to the case where each answer is correct.

1.3.2 Permutations

The rules of the preceding section are often applied when repeated selections are made from one and the same set, and *the order in which the selections are made* is of significance. For instance, if a club has 24 members, of which one is to be elected president and another is to be elected vice-president, there are altogether $24 \cdot 23 = 552$ ways in which the selection can be made. If a third member of the club is to be elected treasurer, and another one is to be elected secretary, the total number of ways in which all four officers can be selected is $24 \cdot 23 \cdot 22 \cdot 21 = 255{,}024$. *After each choice there is one less to choose from for the next selection.*

In general, if r objects are selected from a set of n distinct objects, any particular arrangement of these r objects is referred to as a *permutation.* For instance, 43215 is one of the many possible permutations of the first five positive integers, and 13452 is another; *Idaho, Utah,* and *New Mexico* is a permutation (a particular ordered arrangement) of 3 of the 8 Mountain States; ABG, CHE, and BDF are three of the many possible permutations of 3 of the first 8 letters of the alphabet; and if we were asked to list *all possible permutations* of 2 of the 5 vowels a, e, i, o, u, our answer would be

ae ai ao au ei eo eu io iu ou

ea ia oa ua ie oe ue oi ui uo

The total number of different permutations of r elements selected from a set of n distinct elements is given by the following rule, which is an immediate consequence of Theorem 1.2:

Theorem 1.3

*The number of permutations of r objects selected from a set of n objects is $n(n-1)(n-2)\cdots(n-r+1)$, which will be denoted $P(n, r)$.**

To prove this, we have only to observe that the first selection is made from the whole set of n objects so that $n_1 = n$ in Theorem 1.2, the second selection is made from the $n-1$ objects which remain after the first selection has been made so that $n_2 = n-1$ in Theorem 1.2, the third selection is made from the $n-2$ objects which remain after the first two selections have been made so that $n_3 = n-2$ in Theorem 1.2, ..., and the rth selection is made from the $n-(r-1) = n-r+1$ objects which remain after the first $r-1$ selections have been made so that $n_r = n-r+1$ in Theorem 1.2.

If we use the *factorial notation* according to which $n! = n(n-1)(n-2)\cdots3\cdot2\cdot1$ for any positive integer n and *by definition* $0! = 1$, we can write the formula for the number of permutations of r objects selected from a set of n objects as

$$P(n, r) = \frac{n!}{(n-r)!} \tag{1.3.1}$$

for any positive integer n and $r = 0, 1, 2, \ldots,$ or n. To obtain this result we have only to *multiply and divide* $n(n-1)(n-2)\ldots(n-r+1)$ by $(n-r)!$, and then make use of the fact that $n(n-1)(n-2)\ldots$

* Some authors use ${}_nP_r$, P_r^n, $P_{n,r}$, or $(n)_r$ to denote the number of permutations of r objects selected from a set of n objects.

$(n - r + 1)\cdot(n - r)! = n!$. Thus, the number of permutations of 3 of the 8 Mountain States is

$$P(8, 3) = \frac{8!}{(8-3)!} = \frac{8!}{5!} = \frac{40{,}320}{120} = 336$$

where the values of 8! and 5! were obtained from Table VII. (Of course, with small numbers like this it would have been just as easy to multiply out the product $8\cdot7\cdot6$.)

In the special case where $r = n$, the formula for the number of permutations of n distinct objects taken all together becomes

$$P(n, n) = \frac{n!}{(n-n)!} = \frac{n!}{0!} = n! \qquad (1.3.2)$$

and, referring to Table VII, we find that the starting eleven of a professional football team can be introduced to the public before a game in $11! = 39{,}916{,}800$ different ways.

Throughout our discussion it has been assumed that the n objects from which we select r objects and form permutations are all distinct. This means that our results cannot be used, for example, to determine the number of different ways in which we can arrange the letters in the word "book." If we distinguish for the moment between the two o's by labeling them o_1 and o_2, there are indeed $4! = 24$ different permutations of the symbols b, o_1, o_2, and k. However, if we drop the subscripts, then bo_1ko_2 and bo_2ko_1, for example, both yield boko, and since *each pair of permutations with subscripts yields but one arrangement without subscripts*, the total number of different ways in which the letters in "book" can be arranged is $\frac{24}{2} = 12$. Similarly, *with subscripts* on the e's there are 7! permutations of the letters in the word "receive," but since there are $3! = 6$ permutations of e_1, e_2, and e_3 which lead to the *same* arrangement of the letters in "receive," there are only $\frac{7!}{3!} = 840$ different arrangements of the letters in the word "receive."

Using the same sort of reasoning, we can state more generally that the total number of different arrangements of n symbols of which r_1 are alike, r_2 others are alike, ..., and r_k others are alike is

$$\frac{n!}{r_1!r_2!\cdots r_k!} \qquad (1.3.3)$$

For instance, the number of distinct arrangements of the 10 letters in "statistics," where there are 3 s's, 3 t's, and 2 i's, is $\frac{10!}{3!3!2!} = 50{,}400$.

1.3.3 Combinations

If a person gathering data for a market research organization has to interview 3 of the 20 families living in a certain apartment house, he can select these families in 1,140 ways, but this is *not* the number of permutations of 3 objects selected from a set of 20 objects. If we cared about the order in which these families are interviewed, the answer would be

$$P(20, 3) = 20 \cdot 19 \cdot 18 = 6{,}840$$

but each set of 3 families would then be counted $3! = 6$ times. If we are not interested in the order in which the 3 families are interviewed, there are thus only $\frac{6{,}840}{6} = 1{,}140$ ways in which we can select 3 of the 20 families, and this figure is referred to as *the number of combinations of 3 objects selected from a set of 20 objects.*

Thus, "combination" means the same as "subset," and when we ask for the *number of combinations of r objects selected from a set of n objects,* we are simply asking for the total number of different subsets containing r elements.

In general, there are $r!$ permutations of the elements of a subset of r elements, so that the $P(n,r)$ permutations of r elements selected from a set of n elements contain each subset $r!$ times. Dividing $P(n, r)$ by $r!$, we thus have:

THEOREM 1.4

The number of combinations of r objects selected from a set of n objects is $\frac{n!}{r!(n-r)!}$, *which will be denoted* $\binom{n}{r}$.*

This formula holds for any positive integer n and $r = 0, 1, 2, \ldots,$ or n, and it should be observed that it was obtained by dividing the expression for $P(n, r)$, given in (1.3.1), by $r!$. For reasons to be explained in Section 1.3.4, it is customary to refer to $\binom{n}{r}$ as a *binomial coefficient.*

To consider an applied example, suppose we want to know in how many different ways 6 tosses of a coin can yield 2 heads and 4 tails. This is equivalent to asking in how many different ways one can select the 2 tosses on which heads is to occur, or in other words, how many subsets

* Some authors use $C(n,r)$, ${}_nC_r$, C_r^n, or $C_{n,r}$ to denote the number of combinations of r objects selected from a set of n objects.

of 2 elements can be selected from a set of 6 elements. Applying Theorem 1.4, we thus find that the answer is

$$\frac{6!}{2!4!} = \frac{720}{2 \cdot 24} = 15$$

This result could also have been obtained by the rather tedious process of enumerating the various possibilities, HHTTTT, TTHTHT, HTHTTT, . . ., where H stands for heads and T for tails.

In many instances it is easier (that is, simpler arithmetically) to write the formula for the number of combinations of r objects selected from a set of n objects as

$$\binom{n}{r} = \frac{n(n-1)(n-2)\cdots(n-r+1)}{r!} \tag{1.3.4}$$

where we divided the expression given for $P(n, r)$ in Theorem 1.3 by $r!$. For instance, the number of ways in which a committee of 4 can be selected from among the 120 employees of a company is

$$\binom{120}{4} = \frac{120 \cdot 119 \cdot 118 \cdot 117}{4!} = \frac{197{,}149{,}680}{24} = 821{,}457$$

and the number of ways in which one can select 5 of the 14 counties of the state of Arizona is

$$\binom{14}{5} = \frac{14 \cdot 13 \cdot 12 \cdot 11 \cdot 10}{5!} = \frac{240{,}240}{120} = 2{,}002$$

1.3.4 Binomial Coefficients

If n is a positive integer and we multiply out $(1 + x)^n$ term by term, each term will be the product of x's and 1's, with an x or a 1 coming from each of the n factors $(1 + x)$. For instance, the expansion of $(1 + x)^3$ yields

$$\begin{aligned}(1 + x)(1 + x)(1 + x) &= 1 \cdot 1 \cdot 1 + 1 \cdot 1 \cdot x + 1 \cdot x \cdot 1 + x \cdot 1 \cdot 1 \\ &\quad + 1 \cdot x \cdot x + x \cdot 1 \cdot x + x \cdot x \cdot 1 + x \cdot x \cdot x \\ &= 1 + 3x + 3x^2 + x^3\end{aligned}$$

and it should be noted that the coefficient of x, for example, is 3, because there are 3 ways in which *one* x and *two* 1's can be selected, one from each of the three factors $(1 + x)$. Similarly, the coefficient of x^2 is 3 because there are 3 ways in which *two* x's and *one* 1 can be selected, one from each of the three factors $(1 + x)$, and the constant term and the coefficient of x^3 both equal 1, because there is only one way in which *three* 1's or *three* x's can be selected, one from each of the three factors $(1 + x)$.

More generally, if n is a positive integer and we multiply out $(1 + x)^n$ term by term, the coefficient of x^r equals the number of ways in which r factors x and $n - r$ factors 1 can be selected, one from each of the n factors $(1 + x)$, namely, $\binom{n}{r}$. This explains why we referred to $\binom{n}{r}$ as a *binomial coefficient*—it is the coefficient of x^r in the *binomial expansion* of $(1 + x)^n$. Making use of the fact that by definition $\binom{n}{0} = \frac{n!}{0!(n-0)!} = 1$, we can now state the following theorem:

THEOREM 1.5

$$(1 + x)^n = \sum_{r=0}^{n} \binom{n}{r} x^r \qquad \textit{for any positive integer } n.$$

(In case the reader is not familiar with the Σ notation, he will find a brief explanation at the end of the book.)

The calculation of binomial coefficients can often be simplified by making use of the following theorems:

THEOREM 1.6

$$\binom{n}{r} = \binom{n}{n-r} \qquad \textit{for any positive integer } n \textit{ and } r = 0, 1, \ldots, \textit{or } n.$$

THEOREM 1.7

$$\binom{n}{r} = \binom{n-1}{r} + \binom{n-1}{r-1} \qquad \textit{for any positive integer } n \textit{ and } r = 1, 2, \ldots, \text{or } n-1.$$

To prove Theorem 1.6 we might argue that when we select a subset of r objects from a set of n objects we *leave* a subset of $n - r$ objects, and, hence, there are as many ways of selecting r objects as there are ways of leaving (or selecting) $n - r$ objects. To prove the theorem algebraically we have only to write

$$\binom{n}{n-r} = \frac{n!}{(n-r)![n-(n-r)]!} = \frac{n!}{(n-r)!r!} = \frac{n!}{r!(n-r)!} = \binom{n}{r}$$

Theorem 1.6 implies that if we calculate the binomial coefficients for $r = 0, 1, 2, \ldots, \frac{n}{2}$ when n is *even* and for $r = 0, 1, \ldots, \frac{n-1}{2}$ when n is *odd*, the remaining binomial coefficients can be obtained by making use of the theorem. For instance, if we are given $\binom{4}{0} = 1$, $\binom{4}{1} = 4$, and $\binom{4}{2} = 6$, we can immediately write $\binom{4}{3} = \binom{4}{4-3} = \binom{4}{1} = 4$ and $\binom{4}{4} = \binom{4}{4-4} = \binom{4}{0} = 1$; similarly, if we are given $\binom{5}{0} = 1$, $\binom{5}{1} = 5$, and $\binom{5}{2} = 10$, we can immediately write $\binom{5}{3} = \binom{5}{5-3} = \binom{5}{2} = 10$, $\binom{5}{4} = \binom{5}{5-4} = \binom{5}{1} = 5$, and $\binom{5}{5} = \binom{5}{5-5} = \binom{5}{0} = 1$. It is precisely in this fashion that Theorem 1.6 is used in connection with Table VII at the end of the book. To find $\binom{20}{12}$, for example, we make use of the fact that $\binom{20}{12} = \binom{20}{8}$ and look up $\binom{20}{8}$; similarly, to find $\binom{17}{10}$ we make use of the fact that $\binom{17}{10} = \binom{17}{7}$ and look up $\binom{17}{7}$.

Theorem 1.7 can be proved by expressing the binomial coefficients on both sides of the equation in terms of factorials and then proceeding algebraically, but we shall leave this to the reader in Exercise 3 on page 29; instead, we shall use an alternate method which illustrates a very useful technique. Writing $(1 + x)^n$ as

$$(1 + x)(1 + x)^{n-1} = (1 + x)^{n-1} + x(1 + x)^{n-1}$$

we can prove Theorem 1.7 by equating the coefficient of x^r in $(1 + x)^n$ with that in $(1 + x)^{n-1} + x(1 + x)^{n-1}$. Since the coefficient of x^r in $(1 + x)^n$ is $\binom{n}{r}$ and the coefficient of x^r in $(1 + x)^{n-1} + x(1 + x)^{n-1}$ is the sum of the coefficient of x^r in $(1 + x)^{n-1}$, namely, $\binom{n-1}{r}$, and the coefficient of x^{r-1} in $(1 + x)^{n-1}$, namely, $\binom{n-1}{r-1}$, we obtain

$$\binom{n}{r} = \binom{n-1}{r} + \binom{n-1}{r-1}$$

which completes the proof. Theorem 1.7 has many important applications; in Exercise 9 on page 30, for example, it provides the key for the construction of what is known as *Pascal's triangle*.

To state and prove another theorem about binomial coefficients, let us make the definition that $\binom{n}{r} = 0$ whenever n is a positive integer and r is a positive integer greater than n. (Clearly, there is no way in which we can select a subset from a set which contains *more* elements than the set itself.)

Theorem 1.8

$$\sum_{r=0}^{k} \binom{m}{r}\binom{n}{k-r} = \binom{m+n}{k}$$

Using the same technique as in the proof of Theorem 1.7, let us prove this theorem by equating the coefficients of x^k in the expressions on both sides of the equation

$$(1+x)^{m+n} = (1+x)^m(1+x)^n$$

The coefficient of x^k in $(1+x)^{m+n}$ is $\binom{m+n}{k}$, and the coefficient of x^k in

$$(1+x)^m(1+x)^n = \left[\binom{m}{0} + \binom{m}{1}x + \cdots + \binom{m}{m}x^m\right] \times \left[\binom{n}{0} + \binom{n}{1}x + \cdots + \binom{n}{n}x^n\right]$$

is the *sum* of the products which we obtain by multiplying the constant term of the first factor by the coefficient of x^k in the second factor, the coefficient of x in the first factor by the coefficient of x^{k-1} in the second factor, the coefficient of x^2 in the first factor by the coefficient of x^{k-2} in the second factor, . . . , and the coefficient of x^k in the first factor by the constant term of the second factor. Thus, the coefficient of x^k in $(1+x)^m(1+x)^n$ is

$$\binom{m}{0}\binom{n}{k} + \binom{m}{1}\binom{n}{k-1} + \binom{m}{2}\binom{n}{k-2} + \cdots + \binom{m}{k}\binom{n}{0} = \sum_{r=0}^{k}\binom{m}{r}\binom{n}{k-r}$$

and this completes the proof. Note that in the application of Theorem 1.8 it is not necessary that m and n are greater than k. For instance, for $m = 2$, $n = 3$, and $k = 4$ we obtain

$$\binom{2}{0}\binom{3}{4} + \binom{2}{1}\binom{3}{3} + \binom{2}{2}\binom{3}{2} + \binom{2}{3}\binom{3}{1} + \binom{2}{4}\binom{3}{0} = \binom{5}{4}$$

and since $\binom{3}{4}$, $\binom{2}{3}$, and $\binom{2}{4}$ equal 0 according to the definition on page 28, the equation reduces to

$$\binom{2}{1}\binom{3}{3} + \binom{2}{2}\binom{3}{2} = \binom{5}{4}$$

which checks, since $2 \cdot 1 + 1 \cdot 3 = 5$.

THEORETICAL EXERCISES

1. Show that the total number of subsets of a set of n elements (including the empty set $\varnothing$ and the whole set itself) is 2^n by
   (a) making use of the fact that we must decide for each element whether or not it is to be included in the subset;
   (b) using Theorem 1.5 to evaluate $\binom{n}{0} + \binom{n}{1} + \binom{n}{2} + \cdots + \binom{n}{n}$.

2. In the illustration on page 18 we showed by means of a tree diagram that the total number of different outcomes (sequences of wins and losses) in a "3 out of 5" play-off in sports is 20.
   (a) Counting separately the number of play-offs requiring 3, 4, and 5 games, show that the result is given by $2\left[1 + \binom{3}{2} + \binom{4}{2}\right]$.
   (b) Duplicating the method of part (a), show that the total number of different outcomes in a "4 out of 7" play-off is 70.

3. Prove Theorem 1.7 by expressing all the binomial coefficients in terms of factorials and then simplifying algebraically.

4. Expressing the binomial coefficients in terms of factorials and simplifying algebraically, show that
   (a) $\binom{n}{r} = \frac{n-r+1}{r}\cdot\binom{n}{r-1}$; (c) $n\binom{n-1}{r} = (r+1)\binom{n}{r+1}$.
   (b) $\binom{n}{r} = \frac{n}{n-r}\cdot\binom{n-1}{r}$;

5. Substituting a suitable value of x into the formula of Theorem 1.5, show that
   (a) $\sum_{r=0}^{n} (-1)^r\binom{n}{r} = 0$;
   (b) $\sum_{r=0}^{n} \binom{n}{r}(a-1)^r = a^n$.

6. Repeatedly applying Theorem 1.7, show that $\binom{n}{r} = \sum_{i=1}^{r+1} \binom{n-i}{r-i+1}$.

7. Using Theorem 1.8, show that $\sum_{r=0}^{n} \binom{n}{r}^2 = \binom{2n}{n}$.

8. Show that $\sum_{r=0}^{n} r \binom{n}{r} = n2^{n-1}$ by
   (a) differentiating with respect to x the expressions on both sides of the equation of Theorem 1.5, and then substituting $x = 1$;
   (b) making use of part (b) of Exercise 1 with $n - 1$ substituted for n, multiplying through by n, and then using part (c) of Exercise 4.

9. **PASCAL'S TRIANGLE** If no tables are available, it is sometimes convenient to determine binomial coefficients by means of the following arrangement, called *Pascal's triangle,*

$$
\begin{array}{ccccccccccc}
 & & & & 1 & & 1 & & & & \\
 & & & 1 & & 2 & & 1 & & & \\
 & & 1 & & 3 & & 3 & & 1 & & \\
 & 1 & & 4 & & 6 & & 4 & & 1 & \\
1 & & 5 & & 10 & & 10 & & 5 & & 1 \\
\cdots & \cdots & \cdots & \cdots & \cdots & \cdots & \cdots & \cdots & \cdots & \cdots & \cdots
\end{array}
$$

   where each row begins with a 1, ends with a 1, and each other entry is the *sum* of the nearest two entries in the row immediately above.
   (a) Verify that with this method of construction the rth entry of the nth row is the binomial coefficient $\binom{n}{r-1}$.
   (b) Construct the next two rows of the triangle and read off the coefficients in the binomial expansion of $(1 + x)^6$ and $(1 + x)^7$.

10. **STIRLING'S FORMULA** When n is large, $n!$ can be approximated by means of the expression $\sqrt{2\pi n}\left(\frac{n}{e}\right)^n$, called *Stirling's formula,* where e, the base of natural logarithms, is approximately 2.7183. (A derivation of this formula may be found in the book by W. Feller cited among the References at the end of this chapter.)
   (a) Use Stirling's formula to obtain an approximation for 12!, and find the *percentage error* of this approximation by comparing it with the correct value given in Table VII.

(b) Use Stirling's formula to obtain an approximation for the number of 13-card bridge hands that can be dealt with an ordinary deck of 52 playing cards.

(c) Use Stirling's formula to show that

$$\lim_{n\to\infty} \frac{\binom{2n}{n}\sqrt{\pi n}}{2^{2n}} = 1$$

11. **OCCUPANCY THEORY** In occupancy theory we are concerned with the number of ways in which several distinguishable or indistinguishable objects can be distributed among a given number of individuals (or into a given number of boxes or cells).

(a) Construct a tree diagram or otherwise enumerate the number of ways in which six (indistinguishable) apples can be distributed among three children.

(b) To solve part (a) systematically, we might argue that $A|AAA|AA$, for example, represents the case where the first child gets one apple, the second child gets three, and the third child gets two. Thus, the total number of ways in which we can distribute the apples equals the number of ways in which we can arrange the six A's and the two vertical bars, namely, $\binom{8}{2} = 28$, and this should agree with the answer to part (a). Use a similar argument to find a general formula for the number of ways in which r *indistinguishable* objects can be distributed among n cells.

(c) Show that if there must be at least one object in each cell, the number of ways in which r *indistinguishable* objects can be distributed among n cells is $\binom{r-1}{n-1}$.

12. **OCCUPANCY THEORY, CONTINUED** Find the number of ways in which r *distinguishable* objects can be distributed among n cells, when there is no restriction as to the number of objects allowed or required in any one cell.

13. **BINOMIAL EXPANSIONS WITH NEGATIVE OR NON-INTEGRAL EXPONENTS** If n is not a positive integer or zero, the binomial expansion of $(1 + x)^n$ yields, for $-1 < x < 1$, the infinite series

$$1 + \binom{n}{1}x + \binom{n}{2}x^2 + \binom{n}{3}x^3 + \cdots + \binom{n}{r}x^r + \cdots$$

where $\binom{n}{r} = \frac{n(n-1)\cdots(n-r+1)}{r!}$ for $r = 1, 2, 3, \ldots$. Use this generalized definition of the symbol $\binom{n}{r}$ which, incidentally, agrees with that on page 24 for positive integral values of n, to

(a) evaluate $\binom{\frac{1}{2}}{4}$ and $\binom{-3}{3}$;

(b) show that $\binom{-1}{r} = (-1)^r$;

(c) show that $\binom{-n}{r} = (-1)^r \binom{n+r-1}{r}$ for $n > 0$.

14. **MULTINOMIAL COEFFICIENTS** If we extend the discussion on page 23 to the case where $r_1 + r_2 + \cdots + r_k = n$, namely, the case where a set of n objects consists of r_1 which are alike, r_2 others which are alike, ..., r_k others which are alike, and all of the elements of the set are thus accounted for, expression (1.3.3) leads to the result that the total number of different arrangements is

$$\frac{n!}{r_1!r_2!\cdots r_k!}$$

Symbolically, we denote this $\binom{n}{r_1, r_2, \ldots, r_k}$ and refer to it as a *multinomial coefficient* because it is the coefficient of $x_1^{r_1} \cdot x_2^{r_2} \cdots x_k^{r_k}$ in the expansion of $(x_1 + x_2 + \cdots + x_k)^n$. Use this formula to find

(a) the number of ways in which 10 rolls of a die can yield 2 *ones*, 3 *twos*, no *threes*, 1 *four*, 2 *fives*, and 2 *sixes;*

(b) the number of ways in which one A, three B's, two C's, and one F can be distributed among seven students taking a course in statistics;

(c) the different number of ways in which among 8 persons having dinner together 3 can order chicken, 4 can order steak, and 1 can order lobster.

APPLIED EXERCISES

15. A theatrical producer plans to put \$25,000 into *one* Broadway production each year so long as the number of flops in which he invests does not exceed the number of hits.

(a) Assuming that each production must be either a flop or a hit, draw a tree diagram showing the 14 possible situations that can arise during the first five years the plan is in operation.

(b) In how many of the situations described in part (a) will he continue the plan in the sixth year?

(c) If the producer loses his total investment in a flop and doubles his money in a hit, in how many of the situations described in part (a) will he be \$75,000 ahead?

16. An artist has two paintings in a show which lasts two days.

(a) If we are interested only in how many of his paintings are sold on each day, show by means of a tree diagram that there are altogether six possible outcomes.

(b) How many possible outcomes are there if we care which painting is sold on what day?

17. A department store classifies its delinquent charge accounts according to whether or not they have ever been delinquent before, according to whether the amount owed is less than \$100 or \$100 or more, and according to whether the payments are overdue by less than two months or by two months or more.

(a) Construct a tree diagram showing the various ways in which the department store classifies its delinquent charge accounts.

(b) If there are 40 accounts in each of the categories of part (a) and a courteous reminder is sent to all those whose charge accounts are delinquent for the first time *or* who owe less than \$100, how many of these courteous reminders will the department store have to mail out?

(c) If a warning is sent to all those whose charge accounts are overdue by two months or more, how many warnings will the department store have to mail out?

(d) How many persons with delinquent charge accounts at this department store will receive a courteous reminder as well as a warning?

18. If the five finalists in the Miss Universe contest are Miss Peru, Miss England, Miss U.S.A., Miss Japan, and Miss Italy, in how many ways can the judges choose

(a) the winner and the first runner-up;

(b) the winner, the first runner-up, and the second runner-up?

19. In a primary election, there are three candidates for mayor, five candidates for city treasurer, and four candidates for county attorney. In how many different ways can a voter mark his ballot for these three offices? In how many ways can a person vote if he exercises his option of not voting for any one of the candidates for any or all of these offices?

20. In a certain city, a child watching television on a Saturday morning has the choice between two different cartoon shows from 7:30 to 8, three different cartoon shows from 8 to 8:30, and three different cartoon shows from 8:30 to 9.

(a) In how many ways can one of these children plan his Saturday morning entertainment of watching cartoons continuously from 7:30 to 9?

(b) In how many ways can one of these children plan his Saturday morning entertainment of watching cartoons, if he is allowed to watch only for one hour?

21. A multiple-choice test consists of 12 questions, each permitting a choice of 4 alternatives. In how many different ways can a student check off his answers to these questions?

22. On each business trip, a salesman visits 5 of the 12 major cities in his territory. In how many different ways can he schedule his route (that is, the cities and their order) for such a trip?

23. Assuming that no member can hold more than one office, in how many ways can the 80 members of a labor union elect a president, a vice-president, and a secretary-treasurer?

24. In how many ways can a student arrange his 8 textbooks on a shelf? In how many ways can he do this if he wants to keep his 3 mathematics books together, his 2 physics books together, and his 2 history books together?

25. In how many ways can a television director schedule a sponsor's eight different commercials during the eight time slots allocated to commercials during an hour "special"?

26. In how many ways can the television director of Exercise 25 fill the eight time slots for commercials if

(a) the sponsor has four different commercials, each of which is to be shown twice;

(b) the sponsor has only two commercials, each of which is to be shown four times?

27. On page 21 we showed that a true-false test which consists of 20 questions can be marked in 1,048,576 different ways. In how many ways can they be marked true or false so that

(a) 3 are right and 17 are wrong;

(b) 10 are right and 10 are wrong;

(c) *at least* 18 are right?

28. In a shipment of 18 television sets there is one that is defective. In how many ways can one choose 4 of these television sets so that

(a) the defective set is not included;

(b) the defective set is included?

29. Mr. Jones has 4 suits, 12 ties, and 5 pairs of shoes. In how many ways can he choose 2 suits, 5 ties, and 2 pairs of shoes to take along on a business trip?

30. A committee of 6 is to be chosen from among 4 employers and 20 employees. In how many ways can this be done if
(a) the committee is to include 3 employers and 3 employees;
(b) the committee is to contain at least 2 employers?

31. Using the theory of Exercise 13, evaluate $\sqrt{5}$ by making use of the fact that $5 = 4(1 + \frac{1}{4})$ and using the first four terms of the binomial expansion of $(1 + \frac{1}{4})^{\frac{1}{2}}$.

REFERENCES

A wealth of material on combinatorial methods can be found in

Feller, W., *An Introduction to Probability Theory and its Applications*, Vol. I, 2nd ed. New York: John Wiley & Sons, Inc., 1957,

and in

Whitworth, W. A., *Choice and Chance*, 5th ed. New York: Hafner Publishing Co., Inc., 1959,

which has become a classic in this field. More advanced treatments may be found in:

Beckenbach, E. F., ed., *Applied Combinatorial Mathematics*. New York: John Wiley & Sons, Inc., 1964,

David, F. N., and Barton, D. E., *Combinatorial Chance*. New York: Hafner Publishing Co., Inc., 1962,

and

Riordan, J., *An Introduction to Combinatorial Analysis*. New York: John Wiley & Sons, Inc., 1958.

2

PROBABILITY

2.1 INTRODUCTION

Directly or indirectly, probability plays an important role in all problems of science, business, and everyday life, where decisions and understanding involve an element of uncertainty or risk. Thus, it is unfortunate that the term "probability," itself, is controversial and difficult to define.

Among the different theories of probability, that most widely held is the *frequency concept of probability*, according to which the probability of an event is interpreted as *the proportion of the time that events of the same kind will occur in the long run.** If we say that there is a probability of 0.82 that a jet from New York to Miami, Florida, will arrive on time, this means that such flights should arrive on time about 82 percent of the time. More generally, we say that an event has a probability of, say, 0.90, in the same sense in which we might say that our car will start in cold weather about 90 percent of the time. *We cannot guarantee what will happen on any one try, the car may start and then it may not, but it would be reasonable to bet $9.00 against $1.00 or ninety cents against a dime (namely at odds of 9 to 1) that the car will start at any given try.* By "reasonable" we mean that we would thus win $1.00 (or a dime) about 90 percent of the time, lose $9.00 (or ninety cents) about 10 percent of the time, and, hence, can expect to break even in the long run.

In accordance with the frequency concept of probability, we *estimate* the probability of an event by observing how often (what part of the time) similar events have occurred in the past. For instance, if airline records show that (over a period of time) 574 of 700 jets from New York to Miami, Florida, arrived on time, we *estimate* the probability that any such flight (perhaps, the next one) will arrive on time as $\frac{574}{700} = 0.82$.

An alternate point of view, which is currently gaining favor, is to interpret probabilities as *personal* or *subjective* evaluations. For instance, a businessman may "feel" that the odds for the success of a new venture, say, a new restaurant, are 3 to 2. This means that he would be willing to bet (or consider it fair to bet) $30 against $20, or perhaps $3,000 against

* Another popular concept of probability, based on "equally likely" events, will be discussed later on page 41.

\$2,000, that the venture will succeed. In this way he expresses the *strength of his belief* regarding the uncertainties connected with the success of the new restaurant. This method of dealing with uncertainties works well (and is certainly justifiable) in situations where there is very little direct evidence, and there may be no choice but to rely on pertinent collateral information, "educated" guesses, and perhaps intuition and other subjective factors.

Regardless of how we interpret probabilities and odds, subjectively or objectively (in terms of frequencies or proportions), the mathematical relationship between the two is always the same. If somebody considers it fair or equitable to bet a dollars against b dollars that a given event will occur, he is, in fact, assigning the event the probability $\frac{a}{a+b}$. Thus, the businessman of the preceding paragraph is actually assigning the success of the new restaurant a probability of $\frac{3}{3+2} = 0.60$. To illustrate how probabilities are converted into odds, let us refer back to the example on page 36, which dealt with the question whether or not we could start our car in cold weather. As we pointed out at the time, the probability of 0.90 implies that we should win (the bet that we can start the car) about 90 percent of the time, lose about 10 percent of the time, and, hence, that the proper odds are 9 to 1. In general, if the probability of an event is p, then the odds for its occurrence are p to $1 - p$. Odds are usually quoted as ratios of positive integers (having no common factors), and if the probability of an event is 0.15, for example, we say that the odds for its occurrence are 3 to 17 instead of 0.15 to 0.85. Similarly, if the probability that we shall not have to wait for a table at our favorite restaurant is 0.75, then the odds that we shall not have to wait are 3 to 1.

2.2 THE MATHEMATICS OF PROBABILITY

It is important to note that the mathematical theory of probability and statistics, which is the subject matter of this book, does not depend on philosophical arguments concerning the meaning of "probability"—there is general agreement concerning the postulates and definitions which will be given here and in Chapter 4. Questions of meaning will, of course, arise when it comes to applications, and in that case we shall follow mostly the very widely held frequency interpretation.

2.2.1 The Postulates of Probability

To formulate the postulates of probability we shall continue the practice of denoting events by means of capital letters, and we shall write the probability of event A as $P(A)$, the probability of event B as $P(B)$, and so forth. Furthermore, we shall continue to denote the set of all possible outcomes, namely the *sample space*, by the letter S. As we shall formulate them here, the postulates of probability apply only when the sample space S is *discrete* (finite or countably infinite); the modifications that are required for the continuous case will be discussed in Chapter 4.

POSTULATE 1

The probability of any event is a non-negative real number; that is, $P(A) \geq 0$ *for any subset* A *of* S.

POSTULATE 2

$P(S) = 1$

POSTULATE 3

If $A_1, A_2, A_3, \ldots,$ *is a finite or infinite sequence of disjoint subsets of* S, *then*

$$P(A_1 \cup A_2 \cup A_3 \cup \cdots) = P(A_1) + P(A_2) + P(A_3) + \cdots$$

Thus, probabilities are values of a set function, called a *probability measure*, which must satisfy the three given postulates, and it is of interest to note that Postulate 1 expresses the first of the special properties of measure functions listed on page 14, and that Postulate 3 generalizes the second property to any number of mutually exclusive events.

Postulates *per se* require no proof, but if the resulting theory is to be applied, we must show that the postulates are satisfied when we give probabilities a "real" interpretation. So far as the first postulate is concerned, it is in complete agreement with the frequency interpretation as well as the subjective concept of probability—proportions are always positive or zero, and so long as a and b (the amounts bet for and against the occurrence of an event) are positive, the probability $\frac{a}{a+b}$ cannot be negative.

The second postulate states indirectly that *certainty* is identified with a probability of 1; after all, it is always assumed that *one of the possibilities included in S must occur*, and it is to this certain event that we assign a probability of 1. So far as the frequency interpretation is concerned, a probability of 1 implies that the event will occur 100 percent of the time, or in other words, that it is certain to occur. So far as subjective probabilities are concerned, the surer we are that an event will occur, the "better" odds we should be willing to give—say, 100 to 1, 1,000 to 1, or perhaps even 1,000,000 to 1. The corresponding probabilities are $\frac{100}{100+1}$, $\frac{1{,}000}{1{,}000+1}$, and $\frac{1{,}000{,}000}{1{,}000{,}000+1}$ (or approximately 0.99, 0.999, and 0.999999), and it can be seen that *the "surer" we are that an event will occur, the closer its probability will be to 1.*

Taking Postulate 3 in the simplest case, namely, for two mutually exclusive events A_1 and A_2, we can easily show that it is satisfied by the frequency interpretation. If one event occurs, say, 28 percent of the time, another event occurs 39 percent of the time, and *they cannot both occur at the same time* (that is, they are mutually exclusive), then one or the other will occur $28 + 39 = 67$ percent of the time. So far as subjective probabilities are concerned, the third postulate does not follow from our discussion in Section 2.1, but it is imposed by proponents of the subjective point of view as a "consistency criterion." In other words, if a person's subjective probabilities "behave" in accordance with Postulate 3, he is said to be *consistent;* otherwise, he is said to be *inconsistent* and his probability judgments can be taken only with a good deal of scepticism.

Before we study some of the immediate consequences of the postulates of probability, let us stress the point that *the three postulates do not tell us how to assign probabilities to events, they merely restrict the ways in which it can be done.* For instance, the following are three ways in which we *can* assign probabilities to the four possible and mutually exclusive outcomes A, B, C, and D of an experiment:

(1) $P(A) = \frac{2}{9}$, $P(B) = \frac{1}{3}$, $P(C) = \frac{1}{3}$, $P(D) = \frac{1}{9}$
(2) $P(A) = 0.27$, $P(B) = 0.42$, $P(C) = 0.15$, $P(D) = 0.16$
(3) $P(A) = \frac{1}{4}$, $P(B) = \frac{1}{4}$, $P(C) = \frac{1}{4}$, $P(D) = \frac{1}{4}$

On the other hand,

(4) $P(A) = 0.12$, $P(B) = 0.63$, $P(C) = 0.45$, $P(D) = -0.20$

and

(5) $P(A) = \frac{9}{120}$, $P(B) = \frac{45}{120}$, $P(C) = \frac{37}{120}$, $P(D) = \frac{56}{120}$

are *not* permissible; in case (4) $P(D) = -0.20$ violates Postulate 1, and in case (5) we have $P(S) = P(A \cup B \cup C \cup D) = \frac{9}{120} + \frac{45}{120} + \frac{37}{120} + \frac{56}{120} = \frac{147}{120}$, which violates Postulate 2. *Of course, in actual practice, probabilities are assigned on the basis of past experience, on the basis of a careful analysis of all underlying conditions, or on the basis of assumptions—say, the assumption that all possible outcomes are equiprobable.*

To assign a probability measure to a sample space, it is not necessary to specify the probability of each possible subset. This is fortunate, for a sample space with as few as 20 possible outcomes has already $2^{20} = 1{,}048{,}576$ subsets (see Exercise 1 on page 29), and the number of subsets grows very rapidly when there are 50 possible outcomes, 100 possible outcomes, or more. Instead of listing the probabilities of all possible subsets, we often make use of the following theorem:

THEOREM 2.1

If A is a subset of a discrete sample space S, then $P(A)$ equals the sum of the probabilities of the individual outcomes comprising A.

To prove this theorem, let $E_1, E_2, E_3, \ldots$, be a finite or infinite sequence of subsets of S, representing the individual outcomes comprising A. Thus,

$$A = E_1 \cup E_2 \cup E_3 \cup \cdots$$

and since the individual outcomes, the E's, are by definition mutually exclusive, the third postulate of probability yields

$$P(A) = P(E_1) + P(E_2) + P(E_3) + \cdots$$

which completes the proof. To illustrate how this theorem might be used, let us refer to the sample space of Figure 1.5, representing the 36 possible outcomes for a roll of a pair of dice. If each of the outcomes is assigned a probability of $\frac{1}{36}$, it can immediately be seen that the probability of rolling a 7 is $\frac{6}{36}$ and that the probability of rolling a 2, 3, or 12 is $\frac{4}{36}$. In the first case we add the probabilities of the six points comprising the event "rolling a 7," and in the second case we add the probabilities of the four points comprising the event "rolling a 2, 3, or 12." If a sample space is discrete but *infinite*, probabilities will have to be assigned to the individual outcomes by means of some mathematical rule. For instance, if E_1, E_2, $E_3, \ldots$, represent the infinitely many outcomes of an experiment, their probabilities might be specified by the rule

$$P(E_i) = (\tfrac{1}{2})^i \quad \text{for} \quad i = 1, 2, 3, \ldots$$

As we shall see in Chapter 3, this probability measure would be appropriate if E_i represents the event that a person flipping a balanced coin gets *heads* for the first time on the ith try. Thus, the probability that the first *heads* will come on the third, fourth, or fifth try is $(\frac{1}{2})^3 + (\frac{1}{2})^4 + (\frac{1}{2})^5 = \frac{7}{32}$, and the probability that the first *heads* will come on an *odd-numbered* try is

$$(\tfrac{1}{2})^1 + (\tfrac{1}{2})^3 + (\tfrac{1}{2})^5 + \cdots = \tfrac{2}{3}$$

as can easily be verified by using the formula for the value of an infinite geometric progression. (To be rigorous in a situation like this, the word "sum" in Theorem 2.1 will have to be interpreted so that it includes the value of an infinite series, if necessary.)

In the example dealing with the pair of dice, we could have used to advantage the following theorem which applies whenever the individual outcomes are all equiprobable:

THEOREM 2.2

If an experiment has n equiprobable outcomes and if s of these comprise event A, then $P(A) = \dfrac{s}{n}\cdot$

This theorem follows immediately from Theorem 2.1 and the fact that if there are n possible outcomes which are all equiprobable, each outcome has the probability $\dfrac{1}{n}\cdot$

Theorem 2.2 has many useful applications in problems relating to games of chance, where it can be assumed, for example, that if a deck of cards is properly shuffled each card has the same chance of being drawn, if an "honest" coin is properly flipped *heads* is as likely to come up as *tails*, and if a balanced die is properly rolled each face is as likely to come up as any other. Thus, the probability of drawing a queen from an ordinary deck of 52 playing cards is $\frac{4}{52}$, the probability of getting *heads* with a balanced coin is $\frac{1}{2}$, and the probability of rolling a 3 or a 4 with a balanced die is $\frac{2}{6}$. [Theorem 2.2 is sometimes given as a *definition* of probability; this not only limits the applicability of probability theory to very special kinds of experiments (those with equiprobable outcomes), but it can be criticized on the grounds that there is an obvious *circularity* in defining "probability" in terms of "equiprobable" events.]

2.2.2. Some Elementary Theorems

By using the three postulates of probability, we can derive many further rules which have important applications. Among the immediate consequences of the postulates, we find that *probabilities cannot exceed 1*, that *the empty set* $\emptyset$ *has the probability 0*, and that *the respective probabilities that an event will occur and that it will not occur always add up to 1*. Symbolically,

THEOREM 2.3

$P(A) \leq 1$ for any event A.

THEOREM 2.4

$$P(\varnothing) = 0$$

and

THEOREM 2.5

$$P(A) + P(A') = 1$$

The first of these three theorems (which the reader will be asked to prove in Exercise 1 on page 46) simply expresses the fact that an event cannot occur more than 100 percent of the time, or that the probability $\frac{a}{a+b}$ cannot exceed 1 when a and b are positive amounts bet for and against the occurrence of an event. The second theorem expresses the fact that an impossible event happens 0 percent of the time, and to prove it we have only to make use of the facts that $S \cup \varnothing = S$, that S and $\varnothing$ are disjoint, and write

$$P(S) = P(S \cup \varnothing)$$

$$P(S) = P(S) + P(\varnothing)$$

according to Postulate 3, and, hence, $P(\varnothing) = 0$ by subtraction. It is important to note that it does *not* follow from $P(A) = 0$ that A is necessarily an empty set. In actual practice, we often assign a probability of 0 to an event which, in colloquial terms, would not happen in a million years. For instance, there is the classical example that we assign a probability of 0 to the event that a monkey set loose on a typewriter will type Plato's *Republic* word for word without a mistake. As we shall see in Chapter 4, the fact that $P(A) = 0$ does not imply that A is an empty set is of relevance, particularly, in the continuous case.

To prove Theorem 2.5 we have only to make use of the fact that $A \cup A' = S$ and A and A' are disjoint according to the sixth postulate for sets (on page 11), and write

$$P(S) = P(A \cup A')$$

$$P(S) = P(A) + P(A')$$

according to Postulate 3, and, hence,

$$1 = P(A) + P(A')$$

according to Postulate 2. This result, which is often written in the form $P(A') = 1 - P(A)$, implies according to the frequency interpretation that if an event occurs, say, 37 percent of the time, it does *not* occur 63 percent of the time, or if the probability that it will rain is 0.28, the probability that it will *not* rain is $1 - 0.28 = 0.72$. Subjectively speaking, if a person considers it fair, or equitable, to bet a dollars against b dollars that a given event will occur, he is actually assigning to the event the probability $\frac{a}{a+b}$ and to its non-occurrence the probability $\frac{b}{b+a}$; evidently, these two probabilities also add up to 1.

2.2.3 Further Addition Rules

If A and B are mutually exclusive events, the formula $P(A \cup B) = P(A) + P(B)$, often called the *special rule of addition*, is merely a special case of Postulate 3. The fact that this formula does not necessarily apply when A and B are *not* mutually exclusive is apparent from the following example: if A represents the event that a certain student will pass a chemistry test, B represents the event that he will pass a test in mathematics, $P(A) = 0.80$, and $P(B) = 0.75$, then the probability that he will pass at least one of the two tests is *not* $0.80 + 0.75 = 1.55$. In arriving at this figure we made the mistake of adding in *twice* the probability $P(A \cap B)$ that he will pass in chemistry as well as mathematics, and to compensate for this we should have used the following theorem, often called the *general rule of addition:*

THEOREM 2.6

If A and B are any two subsets of a sample space S, then

$$P(A \cup B) = P(A) + P(B) - P(A \cap B)$$

Thus, if the probability that the student of the preceding example will pass both tests is 0.65, then the probability that he will pass at least one of the two tests is $P(A \cup B) = 0.80 + 0.75 - 0.65 = 0.90$; similarly, if C is the event that a car stopped at a road block has faulty brakes, D is the event that its tires are badly worn, $P(C) = 0.23$, $P(D) = 0.24$, and $P(C \cap D) = 0.09$, then the probability that such a car has faulty brakes and/or badly worn tires is $P(C \cup D) = 0.23 + 0.24 - 0.09 = 0.38$.

To prove Theorem 2.6, let us make use of part (d) of Exercise 3 on page 14, according to which $A \cup B = (A \cap B) \cup (A \cap B') \cup (A' \cap B)$, and the fact that the sets $A \cap B$, $A \cap B'$, and $A' \cap B$ are disjoint (as is apparent from Figure 2.1). Thus, we can write

$$\begin{aligned} P(A \cup B) &= P[(A \cap B) \cup (A \cap B') \cup (A' \cap B)] \\ &= P(A \cap B) + P(A \cap B') + P(A' \cap B) \end{aligned}$$

and, upon *adding and subtracting* $P(A \cap B)$, this becomes

$$\begin{aligned} P(A \cup B) &= [P(A \cap B) + P(A \cap B')] + [P(A' \cap B) + P(A \cap B)] \\ &\qquad - P(A \cap B) \\ &= P(A) + P(B) - P(A \cap B) \end{aligned}$$

In the last step of this proof we made use of part (c) of Exercise 3 on page 14, according to which $A = (A \cap B) \cup (A \cap B')$ and $B = (A' \cap B) \cup (A \cap B)$, and that $A \cap B$ and $A \cap B'$ are mutually exclusive and so are $A' \cap B$ and $A \cap B$; all this is apparent from Figure 2.1.

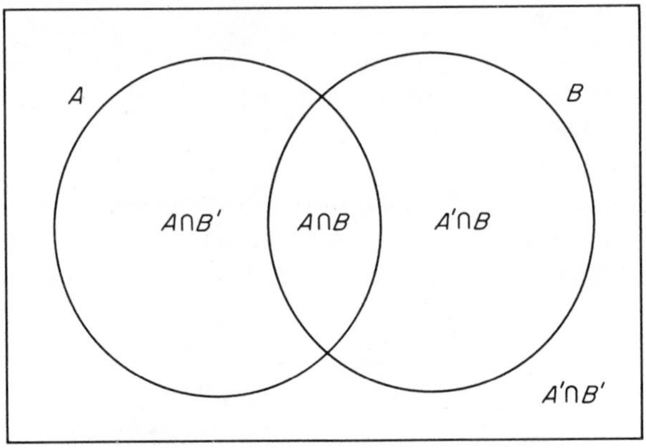

FIGURE 2.1 Venn diagram.

An easier way of looking at this proof is to assign $A \cap B$, $A \cap B'$, and $A' \cap B$ the probabilities a, b, and c as in Figure 2.2, so that it can immediately be seen that $P(A) = b + a$, $P(B) = a + c$, $P(A \cup B) = a + b + c$, and, hence, that

$$\begin{aligned} P(A) + P(B) - P(A \cap B) &= (b + a) + (a + c) - a \\ &= a + b + c \\ &= P(A \cup B) \end{aligned}$$

Using the same idea and assigning the probabilities a, b, c, d, e, f, and g

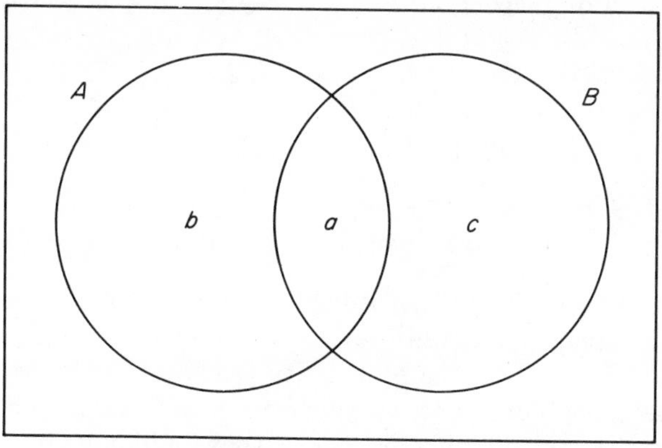

FIGURE 2.2 Venn diagram.

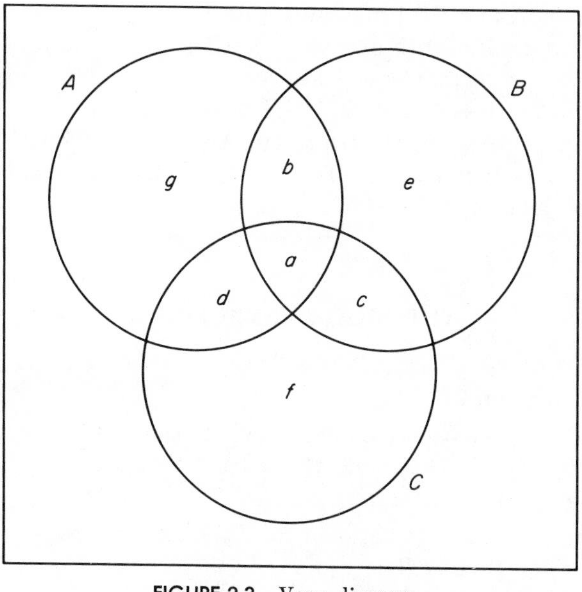

FIGURE 2.3 Venn diagram.

as in the Venn diagram of Figure 2.3, it is easy to verify the following generalization of the *addition rule* of Theorem 2.6:

THEOREM 2.7

If A, B, and C are any three subsets of a sample space S, then

$$P(A \cup B \cup C) = P(A) + P(B) + P(C) - P(A \cap B) \\ - P(A \cap C) - P(B \cap C) + P(A \cap B \cap C)$$

As can be seen by inspection, $P(A) = a + b + d + g$, $P(B) = a + b + c + e$, $P(C) = a + c + d + f$, $P(A \cap B) = a + b$, $P(A \cap C) = a + d$, $P(B \cap C) = a + c$, $P(A \cap B \cap C) = a$, and $P(A \cup B \cup C) = a + b + c + d + e + f + g$, so that $P(A) + P(B) + P(C) - P(A \cap B) - P(A \cap C) - P(B \cap C) + P(A \cap B \cap C)$ equals

$$(a + b + d + g) + (a + b + c + e) + (a + c + d + f) - (a + b) \\ - (a + d) - (a + c) + a = a + b + c + d + e + f + g$$

and, hence, $P(A \cup B \cup C)$. To illustrate the use of Theorem 2.7, suppose that if a person visits his dentist, the probability that he will have his teeth cleaned is 0.44, the probability that he will have a cavity filled is 0.24, the probability that he will have a tooth extracted is 0.21, the probability that he will have his teeth cleaned and a cavity filled is 0.08, the probability that he will have his teeth cleaned and a tooth extracted is 0.11, the probability that he will have a cavity filled and a tooth extracted is 0.07, and the probability that he will have his teeth cleaned, a cavity filled, and a tooth extracted is 0.03. Thus, the probability that he will have his teeth cleaned and/or a cavity filled and/or a tooth extracted is $0.44 + 0.24 + 0.21 - 0.08 - 0.11 - 0.07 + 0.03 = 0.66$. In Exercise 2Δ on page 50 the reader will be asked to show that the probability that a person visiting his dentist will have *one and only one* of these things done to him is 0.46.

THEORETICAL EXERCISES

1. Prove Theorem 2.3.
2. Show that if set B is *contained* in set A (that is, B is a subset of A), then $P(B) \leq P(A)$. [*Hint:* Use part (c) of Exercise 3 on page 14 and the fact that $A \cap B = B$ when B is a subset of A.]
3. Referring to parts (c) and (d) of Exercise 3 on page 14, show that
   (a) $P(A) \geq P(A \cap B)$;
   (b) $P(A) \leq P(A \cup B)$.
4. An experiment has five possible distinct outcomes A, B, C, D, and E. Check whether the following constitute *probability measures* and explain your answers:
   (a) $P(A) = 0.20$, $P(B) = 0.20$, $P(C) = 0.20$, $P(D) = 0.20$, and $P(E) = 0.20$;
   (b) $P(A) = 0.21$, $P(B) = 0.26$, $P(C) = 0.58$, $P(D) = 0.01$, and $P(E) = 0.06$;
   (c) $P(A) = 0.18$, $P(B) = 0.19$, $P(C) = 0.20$, $P(D) = 0.21$, and $P(E) = 0.22$;

(d) $P(A) = 0.10$, $P(B) = 0.30$, $P(C) = 0.10$, $P(D) = 0.60$, and $P(E) = -0.10$;
(e) $P(A) = 0.23$, $P(B) = 0.12$, $P(C) = 0.05$, $P(D) = 0.50$, and $P(E) = 0.08$.

5. If A and B are mutually exclusive events, $P(A) = 0.33$, and $P(B) = 0.42$, find
(a) $P(A')$;
(b) $P(B')$;
(c) $P(A \cup B)$;
(d) $P(A \cap B)$;
(e) $P(A \cap B')$;
(f) $P(A' \cap B')$.

6. Given $P(A) = 0.48$, $P(B) = 0.37$, and $P(A \cap B) = 0.13$, find
(a) $P(A \cup B)$;
(b) $P(A \cap B')$;
(c) $P(A' \cup B')$;
(d) $P(A' \cap B')$.

7. Use Theorem 2.6 to show that
(a) $P(A \cap B) \leq P(A) + P(B)$;
(b) $P(A \cap B) \geq P(A) + P(B) - 1$.

8. Duplicating the method of proof of Theorem 2.7, show that $P(A \cup B \cup C \cup D) = P(A) + P(B) + P(C) + P(D) - P(A \cap B) - P(A \cap C) - P(A \cap D) - P(B \cap C) - P(B \cap D) - P(C \cap D) + P(A \cap B \cap C) + P(A \cap B \cap D) + P(A \cap C \cap D) + P(B \cap C \cap D) - P(A \cap B \cap C \cap D)$. (*Hint:* Divide each of the 7 relevant regions of Figure 2.3 into two parts, one inside D and one outside D and assign to the resulting 14 regions the probabilities $a, b, c, \ldots,$ and n.)

APPLIED EXERCISES

9. Mr. Jones claims that the odds are 10 to 15 that a person driving from New York to San Francisco will have at least one flat tire, while Mr. Martin claims that the odds are 9 to 14. Which of the two assigns this event a higher probability?

10. Asked about his chances of getting a \$500 raise in his annual salary or a \$1,000 raise, a university professor replies that the odds are 1 to 2 that he will get a \$500 raise, 1 to 3 that he will get a \$1,000 raise, and 7 to 5 that he will get either a \$500 raise or a \$1,000 raise. Discuss the consistency of these odds.

11. If an explorer is willing to bet \$50 against \$30 that his expedition will be a success, but not \$75 against \$25, what limits can we put on the probability which he assigns to the success of his expedition?

12. Asked about the chances that business conditions will improve, remain the same, or get worse, an economist replies that the odds are 2 to 1 that they will improve, 4 to 1 that they will *not* remain the same, and 6 to 1 that they will *not* get worse. Discuss the consistency of these odds.

13. Explain why there must be a mistake in each of the following statements:

(a) The probabilities that a secretary will make 0, 1, 2, 3, 4, or 5 *or more* mistakes in typing a report are, respectively, 0.12, 0.25, 0.36, 0.14, 0.09, and 0.07.

(b) The probabilities that 0, 1, 2, or 3 *or more* players will get hurt during a football game are, respectively, 0.08, 0.21, 0.29, and 0.40.

(c) The probability that Mr. Brown will visit his dentist twice in 1972 is 0.18, and the probability that he will visit his dentist at least twice is 0.15.

(d) Mrs. Adams has two daughters; the probability that the older daughter will get married within a year is 0.32, and the probability that both daughters will get married within a year is 0.41.

(e) The probability that a person visiting the San Diego Zoo will see the giraffes is 0.72, the probability that he will see the bears is 0.85, and the probability that he will see them both is 0.51.

[*Hint:* For parts (c), (d), and (e) refer to Exercises 3 and 7.]

14. Supposing that each of the 30 points of the sample space of Exercise 9 on page 15 is assigned the probability $\frac{1}{30}$, find

(a) the probability that at least one of the station wagons is empty;

(b) the probability that each of the two station wagons carries the same number of passengers;

(c) the probability that the larger station wagon carries more passengers than the smaller station wagon;

(d) the probability that together they carry at least six passengers.

15. Suppose that in Exercise 11 on page 16 the probabilities that the dealer will first sell Car 1, Car 2, . . . , or Car 8, are, respectively, 0.15, 0.12, 0.08, 0.21, 0.17, 0.10, 0.04, and 0.13. Find the probabilities that the first car he sells will

(a) have air-conditioning;

(b) have no power steering;

(c) be at most two years old.

16. With reference to the example on page 40 where we assigned outcome E_i the probability $(\frac{1}{2})^i$, verify that this satisfies the second postulate of probability.

17. Sometimes Mrs. Clark has lunch by herself at home, sometimes she has lunch out by herself while shopping, sometimes she goes to luncheon meetings sponsored by organizations to which she belongs, sometimes she is invited out for lunch by friends, and sometimes she skips lunch to lose weight. If the probabilities of these alternatives are respectively, 0.22, 0.19, 0.23, 0.14, and 0.22, find

(a) the probability that she will eat lunch by herself;

(b) the probability that she will not go out for lunch;

(c) the probability that she will have lunch out, but not by herself.

18. The probabilities that a student will receive an A, B, C, D, or F in a final examination in a course in mathematics are, respectively, 0.10, 0.26, 0.44, 0.14, and 0.06. What are the probabilities that he will receive
(a) at least a C;
(b) neither an A nor an F;
(c) at most a C?

19. If each of the 52 cards of an ordinary deck of 52 playing cards has a probability of $\frac{1}{52}$ of being drawn, what is the probability of drawing
(a) a red king;
(b) a black card;
(c) a 7, 8, 9, or 10;
(d) a red queen or a black king?

20. If a hat contains twenty white slips of paper numbered from 1 through 20, ten red slips of paper numbered from 1 through 10, forty yellow slips of paper numbered from 1 through 40, ten blue slips of paper numbered from 1 through 10, and they are thoroughly shuffled so that each slip has a probability of $\frac{1}{80}$ of being drawn, find the probability of drawing
(a) a slip of paper which is either blue or white;
(b) a slip of paper numbered 1, 2, 3, 4, 5, or 6;
(c) a red or yellow slip numbered 1, 2, or 3;
(d) a slip of paper whose number is divisible by 3;
(e) either a white slip numbered higher than 15 or a yellow slip numbered 25 or less.

21. For married couples living in a certain suburb, the probability that the husband will vote in a school board election is 0.21, the probability that his wife will vote in the election is 0.28, and the probability that they will both vote is 0.15. What is the probability that at least one of them will vote?

22. The probability that a person with a fever visiting his doctor will get a shot is 0.46, the probability that he will get a pill is 0.38, and the probability that he will get a shot as well as a pill is 0.22. What is the probability that a person with a fever visiting his doctor will
(a) get a shot and/or a pill;
(b) get a shot but no pill;
(c) get a shot or a pill, but not both;
(d) get neither a shot nor a pill?

23. Among the 180 professors on the faculty of a university, 135 hold Ph.D. degrees, 146 devote at least part of their time to research, and 114 of the 135 with Ph.D. degrees devote at least part of their time to research. If one of these professors is chosen by lot to serve on an administrative committee, find the probabilities that

(a) the one who is chosen holds a Ph.D. degree and/or devotes at least part of his time to research;
(b) the one who is chosen does not hold a Ph.D. degree and does not devote any part of his time to research.

24. With reference to the illustration on page 46, verify the claim that the probability that a person visiting his dentist will have *one and only one* of the things done to him is 0.46.

25. Suppose that if a person visits Disneyland, the probability that he will go on the Jungle Cruise is 0.74, the probability that he will ride the Monorail is 0.70, the probability that he will go on the Matterhorn ride is 0.62, the probability that he will go on the Jungle Cruise and ride the Monorail is 0.52, the probability that he will go on the Jungle Cruise as well as the Matterhorn ride is 0.46, the probability that he will ride the Monorail and go on the Matterhorn ride is 0.44, and the probability that he will go on all three of these rides is 0.34. What is the probability that a person visiting Disneyland will go on at least one of these three rides?

26. Suppose that if a person travels to Europe for the first time, the probability that he will see London is 0.70, the probability that he will see Paris is 0.64, the probability that he will see Rome is 0.58, the probability that he will see Amsterdam is 0.58, the probability that he will see London and Paris is 0.45, the probability that he will see London and Rome is 0.42, the probability that he will see London and Amsterdam is 0.41, the probability that he will see Paris and Rome is 0.35, the probability that he will see Paris and Amsterdam is 0.39, the probability that he will see Rome and Amsterdam is 0.32, the probability that he will see London, Paris, and Rome is 0.23, the probability that he will see London, Paris, and Amsterdam is 0.26, the probability that he will see London, Rome, and Amsterdam is 0.21, the probability that he will see Paris, Rome, and Amsterdam is 0.20, and the probability that he will see all four of these cities is 0.12. What is the probability that a person traveling to Europe for the first time will see at least one of these four cities? (*Hint:* Use the formula of Exercise 8.)

2.3 CONDITIONAL PROBABILITY

As we have defined probability, it is meaningless to speak of the probability of an event unless we specify the sample space to which it belongs. To ask for the probability that a psychologist has an annual income of $12,000 or more is meaningful only if we specify whether we are referring to all professional psychologists in the United States, all those employed in industry, all those holding advanced college degrees, or perhaps all those who belong to a certain professional organization.

Since the choice of the sample space (namely, the set of all possibilities under consideration) is by no means always self-evident, let us introduce the symbol $P(A|S)$ to denote the *conditional probability* of event A relative to the sample space S, or as we often call it "the probability of A given S." Every probability is thus a conditional probability, and we use the abbreviated notation $P(A)$ and drop the word "conditional" only when the tacit choice of S is clearly understood.

To elaborate on the concept of a *conditional probability*, let us consider the following problem: There are 100 qualified applicants for a teaching position in an elementary school, of which some have had at least three years of teaching experience and some have not, some are married and some are single, with the exact breakdown being

	Married	*Single*
At least three years teaching experience	12	24
Less than three years teaching experience	18	46

If we assume that each applicant has the same probability of $\frac{1}{100}$ of being selected (say, because there is no time to screen their credentials), E denotes the event that the applicant who is selected will have had at least three years of teaching experience, and M denotes the event that the applicant who is selected is married, the probabilities associated with the various possibilities are as shown in the Venn diagram of Figure 2.4. We can thus write $P(E) = 0.24 + 0.12 = 0.36$, for example, $P(M) = 0.12 + 0.18 = 0.30$, and $P(E \cap M) = 0.12$, with the tacit understanding that we are referring to the sample space S in which each of the 100 applicants has the probability $\frac{1}{100}$ of getting the job.

Now suppose that for some reason the selection is to be limited to applicants who are married, and that we are interested in the probability $P(E|M)$, namely, the probability that under these conditions the person who gets the job will have had at least three years of teaching experience. This reduces the sample space to 30 possibilities, and if we assume that each of the remaining applicants still has an equal chance (namely, a probability of $\frac{1}{30}$) of getting the job, it follows immediately that $P(E|M) = \frac{12}{30} = 0.40$. Note that this *exceeds* $P(E) = 0.36$, the probability that a person selected from the original 100 applicants will have had at least three years of teaching experience. On the other hand, in Exercise 7 on page 64 the reader will be asked to verify that $P(E|M') = \frac{24}{70}$ or approximately 0.34, and all this simply expresses the fact that there are relatively fewer experienced applicants among those who are single than among those who are married.

When we restricted the selection to applicants who are married, we actually reduced the sample space of Figure 2.4 to that part which is

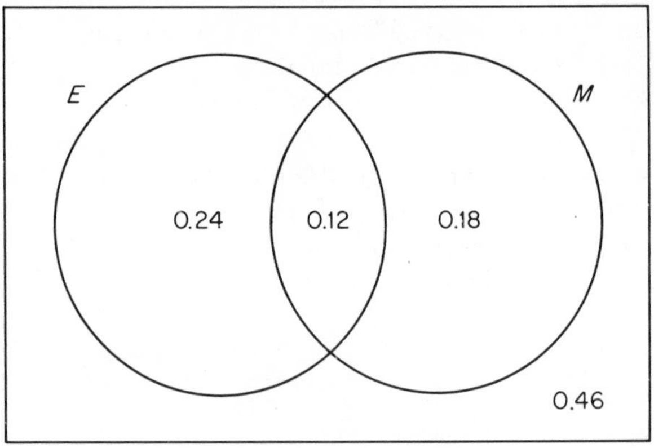

FIGURE 2.4 Venn diagram.

shown in Figure 2.5. Of course, the probabilities assigned to the two regions have changed—in Figure 2.5 they have to add up to 1—but they are still proportional. In fact, the two probabilities of Figure 2.5 could have been obtained by dividing the corresponding probabilities of Figure 2.4 by $P(M) = 0.30$. Generalizing from this observation, let us now make the following definition:

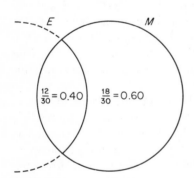

FIGURE 2.5 Reduced sample space.

If A and B are two subsets of a sample space S and if $P(B) \neq 0$, then the conditional probability of A relative to B is given by

$$P(A|B) = \frac{P(A \cap B)}{P(B)}$$

As in our numerical illustration, we divided by $P(B)$ so that in the *reduced* sample space, the set B, the probabilities will add up to 1. This can be seen from the diagrams of Figures 2.6 and 2.7, paralleling those of Figures 2.4 and 2.5, and in Exercise 1 on page 63 the reader will be asked to

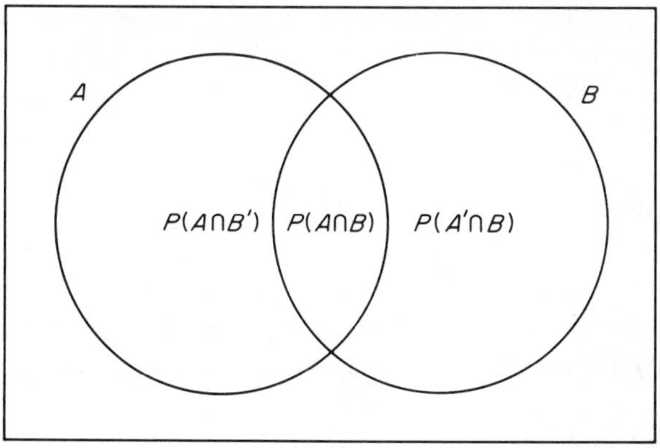

FIGURE 2.6 Venn diagram.

prove formally that with this definition $P(A|B)$ and $P(A'|B)$ add up to 1. It should also be observed that our definition of conditional probabilities is quite general; it does *not* depend on the assumption of equally likely possibilities as in our illustration.

To consider an example in which we are not dealing with equiprobable outcomes, suppose that a manufacturer of certain airplane parts knows from past experience that (1) the probability that an order will be ready for shipment on time is 0.80 and (2) the probability that an order will be ready for shipment on time *and* that it will also be delivered on time is 0.72. What we would like to know is the probability that such an order will be delivered on time *given that it was ready for shipment on time.* If we let R stand for the event that an order is ready for shipment on time and D for the event that it is delivered on time, we can write the given information as $P(R) = 0.80$ and $P(R \cap D) = 0.72$, and it follows that

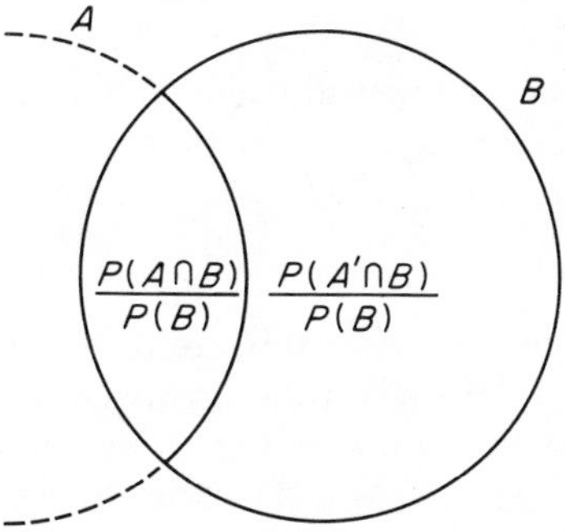

FIGURE 2.7 Reduced sample space.

$$P(D|R) = \frac{P(R \cap D)}{P(R)} = \frac{0.72}{0.80} = 0.90$$

Thus, 90 percent of these shipments are delivered on time provided that they are shipped on time. [Note that $P(R|D)$, the probability that a shipment which is delivered on time *was also ready for shipment on time*, cannot be determined without further information; for this purpose we would also have to know $P(D)$.]

2.3.1 Multiplication Rules

An immediate consequence of our definition of conditional probabilities is expressed in the following theorem, which is often called the *general rule of multiplication:**

THEOREM 2.8

If A and B are any two subsets of a sample space S, then

$$P(A \cap B) = P(B) \cdot P(A|B) \quad \textit{provided} \quad P(B) \neq 0$$
$$P(A \cap B) = P(A) \cdot P(B|A) \quad \textit{provided} \quad P(A) \neq 0$$

To prove the second part of this theorem we have only to interchange A and B in the first part and make use of the fact that $A \cap B = B \cap A$ for any two sets A and B. In words, Theorem 2.8 states that the probability that events A and B will *both* occur is the product of the probability that one of the events will occur and the conditional probability that the other event will occur given that the first event has occurred (occurs, or will occur). For instance, the probability of having the misfortune of picking two defective television tubes in succession from a shipment of 240 television tubes among which 15 are defective is $\frac{15}{240} \cdot \frac{14}{239} = \frac{7}{1,912}$ (since there are only 14 defective tubes among the 239 which remain after one defective tube has been picked). This assumes, incidentally, that we are "sampling without replacement," namely, that the first tube is not replaced before the second tube is picked, and that in each case all of the tubes from which we take a pick have the same chance of being selected. Under the same conditions, the probability of drawing two aces in succession from an ordinary deck of 52 playing cards is $\frac{4}{52} \cdot \frac{3}{51} = \frac{1}{221}$, but if the first card is replaced before the second is drawn the corresponding probability is $\frac{4}{52} \cdot \frac{4}{52} = \frac{1}{169}$. In both of these examples there is a definite *temporal order* between the two events A and B, but this need not be the case when we write $P(A|B)$ or $P(B|A)$; for instance, we could ask for the probability that the first card drawn is an ace given that the second card drawn (without replacement) is an ace—the answer is also $\frac{3}{51}$.

* A corresponding "special" rule will be given in Section 2.3.2.

Theorem 2.8 can easily be generalized so that it applies to more than two events; for instance, for three events we have:

THEOREM 2.9

If A, B, and C are any three subsets of a sample space S, then

$$P(A \cap B \cap C) = P(A)\cdot P(B|A)\cdot P(C|A \cap B) \quad \textit{provided } P(A \cap B) \neq 0$$

Since the letters A, B, and C can be interchanged, this theorem states *in words* that the probability that three events will *all* occur is the product of the probability that one of the events will occur, the conditional probability that one of the other events will occur given that the first event has occurred (occurs, or will occur), and the conditional probability that the remaining event will occur given that the other two events have occurred (are occurring, or will occur). To prove Theorem 2.9 let us substitute D for $A \cap B$, so that, repeatedly applying Theorem 2.8, we can write

$$\begin{aligned} P(A \cap B \cap C) &= P(D \cap C) \\ &= P(D)\cdot P(C|D) \\ &= P(A \cap B)\cdot P(C|A \cap B) \\ &= P(A)\cdot P(B|A)\cdot P(C|A \cap B) \end{aligned}$$

To illustrate the use of this formula, suppose that in Portland, Oregon, the probability that a rainy fall day is followed by a rainy day is 0.80 and the probability that a sunny fall day is followed by a rainy day is 0.60. Then, the probability that a rainy fall day is followed by a rainy day, a sunny day, and another rainy day is $0.80\cdot(1 - 0.80)\cdot 0.60 = 0.096$, and the probability that a rainy fall day is followed by two sunny days and then another rainy day is $(1 - 0.80)\cdot(1 - 0.60)\cdot 0.60 = 0.048$. Also, the probability of drawing (without replacement) three aces in succession from an ordinary deck of 52 playing cards is $\frac{4}{52}\cdot\frac{3}{51}\cdot\frac{2}{50} = \frac{1}{5{,}525}$. A further generalization of Theorems 2.8 and 2.9, applicable to the occurrence of *four* events, will be given in Exercise 2 on page 63.

2.3.2 Independent Events

Informally speaking, event A is said to be *independent* of event B if the probability of its occurrence is in no way affected by the occurrence or the non-occurrence of event B. For instance, if we draw cards from an ordinary deck of 52 playing cards and each card is replaced before the next one is drawn, the outcomes of successive draws are all independent—the probability of getting an ace, for example, is $\frac{4}{52}$ in each case. To express this idea formally, let us now make the following definition:

Event A is independent of event B if and only if $P(A|B) = P(A)$.

An immediate consequence of this definition is that *if A is independent of B then B is also independent of A.* To prove this we have only to make use of the fact that $P(B) \cdot P(A|B)$ and $P(A) \cdot P(B|A)$ both equal $P(A \cap B)$ according to Theorem 2.8, so that we can write

$$P(B) \cdot P(A|B) = P(A) \cdot P(B|A)$$

Then, substituting $P(A)$ for $P(A|B)$ we get $P(B) \cdot P(A) = P(A) \cdot P(B|A)$, and upon dividing by $P(A)$ we arrive at the result that $P(B) = P(B|A)$, namely, that B is also independent of A. Thus, whenever A is independent of B or vice versa, we can simply say that *A and B are independent.* When two events are not independent, we say that they are *dependent;* for instance, getting aces in two successive draws (without replacement) from an ordinary deck of 52 playing cards are *dependent events.*

In the special case where A and B are independent events, the formulas of Theorem 2.8 reduce to what is called the *special rule of multiplication,* namely,

THEOREM 2.10

If A and B are independent events, then

$$P(A \cap B) = P(A) \cdot P(B)$$

To obtain this result we have only to substitute $P(A)$ for $P(A|B)$ in the first of the two formulas of Theorem 2.8, or $P(B)$ for $P(B|A)$ in the second. Actually, the "if" in Theorem 2.10 can be replaced by "if and only if," for $P(A \cap B) = P(A) \cdot P(B)$ together with the first formula of Theorem 2.8 yields $P(A) = P(A|B)$, which means that A and B are independent. Thus, $P(A \cap B) = P(A) \cdot P(B)$ can be used as a *criterion* for deciding whether or not two events A and B are independent.

To illustrate the use of Theorem 2.10, let us calculate the probability of getting two *heads* in a row with a balanced coin, the probability of getting two 7's in a row with a pair of "honest" dice, and the probability that two totally unrelated persons will choose blue as their favorite color, given that for any one person the probability of his choosing blue is 0.24. All this is very simple, for the two *heads* we get $\frac{1}{2} \cdot \frac{1}{2} = \frac{1}{4}$, for the two 7's we get $\frac{1}{6} \cdot \frac{1}{6} = \frac{1}{36}$ (see page 40), and for the two persons choosing blue we get $(0.24)(0.24) = 0.0576$.

In order to extend the concept of independence to more than two events, let us now make the following definition:

> *$A_1, A_2, \ldots,$ and A_k are independent events if and only if the probability of the intersection of any 2, 3, ..., or k of them equals the product of their respective probabilities.*

Thus, three events A, B, and C are independent if and only if $P(A \cap B) = P(A) \cdot P(B)$, $P(A \cap C) = P(A) \cdot P(C)$, $P(B \cap C) = P(B) \cdot P(C)$, and $P(A \cap B \cap C) = P(A) \cdot P(B) \cdot P(C)$.

It is of interest to note that three or more events can be *pairwise independent* without being independent. Suppose, for instance, that a large living room has three separate switches which control the ceiling lights and that we are concerned *only* with those cases (positions of the switches) where the lights are on. The lights will be on when all three switches are "up," and hence also when one of the switches is "up" and the other two are "down." If A denotes the event that the first switch is "up," B denotes the event that the second switch is "up," and C denotes the event that the third switch is "up," the Venn diagram of Figure 2.8 shows the probabilities associated with the different switches being "up" or "down," assuming that each switch is as likely to be "up" as "down," and that *the ceiling lights are on.* As can be seen from this diagram, $P(A) = P(B) = P(C) = \frac{1}{2}$, $P(A \cap B) = P(A \cap C) = P(B \cap C) = \frac{1}{4}$, and $P(A \cap B \cap C) = \frac{1}{4}$, so that A and B are independent, A and C are independent, B and C are independent, but A, B, and C are *not* independent because $P(A \cap B \cap C) = \frac{1}{4}$ while $P(A) \cdot P(B) \cdot P(C) = \frac{1}{8}$. It can also happen that $P(A \cap B \cap C) = P(A) \cdot P(B) \cdot P(C)$ without A, B, and C being independent—this the reader will be asked to verify in Exercise 4 on page 63 with reference to the Venn diagram and the probabilities of Figure 2.9.

2.3.3 Bayes' Rule

There are many problems in which the ultimate outcome of an experiment depends on what happens in various intermediate stages. To consider the simplest case where there is *one* intermediate stage consisting of *two* alternatives, suppose that we are concerned with the completion of a highway construction job, which may be delayed because of a strike. Suppose, furthermore, that the probability that there will be a strike is 0.60, and the probability that the job will be completed on time is 0.85

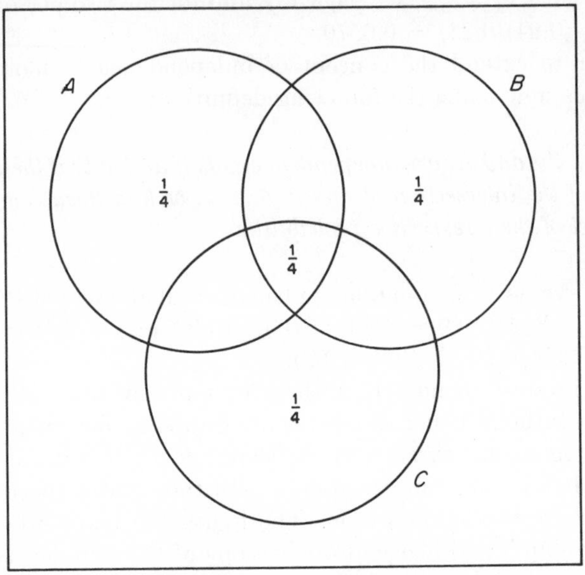

FIGURE 2.8 Venn diagram.

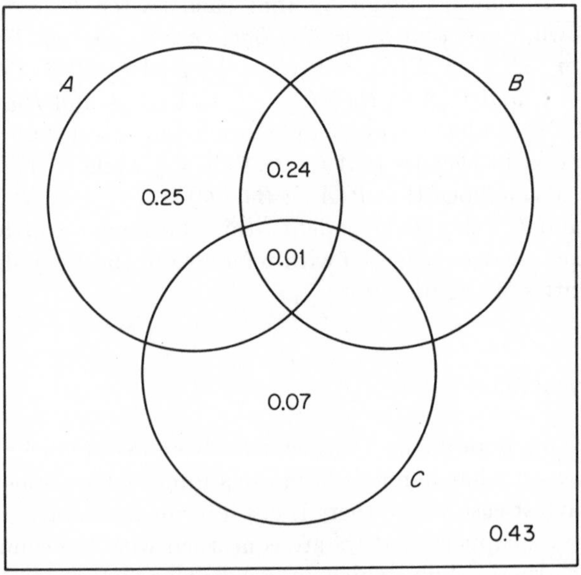

FIGURE 2.9 Venn diagram.

if there is no strike, 0.35 if there is a strike. *What we would like to know is the probability that the job will be completed on time.* If we let A represent the event that the job will be completed on time and B the event that there will be a strike, the information which we have can be written as $P(B) = 0.60$, $P(A|B) = 0.35$, and $P(A|B') = 0.85$. Then, using part (c) of Exercise 3 on page 14, we can write

$$A = (A \cap B) \cup (A \cap B')$$

where $A \cap B$ and $A \cap B'$ are mutually exclusive events. We thus get

$$\begin{aligned} P(A) &= P[(A \cap B) \cup (A \cap B')] \\ &= P(A \cap B) + P(A \cap B') \\ &= P(B) \cdot P(A|B) + P(B') \cdot P(A|B') \end{aligned}$$

and substitution of the given numerical values yields

$$P(A) = (0.60)(0.35) + (1 - 0.60)(0.85) = 0.55$$

An immediate generalization of this method to the case where the intermediate stage permits k alternatives (whose occurrence is denoted $B_1, B_2, \ldots, B_k$) is taken care of by the following theorem, sometimes called the *rule of elimination:*

THEOREM 2.11

If $B_1, B_2, \ldots,$ and B_k constitute a set of mutually exclusive events of which one must occur and none has a zero probability, then for any event A

$$P(A) = \sum_{i=1}^{k} P(B_i) \cdot P(A|B_i)$$

A formal proof of this theorem consists, essentially, of the same steps which we used above to prove the special case where $k = 2$, and it will be left to the reader in Exercise 6 on page 63. More instructive, perhaps, is to visualize the situation by means of the tree diagram of Figure 2.10, where the probability of the final outcome A is given by the *sum* of the *products* of the probabilities corresponding to the two parts of each branch.

To illustrate the use of Theorem 2.11, suppose that the probabilities that a brewery decides to sponsor the televising of football games, a soap opera, a game show, or a news program are, respectively, 0.20, 0.40, 0.30, and 0.10. If they decide on the football games, the probability that they will get a high rating is 0.85; if they decide on a soap opera, the proba-

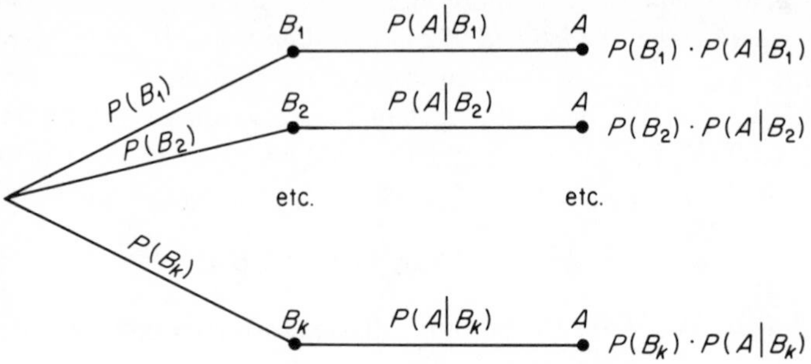

FIGURE 2.10 Tree diagram.

bility that they will get a high rating is 0.45; if they decide on a game show, the probability that they will get a high rating is 0.35; and if they decide on a news program, the probability that they will get a high rating is 0.15. If we let A denote their getting a high rating and B_1, B_2, B_3, and B_4 their choosing the respective kinds of programs, we have $P(B_1) = 0.20$, $P(B_2) = 0.40$, $P(B_3) = 0.30$, $P(B_4) = 0.10$, $P(A|B_1) = 0.85$, $P(A|B_2) = 0.45$, $P(A|B_3) = 0.35$, $P(A|B_4) = 0.15$, and substitution into the formula yields

$$\begin{aligned} P(A) &= (0.20)(0.85) + (0.40)(0.45) + (0.30)(0.35) + (0.10)(0.15) \\ &= 0.47 \end{aligned}$$

for the probability that they will get a high rating. To picture what is going on, it will help to draw a tree diagram like that of Figure 2.11, where the final result is given by the sum of the probabilities associated with the four branches leading to A.

Theorem 2.11 can easily be generalized to situations involving several intermediate states. For instance, it can be shown that:

THEOREM 2.12

If B_1, B_2, . . ., and B_k constitute a set of mutually exclusive events of which one must occur and C_1, C_2, . . ., and C_m constitute another set of mutually exclusive events of which one must occur, then for any event A

$$P(A) = \sum_{i=1}^{k}\sum_{j=1}^{m} P(B_i)\cdot P(C_j|B_i)\cdot P(A|B_i \cap C_j)$$

To illustrate the use of this formula with an example from quality control inspection, suppose that B_1, B_2, and B_3 represent a missile component's being perfect, good, or inferior, that C_1 and C_2 represent its being properly or improperly installed, and that A represents its performing well when the missile is fired. If the various probabilities are as shown in Figure 2.12,

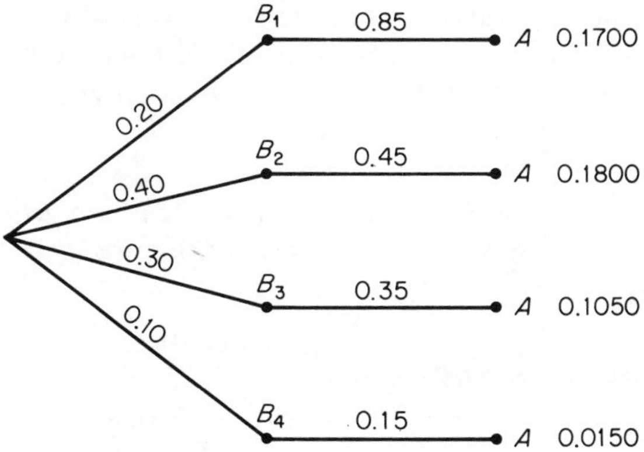

FIGURE 2.11 Tree diagram.

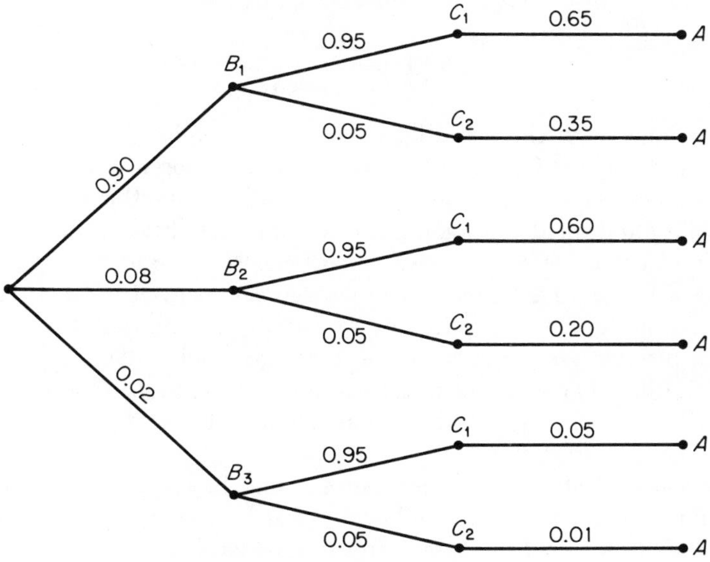

FIGURE 2.12 Tree diagram.

we find that the probability that the component will perform well when the missile is fired is

$$\begin{aligned} P(A) &= (0.90)(0.95)(0.65) + (0.90)(0.05)(0.35) + (0.08)(0.95)(0.60) \\ &\quad + (0.08)(0.05)(0.20) + (0.02)(0.95)(0.05) + (0.02)(0.05)(0.01) \\ &= 0.62 \end{aligned}$$

rounded to two decimals.

Referring back to the example on page 59, suppose now that we are interested in an entirely different question: *Discovering later on that the brewery's television program got a high rating, we want to know the probability that they had decided to sponsor a soap opera.* To answer this kind of question, we need the following theorem, called *Bayes' rule:*

THEOREM 2.13

If B_1, B_2, ..., and B_k constitute a set of mutually exclusive events of which one must occur and none has a zero probability, then for any event A for which $P(A) \neq 0$

$$P(B_r|A) = \frac{P(B_r) \cdot P(A|B_r)}{\sum_{i=1}^{k} P(B_i) \cdot P(A|B_i)}$$

for $r = 1, 2, \ldots,$ or k.

To prove this theorem we have only to write

$$P(B_r|A) = \frac{P(B_r \cap A)}{P(A)}$$

in accordance with the definition of conditional probabilities on page 52, and then substitute for $P(A)$ the expression of Theorem 2.11.

Although Bayes' rule follows quite readily from the postulates of probability and the definition of conditional probabilities, it has been the subject of extensive controversy. There can be no question about the *validity* of Bayes' rule, but considerable arguments have been raised about the interpretation of the "prior" or "a priori" probabilities $P(B_i)$. In his original formulation, Bayes assumed that the probabilities $P(B_i)$ are all equal and, hence, cancel out in the formula for $P(B_r|A)$. This, of course, limits the applicability of the theorem and exposes it to all sorts of criticism.

A good deal of the mysticism surrounding Bayes' rule is due to the fact that it entails a "backward" or "inverse" sort of reasoning, namely, reasoning "from effect to cause." In the question raised at the top of this page we had discovered that the brewery's television program got a high rating, and wanted to know whether this was "caused" by their choosing a soap opera. To answer this question, we can now substitute the appropriate probabilities into the formula of Theorem 2.13, getting

$$\begin{aligned} P(B_2|A) &= \frac{(0.40)(0.45)}{(0.20)(0.85) + (0.40)(0.45) + (0.30)(0.35) + (0.10)(0.15)} \\ &= \frac{0.18}{0.47} \\ &= 0.38 \end{aligned}$$

rounded to two decimals. With reference to the tree diagram of Figure 2.11, this answer is the *ratio* of the probability of the branch which leads to A via B_2 to the *sum* of the probabilities of all four branches of the diagram.

To consider another example, suppose that there is a fifty-fifty chance that Tom's ex-wife will show up at the New Year's party to which he is planning to go. If she does not show up at the party, the odds are 3 to 1 that Tom will have a good time, but if she does show up, the odds are 4 to 1 that he will *not* have a good time. If someone tells us later that Tom did have a good time, *what is the probability that his ex-wife did not show up?* Letting A denote Tom's having a good time at the party, B_1 his ex-wife's showing up, and B_2 his ex-wife's not showing up, we have $P(B_1) = 0.50$, $P(B_2) = 0.50$, $P(A|B_1) = 0.20$, and $P(A|B_2) = 0.75$, so that

$$P(B_2|A) = \frac{(0.50)(0.75)}{(0.50)(0.20) + (0.50)(0.75)} = 0.79$$

rounded to two decimals. Further uses of Bayes' rule will be taken up in Chapter 9.

THEORETICAL EXERCISES

1. Referring to the definition of conditional probabilities on page 52, show that $P(A|B) + P(A'|B) = 1$, provided $P(B) \neq 0$.
2. Duplicating the method of proof of Theorem 2.9, show that $P(A \cap B \cap C \cap D) = P(A) \cdot P(B|A) \cdot P(C|A \cap B) \cdot P(D|A \cap B \cap C)$ provided $P(A \cap B \cap C) \neq 0$.
3. Show that if A is independent of B then A is also independent of B', namely, that $P(A|B) = P(A)$ implies $P(A|B') = P(A)$, provided $P(B) \neq 1$. [*Hint:* Make use of the fact that $A = (A \cap B) \cup (A \cap B')$ where $A \cap B$ and $A \cap B'$ are mutually exclusive.]
4. Use the Venn diagram of Figure 2.9 to demonstrate that $P(A \cap B \cap C) = P(A) \cdot P(B) \cdot P(C)$ does not necessarily imply that A, B, and C are independent.
5. Show that $2^k - k - 1$ conditions must be satisfied for k events to be independent.
6. Prove Theorem 2.11, making use of the following generalization of the first distributive law on page 11:

$$A \cap (B_1 \cup B_2 \cup \cdots \cup B_k) = (A \cap B_1) \cup (A \cap B_2) \cup \cdots \cup (A \cap B_k)$$

APPLIED EXERCISES

7. Verify for the example on page 51 that $P(E|M') = \frac{24}{70}$, or approximately 0.34.
8. With reference to Exercise 21 on page 49, what is the probability that a husband will vote in the given school board election given that his wife will vote?
9. With reference to Exercise 22 on page 49, what is the probability that a person with a fever visiting his doctor who gets a shot will also get a pill?
10. With reference to Exercise 23 on page 49, find the probabilities that
    (a) the professor who is chosen will hold a Ph.D. degree given that he devotes part of his time to research;
    (b) the professor who is chosen will devote at least part of his time to research given that he does *not* hold a Ph.D. degree.
11. With reference to Exercise 25 on page 50 find the probabilities that a person who visits Disneyland will
    (a) ride the Monorail given that he will go on the Jungle Cruise;
    (b) go on the Matterhorn ride given that he will go on the Jungle Cruise *and* ride the Monorail;
    (c) *not* go on the Jungle Cruise given that he will ride the Monorail and/or go on the Matterhorn ride;
    (d) go on the Matterhorn ride *and* the Jungle Cruise given that he will *not* ride the Monorail.

    (*Hint:* Draw a Venn diagram and fill in the various probabilities given in Exercise 25 on page 50.)
12. With reference to the example on page 55, find the probabilities that in Portland, Oregon, a rainy fall day is
    (a) followed by three more rainy days;
    (b) followed by another rainy day and then two sunny days;
    (c) followed by two rainy days and then two sunny days.

    [*Hint:* In part (c) use the formula of Exercise 2.]
13. Use the formula of Exercise 2 to find the probability of randomly choosing (without replacement) *four* healthy guinea pigs from a cage containing 20 guinea pigs of which 15 are healthy and 5 are diseased.
14. Verify that for the situation described in Figure 2.13, events A, B, and C are not independent. Are any two of them pairwise independent?
15. Verify that for the situation described in Figure 2.14, events A, B, C, and D are independent. Note that the region which represents event A consists of two circles, and so do those regions representing events B and C.

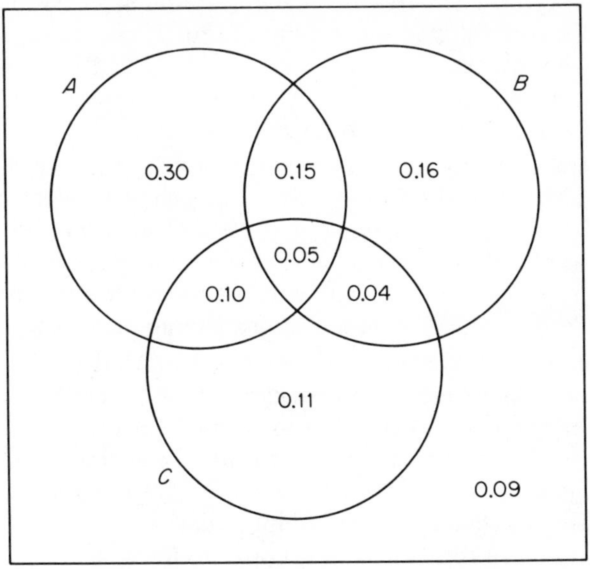

FIGURE 2.13 Venn diagram for Exercise 14.

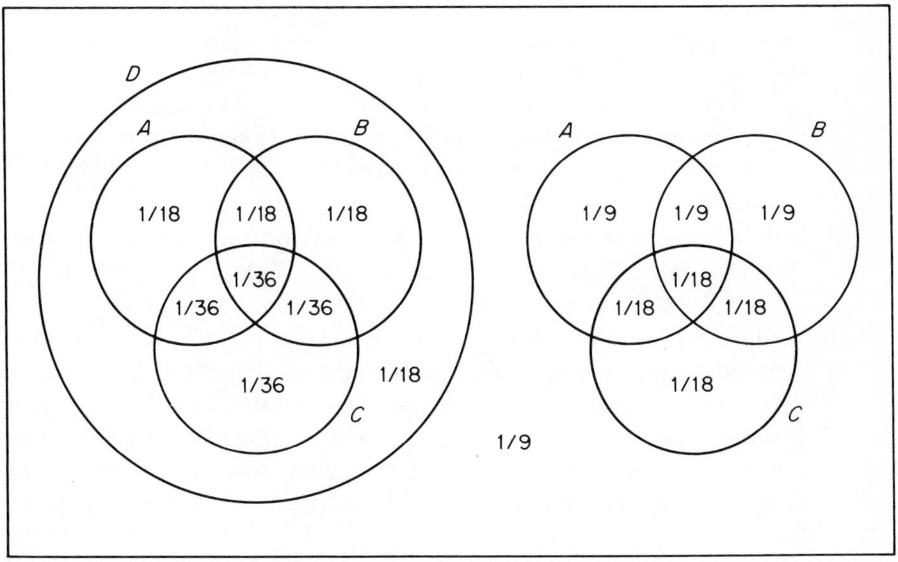

FIGURE 2.14 Diagram for Exercise 15.

16. Roy has enrolled as a freshman at a certain university and the probability that he will get a scholarship is 0.30. If he gets the scholarship, the probability that he will graduate is 0.85, and if he does not get the scholarship, the probability that he will graduate is only 0.45. What is the probability that he will graduate?

17. In a certain community, 10 percent of all adults over fifty have diabetes. If a certain doctor *correctly* diagnoses 95 percent of all persons with diabetes as having the disease and *incorrectly* diagnoses 2 percent of all persons without diabetes as having the disease, what is the probability that he will diagnose an adult over 50 (randomly chosen from the given community) as having diabetes?

18. A toy store employs three gift wrappers at Christmas time. Ruth, who wraps 38 percent of all packages, fails to remove the price tag 2 percent of the time; Helen, who wraps 22 percent of all packages, fails to remove the price tag 8 percent of the time; and Jean, who wraps the remaining packages, fails to remove the price tag 5 percent of the time. What is the probability that the price tag will not be removed from a toy gift wrapped at this store?

19. The manager of a restaurant knows that the odds are 2 to 1 that a customer will not have a cocktail before dinner. If he has a cocktail, the odds are 3 to 2 that he will not order steak, and if he does not have a cocktail, the odds are 3 to 1 that he will not order steak. If a customer has a cocktail and a steak, the odds are 9 to 1 that he will not have dessert; if he has a cocktail but does not order steak, the odds are 7 to 1 that he will not have dessert; if he does not have a cocktail but orders steak, the odds are 3 to 1 that he will not have dessert; and if he does not have a cocktail and does not order steak, the odds are 2 to 1 that he will not have dessert. What is the probability that any one customer of this restaurant will order dessert?

20. With reference to Exercise 16, suppose that we hear years later that Roy graduated from the given university. What is the probability that he did get the scholarship?

21. With reference to Exercise 17, suppose that an adult over fifty living in the given community is diagnosed by the doctor as having diabetes. What is the probability that he actually has this disease?

22. With reference to Exercise 18, suppose that a customer discovers later that a toy he had gift wrapped at the given store did not have the price tag removed. What is the probability that it was wrapped by Ruth?

23. With reference to Exercise 19, find the probabilities that
 (a) a customer who orders dessert had a cocktail before dinner;
 (b) a customer who does not order dessert had a steak;

(c) a customer who orders steak did not have a cocktail before dinner;
(d) a customer who orders steak and dessert also had a cocktail before dinner.

(*Hint:* Draw a tree diagram, if necessary.)

REFERENCES

Among the numerous textbooks on probability published in recent years, one of the most popular is the book by W. Feller listed on page 35. More elementary treatments may be found in

Burford, R. L., *Introduction to Finite Probability.* Columbus, Ohio: Charles E. Merrill Books, Inc., 1967.

Dixon, J. R., *A Programmed Introduction to Probability.* New York: John Wiley & Sons, Inc., 1964.

Goldberg, S., *Probability—An Introduction.* Englewood Cliffs, N.J.: Prentice-Hall, Inc., 1960.

More advanced treatments are given in many texts, for instance, in

McCord, J. R., and Moroney, R. M., *Introduction to Probability Theory.* New York: The Macmillan Company, 1964.

Papoulis, A., *Probability, Random Variables, and Stochastic Processes.* New York: McGraw-Hill Book Company, 1965.

Parzen, E., *Modern Probability Theory and Its Applications.* New York: John Wiley & Sons, Inc., 1960.

Interesting discussions of philosophical questions relating to probability may be found in

Borel, E., *Elements of the Theory of Probability.* Englewood Cliffs, N.J.: Prentice-Hall, Inc., 1965,

Good, I. J., *The Estimation of Probabilities: An Essay on Modern Bayesian Methods.* Cambridge, Mass.: Massachusetts Institute of Technology Press, 1965,

Nagel, E., *Principles of the Theory of Probability.* Chicago: University of Chicago Press, 1939,

Savage, L. J., *The Foundations of Statistics.* New York: John Wiley & Sons, Inc., 1954,

and in the works of Rudolf Carnap, Hans Reichenbach, and Richard von Mises.

3

PROBABILITY FUNCTIONS

3.1 RANDOM VARIABLES

In most applications of probability theory we are interested only in a particular aspect (or in two or three particular aspects) of the outcome of an experiment. When rolling a pair of dice, for example, we are generally interested only in the total, and not in the outcome for each die; when interviewing a randomly selected housewife, we may be interested in the size of her family or her husband's income, but not in her age or her

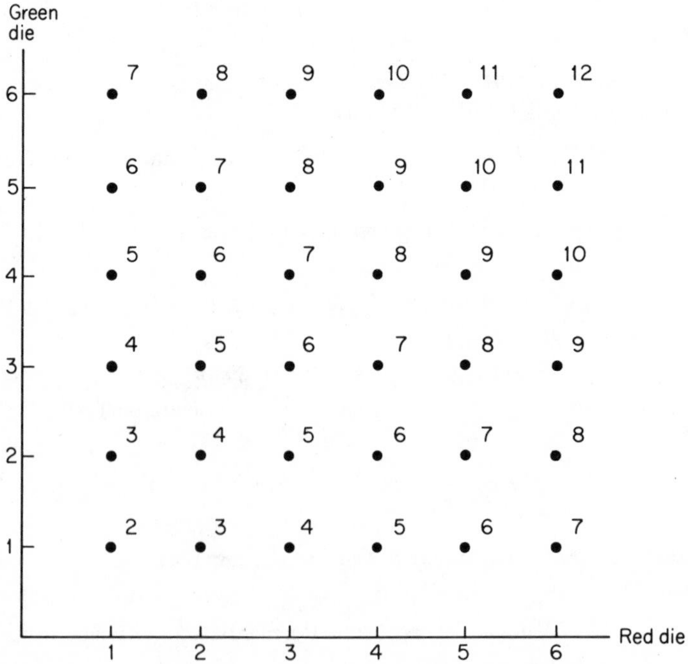

FIGURE 3.1 Outcomes for two dice.

weight; and when sampling mass-produced light bulbs, we may be interested in their durability or their brightness, but not in their price.

In each of these examples we are interested in a number which is associated with the outcome of a chance experiment, namely, the value which is taken on by a so-called *random variable.* In the language of probability and statistics, the total which we obtain with a pair of dice is a random variable, the size of the family of a randomly selected housewife and her husband's income are random variables, and so are the durability and the brightness of a light bulb randomly picked for quality inspection.

To be more explicit, let us consider Figure 3.1, which (like Figure 1.5 on page 4) pictures the sample space for an experiment in which we roll a pair of dice, and let us assume (as on page 40) that each of the 36 possible outcomes has a probability of $\frac{1}{36}$. Note, however, that in contrast to Figure 1.5, we have attached a number to each point: for instance, we attached the number 2 to the point (1, 1), the number 6 to the point (1, 5), the number 8 to the point (6, 2), the number 11 to the point (5, 6), and so forth. Evidently, we associated with each point the corresponding *total* rolled with the two dice.

Since "associating a number with each point (element) of a sample space" is merely another way of saying that we are "defining a function over the points of a sample space," let us now make the following definition:

> *If S is a sample space with a probability measure and if* **x** *is a real-valued function defined over the points of S, then* **x** *is called a random variable.**

In this book we shall always write random variables symbolically in boldface type. This practice is relatively new and not universally accepted (some authors use capital letters to denote random variables), but it is consistent with the notation of modern mathematics, where we write **f**, for example, for the function whose values are given by $y = f(x)$, **g** for the function whose values are given by $y = g(x)$, and so forth. When taking notes or writing on a blackboard, the reader will find it convenient to indicate random variables by underlining the respective symbols, perhaps with wiggly lines, the usual type-setting symbol for boldface type.

The fact that the above definition is limited to real-valued functions does not impose any undue restrictions. If the numbers which we want to assign to the outcomes of an experiment are complex, we can always look upon the real and the imaginary parts separately as values taken on by *two* random variables. Also, if we want to describe the outcomes of an

* Instead of "random variable," the terms "chance variable," "stochastic variable," and "variate" are also used in some books.

experiment qualitatively, say, by giving the color of a person's hair, we can arbitrarily make the descriptions real-valued by *coding* the various colors—perhaps represent them with the numbers 1, 2, 3, etc.

For the time being we shall limit our discussion to *discrete* sample spaces, and, hence, to *discrete random variables,* namely, random variables whose range is finite or countably infinite (see page 6). Continuous sample spaces and continuous random variables will be taken up in Chapter 4.

3.2 PROBABILITY FUNCTIONS

Perhaps the most pertinent feature of discrete random variables is that the probabilities specified for the elements of S, the discrete sample space over which a random variable is defined, automatically provide the probabilities that the random variable will take on any given set of values within its range. For instance, having assigned probabilities of $\frac{1}{36}$ to the elements of the sample space of Figure 3.1, we immediately find that the probabilities with which the random variable *the total obtained with the pair of dice* takes on the values 2, 3, 4, . . . , 11, and 12 are as shown in the following table:

Total Obtained with the Dice	*Probability*
2	$\frac{1}{36}$
3	$\frac{2}{36}$
4	$\frac{3}{36}$
5	$\frac{4}{36}$
6	$\frac{5}{36}$
7	$\frac{6}{36}$
8	$\frac{5}{36}$
9	$\frac{4}{36}$
10	$\frac{3}{36}$
11	$\frac{2}{36}$
12	$\frac{1}{36}$

To obtain these results we merely added the probabilities of the respective points for which the total for the two dice equaled 2, 3, 4, . . . , 11, and 12.

The correspondence which is displayed by means of the above table illustrates what we mean by a *probability function.* In general,

A probability function is a function which assigns a probability $f(x)$ to each real number x within the range of a discrete random variable **x**.

To distinguish between the probability functions of different random variables **x**, we sometimes use the symbols $g(x)$, $h(x)$, $f_1(x)$, $\varphi(x)$, ..., instead of $f(x)$, and if we are dealing with two random variables **y** and **z**, we might write the values of their respective probability functions as $f(y)$ and $g(z)$, Note also that the values of a probability function, being probabilities, must be non-negative real numbers less than or equal to 1, and that $\Sigma f(x)$, summed over all values of x within the domain of the random variable **x**, must equal 1.

Whenever possible, we try to express probability functions by means of formulas, or equations, which enable us to calculate the probabilities associated with values of random variables by direct substitution. For the total obtained with a pair of dice we might, thus, write

$$f(x) = \frac{6 - |x - 7|}{36} \qquad \text{for } x = 2, 3, \ldots, 11, 12$$

where the absolute value $|x - 7|$ equals $x - 7$ or $7 - x$, whichever is positive or zero. For instance, for $x = 3$ we get $f(3) = \frac{6 - |3 - 7|}{36} = \frac{6 - 4}{36} = \frac{2}{36}$, for $x = 8$ we get $f(8) = \frac{6 - |8 - 7|}{36} = \frac{6 - 1}{36} = \frac{5}{36}$, and both of these values agree with the ones shown in the table on page 70.

Since there are many problems in which it is of interest to know the probability that the value of a random variable is less than or equal to some real number x, let us write the probability that **x** takes on a value less than or equal to x as

$$F(x) = \sum_{t \leq x} f(t) \qquad \text{for } -\infty \leq x \leq \infty \tag{3.2.1}$$

where $f(t)$ is the probability that **x** takes on the value t, and the summation extends over all values within the range of **x** that are less than or equal to x. It is customary to refer to the function defined by means of (3.2.1) as the *distribution function* (or the *cumulative distribution*) of the random variable **x**. Note that in (3.2.1) the domain of the distribution function is the set of all real numbers, and not merely those within the range of **x**.

Referring again to the total obtained with a pair of dice and the table on page 70, it can easily be seen that the distribution function of this random variable is given by

$$F(x) = \begin{cases} 0 & \text{for } x < 2 \\ \frac{1}{36} & \text{for } 2 \leq x < 3 \\ \frac{3}{36} & \text{for } 3 \leq x < 4 \\ \frac{6}{36} & \text{for } 4 \leq x < 5 \\ \frac{10}{36} & \text{for } 5 \leq x < 6 \\ \frac{15}{36} & \text{for } 6 \leq x < 7 \\ \frac{21}{36} & \text{for } 7 \leq x < 8 \\ \frac{26}{36} & \text{for } 8 \leq x < 9 \\ \frac{30}{36} & \text{for } 9 \leq x < 10 \\ \frac{33}{36} & \text{for } 10 \leq x < 11 \\ \frac{35}{36} & \text{for } 11 \leq x < 12 \\ 1 & \text{for } 12 \leq x \end{cases}$$

The graph of this distribution functions is shown in Figure 3.2, and it should be observed that at all points of discontinuity this *step-function* assumes the greater of the two values.

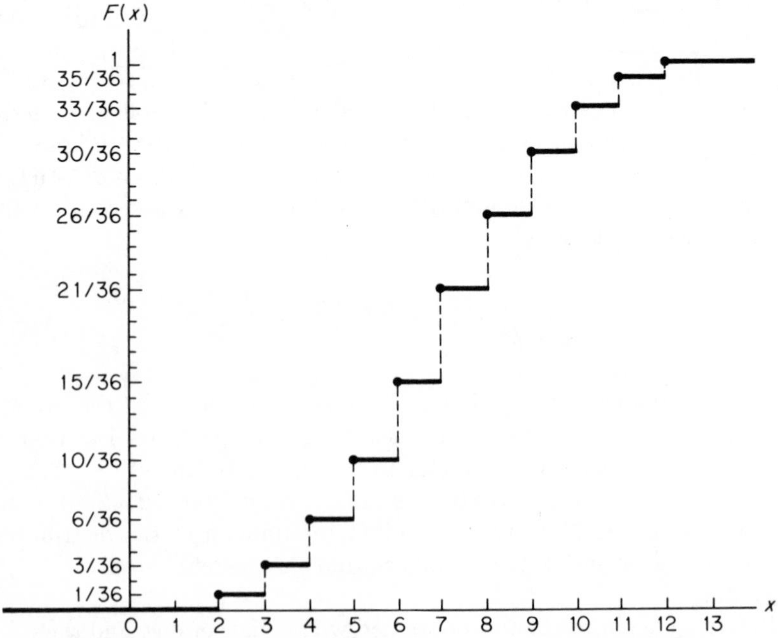

FIGURE 3.2 Distribution function for total obtained with a pair of dice.

3.3 SPECIAL PROBABILITY DISTRIBUTIONS

In the literature of probability and statistics, the terms "probability function" and "probability distribution" are often used interchangeably, although some writers make the distinction that the term "probability distribution" applies to *all* the probabilities associated with a random variable, and not only those given directly by its probability function. (The following illustrates this rather fine distinction: The probability of getting "7 or 11" with a pair of dice is part of the *probability distribution* of the total we obtain with a pair of dice, but it is *not* a value of its *probability function*. Of course, if we know the probability function, this probability can easily be calculated; in fact, the table on page 70 shows that the answer is $\frac{6}{36} + \frac{2}{36} = \frac{8}{36}$.)

The remainder of this chapter will be devoted to some important special kinds of probability distributions and their applications. Further mathematical properties of these probability distributions (as well as probability distributions in general) will be discussed in Chapter 5.

3.3.1 The Discrete Uniform Distribution

This probability distribution arises when a random variable can take on k different values with *equal probabilities*. In the special case where the values which the random variable can take on are $x = 1, 2, \ldots, k$, its probability function is

$$f(x) = \frac{1}{k} \qquad \text{for } x = 1, 2, \ldots, k \tag{3.3.1}$$

For instance, for the number of points rolled with a balanced die we have $f(1) = f(2) = f(3) = f(4) = f(5) = f(6) = \frac{1}{6}$.

3.3.2 The Bernoulli Distribution

This probability distribution applies when an experiment has two possible outcomes (often referred to as "failure" and "success"). If the probabilities of failure and success are, respectively, $1 - \theta$ and θ, and we *code* these two outcomes as 0 (*zero* successes) and 1 (*one* success), the probability function of the *Bernoulli distribution* can be written as

$$f(x; \theta) = \theta^x(1 - \theta)^{1-x} \qquad \text{for } x = 0, 1 \tag{3.3.2}$$

which expresses by means of *one* formula that $f(0; \theta) = 1 - \theta$ and $f(1; \theta) = \theta$.

In (3.3.2) we used the notation $f(x; \theta)$ to indicate explicitly that the Bernoulli distribution depends on the *parameter* θ. This term is used here as in most other branches of mathematics—the parameter θ is a *constant* when we refer to a *specific* Bernoulli distribution, but it can, and generally will, take on different values in different problems. For instance, $\theta = 0.50$ when we flip an "honest" coin, and 0 and 1 represent *tails* and *heads*, respectively; or $\theta = 0.02$ when we randomly select one fishing reel from a very large shipment containing two percent defectives, and 0 represents our getting a good reel while 1 represents our getting a bad one. Note that in this last example our getting a *good* reel is a "failure" while our getting a *bad* reel is a "success." This inconsistency is explained by the fact that this whole terminology is a holdover from the days when probability theory was applied only to games of chance (and one player's failure was the other's success). Also for this reason, we refer to an experiment to which the Bernoulli distribution applies as a "Bernoulli trial," or simply a "trial," and to the repetition of such experiments as "repeated trials."

3.3.3 The Binomial Distribution

Repeated Bernoulli trials play a very important role in probability and statistics, especially when the parameter θ (the probability of a "success") is the same for each trial, and the trials are all independent. In this section we shall be concerned only with the *total number of "successes"* in n trials, but as we shall see in Sections 3.3.5 and 3.3.6, there are other random variables of interest that are related to repeated Bernoulli trials.

The theory which we shall discuss in this section has many applications; for instance, it applies when we want to know the probability of getting 5 heads in 12 flips of a coin, the probability that 7 of 10 persons will recover from a given disease, or the probability that 35 of 80 persons will respond to a mail questionnaire. However, this is true only if each of the 10 persons has the same chance of recovering from the disease and their recoveries are independent (say, they are not all treated by the same doctor in the same hospital), and if the probability of getting a reply to the mail questionnaire is the same for each of the 80 persons and there is independence (say, no two of them belong to the same household).

To derive a formula for the probability of "x successes in n trials" under the stated conditions, we have only to observe that the probability of getting x successes and $n - x$ failures *in a specific order* is $\theta^x(1 - \theta)^{n-x}$.

There is one factor θ for each success, one factor $1 - \theta$ for each failure, and the x factors θ and $n - x$ factors $1 - \theta$ are all multiplied together by virtue of the assumption that the n trials are independent. Since this probability applies to any sequence of n trials in which there are x successes and $n - x$ failures, the desired probability for x successes in n trials *in any order* is simply the product of $\theta^x(1 - \theta)^{n-x}$ and $\binom{n}{x}$, the number of ways in which x successes can occur among the n trials. Hence, the probability for x successes in n trials is given by

$$b(x; n, \theta) = \binom{n}{x}\theta^x(1 - \theta)^{n-x} \qquad \text{for } x = 0, 1, 2, \ldots, n \quad (3.3.3)$$

We refer to the probability distribution of this random variable, the total number of successes in n trials, as the *binomial distribution*, and we used the symbol $b(x; n, \theta)$ to indicate that the values of the probability function depend on the parameters n and θ (the number of trials and the constant probability of success for each trial). Incidentally, the name "binomial distribution" derives from the fact that the values of $b(x; n, \theta)$ for $x = 0, 1, 2, \ldots$, and n are the successive terms of the *binomial expansion* of $[(1 - \theta) + \theta]^n$; this verifies also that the sum of the probabilities equals 1, as it should.

To illustrate the use of (3.3.3), let us calculate the first two of the probabilities mentioned in the second paragraph of this section. For the probability of 5 heads and 7 tails in 12 flips of a balanced coin, we substitute $x = 5$, $n = 12$, and $\theta = \frac{1}{2}$, getting

$$b(5; 12, \tfrac{1}{2}) = \binom{12}{5}(\tfrac{1}{2})^5(1 - \tfrac{1}{2})^7$$

and, looking up the value of $\binom{12}{5}$ in Table VII, we find that the result is $\frac{99}{512}$ or approximately 0.19. Similarly, to find the probability that 7 of 10 persons will recover from the given disease, let us suppose that for each person the probability of recovering from the disease is $\frac{4}{5}$. Then, substituting $x = 7$, $n = 10$, and $\theta = \frac{4}{5}$ into (3.3.3), we get

$$b(7; 10, \tfrac{4}{5}) = \binom{10}{7}(\tfrac{4}{5})^7(1 - \tfrac{4}{5})^3$$

and, looking up the value of $\binom{10}{7}$ in Table VII, we find that the result is $\frac{1{,}966{,}080}{9{,}765{,}625}$ or approximately 0.20.

Figures 3.3 and 3.4 show two different kinds of graphical presentations of probability distributions. The first is called a *histogram;* the height of each rectangle represents (is proportional to) the corresponding probability, and by letting 0 be represented by the interval from $-\frac{1}{2}$ to $\frac{1}{2}$, 1

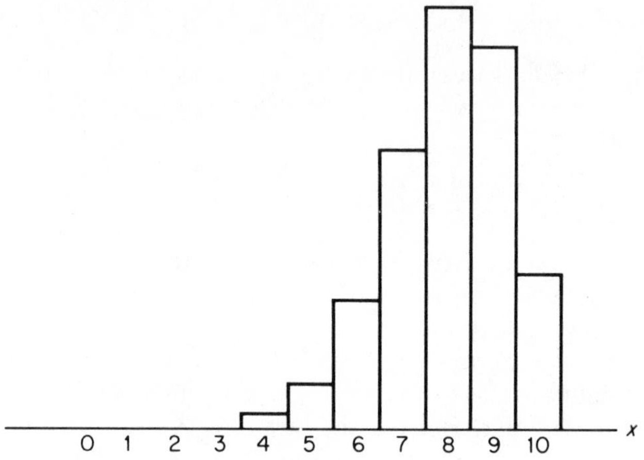

FIGURE 3.3 Binomial distribution with $n = 10$ and $\theta = \frac{4}{5}$.

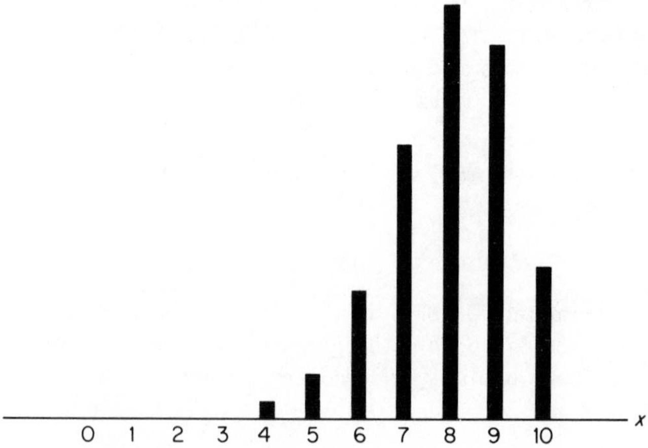

FIGURE 3.4 Binomial distribution with $n = 10$ and $\theta = \frac{4}{5}$.

by the interval from $\frac{1}{2}$ to $1\frac{1}{2}$, 2 by the interval from $1\frac{1}{2}$ to $2\frac{1}{2}$, . . . , we are so to speak "spreading" the values of the discrete random variable (in this case, the number of persons who recover) over a continuous scale. The same idea will be used later, when we shall approximate the graphs of binomial distributions (and other probability distributions) with continuous curves. The graph of Figure 3.4 is called a *bar chart;* each probability is represented by (is proportional to) the height of the corresponding bar.

Had we tried to calculate the third probability asked for on page 74 by substituting $x = 35$, $n = 80$, and, say, $\theta = 0.15$, into (3.3.3), the direct evaluation of $b(35; 80, 0.15)$ would have required a prohibitive amount of work. In actual practice, binomial probabilities are rarely calculated directly, for they have been tabulated extensively for various values of θ—by the National Bureau of Standards for $n = 2$ to $n = 49$, and by H. G. Romig (see page 100) for $n = 50$ to $n = 100$. In this book, Table I beginning on page 425 gives the values of $b(x; n, \theta)$ for $n = 1$ to $n = 20$ and $\theta = 0.05, 0.10, 0.15, \ldots, 0.45$, and 0.50. To make use of this table when θ is greater than 0.50, we have only to employ the identity

$$b(x; n, \theta) = b(n - x; n, 1 - \theta) \tag{3.3.4}$$

which the reader will be asked to prove in Exercise 4 on page 84. Thus, to find $b(11; 18, 0.70)$, for example, we look up $b(7; 18, 0.30)$, getting 0.1376.

There are also several ways in which binomial probabilities can be approximated when n is large. One of these will be discussed in Section 3.3.7, and another in Section 5.4. A method by which one can directly determine *sums* of binomial probabilities is referred to at the end of this chapter, but as it requires knowledge of certain topics of advanced calculus, it will not be discussed in this text.

3.3.4 The Hypergeometric Distribution

In Chapter 2 we discussed sampling (selecting objects) with and without replacement in order to illustrate the multiplication rules for independent and dependent events. To develop a formula analogous to that of the binomial probability function which applies to sampling *without replacement,* in which case successive trials are *not independent,* let us consider a set having $a + b$ elements of which a are labeled "success" and b are labeled "failure." As in Section 3.3.3, we are interested in the probability of getting x successes in n trials, namely, *the probability of choosing a subset of n elements of which x are labeled "success" and $n - x$ are labeled "failure."* Although we could assume that the elements are chosen one at a time without replacement, and that in each case all of the remaining elements have the same chance of being selected, it will be more convenient to make the *equivalent assumption* (see Exercise 7 on page 84) that each subset of n elements has the same probability of being selected.

Since there are $\binom{a+b}{n}$ ways in which a subset of n elements can be selected from a set of $a + b$ elements, each element of the sample space (representing the selection of a particular subset of n elements) has the probability $\frac{1}{\binom{a+b}{n}}$. Also, since $\binom{a}{x}$ is the number of ways in which a subset of x elements can be selected from the set of a elements labeled "success," and $\binom{b}{n-x}$ is the number of ways in which a subset of $n - x$ elements can be selected from the set of b elements labeled "failure," there are altogether $\binom{a}{x}\binom{b}{n-x}$ ways in which x elements labeled "success" and $n - x$ elements labeled "failure" can be selected from the entire set of $a + b$ elements. Thus, the special rule for equiprobable events on page 41 leads to the result that under the stated conditions the probability of "x successes in n trials" is

$$h(x; n, a, b) = \frac{\binom{a}{x}\binom{b}{n-x}}{\binom{a+b}{n}} \quad \text{for } x = 0, 1, 2, \ldots, n \qquad (3.3.5)$$

subject to the restriction that x cannot exceed a and $n - x$ cannot exceed b. This probability function defines the *hypergeometric distribution*, and we used the symbol $h(x; n, a, b)$ to indicate that its values depend on the parameters n, a, and b.

To illustrate the use of (3.3.5), suppose we want to determine the probability that the Internal Revenue Service will catch 3 income tax returns with illegitimate deductions, if it randomly selects 6 returns from among 20 income tax returns of which 8 contain illegitimate deductions. Substituting $x = 3$, $n = 6$, $a = 8$, and $b = 12$ into (3.3.5), we get

$$h(3; 6, 8, 12) = \frac{\binom{8}{3}\binom{12}{3}}{\binom{20}{6}} = \frac{56 \cdot 220}{38{,}760} = 0.318$$

and in Exercise 26 on page 87 the reader will be asked to show that the probability of their catching fewer than 3 of the income tax returns with illegitimate deductions is 0.545.

To consider another example, suppose that among 120 applicants for a job only 80 are actually qualified. If 5 of these applicants are randomly selected for an "in depth" interview, what is the probability that only 2 of the 5 are actually qualified for the job? Substituting $x = 2$, $n = 5$, $a = 80$, and $b = 40$ into (3.3.5), we get

$$h(2; 5, 80, 40) = \frac{\binom{80}{2}\binom{40}{3}}{\binom{120}{5}} = \frac{300{,}200}{1{,}832{,}481}$$

or approximately 0.164. Had we made the *mistake* of using the binomial distribution with $n = 5$ and $\theta = \dfrac{a}{a+b} = \dfrac{80}{80+40} = \dfrac{2}{3}$ in this example, we would have obtained

$$b(2; 5, \tfrac{2}{3}) = \binom{5}{2}(\tfrac{2}{3})^2(1 - \tfrac{2}{3})^3 = \tfrac{40}{243}$$

or approximately 0.165, which is surprisingly close to the correct value of 0.164. Indeed, we often use the binomial distribution with $\theta = \dfrac{a}{a+b}$ to approximate hypergeometric probabilities so long as n does not constitute more than 5 percent of $a + b$.

3.3.5 The Geometric Distribution

As we suggested on page 74, the total number of "successes" is not the only random variable which may be of interest in a series of Bernoulli trials. For instance, if n is not fixed, we may be interested in *the number of the trial on which there is the first success.* This would certainly be of concern to someone who gambles with limited funds, but, strangely enough, the same question arises in many other (seemingly unrelated) areas. An applicant for a driver's license may want to know the probability that he will fail the road test four times before he finally passes on the fifth try, a manufacturer of light switches may want to know the probability that one of his switches can be turned on and off 1,000 times before it finally fails, and a burglar may want to know the probability that he will get away with 3 "jobs" before he finally gets caught.

Probabilities like these are easy to determine provided the trials are independent and the probability of success has the same value θ for each trial. We can then argue that the probability of getting $x - 1$ failures in a row is $(1 - \theta)^{x-1}$, the probability of finally getting a success on the xth trial is θ, and, hence, that the desired probability of getting the first success on the xth trial is

$$g(x; \theta) = \theta(1 - \theta)^{x-1} \qquad \text{for } x = 1, 2, 3, 4, \ldots \tag{3.3.6}$$

A random variable having this probability function is said to have a *geometric distribution.* Its parameter θ is the constant probability of success for each trial, and it owes its name to the fact that for successive values of x the probabilities constitute a geometric progression.

To return to the first application referred to above, suppose that the probability that an applicant for a driver's license will pass the road test on any given try is 0.60. Then the probability that an applicant will eventually get his license on the fifth try is

$$g(5; 0.60) = 0.60(1 - 0.60)^4$$

which is approximately 0.015. Of course, this assumes that if a person has already failed the test 4 times, the probability that he will get his license on the fifth try is still 0.60 (and there may be some doubts about the validity of this assumption). To consider another example, suppose that the probability that a burglar will get caught at any given "job" is 0.30. Then, the probability that he will get caught on his fourth "job" is

$$g(4; 0.30) = 0.30(1 - 0.30)^3$$

which is approximately 0.103.

3.3.6 The Negative Binomial Distribution

An important difference between the geometric distribution and all of the others we had discussed before is that the domain of its probability function is *countably infinite.* This will be the case also for the probability distributions which we shall discuss in this section and in Section 3.3.7. The probability distribution which we shall consider in this section is similar, conceptually, to the geometric distribution—the only difference is that we shall now concern ourselves with the kth success instead of the first. Arguing as in the preceding section, we can say that the probability of getting $k - 1$ successes in the first $x - 1$ trials is

$$b(k - 1; x - 1, \theta) = \binom{x-1}{k-1} \theta^{k-1}(1 - \theta)^{x-k}$$

and since the probability of getting another success on the xth trial is θ, it follows that the desired probability of getting the kth success on the xth trial is

$$f(x; k, \theta) = \binom{x-1}{k-1}\theta^k(1-\theta)^{x-k} \quad \text{for } x = k, k+1, k+2, \ldots \tag{3.3.7}$$

A random variable having this probability function is said to have the *negative binomial distribution.* Its parameters are θ and k, and it owes its name to the fact that for $x = k$, $k + 1$, $k + 2$, $\ldots$, the values of $f(x; k, \theta)$ are the successive terms of the binomial expansion of $\left(\frac{1}{\theta} - \frac{1-\theta}{\theta}\right)^{-k}$. (Binomial expansions with negative exponents are explained in the book by W. Feller listed on page 35.)

To illustrate the use of (3.3.7), suppose that the probability for a child exposed to a certain contagious disease to catch it is 0.20, and that we are interested in the probability that the twelfth child exposed to the disease will be the third to catch it. Substituting $x = 12$, $k = 3$, and $\theta = 0.20$ into (3.3.7), we get

$$f(12; 3, 0.20) = \binom{11}{2}(0.20)^3(1-0.20)^9$$

which is approximately 0.059.

The determination of negative binomial probabilities can be simplified greatly by making use of the identity

$$f(x; k, \theta) = \frac{k}{x}\cdot b(k; x, \theta) \tag{3.3.8}$$

which the reader will be asked to verify in Exercise 11 on page 85. Looking up the binomial probability $b(3; 12, 0.20)$ in Table I and getting 0.2362, we could thus have solved the above problem much more simply by writing

$$f(12; 3, 0.20) = \tfrac{3}{12}\cdot(0.2362) = 0.059$$

3.3.7 The Poisson Distribution

When n is large, the direct calculation of binomial probabilities with the use of formula (3.3.3) can involve a prohibitive amount of work. For instance, to calculate the probability that 13 of 3,000 persons attending a sporting event on a hot summer day will suffer from heat exhaustion, we have to determine $\binom{3{,}000}{13}$, and if the probability that any one of them will suffer from heat exhaustion is 0.005, we also have to calculate $(0.005)^{13}(1 - 0.005)^{2,987}$.

In this section we shall introduce a probability distribution which can be used to approximate probabilities of this kind, namely, binomial probabilities for which n is very large and θ is very small. More rigorously, we shall investigate the limiting form of the binomial distribution when $n \to \infty$, $\theta \to 0$, while $n\theta$ remains constant. Letting this constant be λ (*lambda*), that is, $n\theta = \lambda$, we can write (3.3.3) as $\binom{n}{x}\left(\frac{\lambda}{n}\right)^x\left(1 - \frac{\lambda}{n}\right)^{n-x}$, and, making use of the fact that

$$\binom{n}{x} = \frac{n(n-1)(n-2)\cdots(n-x+1)}{x!}$$

we get

$$b(x; n, \theta) = \frac{n(n-1)(n-2)\cdots(n-x+1)}{x!}\left(\frac{\lambda}{n}\right)^x\left(1 - \frac{\lambda}{n}\right)^{n-x}$$

Then, if we divide one of the x factors n in $\left(\frac{\lambda}{n}\right)^x$ into each factor of the product $n(n-1)(n-2)\cdots(n-x+1)$ and write

$$\left(1 - \frac{\lambda}{n}\right)^{n-x} \quad \text{as} \quad \left[\left(1 - \frac{\lambda}{n}\right)^{-n/\lambda}\right]^{-\lambda}\left(1 - \frac{\lambda}{n}\right)^{-x}$$

we obtain

$$\frac{1\left(1 - \frac{1}{n}\right)\left(1 - \frac{2}{n}\right)\cdots\left(1 - \frac{x-1}{n}\right)}{x!}(\lambda)^x\left[\left(1 - \frac{\lambda}{n}\right)^{-n/\lambda}\right]^{-\lambda}\left(1 - \frac{\lambda}{n}\right)^{-x}$$

Finally, if we let $n \to \infty$ while x and λ remain fixed, we find that

$$1\left(1 - \frac{1}{n}\right)\left(1 - \frac{2}{n}\right)\cdots\left(1 - \frac{x-1}{n}\right) \to 1$$

$$\left(1 - \frac{\lambda}{n}\right)^{-x} \to 1$$

$$\left(1 - \frac{\lambda}{n}\right)^{-n/\lambda} \to e$$

and, hence, that the limiting distribution becomes

$$p(x; \lambda) = \frac{\lambda^x e^{-\lambda}}{x!} \qquad \text{for} \qquad x = 0, 1, 2, \ldots \tag{3.3.9}$$

A random variable having this probability function is said to have the *Poisson distribution*, which is named after the French mathematician

Simeon Poisson (1781–1840); the significance of its parameter λ will be discussed in Chapter 5.

To illustrate the Poisson approximation of the binomial distribution, let us compare the values of the binomial distribution with $n = 48$ and $\theta = 0.05$ (obtained from the National Bureau of Standards table referred to on page 100) with the values of the Poisson distribution with $\lambda = 48(0.05) = 2.4$ (obtained from Table II at the end of this book). Rounding both sets of probabilities to three decimals, we get

x	*Binomial Probabilities*	*Poisson Probabilities*
0	0.085	0.091
1	0.215	0.218
2	0.266	0.261
3	0.215	0.209
4	0.127	0.125
5	0.059	0.060
6	0.022	0.024
7	0.007	0.008
8	0.002	0.002
...	...	...

and it can be seen that the differences are quite small. Generally speaking, the Poisson distribution will provide a good approximation to binomial probabilities when n is at least 20 and θ is at most 0.05; when n is at least 100, the approximation will generally be excellent provided $n\theta$ does not exceed 10.

THEORETICAL EXERCISES

1. Verify that $f(x) = \dfrac{2x}{k(k+1)}$ for $x = 1, 2, 3, \ldots, k$ can serve as the probability function of a random variable.

2. Find an expression for c, if $f(x) = cx^2$ for $x = 1, 2, \ldots, k$ is to be a probability function. (*Hint:* Refer to the Appendix at the end of the book.)

3. If $F(x)$ is defined as in (3.2.1) as the probability that the random variable $\mathbf{x}$ takes on a value less than or equal to x, show that
(a) $F(-\infty) = 0$;
(b) $F(\infty) = 1$;
(c) $F(a) \leq F(b)$ if $a < b$.

4. Verify the identity $b(x; n, \theta) = b(n - x; n, 1 - \theta)$.

5. The calculation of binomial probabilities is often simplified by first determining $b(0; n, \theta)$ with the use of (3.3.3) and then obtaining the other probabilities by means of the *recursion formula*

$$b(x + 1; n, \theta) = \frac{\theta(n - x)}{(x + 1)(1 - \theta)} \cdot b(x; n, \theta)$$

Verify this formula and use it to calculate the values of the binomial probability function for $n = 8$ and $\theta = 0.25$. (Check the results in Table I.)

6. Use the recursion formula of Exercise 5 to show that for $\theta = \frac{1}{2}$ the binomial probability function has a maximum at $x = \frac{n}{2}$ when n is even, and maxima at $x = \frac{n - 1}{2}$ and $x = \frac{n + 1}{2}$ when n is odd.

7. Show that the two assumptions which we made on page 77 about the selection of n elements from a set having $a + b$ elements are actually equivalent.

8. The calculation of hypergeometric probabilities is often simplified by first determining $h(0; n, a, b)$ with the use of (3.3.5) and then obtaining the other probabilities by means of the recursion formula

$$h(x + 1; n, a, b) = \frac{(n - x)(a - x)}{(x + 1)(b - n + x + 1)} \cdot h(x; n, a, b)$$

Verify this formula and use it to calculate the values of the hypergeometric probability function for $n = 6$, $a = 8$, and $b = 12$.

9. The determination of geometric probabilities is often simplified by making use of the identity

$$g(x; \theta) = \frac{1}{x} \cdot b(1; x, \theta)$$

and looking up $b(1; x, \theta)$ in a table of binomial probabilities. Verify this identity and use it (and Table I) to evaluate

(a) $g(12; 0.15)$;

(b) $g(10; 0.35)$;

(c) $g(20; 0.80)$;

(d) $\sum_{x=1}^{5} g(x; 0.50)$.

10. Find the equation of the distribution function of the geometric distribution, and use it to verify part (d) of Exercise 9.

11. Verify the identity (3.3.8) on page 81.

12. Verify that the sum of the Poisson probabilities (3.3.9) is, indeed, equal to 1.

13. Find a recursion formula for the Poisson distribution which expresses $p(x + 1; \lambda)$ in terms of $p(x; \lambda)$, and use it to verify the values given in Table II for $\lambda = 2$. (Use $e^{-2} = 0.1353$.)

14. Approximate the binomial probability $b(3; 100, 0.10)$ by looking up the Poisson probability $p(3; 10)$ in Table II, and compare the result with the exact probability calculated with the use of logarithms in accordance with (3.3.3).

15. The Poisson distribution has many important applications which have no connection with the binomial distribution. Suppose, for instance, that $f(x, t)$ is the probability of getting x successes during a time interval of length t when (i) the probability of a success during a *very small* time interval from t to $t + \Delta t$ is $\alpha \cdot \Delta t$, (ii) the probability of more than one success occurring during such a time interval is negligible, and (iii) the probability of a success during such a time interval does not depend on what happened prior to time t.

(a) Show that under these conditions

$$f(x, t + \Delta t) = f(x, t)[1 - \alpha \cdot \Delta t] + f(x - 1, t)\alpha \cdot \Delta t$$

and, hence, that

$$\frac{d[f(x, t)]}{dt} = \alpha[f(x - 1, t) - f(x, t)]$$

(b) Show by direct substitution that a solution of this infinite system of differential equations (there is one for each value of x) is given by the Poisson distribution with $\lambda = \alpha t$.

Applications where these assumptions are met may be found in Exercises 35 through 38 below.

APPLIED EXERCISES

16. Using (3.3.3) find the probability that 3 of 8 hibiscus plants will survive a frost, given that any such plant will survive a frost with a probability of 0.30.

17. A safety engineer claims that 1 in 10 automobile accidents is due to driver fatigue. Using (3.3.3) find the probabilities that among 5 accidents 0, 1, 2, 3, 4, and 5 are due to driver fatigue, and draw a histogram of this probability function.

18. When the mail room clerk of a publishing house is in a hurry to go home, the probability that he will make a mistake in an address is 0.25. If he has to send out 6 shipments just before closing time, find the probability that he will make a mistake in addressing *at least* 2 of these shipments
(a) using (3.3.3);
(b) referring to Table I.
(*Hint:* Subtract from 1 the probabilities of his making 0 or 1 mistakes.)

19. It is known from experience (according to some "authority") that 40 percent of all housewives who let a door-to-door magazine salesman into their homes will end up ordering a magazine. Find the probability that *at most* 3 of 12 housewives who let such a salesman into their homes will end up ordering a magazine,
(a) using (3.3.3);
(b) referring to Table I.

20. A doctor knows from experience that 15 percent of the patients who are given a certain medicine will have undesirable side effects. Use Table I to find the probabilities that among 18 patients who are given this medicine,
(a) none will have any undesirable side effects;
(b) exactly 3 will have undesirable side effects;
(c) at least 5 will have undesirable side effects;
(d) at most 8 will have undesirable side effects.

21. A study has shown that in a given large city 70 percent of all families in a certain income group own at least two television sets. Use Table I and (3.3.4) to find the probabilities that among 20 such families (randomly selected for a survey)
(a) all will have at least two television sets;
(b) anywhere from 10 to 15, inclusive, will have at least two television sets;
(c) at most 10 will have at least two television sets;
(d) at least 18 will have at least two television sets.

22. In planning the operations of a new department store, one expert claims that only 1 out of 4 salesladies can be expected to stay with the store for more than a year, while a second expert claims that it would be more correct to say 1 out of 5. In the past, the two experts have been about equally reliable, so that in the absense of any direct information we would assign their judgments equal weight, that is, we would assign $\theta = \frac{1}{4}$ and $\theta = \frac{1}{5}$ equal *prior probabilities* of 0.50

(assuming that one or the other must be right). Use Bayes' rule, Theorem 2.13, to find the *posterior probabilities* which we would assign to these two values of θ if it were found that among 15 salesladies hired for the store only 2 stayed for more than a year. (Refer to Table I.)

23. Mr. Adam, Mr. Black, and Mr. Clark are planning to open a chain of restaurants; according to Mr. Adam the probability that any one of them will show a profit during the first year is 0.10, and according to Mr. Black and Mr. Clark this probability is 0.05 and 0.15, respectively. An impartial expert feels that Mr. Adams is three times more likely to be right than Mr. Black, and that Mr. Black is twice as likely to be right as Mr. Clark. If a survey shows that among 20 such restaurants only one showed a profit during the first year, find
(a) the *prior probabilities* which the impartial expert is assigning to $\theta = 0.10$, $\theta = 0.05$, and $\theta = 0.15$ (assuming that these are the only three possibilities);
(b) the *posterior probabilities* which we get with the use of Bayes' rule, Theorem 2.13, for these three values of θ, if we also consider the information that in the survey only one of 20 such restaurants showed a profit during the first year. (Refer to Table I.)

24. A collection of 12 Spanish gold doubloons contains 4 counterfeits. If 3 of these coins are randomly selected to be sent to a dealer, find the probabilities that
(a) none of the coins will be counterfeits;
(b) exactly one of the coins will be a counterfeit;
(c) exactly two of the coins will be counterfeits;
(d) all three of the coins will be counterfeits.
Also draw a *histogram* of this probability function.

25. A quality control engineer inspects a random sample of 2 toasters from each incoming lot of 24 toasters, and accepts the lot only if both are in good working condition; otherwise all of the toasters are inspected with the cost charged to the vendor. What is the probability that such a lot will be accepted without further inspection if
(a) it contains 5 toasters which are not in good working condition;
(b) it contains 10 toasters which are not in good working condition;
(c) it contains 15 toasters which are not in good working condition?

26. With reference to the example on page 78, verify that the probability of the I. R. S. agent catching fewer than 3 of the income tax returns with illegitimate deductions is 0.545.

27. What is the probability that a five-card poker hand dealt from an ordinary deck of 52 playing cards contains exactly one ace? How big an error would we make if we approximated this probability with the binomial probability of getting 1 success in 5 trials when $\theta = \frac{1}{13}$?

28. Among 75 sites considered for army installations, 15 are in California. If the army randomly selects 6 of these sites, what is the probability that none will be in California? How big an error would we make if we approximated this probability with the binomial probability of zero successes in 6 trials when $\theta = \frac{1}{5}$?

29. If the probability that a person will believe a rumor about the retirement of a certain politician is 0.25, what are the probabilities that
(a) the sixth person to hear the rumor will be the first one to believe it;
(b) the twelfth person to hear the rumor will be the fourth to believe it;
(c) the eighteenth person to hear the rumor will be the sixth to believe it?
[*Hint:* Use (3.3.8).]

30. An expert shot hits a distant target 95 percent of the time. What are the probabilities that he will miss the target
(a) for the first time on the fifteenth shot;
(b) for the second time on the eighteenth shot?

31. In a "torture test" a light switch is turned on and off until it fails. If the probability that the switch will fail any time it is turned on or off is 0.001, what is the probability that the switch will fail *after* it has been turned on or off 1,200 times. Assume that the conditions underlying the geometric distribution are met. (*Hint:* Use the formula for the value of an infinite geometric progression and logarithms.)

32. If the probabilities of having a male or female offspring are both 0.50, find the probabilities that
(a) a family's fifth child is their first son;
(b) a family's sixth child is their second daughter;
(c) a family's fifth child is their third or fourth son.

33. Use the Poisson approximation (and Table II) to determine the following binomial probabilities:
(a) If 2 percent of the fuses delivered to an arsenal are defective, what is the probability that a (random) sample of 400 of them will contain exactly 6 defectives?
(b) If 4 percent of all passengers traveling on a certain airline have difficulty with their luggage, what is the probability that among 150 persons traveling on this air line *at least* 5 have difficulty with their luggage?
(c) If 8 percent of all drivers in a certain city get at least one parking ticket a year, what is the problem that among 75 drivers (ran-

domly selected in this city) *more than* 10 received at least one parking ticket during the last year?

34. Use the Poisson distribution (and Table II) to approximate the following binomial probabilities:
(a) If 3 percent of the school children in a very large city have an I.Q. over 150, what is the probability that if 60 of them are (randomly) selected, exactly 2 will have an I.Q. over 150?
(b) If the probability that a tourist visiting the Grand Canyon will be bitten by a rattlesnake is 0.00002, what is the probability that among 75,000 tourists visiting the Grand Canyon *at least* 2 will be bitten by a rattlesnake?
(c) If the probability that a car will have a flat tire while driving through the Holland Tunnel is 0.0004, what is the probability that of 10,000 cars passing through the Holland Tunnel *fewer than* 3 will have a flat tire?

35. If the number of fish a person catches per hour at Woods Canyon Lake is a random variable having the Poisson distribution with $\lambda = 1.8$, find the probability that a person fishing there for an hour
(a) will not catch any fish at all;
(b) will catch at least four;
(c) will catch at most two.

36. If the number of incoming calls reaching the switchboard of a large company per minute is a random variable having the Poisson distribution with $\lambda = 6.2$, find the probability that during any given minute the switchboard of the company will receive
(a) anywhere from 5 through 10 calls, inclusive;
(b) more than 8 calls;
(c) fewer than 3 calls.

37. If the number of accidents on the Pasadena Freeway between 9 A.M. and 10 A.M. on a Sunday morning is a random variable having the Poisson distribution with $\lambda = 7.3$, find the probabilities that there will be
(a) exactly 6 accidents,
(b) fewer than 4 accidents,
(c) more than 12 accidents,
(d) anywhere from 5 to 10 accidents, inclusive,
on this freeway between 9 A.M. and 10 A.M. on a Sunday morning.

38. Three scientists (who are considered about equally reliable in their claims) make the conflicting statements that the number of gamma rays emitted per second by a certain radioactive substance is a random variable having the Poisson distribution with $\lambda = 1.2$, $\lambda = 2.1$, and $\lambda = 3.2$, respectively. If it is found in an experiment that the substance actually emits four gamma rays in one second, use Bayes' rule, Theorem 2.13, to find the *posterior probabilities* of the three scientists' claims (assuming that no other values of λ are possible).

3.4 MULTIVARIATE DISTRIBUTIONS

On page 69 we defined a random variable as a real-valued function defined over a sample space with a probability measure, and it stands to reason that we can define many different random variables over one and the same sample space. With reference to the sample space of Figure 3.1, for example, we considered only the random variable whose values were the totals which we obtained with the pair of dice, but we could also have considered the random variable whose values are the products of the numbers obtained with the two dice, the random variable whose values are the differences between the numbers obtained with the red die and the green die, the random variable whose values are 0, 1, or 2 depending on the number of dice which came up 2, and so forth. Closer to life, an experiment may consist of randomly selecting one of the 345 students attending an elementary school, and the principal may be interested in his I.Q., the school nurse in his weight, his teacher in the number of days he has been absent, and so forth.

In this section we shall be concerned with situations where we are interested *at the same time* in two or more random variables defined over one and the same sample space. Starting with two random variables $\mathbf{x}$ and $\mathbf{y}$, the probability that $\mathbf{x}$ takes on the value x *and* $\mathbf{y}$ takes on the value y will be denoted $f(x, y)$, and we refer to this probability as a value of the *joint probability function* (*joint probability distribution*, or simply the *joint distribution*) of the two random variables $\mathbf{x}$ and $\mathbf{y}$.

To illustrate, let us refer again to the sample space of Figure 1.5 with a probability of $\frac{1}{36}$ assigned to each point, and let us suppose that someone wants to know how many dice come up 1, while someone else wants to know how many dice come up either 5 or 6. The values of these two random variables are indicated in Figure 3.5, where we attached two numbers to each of the 36 points—the first number is the value of the random variable $\mathbf{x}$, the number of dice which come up 1, and the second number is the corresponding value of the random variable $\mathbf{y}$, the number of dice which come up either 5 or 6.

To obtain the probability $f(x, y)$ for any pair of values of x and y, we have only to add the probabilities of those points of the sample space to which these particular values of the two random variables are assigned.

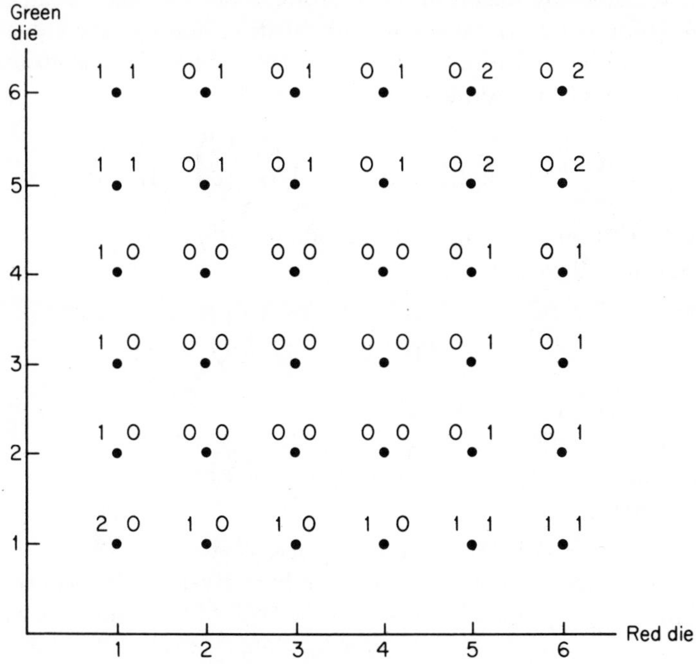

FIGURE 3.5 Outcomes for two dice.

For instance, $f(0, 0) = \frac{9}{36}$ in our example because there are 9 points of the sample space at which x and y both equal 0. Similarly, $f(1, 0) = \frac{6}{36}$ because there are 6 points of the sample space at which x equals 1 and y equals 0, and $f(1, 2) = 0$ because it is impossible for x to equal 1 and y to equal 2. Continuing in this way, it can easily be verified that the values of the joint probability function of the given random variables are as shown *inside* the following table:

		x 0	1	2	
	0	$\frac{9}{36}$	$\frac{6}{36}$	$\frac{1}{36}$	$\frac{16}{36}$
y	1	$\frac{12}{36}$	$\frac{4}{36}$	0	$\frac{16}{36}$
	2	$\frac{4}{36}$	0	0	$\frac{4}{36}$
		$\frac{25}{36}$	$\frac{10}{36}$	$\frac{1}{36}$	

As in the *univariate case,* that is, in the case of one random variable, it is generally desirable to express the joint probability function of two random variables by means of an equation which enables us to calculate the probabilities $f(x, y)$ by simply substituting appropriate values for x and y. As will be shown in Section 3.4.1, the values of the *bivariate probability function* of our example are given by

$$f(x, y) = \frac{2!}{x!y!(2 - x - y)!}\left(\frac{1}{6}\right)^x\left(\frac{2}{6}\right)^y\left(\frac{3}{6}\right)^{2-x-y}$$

for $x = 0, 1, 2$, and $y = 0, 1, 2$, provided $x + y \leq 2$. For instance, for $x = 0$ and $y = 0$ we get

$$f(0, 0) = \frac{2!}{0!0!2!}\left(\frac{1}{6}\right)^0\left(\frac{2}{6}\right)^0\left(\frac{3}{6}\right)^2 = \frac{9}{36}$$

and for $x = 1$ and $y = 0$ we get

$$f(1, 0) = \frac{2!}{1!0!1!}\left(\frac{1}{6}\right)^1\left(\frac{2}{6}\right)^0\left(\frac{3}{6}\right)^1 = \frac{6}{36}$$

which both agree with the results obtained above.

It is of interest to note that the numbers given in the *bottom margin* of the table for the probabilities $f(x, y)$, the *totals* of the respective columns, are, in fact, the probabilities that the random variable $\mathbf{x}$ will take on the values 0, 1, and 2. In other words, the column totals give the values of the probability function of the random variable $\mathbf{x}$, which happen to be the binomial probabilities $b(x; 2, \frac{1}{6})$. By the same token, the row totals, the values given in the *right-hand margin* of the table, are the values of the probability function of the random variable $\mathbf{y}$, which happen to be the binomial probabilities $b(y; 2, \frac{2}{6})$. (Clearly, the probability of rolling a 1 with a die is $\theta = \frac{1}{6}$ and the probability of rolling a 5 or a 6 is $\theta = \frac{2}{6}$.)

In general, if we are given all the joint probabilities $f(x, y)$ for a pair of random variables $\mathbf{x}$ and $\mathbf{y}$, the probability function of $\mathbf{x}$ alone is given by

$$g(x) = \sum_y f(x, y) \tag{3.4.1}$$

where the summation extends over all values within the range of $\mathbf{y}$ for which $f(x, y)$ exists, and the probability function of $\mathbf{y}$ alone is given by

$$h(y) = \sum_x f(x, y) \tag{3.4.2}$$

where the summation extends over all values within the range of $\mathbf{x}$ for which $f(x, y)$ exists. When the probability functions of $\mathbf{x}$ and $\mathbf{y}$ are thus obtained from their joint distribution, we refer to them, respectively, as the *marginal probability function* (or the *marginal distribution*) of **x** and the *marginal probability function* (or the *marginal distribution*) of **y**.

When we are dealing with two random variables, it may also be of interest to know the probability that **x** takes on the value x given that **y** takes on the value y. Writing this conditional probability as $\varphi(x|y)$, we have

$$\varphi(x|y) = \frac{f(x, y)}{h(y)} \qquad \text{provided } h(y) \neq 0 \tag{3.4.3}$$

in accordance with the definitionon page 52, and it is customary to refer to the probabilities $\varphi(x|y)$ as values of the *conditional probability function* (or the *conditional distribution*) of the random variable **x** given that **y** takes on the value y. By the same token, we refer to the probabilities

$$\pi(y|x) = \frac{f(x, y)}{g(x)} \qquad \text{provided } g(x) \neq 0 \tag{3.4.4}$$

as values of the *conditional probability function* (or the *conditional distribution*) of the random variable **y** given that **x** takes on the value x. With reference to our numerical example, we thus find that the probability that one of the dice will come up 1 *given* that neither of the dice will come up 5 or 6 is

$$\varphi(1|0) = \frac{f(1, 0)}{h(0)} = \frac{\frac{6}{36}}{\frac{16}{36}} = \frac{3}{8}$$

and the probability that one of the dice will come up 5 or 6 *given* that one of the dice (the other, of course) will come up 1 is

$$\pi(1|1) = \frac{f(1, 1)}{g(1)} = \frac{\frac{4}{36}}{\frac{10}{36}} = \frac{2}{5}$$

As we indicated in the beginning of this section, any number of random variables $\mathbf{x}_1, \mathbf{x}_2, \ldots, \mathbf{x}_n$ can be defined over one and the same sample space, and we refer to the probability $f(x_1, x_2, \ldots, x_n)$ that $\mathbf{x}_1$ takes on the value x_1, $\mathbf{x}_2$ takes on the value x_2, ..., $\mathbf{x}_n$ takes on the value x_n as a value of the *joint probability function* (or the *joint distribution*) of the n random variables. For instance, if we consider n flips of a balanced coin and let $\mathbf{x}_1$ be the number of heads (0 or 1) obtained on the first flip, $\mathbf{x}_2$ the number of heads obtained on the second flip, ..., and $\mathbf{x}_n$ the number of heads obtained on the nth flip, we can write

$$f(x_1, x_2, \ldots, x_n) = (\tfrac{1}{2})^n$$

where each x can take on the value 0 or 1.

When we are dealing with more than two random variables, we can again speak of the marginal distribution or the conditional distribution of a random variable, but we can also speak of *joint marginal distributions* and *joint conditional distributions.* For instance, in the *trivariate case* where we have the three random variables $\mathbf{x}_1$, $\mathbf{x}_2$, and $\mathbf{x}_3$, and the values of their joint probability function are denoted $f(x_1, x_2, x_3)$, the marginal distribution of $\mathbf{x}_1$ is given by

$$g(x_1) = \sum_{x_2} \sum_{x_3} f(x_1, x_2, x_3) \tag{3.4.5}$$

where the summation extends over all pairs of values within the ranges of $\mathbf{x}_2$ and $\mathbf{x}_3$ for which $f(x_1, x_2, x_3)$ is defined. However, we can also speak of the *joint marginal distribution,* say, of $\mathbf{x}_1$ and $\mathbf{x}_2$, whose values $m(x_1, x_2)$ are given by

$$m(x_1, x_2) = \sum_{x_3} f(x_1, x_2, x_3) \tag{3.4.6}$$

where the summation extends over all values within the range of $\mathbf{x}_3$ for which $f(x_1, x_2, x_3)$ is defined. Thus, $m(x_1, x_2)$ is the probability that $\mathbf{x}_1$ takes on the value x_1 and $\mathbf{x}_2$ takes on the value x_2 without any reference to $\mathbf{x}_3$.

So far as conditional probabilities are concerned, the *conditional probability function* of $\mathbf{x}_3$ given that $\mathbf{x}_1$ takes on the value x_1 and $\mathbf{x}_2$ takes on the value x_2 is given by

$$\varphi(x_3|x_1, x_2) = \frac{f(x_1, x_2, x_3)}{m(x_1, x_2)} \qquad \text{provided } m(x_1, x_2) \neq 0 \tag{3.4.7}$$

and we can also speak of the *joint conditional probability function* of $\mathbf{x}_2$ and $\mathbf{x}_3$ given that $\mathbf{x}_1$ takes on the value of x_1, whose equation is

$$\psi(x_2, x_3|x_1) = \frac{f(x_1, x_2, x_3)}{g(x_1)} \qquad \text{provided } g(x_1) \neq 0 \tag{3.4.8}$$

As can easily be seen, there are numerous other possibilities for marginal and conditional probabilities, especially when the number of random variables is greater than three.

When we are dealing with two or more random variables, questions of *independence* are also of great importance. In general, if $f(x_1, x_2, \ldots, x_n)$ and $f_i(x_i)$ are, respectively, values of the joint distribution of n random variables $\mathbf{x}_1, \mathbf{x}_2, \ldots, \mathbf{x}_n$, and the marginal distribution of the random variable $\mathbf{x}_i$, we say that *the n random variables are independent if and only if*

$$f(x_1, x_2, \ldots, x_n) = \prod_{i=1}^{n} f_i(x_i) \tag{3.4.9}$$

for all values within the range of these random variables for which the probability $f(x_1, x_2, \ldots, x_n)$ *is defined.* Referring again to the example on page 91, it is easy to see that the two random variables are *not* independent. For instance, $f(0, 1) = \frac{12}{36}$, which does *not* equal the product of the marginal probabilities $g(0) = \frac{25}{36}$ and $h(1) = \frac{16}{36}$.

3.4.1 The Multinomial Distribution

An immediate generalization of the binomial distribution arises when each trial has more than two possible outcomes, the probabilities of the respective outcomes are the same for each trial, and the trials are all independent. This would apply, for example, to repeated rolls of a die, where there are 6 alternatives and each alternative has the probability $\frac{1}{6}$; it would also apply to interviews conducted by an opinion poll, where each person is asked whether he is for a candidate, against him, or undecided; and it would apply when samples of manufactured products are rated excellent, above average, average, or inferior.

Thus, let us consider the case where there are n independent trials, with each trial permitting k mutually exclusive outcomes whose respective probabilities are $\theta_1, \theta_2, \ldots, \theta_k$, and $\sum_{i=1}^{k} \theta_i = 1$. The random variables with which we shall be concerned are $\mathbf{x}_1, \mathbf{x}_2, \ldots$, and $\mathbf{x}_k$, where $\mathbf{x}_i$ is the number of times the ith kind of outcome (the ith kind of "success") occurs in n trials, and the probability we shall want to determine is $f(x_1, x_2, \ldots, x_k)$, namely, the probability that $\mathbf{x}_1$ will take on the value x_1, $\mathbf{x}_2$ will take on the value x_2, ..., and $\mathbf{x}_k$ will take on the value x_k, where $\sum_{i=1}^{k} x_i = n$. Proceeding as on page 75, let us first observe that the probability of getting x_1 "successes" of the first kind, x_2 "successes" of the second kind, ..., and x_k "successes" of the kth kind *in a specific order* is

$$\theta_1^{x_1} \cdot \theta_2^{x_2} \cdots \theta_k^{x_k} \tag{3.4.10}$$

There is one factor θ_1 for each success of the first kind, one factor θ_2 for each success of the second kind, ..., one factor θ_k for each success of the kth kind, and these $x_1 + x_2 + \cdots + x_k = n$ factors are all multiplied together by virtue of the assumption that the n trials are independent. Since (3.4.10) applies to any sequence of n trials in which there are x_1 successes of the first kind, x_2 successes of the second kind, ..., and x_k successes of the kth kind, the desired probability $f(x_1, x_2, \ldots, x_k)$ is simply the product of (3.4.10) and the number of ways in which this particular assortment of successes can occur in n trials. Since this is

$$\frac{n!}{x_1!x_2!\cdots x_k!} \tag{3.4.11}$$

according to (1.3.3) on page 23, it follows that the joint probability function of the random variables $\mathbf{x}_1, \mathbf{x}_2, \ldots, \mathbf{x}_k$ is given by

$$f(x_1, x_2, \ldots, x_k) = \frac{n!}{x_1!x_2!\cdots x_k!}\cdot \theta_1^{x_1}\cdot\theta_2^{x_2}\cdots\theta_k^{x_k} \tag{3.4.12}$$

for $x_i = 0, 1, \ldots, n$, for each i, subject to the restriction that $\sum_{i=1}^{k} x_i = n$. This joint distribution is called the *multinomial distribution*, and it owes its name to the fact that for various values of the x_i we obtain corresponding terms of the multinomial expansion of $(\theta_1 + \theta_2 + \cdots + \theta_k)^n$.

To give an illustration, suppose that a supermarket carries four grades of ground beef, and that the probabilities that a housewife will choose the poorest, third best, second best, and best kinds are, respectively, 0.10, 0.30, 0.40, and 0.20. To obtain the probability that among 12 randomly-chosen housewives buying ground beef at this market 1 will choose the poorest kind 3 will choose the third best kind, 6 will choose the second best kind, and 2 will choose the best kind, we simply substitute $x_1 = 1$, $x_2 = 3$, $x_3 = 6$, $x_4 = 2$, $n = 12$, $\theta_1 = 0.10$, $\theta_2 = 0.30$, $\theta_3 = 0.40$, and $\theta_4 = 0.20$ into (3.4.12), and we get

$$\begin{aligned} f(1, 3, 6, 2) &= \frac{12!}{1!3!6!2!}(0.10)^1(0.30)^3(0.40)^6(0.20)^2 \\ &= 0.0245 \end{aligned}$$

THEORETICAL EXERCISES

1. Given the joint probability function of two random variables $\mathbf{x}$ and $\mathbf{y}$, whose values $f(x, y)$ are $f(1, 1) = \frac{6}{30}$, $f(1, 2) = \frac{1}{30}$, $f(1, 3) = \frac{1}{30}$, $f(2, 1) = \frac{4}{30}$, $f(2, 2) = \frac{5}{30}$, $f(2, 3) = \frac{1}{30}$, $f(3, 1) = \frac{2}{30}$, $f(3, 2) = \frac{4}{30}$, and $f(3, 3) = \frac{6}{30}$, find
   (a) the values of $g(x)$, the marginal distribution of $\mathbf{x}$;
   (b) the values of $h(y)$, the marginal distribution of $\mathbf{y}$;
   (c) the values of $\varphi(x|2)$, the conditional distribution of $\mathbf{x}$ given that $\mathbf{y}$ takes on the value 2;

(d) the values of $\pi(y|1)$, the conditional distribution of **y** given that **x** takes on the value 1.

Also check whether the two random variables are independent.

2. Given the joint probability function

$$f(x, y) = \tfrac{1}{30}(x + y) \qquad \text{for } x = 0, 1, 2, 3, \text{ and } y = 0, 1, 2$$

show that

(a) the marginal distribution of **x** is

$$g(x) = \tfrac{1}{10}(x + 1) \qquad \text{for } x = 0, 1, 2, 3$$

(b) the marginal distribution of **y** is

$$h(y) = \frac{2y + 3}{15} \qquad \text{for } y = 0, 1, 2$$

Also find

(c) an expression for $\varphi(x|y)$, the conditional distribution of **x** given that **y** takes on the value y, and use it to evaluate $\varphi(2|2)$, $\varphi(0|1)$, and $\varphi(3|0)$;

(d) an expression for $\pi(y|x)$, the conditional distribution of **y** given that **x** takes on the value x, and use it to evaluate $\pi(1|3)$, $\pi(0|0)$, and $\pi(2|2)$.

3. Given the joint probability function

$$f(x, y, z) = \frac{x \cdot y \cdot z}{108}$$

for $x = 1, 2, 3$, $y = 1, 2, 3$, $z = 1, 2$, find

(a) an expression for $m(x, y)$, the joint marginal distribution of **x** and **y**;

(b) an expression for $n(x, z)$, the joint marginal distribution of **x** and **z**;

(c) an expression for $g(x)$, the marginal distribution of **x**;

(d) an expression for $\varphi(z|x, y)$, the conditional distribution of **z** given that **x** takes on the value x and **y** takes on the value y;

(e) an expression for $\psi(y, z|x)$, the joint conditional distribution of **y** and **z** given that **x** takes on the value x.

4. Given the random variables $\mathbf{x}_1, \mathbf{x}_2, \ldots, \mathbf{x}_n$, the probability that $\mathbf{x}_1$ takes on a value less than or equal to x_1, $\mathbf{x}_2$ takes on a value less than or equal to x_2, ..., and $\mathbf{x}_n$ takes on a value less than or equal to x_n is denoted $F(x_1, x_2, \ldots, x_n)$, and it is referred to as a value of the *joint distribution function* of the n random variables. With reference to the example on page 91, find (a) $F(1, 1)$, (b) $F(3, 5)$, (c) $F(-2, 1)$, (d) $F(\frac{3}{2}, \frac{1}{2})$; (e) $F(5, 0)$; (f) $F(2, -\frac{1}{3})$.

5. With reference to the example on page 91, find the probabilities that
(a) neither die will come up 1 given that neither die is going to come up 5 or 6;
(b) neither die will come up 5 or 6 given that neither die is going to come up 1;
(c) one die will come up 1 given that one die is going to come up 5 or 6;
(d) neither die will come up 5 or 6 given that one die is going to come up 1.

6. Verify that the values of the bivariate probability function on page 92 are, in fact, multinomial probabilities $f(x, y, 2 - x - y)$ with $n = 2$, $\theta_1 = \frac{1}{6}$, $\theta_2 = \frac{2}{6}$, and $\theta_3 = \frac{3}{6}$.

7. Verify (3.4.11) on page 96 by making use of the fact that the x_1 trials on which there are successes of the first kind can be chosen from among the n trials in $\binom{n}{x_1}$ different ways, that the x_2 trials on which there are successes of the second kind can then be chosen from among the remaining $n - x_1$ trials in $\binom{n - x_1}{x_2}$ different ways, that the x_3 trials on which there are successes of the third kind can then be chosen from among the remaining $n - x_1 - x_2$ trials in $\binom{n - x_1 - x_2}{x_3}$ different ways, and so forth.

8. **MULTIVARIATE GENERALIZATION OF THE HYPERGEOMETRIC DISTRIBUTION** Consider a set of N elements among which a_1 are regarded successes of the first kind, a_2 are regarded successes of the second kind, ..., and a_k are regarded successes of the kth kind, so that $\sum_{i=1}^{k} a_i = N$. If each subset of n elements has the same probability of $1/\binom{N}{n}$ of being selected, find an expression for the probability that such a subset will contain x_1 successes of the first kind, x_2 successes of the second kind, ..., and x_k successes of the kth kind, where $\sum_{i=1}^{k} x_k = n$. Applications of this multivariate distribution may be found in Exercises 14 and 15 below.

9. Given that the random variables $\mathbf{x}_1, \mathbf{x}_2, \ldots, \mathbf{x}_k$ have the multinomial distribution (3.4.12), explain why the marginal distribution of $\mathbf{x}_i$ for $i = 1, 2, \ldots, k$ is a binomial distribution with the parameters n and θ_i.

APPLIED EXERCISES

10. The probability that a certain kind of battery will last fewer than 80 hours of continuous use is 0.50, the probability that it will last anywhere from 80 hours to 120 hours is 0.40, and the probability that it will last more than 120 hours is 0.10. What is the probability that among 10 such batteries 4 will last fewer than 80 hours, 4 will last anywhere from 80 to 120 hours, and 2 will last more than 120 hours?

11. In a city which has two television stations, the probability that a person contacted on a Saturday evening by phone is watching Station X is 0.10, the probability that he is watching Station Y is 0.20, and the probability that he is not watching television at all is 0.70. What is the probability that among 12 persons contacted on a Saturday evening by phone, 2 will be watching Station X, 4 will be watching Station Y, and 6 will not be watching television at all?

12. Suppose that 60 percent of all state income tax returns filed in a given state are correct, 20 percent contain only mistakes favoring the taxpayer, 10 percent contain only mistakes favoring the state, and 10 percent contain both kinds of mistakes. What is the probability that among nine of these state income tax returns randomly selected for audit five are correct, one contains only mistakes favoring the taxpayer, one contains only mistakes favoring the state, and two contain both kinds of mistakes?

13. If the probabilities that a one-car accident results in no injury, minor injuries, severe injuries, or fatal injuries to the driver are, respectively, 0.25, 0.40, 0.30, and 0.05, find the probability that in eight such accidents the driver is not injured in one, has minor injuries in three, has severe injuries in two, and is fatally injured in two.

14. A roll of 20 silver dollars minted in 1903 contains 12 from the Philadelphia mint, 7 from the New Orleans mint, and 1 from the San Francisco mint. If 5 of these silver dollars are picked at random, find the probabilities of getting
    (a) 4 from the Philadelphia mint and 1 from the New Orleans mint;
    (b) 3 from the Philadelphia mint and 2 from the New Orleans mint;
    (c) 3 from the Philadelphia mint, 1 from the New Orleans mint, and 1 from the San Francisco mint.

    (*Hint:* Use the theory of Exercise 8.)

15. A panel of prospective jurors consists of 5 married men, 3 single men, 8 married women, and 4 single women. If each subset of 12 of these

persons has the same chance of being selected, what is the probability that a jury of 12 chosen from this panel will include

(a) 3 married men, 1 single man, 5 married women, and 3 single women;

(b) 4 married men, 6 married women, and 2 single women?

(*Hint:* Use the theory of Exercise 8.)

REFERENCES

Binomial probabilities for $n = 2$ to $n = 49$ may be found in

Tables of the Binomial Probability Distribution, National Bureau of Standards Applied Mathematics Series No. 6, Washington, D.C.: U.S. Government Printing Office, 1950,

and for $n = 50$ to $n = 100$ in

Romig, H. G., *50–100 Binomial Tables*, New York: John Wiley & Sons, Inc., 1953.

The most widely used table of Poisson probabilities is

Molina, E. C., *Poisson's Exponential Binomial Limit*, New York: D. Van Nostrand Co., Inc., 1943.

The use of *incomplete beta functions* and *incomplete gamma functions* in evaluating sums of binomial or Poisson probabilities is discussed in the book by W. Feller listed on page 35.

4

PROBABILITY DENSITIES

4.1 THE CONTINUOUS CASE

In Chapters 2 and 3 we limited our discussion to *discrete* sample spaces and, hence, *discrete* random variables (namely, to experiments in which the number of outcomes was finite or countably infinite). To illustrate the difficulties that can arise when there is a continuum of possibilities, suppose that an accident occurs on a freeway which is 120 miles long, and that we are interested in the probability that it occurred at a given location, or perhaps on a given stretch of the road. The outcomes of this "experiment" constitute a continuum of points, those on the continuous interval from 0 to 120, and we shall assume, for the sake of argument, that the probability that an accident will occur on any interval of length L (miles) is $\frac{L}{120}$. Note that this arbitrary assignment of probabilities is consistent with Postulates 1 and 2 on page 38, for the probabilities $\frac{L}{120}$ are all nonnegative and $P(S) = \frac{120}{120} = 1$. Of course, this assignment of probabilities applies only to events represented by *intervals* on the line segment from 0 to 120, and even though we could use Postulate 3 to consider events which are represented by the union of any finite or infinite sequence of intervals, *this would be as far as we can go.* The point we are trying to make is that in the continuous case the postulates of probability remain the same, but they will not apply to *every* subset of the sample space; in other words, in the continuous case *we must restrict the meaning of the term "event."* Practically speaking, this is of very little consequence, if any—we simply do not assign probabilities to some rather abstruse subsets of the sample space which cannot be expressed as the unions or intersections of finitely many or countably many intervals.

In Section 3.1 we introduced the concept of a random variable as a real-valued function defined over the points of a sample space, and in Figure 3.1 on page 68 we illustrated this by assigning the *total* rolled with the two dice to each of the 36 points of the sample space to which we had assigned probabilities of $\frac{1}{36}$. In the continuous case, where random variables can assume values on a continuous scale, the procedure is very much the same. In the one-variable case, the outcomes of an experiment are represented by the points on a line segment or a line, and the values of a random variable are numbers appropriately assigned to the points by some rule or equation. When the value of a random variable is given *directly* by a measurement or observation, we generally do not bother to differentiate between the value of the random variable (the measurement which we obtain) and the outcome of the experiment (the corresponding point on the real axis). Thus, if an experiment consists of determining the actual content of an 8-ounce jar of instant coffee, the result itself, say, 7.93 ounces, is the value of the random variable with which we are concerned, and there is no real need to add that the sample space consists of a certain interval of points on the positive real axis.

4.2 PROBABILITY DENSITIES

In Chapter 3 we introduced the concept of a *distribution function* by defining its values $F(x)$ as the probabilities that a random variable $\mathbf{x}$ takes on a value less than or equal to x, for $-\infty \leq x \leq \infty$. At the time we restricted ourselves to discrete random variables, so that the graphs of their distribution functions were step-functions like the one shown in Figure 3.2 on page 72; another example is shown in Figure 4.1.

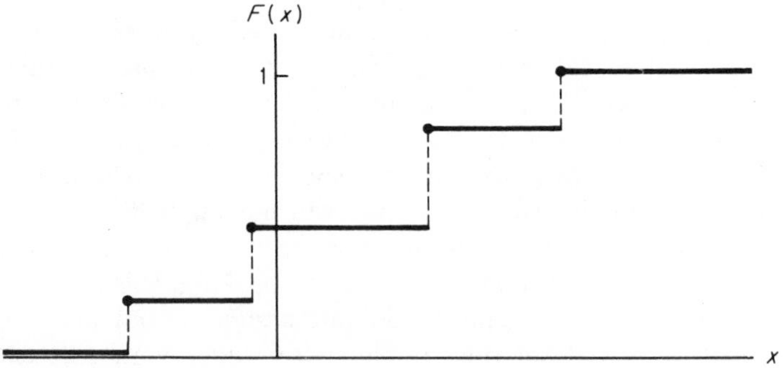

FIGURE 4.1 Distribution function of discrete random variable.

Now let us consider the case where the range of a random variable $\mathbf{x}$ consists of the set of all real numbers and the graph of its distribution function is a continuous curve somewhat like that of Figure 4.2. Although

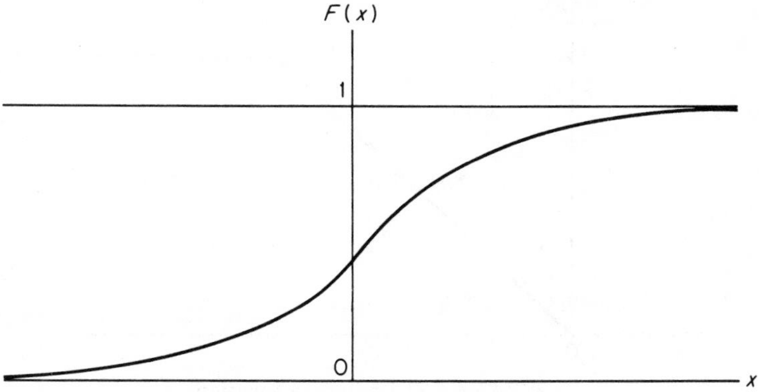

FIGURE 4.2 Continuous distribution function.

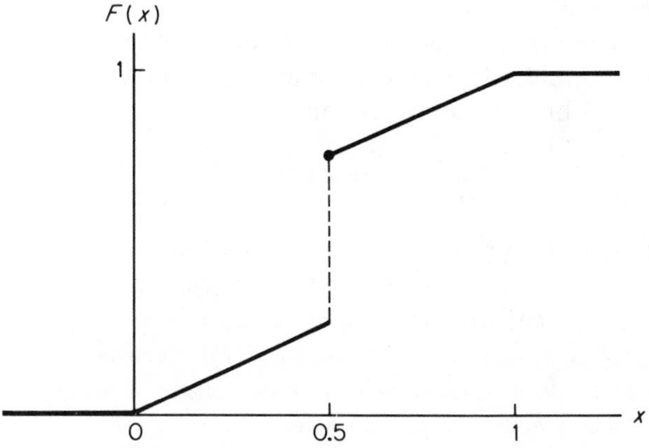

FIGURE 4.3 Discontinuous distribution function.

there are other possibilities (as is illustrated in Figure 4.3), *we shall limit ourselves to distribution functions which are continuous and whose derivative* $F'(x) = f(x)$ *exists for all but a finite set of values of* x. This excludes the distribution function whose graph is shown in Figure 4.3 (see also Exercise 5 on page 109), but it does *not* exclude the distribution function whose graph is shown in Figure 4.4. In the first case there is a discontinuity at $x = \frac{1}{2}$, and in the second case the function is continuous and fails to have a derivative only at $x = 0$ and $x = 1$.

An important consequence of the fact that we are limiting ourselves to continuous distribution functions is that *the probability that a random variable* $\mathbf{x}$ *will take on any specific value* x *is always zero.* To prove this we

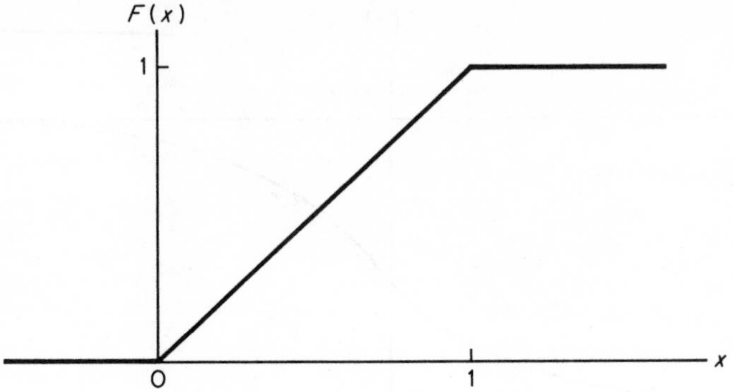

FIGURE 4.4 Distribution function.

have only to observe that the probability that the value of x will actually *equal* x is less than or equal to $F(x) - F(x - \delta)$ for any $\delta > 0$. (In other words, the desired probability is less than or equal to the probability that the random variable will take on a value on the interval from $x - \delta$ to x, excluding $x - \delta$ but including x.) Since

$$\lim_{\delta \to 0} [F(x) - F(x - \delta)] = 0$$

by virtue of the assumption that the distribution function is continuous, we have thus shown what we set out to prove. *The fact that in the continuous case the probability for any given value x is zero should not be viewed with alarm, however, as it does not mean that the corresponding event cannot take place. After all, a random variable must take on some value even though all the possibilities have zero probabilities.*

Since the properties of distribution functions referred to in Exercise 3 on page 83 hold also for the continuous case, that is, $F(-\infty) = 0$, $F(\infty) = 1$, and $F(a) \leq F(b)$ when $a < b$, we can use the fundamental theorem of integral calculus and write

$$\int_{-\infty}^{x} f(t)\, dt = F(x) - F(-\infty) = F(x) \qquad (4.2.1)$$

Thus, the probability that a random variable $\mathbf{x}$ takes on a value less than or equal to x is given by the integral

$$\int_{-\infty}^{x} f(t)\, dt$$

where $f(t)$ is the value of the derivative of the distribution function of $\mathbf{x}$ at t.

An immediate consequence of (4.2.1) is that *for any two constants a and b, with a < b, the probability that* **x** *takes on a value on the interval from a to b is given by*

$$F(b) - F(a) = \int_{-\infty}^{b} f(x)\,dx - \int_{-\infty}^{a} f(x)\,dx$$
$$= \int_{a}^{b} f(x)\,dx \qquad (4.2.2)$$

and, hence, by the area under the curve shaded in Figure 4.5. Actually, $F(b) - F(a)$ is the probability that **x** takes on a value greater than a and

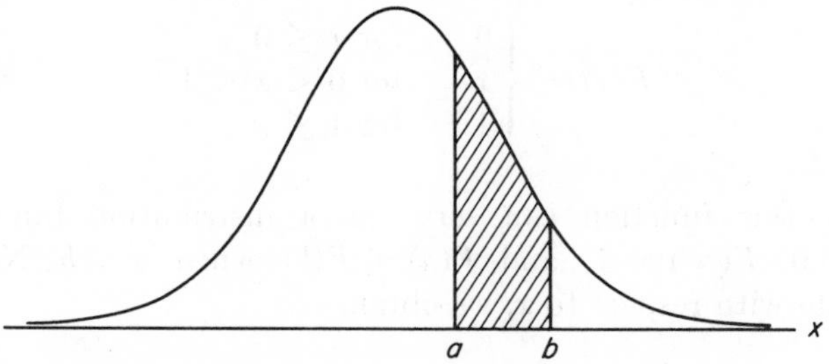

FIGURE 4.5 Probability given by area under curve.

less than or equal to b, but by virtue of the proof on the preceding page it does not matter whether we include $x = a$, or for that matter $x = b$.

Also, it does not matter, of course, whether we write the integral as $\int_a^b f(x)\,dx$ or as $\int_a^b f(t)\,dt$, so long as it is remembered that the function **f** (which is being integrated) is the derivative of the distribution function of the random variable. Incidentally, there is a revealing analogy between this function **f** and the concept of a density function as it is used in physics. Consider a rod of *variable density* which has a uniform cross section of 1 square inch and the density $\rho(x)$, expressed in pounds per cubic inch, at the distance x from one end (see Figure 4.6). To obtain the *weight* of any piece of the rod, say, the section shaded in Figure 4.6, we

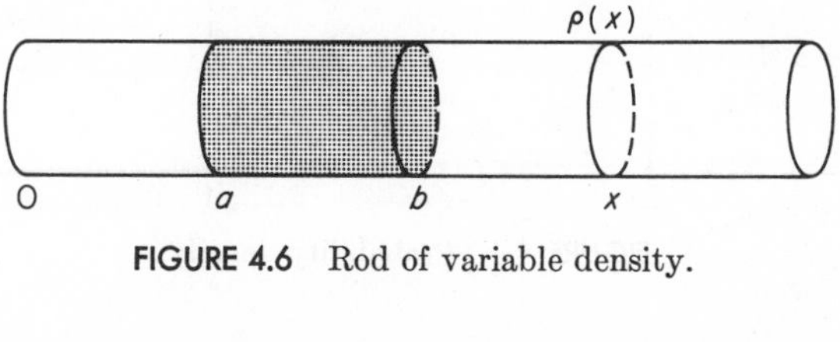

FIGURE 4.6 Rod of variable density.

have to integrate the density function between the appropriate limits, and hence get

$$\int_a^b \rho(x)\, dx$$

In view of this close analogy between *density* and *weight* in physics and the *function* **f** which we integrate in (4.2.2) and *probability* in statistics, it is customary to refer to this function **f** as a *probability density function*, or simply a *probability density*.

To give an example of a distribution function and the corresponding probability density, let us refer to the distribution function pictured in Figure 4.4, which is given by

$$F(x) = \begin{cases} 0 & \text{for } x \leq 0 \\ x & \text{for } 0 < x < 1 \\ 1 & \text{for } 1 \leq x \end{cases}$$

Note that this function can serve as a distribution function since $F(-\infty) = 0$, $F(\infty) = 1$, and $F(a) \leq F(b)$ when $a < b$. Now, if we differentiate with respect to x, we obtain

$$f(x) = \begin{cases} 0 & \text{for } x < 0 \\ 1 & \text{for } 0 < x < 1 \\ 0 & \text{for } 1 < x \end{cases}$$

and we find that this probability density, whose graph is shown in Figure 4.7, is defined for all values of x except 0 and 1. Actually, it does not matter how $f(x)$ is defined at these two points, but there are certain

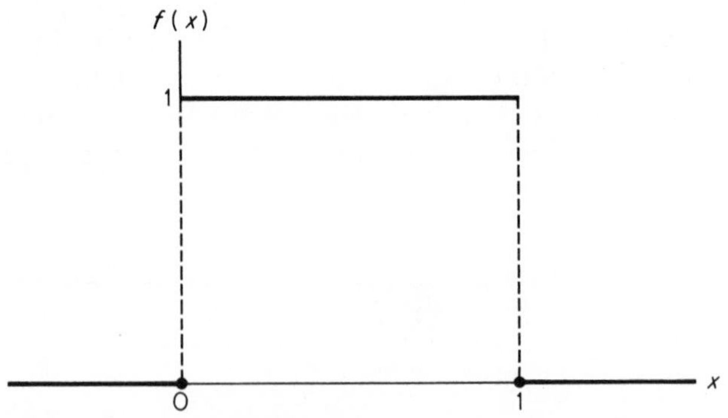

FIGURE 4.7 Probability density.

advantages (which will be explained on page 120) to putting $f(0)$ and $f(1)$ both equal to 0. Thus, we can rewrite the probability density as

$$f(x) = \begin{cases} 1 & \text{for } 0 < x < 1 \\ 0 & \text{elsewhere} \end{cases}$$

where "elsewhere" means all other real values of x.

Using either this probability density or the distribution function with which we began, we can now determine all sorts of probabilities concerning a random variable having the given distribution. For instance, the probability that it takes on a value less than 0.4 is

$$\int_0^{0.4} f(x)\,dx = \int_0^{0.4} 1\,dx = x\Big]_0^{0.4} = 0.4 - 0 = 0.4$$

or more directly $F(0.4) = 0.4$; similarly, the probability that it takes on a value on the interval from 0.6 to 0.9 is

$$\int_{0.6}^{0.9} f(x)\,dx = \int_{0.6}^{0.9} 1\,dx = x\Big]_{0.6}^{0.9} = 0.9 - 0.6 = 0.3$$

or (again) more directly $F(0.9) - F(0.6) = 0.9 - 0.6 = 0.3$. Obviously, when we use the distribution function *directly* the work is much simpler, because the required integration has already been performed.

Since a probability density **f**, integrated between any two constants a and b (with $a < b$), gives the probability that a random variable takes on a value on the interval from a to b, **f** cannot be just any real-valued integrable function. However, if we impose the conditions that

(1) $f(x) \geq 0$ for all x within the domain of **f**,

(2) $\int_{-\infty}^{\infty} f(x)\,dx = 1$

the postulates of probability will be satisfied. The first condition assures that probabilities cannot be negative, and the second condition asserts that $P(S)$, the probability of the entire sample space, equals 1.* So far as the third postulate of probability is concerned, we have only to observe that for non-overlapping intervals (mutually exclusive outcomes) we add the corresponding integrals.

To illustrate, let us check whether

$$f(x) = \begin{cases} 3e^{-3x} & \text{for } x > 0 \\ 0 & \text{for } x \leq 0 \end{cases}$$

* So long as we are limiting ourselves to distribution functions that are continuous, it actually would not hurt if $f(x)$ were negative for a few values of x. This would not affect any probabilities, but (so far as applications are concerned) it would not make any sense.

can serve as the probability density of a random variable. Since $3e^{-3x}$ cannot be negative, all we have to check is whether

$$\int_0^\infty 3e^{-3x}\,dx = 1$$

Actually performing this integration, we get

$$\int_0^\infty 3e^{-3x}\,dx = -e^{-3x}\Big]_0^\infty = 0 - (-1) = 1$$

and it should be noted that the first step of this integration can also be used to show that for $x > 0$ the corresponding distribution function is given by

$$F(x) = -e^{-3x}\Big]_0^x = 1 - e^{-3x}$$

while elsewhere (for $x \leq 0$) $F(x) = 0$.

THEORETICAL EXERCISES

1. If the distribution function of a random variable is given by

$$F(x) = \begin{cases} 1 - \dfrac{4}{x^2} & \text{for } x > 2 \\ 0 & \text{for } x \leq 2 \end{cases}$$

find
(a) the probabilities that it will take on a value which is (i) less than 4, (ii) between 2 and 5, and (iii) greater than 8;
(b) the probability density which corresponds to this distribution function, letting its value equal zero wherever it is undefined.
Also sketch the graphs of the distribution function and the probability density.

2. If the distribution function of a random variable is given by

$$F(x) = \begin{cases} 1 - (1 + x)e^{-x} & \text{for } x > 0 \\ 0 & \text{for } x \leq 0 \end{cases}$$

find
(a) the probabilities that it will take on a value which is (i) less than 1, (ii) between 2 and 4, and (iii) greater than 3;
(b) the probability density which corresponds to this distribution function. Are there any points at which it is undefined?
Also sketch the graphs of the distribution function and the probability density.

3. If the probability density of a random variable is given by

$$f(x) = \begin{cases} k(1 - x) & \text{for } 0 < x < 1 \\ 0 & \text{elsewhere} \end{cases}$$

find

(a) the value of k;

(b) the distribution function which corresponds to this probability density;

(c) the probability that a random variable having this probability density will take on a value less than 0.5.

Also sketch the graphs of the probability density and the distribution function.

4. If the probability density of a random variable is given by

$$f(x) = \begin{cases} cxe^{-x^2} & \text{for } x > 0 \\ 0 & \text{for } x \leq 0 \end{cases}$$

find

(a) the value of c;

(b) the distribution function which corresponds to this probability density.

5. The discontinuous distribution function of Figure 4.3 could be that of a random variable whose value is determined as follows: First we flip an "honest" coin and if it comes up *heads* we assign the random variable the value $\frac{1}{2}$; if it comes up *tails* we choose a value of a random variable having the probability density pictured in Figure 4.7 and let *it* be the value of the original random variable. Find the equation of this distribution function.

APPLIED EXERCISES

6. The actual amount of coffee (in ounces) in an 8-ounce jar filled by a certain machine is a random variable having the probability density

$$f(x) = \begin{cases} 0 & x \leq 7.9 \\ 5 & 7.9 < x < 8.1 \\ 0 & x \geq 8.1 \end{cases}$$

Find the probabilities that an 8-ounce jar filled by this machine will contain

(a) at most 7.93 ounces of coffee;

(b) anywhere from 7.98 to 8.02 ounces of coffee;

(c) at least 7.95 ounces of coffee.

7. The amount of time that a flight from Phoenix to Tucson is early or late is a random variable having the probability density

$$f(x) = \begin{cases} \frac{1}{288}(36 - x^2) & \text{for } -6 < x < 6 \\ 0 & \text{elsewhere} \end{cases}$$

where x is in minutes, negative values are indicative of the flight's being early, and positive values are indicative of its being late. Find the probability that one of these flights will be
(a) at least a minute early;
(b) at least 3 minutes late;
(c) anywhere from 2 to 4 minutes late;
(d) *exactly* 5 minutes early.

8. The mileage (in thousands of miles) which car owners get with a certain kind of tire is a random variable having the probability density

$$f(x) = \begin{cases} \frac{1}{20} e^{-\frac{x}{20}} & \text{for } x > 0 \\ 0 & \text{for } x \leq 0 \end{cases}$$

Find the probabilities that one of these tires will last
(a) at most 12,000 miles;
(b) anywhere from 18,000 to 24,000 miles;
(c) at least 30,000 miles.

4.3 SPECIAL PROBABILITY DENSITIES

As we pointed out in Section 3.3, a discrete random variable which has a certain probability *function* is also said to have that probability *distribution;* for instance, if the probabilities associated with a random variable are given by the binomial probability *function* (say, with $n = 20$ and $\theta = \frac{1}{2}$), it is said to have this binomial *distribution*. The same kind of terminology is used in the continuous case where we shall say that a random variable has a certain *distribution* (say, the *uniform distribution* of Section 4.3.1) when it has the corresponding probability *density*.

In this section we shall introduce several probability densities which are of basic importance; two others are given in the exercises which follow this section, and three probability densities that are of special importance in the theory of sampling will be taken up in Chapter 7. Further mathematical properties of all these special probability densities (as well as probability densities in general) and the significance of their *parameters* will be discussed in Chapter 5.

4.3.1 The Uniform Distribution

The probability density whose graph is shown in Figure 4.7 and the one of Exercise 6 on page 109 are both special cases of the *uniform density* given by

$$f(x) = \begin{cases} \dfrac{1}{\beta - \alpha} & \text{for } \alpha < x < \beta \\ 0 & \text{elsewhere} \end{cases} \tag{4.3.1}$$

where α and β are real constants with $\alpha < \beta$. Although the uniform density has some direct applications, one of which will be discussed in Section 4.4, its main value lies in the fact that due to its simplicity it readily lends itself to the task of illustrating various aspects of statistical theory.

4.3.2 The Exponential Distribution

The probability density which we used in the last illustration of Section 4.2, the one on page 107, was a special case of the *exponential density*

$$f(x) = \begin{cases} \dfrac{1}{\theta} e^{-\frac{x}{\theta}} & \text{for } x > 0 \\ 0 & \text{for } x \leq 0 \end{cases} \tag{4.3.2}$$

where θ is a positive constant. As can easily be verified, the integral of this function, from $-\infty$ to ∞, is equal to 1.

To illustrate how an exponential distribution might arise in actual practice, let us refer to the situation described in Exercise 15 on page 85, where we were interested in the probability of getting x successes during a time interval of length t under the conditions that (i) the probability of a success during a *very small* time interval Δt is $\alpha \cdot \Delta t$, (ii) the probability of more than one success during such a time interval is negligible, and (iii) the probability of a success during such a time interval does not depend on what happened before. Now let us determine the probability density of the random variable $\mathbf{y}$, the time it takes under these conditions until there is the *first success*. Making use of the result of that exercise and formula (3.3.9) for the *Poisson distribution*, we find that the probability of 0 successes during a time interval of length y is

$$p(0; \alpha y) = \frac{(\alpha y)^0 e^{-\alpha y}}{0!} = e^{-\alpha y} \qquad \text{for } \alpha > 0 \text{ and } y > 0$$

It follows that the probability of getting the first success *prior* to time y is

$$F(y) = 1 - e^{-\alpha y} \qquad \text{for } y > 0$$

and we have thus found the *distribution function* of the random variable $\mathbf{y}$. Differentiating with respect to y, we then get the probability density

$$f(y) = \alpha e^{-\alpha y} \qquad \text{for } y > 0$$

which is easily identified as an exponential density with $\theta = \frac{1}{\alpha}$ [after we add that $f(y) = 0$ for $y \leq 0$, since the first success cannot very well occur before the whole process starts, and it does not matter how we define $f(0)$].

The exponential distribution applies not only to the occurrence of the first success in a *Poisson process* (which is what we call a situation like that described in Exercise 15 on page 85), but by virtue of condition (iii) it also applies to the "waiting time" *between* successes. For instance, in Exercise 35 on page 89 it would be reasonable to treat the time between successive catches as a random variable having an exponential distribution with $\theta = \frac{1}{1.8}$, and the probability that it will take, say, less than 20 minutes between successive catches is

$$\int_0^{\frac{1}{3}} 1.8e^{-1.8x}\,dx = -e^{-1.8x}\Big]_0^{\frac{1}{3}} = -e^{-0.6} + 1$$

which is approximately 0.45. (Note that the units are *hours* in this example, so that we had $\alpha \cdot 1 = 1.8$ and, hence, $\theta = \frac{1}{1.8}$, and the upper limit of integration was $\frac{20}{60} = \frac{1}{3}$.)

4.3.3 The Gamma Distribution

An important generalization of the exponential distribution is the one whose probability density is given by

$$f(x) = \begin{cases} kx^{\alpha-1}e^{-\frac{x}{\beta}} & \text{for } x > 0 \\ 0 & \text{for } x \leq 0 \end{cases}$$

where $\alpha > 0$, $\beta > 0$, and k must be such that the integral of the function from $-\infty$ to ∞ is equal to 1. [Evidently, this *is* a generalization of the exponential distribution since it reduces to (4.3.2) when $\alpha = 1$ and $\beta = \theta$.] To evaluate k, let us make the substitution $y = \frac{x}{\beta}$, which yields

$$\int_0^\infty kx^{\alpha-1}e^{-\frac{x}{\beta}}\,dx = k\beta^\alpha \int_0^\infty y^{\alpha-1}e^{-y}\,dy$$

The integral thus obtained depends on α alone, and it *defines* the well-known *gamma function*

$$\Gamma(\alpha) = \int_0^\infty y^{\alpha-1}e^{-y}\,dy \qquad \text{for } \alpha > 0 \tag{4.3.3}$$

which is discussed in great detail in textbooks on advanced calculus. If we integrate by parts, which will be left to the reader in Exercise 3 on page 116, it can be shown that the gamma function satisfies the recursion formula

$$\Gamma(\alpha) = (\alpha - 1)\cdot\Gamma(\alpha - 1) \tag{4.3.4}$$

from which it follows immediately that

$$\Gamma(\alpha) = (\alpha - 1)! \tag{4.3.5}$$

when α is a *positive integer*.

Returning now to the problem of evaluating k, we equate the above integral to 1, getting

$$\int_0^\infty kx^{\alpha-1}e^{-\frac{x}{\beta}}\,dx = k\beta^\alpha\Gamma(\alpha) = 1$$

and, hence,

$$k = \frac{1}{\beta^\alpha\Gamma(\alpha)}$$

Thus, the original probability density, that of the so-called *gamma distribution*, can be written as

$$f(x) = \begin{cases} \dfrac{1}{\beta^\alpha\Gamma(\alpha)}\,x^{\alpha-1}e^{-\frac{x}{\beta}} & \text{for } x > 0 \\ 0 & \text{for } x \le 0 \end{cases} \tag{4.3.6}$$

where $\Gamma(\alpha)$ will have to be looked up in a special table (not in this book) when α is not a positive integer. To give the reader some idea about the shape of the graphs of gamma densities, those for several special values of α and β are shown in Figure 4.8.

Special cases of the gamma distribution play an important role in the theory of statistics. We already saw that for $\alpha = 1$ and $\beta = \theta$ we obtain the exponential distribution of Section 4.3.2; for $\alpha = \frac{n}{2}$ and $\beta = 2$ we

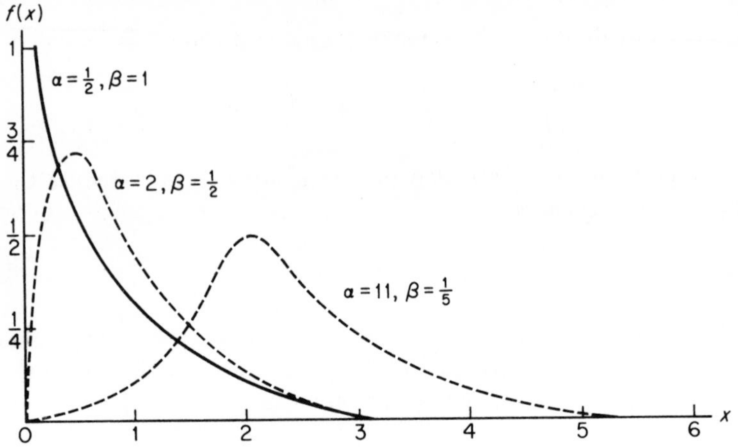

FIGURE 4.8 Gamma distributions.

obtain the so-called *chi-square distribution*, which we shall study in considerable detail in Chapter 7. In addition to the applications given in the exercises on page 119, an important application of the gamma distribution to a problem of theoretical physics is given in Exercise 8 on page 127.

4.3.4 The Beta Distribution

The uniform distribution $f(x) = 1$ for $0 < x < 1$ and $f(x) = 0$ elsewhere as well as the distribution of Exercise 3 on page 109 are special cases of the *beta distribution*, whose density is given by

$$f(x) = \begin{cases} \dfrac{\Gamma(\alpha + \beta)}{\Gamma(\alpha) \cdot \Gamma(\beta)}\, x^{\alpha-1}(1 - x)^{\beta-1} & \text{for } 0 < x < 1 \\ 0 & \text{elsewhere} \end{cases} \qquad (4.3.7)$$

where $\alpha > 0$ and $\beta > 0$. (Note that we obtain this special uniform distribution when $\alpha = 1$ and $\beta = 1$, and the distribution of Exercise 3 on page 109 when $\alpha = 1$ and $\beta = 2$.) In recent years, the beta distribution has found important applications in *Bayesian statistics*, where probabilities are sometimes looked upon as random variables, and there is, therefore, a need for a fairly "flexible" probability density which assumes nonzero values only on the interval from 0 to 1. By "flexible" we mean that the beta distribution can assume a great variety of different shapes, as the reader will be asked to verify in Exercise 9 on page 117. The use of the beta distribution in Bayesian statistics will be discussed in detail in Chapter 9.

4.3.5 The Normal Distribution

The distribution which will be introduced in this section is in many ways the cornerstone of modern statistical theory. It was studied first in the eighteenth century when scientists observed an astonishing degree of regularity in errors of measurement. They found that the patterns (distributions) which they observed were closely approximated by a continuous curve which they referred to as the "normal curve of errors" and attributed to the laws of chance. The mathematical properties of this continuous distribution and its theoretical basis were first investigated by Abraham de Moivre (1667–1745), Pierre Laplace (1749–1827), and Karl Gauss (1777–1855). In honor of Gauss, this distribution is sometimes referred to as the *Gaussian distribution*, but the term *normal distribution* is much more widely used. The equation of the *normal probability density*, whose graph (shaped like the cross section of a bell) is shown in Figure 4.9, is

$$f(x) = \frac{1}{\beta\sqrt{2\pi}}\, e^{-\frac{1}{2}\left(\frac{x-\alpha}{\beta}\right)^2} \qquad \text{for } -\infty < x < \infty \tag{4.3.8}$$

where α and β are real constants and β must be positive. The significance of the *parameters* α and β will be discussed in detail in Section 5.3.6.

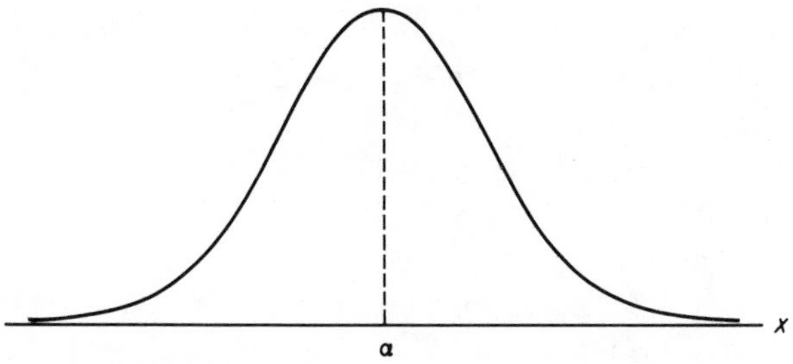

FIGURE 4.9 Graph of normal distribution.

To verify that (4.3.8) is, indeed, a probability density, we have only to show that the integral of the function from $-\infty$ to ∞ is equal to 1, for the values of the function obviously cannot be negative. Making the substitution $u = \dfrac{x-\alpha}{\beta}$, we get

$$\int_{-\infty}^{\infty} \frac{1}{\beta\sqrt{2\pi}}\, e^{-\frac{1}{2}\left(\frac{x-\alpha}{\beta}\right)^2} dx = \frac{1}{\sqrt{2\pi}} \int_{-\infty}^{\infty} e^{-\frac{1}{2}u^2}\, du = \frac{1}{\sqrt{2\pi}} \cdot I$$

and it remains to be shown that the value of the integral I is $\sqrt{2\pi}$. To this end, let us multiply I by itself, writing v in the second integral instead of u, so that we get

$$I^2 = \int_{-\infty}^{\infty} e^{-\frac{1}{2}u^2}\,du \cdot \int_{-\infty}^{\infty} e^{-\frac{1}{2}v^2}\,dv = \int_{-\infty}^{\infty}\int_{-\infty}^{\infty} e^{-\frac{1}{2}(u^2+v^2)}\,du\,dv$$

Then, if we change to polar coordinates, that is, if we let $u = r\cdot\cos\theta$ and $v = r\cdot\sin\theta$, we obtain

$$I^2 = \int_0^{2\pi}\int_0^{\infty} re^{-\frac{1}{2}r^2}\,dr\,d\theta = \int_0^{2\pi} d\theta \cdot \int_0^{\infty} re^{-\frac{1}{2}r^2}\,dr$$

where the first integral equals 2π and straightforward methods of elementary calculus will show that the second integral equals 1. Thus, $I^2 = 2\pi$, $I = \sqrt{2\pi}$, and this completes the proof.

A physical situation which leads to a normal distribution will be given in Exercise 12 on page 138, but otherwise applications will not be discussed until we shall study the distribution further in Chapter 5. Also, most of Chapter 7 will be devoted to the normal distribution and distributions closely related to it.

THEORETICAL EXERCISES

1. Show that if a random variable has the uniform density (4.3.1), the probability that it will take on a value less than $\alpha + p(\beta - \alpha)$ is equal to p.
2. Show that if a random variable has the exponential density (4.3.2), the probability that it will take on a value less than $-\theta\cdot\ln(1 - p)$ equals p for $0 \leq p < 1$.
3. Use integration by parts to show that the gamma function satisfies the recursion formula $\Gamma(\alpha) = (\alpha - 1)\cdot\Gamma(\alpha - 1)$ for $\alpha > 1$.
4. Show that the integral for the gamma function can also be written as

$$\Gamma(\alpha) = 2^{1-\alpha}\cdot\int_0^{\infty} z^{2\alpha-1}e^{-\frac{1}{2}z^2}\,dz \qquad \text{for } \alpha > 0$$

by letting $y = \frac{1}{2}z^2$ in (4.3.3). Use this form of the gamma function and the result obtained earlier on this page to show that $\Gamma(\frac{1}{2}) = \sqrt{\pi}$.

5. Find the probabilities that the value of a random variable will exceed 4, if it has a gamma distribution with
(a) $\alpha = 2$ and $\beta = 2$;
(b) $\alpha = 3$ and $\beta = 4$.

6. Show that the gamma density (4.3.6) with $\alpha > 1$ has a relative maximum at $x = \beta(\alpha - 1)$. What happens when $0 < \alpha < 1$ and when $\alpha = 1$?

7. Verify that the integral of the beta density (4.3.7) from 0 to 1 is equal to 1 for
   (a) $\alpha = 2$ and $\beta = 4$;
   (b) $\alpha = 3$ and $\beta = 3$.

8. Show that the beta density (4.3.7) with $\alpha > 1$ and $\beta > 1$ has a relative maximum at $x = \dfrac{\alpha - 1}{\alpha + \beta - 2}$.

9. Sketch the graphs of the beta densities for which
   (a) $\alpha = 2$ and $\beta = 2$;
   (b) $\alpha = \frac{1}{2}$ and $\beta = 1$;
   (c) $\alpha = 2$ and $\beta = \frac{1}{2}$;
   (d) $\alpha = 2$ and $\beta = 5$.
   [*Hint:* To evaluate $\Gamma(1\frac{1}{2})$ and $\Gamma(2\frac{1}{2})$ make use of (4.3.4) and the result of Exercise 4.]

10. Show that the normal probability density (4.3.8) has a relative maximum at $x = \alpha$ and inflection points at $x = \alpha - \beta$ and $x = \alpha + \beta$.

11. A random variable is said to have the *Cauchy distribution* if its density is given by

$$f(x) = \frac{k}{a^2 + x^2} \qquad \text{for } -\infty < x < \infty$$

   where $a > 0$. Express k in terms of a.

12. A random variable is said to have the *Weibull distribution* if its density is given by

$$f(x) = \begin{cases} kx^{\beta-1}e^{-\alpha x^\beta} & \text{for } x > 0 \\ 0 & \text{for } x \leq 0 \end{cases}$$

   where $\alpha > 0$ and $\beta > 0$. Express k in terms of α and β.

13. If a random variable $\mathbf{t}$ is the *time to failure* of a commercial product (say, the tires of Exercise 8 on page 110) and the values of its probability density and probability distribution are $f(t)$ and $F(t)$, its *failure rate* at any time t is said to be $\dfrac{f(t)}{1 - F(t)}$. Thus, the failure rate is the value of the probability density at time t divided by the probability that the product has not failed prior to time t.
   (a) Show that if the time to failure has an exponential distribution, the failure rate is constant.
   (b) Show that for the *Weibull distribution* of Exercise 12 the value of the failure rate at time t is $\alpha\beta t^{\beta-1}$.

14. PEARSON CURVES It was shown by Karl Pearson, one of the founders of modern statistics, that the differential equation

$$\frac{1}{f(x)} \cdot \frac{d[f(x)]}{dx} = \frac{d - x}{a + bx + cx^2}$$

yields most of the important distributions of statistics for appropriate values of the constants a, b, c, and d. Verify that the differential equation gives

(a) the normal distribution when $b = c = 0$ and $a > 0$;

(b) the exponential distribution when $a = c = d = 0$ and $b > 0$;

(c) the gamma distribution when $a = c = 0$, $b > 0$ and $d > -b$;

(d) the beta distribution when $a = 0$, $b = -c$, $\frac{d-1}{b} < 1$, and $\frac{d}{b} > -1$.

APPLIED EXERCISES

15. Given a line AB, whose midpoint is C and whose length is a. If a point X is chosen on this line so that $\mathbf{x}$, its distance from A, has the uniform density $f(x) = \frac{1}{a}$ for $0 < x < a$ [and $f(x) = 0$ elsewhere], what is the probability that AX, BX, and AC will form a triangle?

16. In Exercise 36 on page 89 it would be reasonable to say that the time between successive calls reaching the switchboard is a random variable having an exponential distribution with $\theta = \frac{1}{6.2}$, where the units are in minutes.

(a) What is the probability that the switchboard operator will have a "breathing spell" of less than 1.0 minutes between successive calls?

(b) What is the probability that the switchboard operator will have a "breathing spell" of at least 0.5 minutes between successive calls?

17. With reference to Exercise 37 on page 89, find the probability that there will be at least 24 minutes between accidents on the Pasadena Freeway during the given period of time.

18. If family income in a certain urban area (in thousands of dollars) can be looked upon as a random variable having a gamma distribution with $\alpha = 2$ and $\beta = 4$, find the probabilities that a family randomly selected from this area will have

(a) an income of less than \$4,000;
(b) an income anywhere from \$6,000 to \$12,000;
(c) an income greater than \$20,000.

19. In a certain city, the daily consumption of electric power (in millions of kilowatt-hours) can be treated as a random variable having a gamma distribution with $\alpha = 3$ and $\beta = 2$. If the power plant of this city has a daily capacity of 12 million kilowatt-hours, what is the probability that this power supply will be inadequate on any given day?

20. If the annual proportion of erroneous income tax returns filed with the I. R. S. can be looked upon as a random variable having a beta distribution with $\alpha = 2$ and $\beta = 9$, what is the probability that in any given year there will be fewer than 10 percent erroneous returns?

21. If the proportion of donut shops which make a profit during their first year of operation is a random variable having a beta distribution with $\alpha = 3$ and $\beta = 2$, what is the probability that at least 80 percent of the new donut shops belonging to a very large chain will make a profit during their first year of operation?

22. Suppose that the service life (in hours) of a semiconductor is a random variable having the Weibull distribution of Exercise 12 with $\alpha = 0.025$ and $\beta = 0.50$. What is the probability that such a semiconductor will still be in operating condition after 4,000 hours?

4.4 CHANGE OF VARIABLE

Let us now study the problem of obtaining the probability distribution of a random variable **y**, given the probability distribution of a random variable **x** and a relationship of the form $y = h(x)$ between the values of **x** and those of **y**. In the discrete case there is no real problem, for all we have to do is make the appropriate substitutions. To illustrate, suppose that the random variable **x** has a binomial distribution with $\theta = \frac{1}{4}$ and $n = 5$, and that we are interested in the random variable **y** whose values are related to those of **x** by means of the equation $y = \dfrac{1}{1+x}$, which can also be written as $x = \dfrac{1}{y} - 1$. Thus,

$$f(x) = \binom{5}{x}\left(\frac{1}{4}\right)^x\left(\frac{3}{4}\right)^{5-x} \qquad \text{for } x = 0, 1, 2, 3, 4, 5$$

and if we let $g(y)$ denote the values of the probability function of **y**, we get

$$g(y) = f(x) = f\left(\frac{1}{y} - 1\right) = \binom{5}{\frac{1}{y} - 1}\left(\frac{1}{4}\right)^{\frac{1}{y}-1}\left(\frac{3}{4}\right)^{6-\frac{1}{y}}$$

$$\text{for } y = 1, \tfrac{1}{2}, \tfrac{1}{3}, \tfrac{1}{4}, \tfrac{1}{5}, \tfrac{1}{6}$$

Note that *the probabilities have remained unchanged*—the only difference is that they are now associated with the various values of **y** instead of the corresponding values of **x**. This is all there is to it, so long as the relationship $y = h(x)$ is one-to-one; if it is not one-to-one, we will have to be careful, as is illustrated in Exercise 3 on page 126.

In the continuous case, we shall assume that the function given by $y = h(x)$ is differentiable and either increasing or decreasing for all values within the range of **x**, so that the inverse function, given by $x = H(y)$, exists for all of the corresponding values of **y**, and that it is differentiable except when $h'(x) = 0$. Under these conditions, we can prove the following theorem:

THEOREM 4.1

If the probability density of **x** *is given by* $f(x)$ *and the function given by* $y = h(x)$ *is differentiable and either increasing or decreasing for all values within the range of* **x** *for which* $f(x) \neq 0$,* *then the probability density of* **y** *is given by*

$$g(y) = f(x) \cdot \left|\frac{dx}{dy}\right|, \qquad \frac{dy}{dx} \neq 0 \tag{4.4.1}$$

In what follows, we shall prove this theorem for the case where the function given by $y = h(x)$ is *increasing;* the second case, where the function is assumed to be *decreasing* will be left to the reader in Exercise 4 on page 126. Note that we distinguished between the probability densities of **x** and **y** by writing their values as $f(x)$ and $g(y)$; correspondingly, we shall write the values of their distribution functions as $F(x)$ and $G(y)$.

Using this notation, we can write the probability that **x** will take on a value less than or equal to a as

$$F(a) = \int_{-\infty}^{a} f(x)\, dx$$

and since [under the assumption that the function given by $y = h(x)$ is increasing] this equals the probability that **y** will take on a value less

* Note that this is why we let $f(0) = f(1) = 0$ on page 107, and in general did not include the endpoints of the intervals for which probability densities are non-zero. For instance, in the illustration on page 121, $y = \sqrt{x}$ does *not* have a derivative at $x = 0$.

than or equal to $h(a)$, we also have

$$G[h(a)] = \int_{-\infty}^{a} f(x)\,dx$$

Now, if we perform the change of variable $y = h(x)$ and, hence, $x = H(y)$ in the last integral, we get

$$G[h(a)] = \int_{-\infty}^{h(a)} f[H(y)]\cdot H'(y)\,dy$$

for any real number $h(a)$ within the range of $\mathbf{y}$. Hence, the *integrand* of this last integral gives the probability density of $\mathbf{y}$ so long as $H'(y)$ exists, and we can write

$$g(y) = f[H(y)]H'(y), \qquad \frac{1}{H'(y)} \neq 0$$

or

$$g(y) = f(x)\cdot\frac{dx}{dy}, \qquad \frac{dy}{dx} \neq 0$$

since $H(y) = x$ and $H'(y) = \frac{dx}{dy}$. The absolute value signs of (4.4.1) were not needed in this case because it was assumed that the function given by $y = h(x)$ is *increasing;* however, they will be needed in the case which the reader is asked to prove in Exercise 4 on page 126, namely, when the function given by $y = h(x)$ is *decreasing.*

To illustrate the use of Theorem 4.1, let us consider a random variable $\mathbf{x}$ which has the exponential density (4.3.2) with $\theta = 1$, and let us determine the probability density of the random variable $\mathbf{y}$ whose values are given by $y = \sqrt{x}$ (where $\sqrt{x}$ is the positive square root of x) and, hence, $x = y^2$. Since the probability density of x is given by

$$f(x) = \begin{cases} e^{-x} & \text{for } x > 0 \\ 0 & \text{for } x \leq 0 \end{cases}$$

and $\frac{dx}{dy} = 2y$, substitution into (4.4.1) yields

$$g(y) = e^{-y^2}\cdot 2y = 2ye^{-y^2} \qquad \text{for } y > 0$$

Also, we let $g(y) = 0$ for $y \leq 0$, since the probability of getting a value of $\mathbf{x}$ less than 0 is *zero*, and the same must be true also for $\mathbf{y}$. Note that the result we have obtained is a *Weibull distribution* (see Exercise 12 on page 117), and that the two shaded regions of Figure 4.10 both represent probabilities of 0.35. As in the discrete case, probabilities remain the same, but they pertain to different values (intervals) of the corresponding random variables. Whereas the one on the left represents the probability that $\mathbf{x}$ takes on a value on the interval from 1 to 4, the one on the right

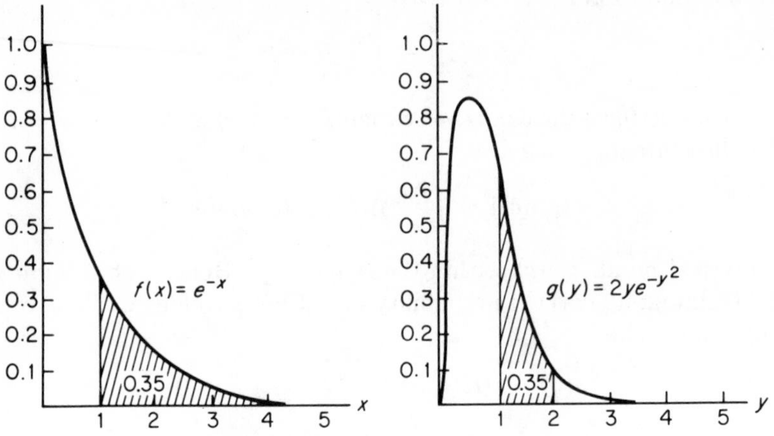

FIGURE 4.10 Change of variable.

represents the probability that **y** correspondingly takes on a value on the interval from 1 to 2.

The following is another instructive illustration of Theorem 4.1. Suppose that the double arrow of Figure 4.11 is spun so that the random variable θ has the uniform density

$$f(\theta) = \begin{cases} \dfrac{1}{\pi} & \text{for } -\dfrac{\pi}{2} < \theta < \dfrac{\pi}{2} \\ 0 & \text{elsewhere} \end{cases}$$

and that we want to determine the probability density of **x**, the coordinate of the point to which the arrow is pointing in Figure 4.11. As is apparent from the diagram, the relationship between x and θ is given by $x = a \cdot \tan \theta$, so that

$$\frac{d\theta}{dx} = \frac{a}{a^2 + x^2}$$

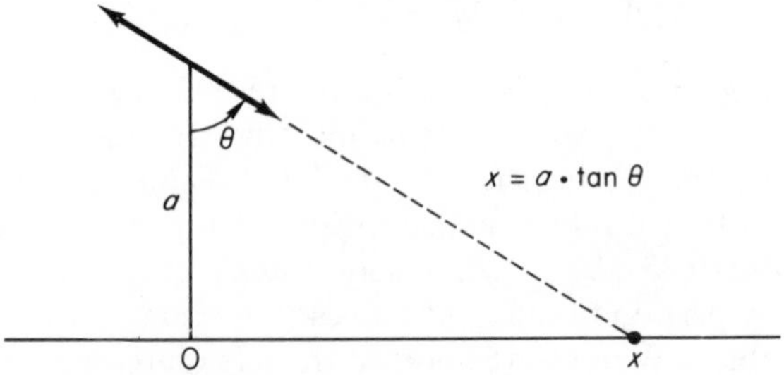

FIGURE 4.11 Change of variable.

and according to Theorem 4.1

$$g(x) = \frac{1}{\pi} \cdot \frac{a}{a^2 + x^2} \qquad \text{for } -\infty < x < \infty$$

Note that this is the *Cauchy distribution* introduced in Exercise 11 on page 117; in fact, our result provides the answer to that exercise. (Of course, $a > 0$ was tacitly assumed in Figure 4.11.)

When the assumptions underlying Theorem 4.1 are *not* met, we can be in serious difficulties, although sometimes there is an easy escape. Suppose, for instance, that the random variable **x** has the normal distribution (4.3.8) with $\alpha = 0$ and $\beta = 1$, namely

$$f(x) = \frac{1}{\sqrt{2\pi}} e^{-\frac{1}{2}x^2} \qquad \text{for } -\infty < x < \infty$$

and that we want to find the probability density of the random variable **y**, where $y = x^2$. Since this function is increasing for positive values of x and decreasing for negative values of x, the conditions of Theorem 4.1 are not met, but we can eliminate this difficulty by working with the random variable **z** whose values are given by $z = |x|$, namely, the corresponding absolute values of x. Since the given normal distribution is symmetrical about $x = 0$, that is, $f(-x) = f(x)$, we can obtain the probability density of **z**, so to speak, by folding the first diagram of Figure 4.12 along the dotted line and overlapping the two parts. We thus get the second

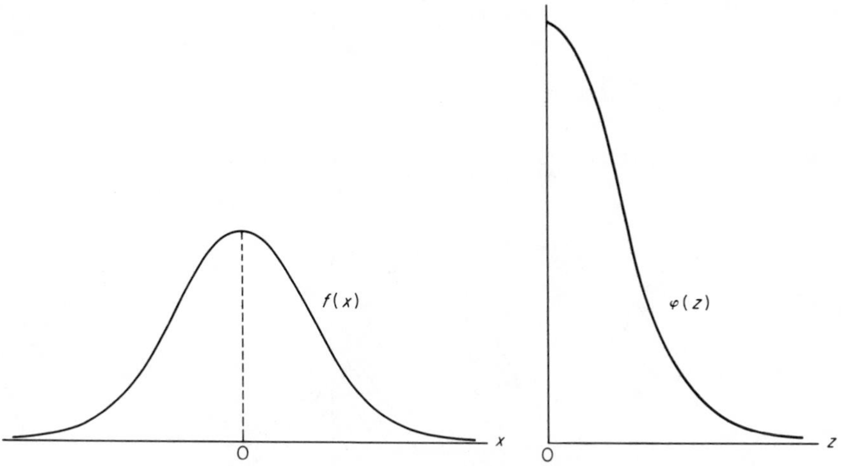

FIGURE 4.12 Change of variable.

diagram of Figure 4.12 and the probability density

$$\varphi(z) = \begin{cases} \dfrac{2}{\sqrt{2\pi}} e^{-\frac{1}{2}z^2} & \text{for } z > 0 \\ 0 & \text{for } z \le 0 \end{cases}$$

Now $y = z^2$ is increasing for *all* values of z for which its probability density is non-zero, and since $\dfrac{dz}{dy} = \dfrac{1}{2\sqrt{y}}$, Theorem 4.1 yields

$$g(y) = \frac{2}{\sqrt{2\pi}} e^{-\frac{y}{2}} \cdot \frac{1}{2\sqrt{y}} = \frac{1}{\sqrt{2\pi}} y^{-\frac{1}{2}} e^{-\frac{y}{2}} \qquad \text{for } y > 0$$

which is readily identified as a *gamma density* (4.3.6) with $\alpha = \frac{1}{2}$ and $\beta = 2$. Of course, we have to add that $g(y) = 0$ for $y \le 0$ as in the illustration on page 121.

Finally, let us study an application of Theorem 4.1 which is not only of theoretical importance, but widely used in the *simulation* of random variables having continuous distributions. Let us consider an arbitrary random variable **x** whose probability density has the values $f(x)$, and let us determine the probability density of the random variable **y** whose values are given by

$$y = \int_{-\infty}^{x} f(x)\, dx = F(x) \tag{4.4.2}$$

As can be seen from Figure 4.13, the value of **y** corresponding to any value of **x** is thus given by the area under the graph of $f(x)$ to the left of x. Differentiating $y = F(x)$ with respect to x, we get

$$\frac{dy}{dx} = F'(x) = f(x)$$

and according to Theorem 4.1 we have

$$g(y) = f(x) \cdot \frac{1}{f(x)} = 1 \qquad \text{for } 0 < y < 1$$

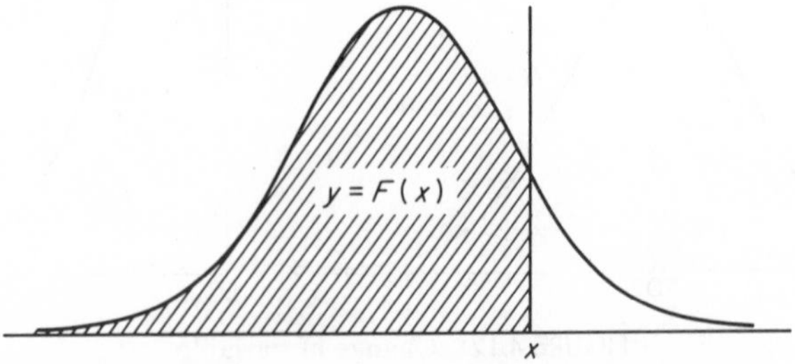

FIGURE 4.13 Probability integral transformation.

The probability density of $\mathbf{y}$ is, thus, the *uniform density* (4.3.1) with $\alpha = 0$ and $\beta = 1$. The change of variable which we have performed here is often referred to as the *probability integral transformation.*

Random variables having the uniform density $f(y) = 1$ for $0 < y < 1$ and $f(y) = 0$ elsewhere can easily be simulated with the use of random numbers (like those of Table IX at the end of the book). For instance, to obtain values of such a random variable rounded to three decimals, we take any three columns of the table, start with some randomly chosen row, and put a decimal point to the left of the first digit. Thus, if we use the eleventh, twelfth, and thirteenth columns of the table on page 445 starting with the 16th row, we obtain .188, .509, .773, .418, .544, .564, .587, .777, .017, .316, . . . , as a *sample* of the values of a random variable having the given uniform distribution. This is justified on the grounds that in the construction of the table each digit has a probability of $\frac{1}{10}$ of being either 0, 1, 2, . . . , 8, or 9. (Tables of random numbers are usually generated by means of computers and then subjected to stringent tests of randomness to assure that the above assumption is met.) Of course, we can round the values of our random variable to as many decimals as we want by using correspondingly many columns of random digits.

If we combine this technique with the change of variable referred to above as the probability integral transformation, we can easily simulate (that is, sample the values of) any random variable. To this end we plot the graph of the distribution function of the random variable $\mathbf{x}$ which we want to simulate, use random numbers to obtain values of the random variable $\mathbf{y}$ described in the preceding paragraph, plot these values on the

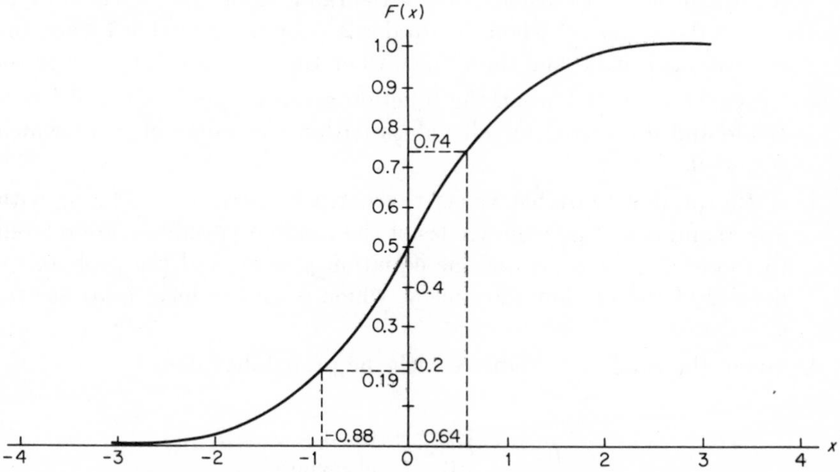

FIGURE 4.14 Distribution function of normal distribution.

vertical scale, and read off the corresponding values of **x** as in Figure 4.14, where we plotted the distribution function of a random variable having the normal density (4.3.8) with $\alpha = 0$ and $\beta = 1$. Using the thirty-first and thirty-second columns of the table of random numbers on page 444 starting with the twenty-sixth row, we obtain .74, .19, .22, .33, .23, .67, . . ., and as can be seen from Figure 4.14, the values of **x** which correspond to the first two values of **y** are 0.64 and -0.88. In Exercise 9 on page 127 the reader will be asked to find the values of **x** which correspond to the next four values of **y**.

THEORETICAL EXERCISES

1. If the random variable **x** has the geometric distribution (3.3.6) and the values of the random variable **y** are related to those of **x** by means of the equation $y = 3 - 2x$, find the probability distribution of **y**.
2. If the random variable **x** has the hypergeometric distribution (3.3.5) and the values of the random variable **y** are given by the number of "successes" *minus* the number of "failures," find the probability distribution of **y**.
3. If the random variable **x** is the total we obtain with a pair of balanced dice (for which the probability distribution is given on page 70), find
   (a) the probability distribution of the random variable **y** which takes on the value 0 when the total is *even* and the value 1 when the total is *odd;*
   (b) the probability distribution of the random variable **z** which takes on the value -1 when the total is 2, 3, or 12, the value 1 when the total is 7 or 11, and the value 0 when the total is 4, 5, 6, 8, 9, or 10.
4. Prove Theorem 4.1 when the function given by $y = h(x)$ is differentiable and *decreasing* for all values within the range of x for which $f(x) \neq 0$.
5. If the random variable **x** has the normal distribution (4.3.8) with $\alpha = 0$ and $\beta = 1$ and the values of the random variable **y** are related to those of **x** by means of the equation $y = e^x$, find the probability density of the random variable **y**, which is said to have a *log-normal distribution.*
6. Given the random variable **x** with the probability density

$$f(x) = \begin{cases} \frac{1}{2}x & \text{for } 0 < x < 2 \\ 0 & \text{elsewhere} \end{cases}$$

find the probability density of the random variable $\mathbf{y}$, whose values are related to those of $\mathbf{x}$ by means of the equation $y = x^2$. Also plot the graphs of these two probability densities.

7. Given the random variable $\mathbf{x}$ with the probability density

$$f(x) = \begin{cases} \dfrac{kx^3}{(1+2x)^6} & \text{for } x > 0 \\ 0 & \text{for } x \leq 0 \end{cases}$$

where k is an appropriate constant, find the probability density of the random variable $\mathbf{y}$, whose values are related to those of $\mathbf{x}$ by means of the equation $y = \dfrac{2x}{1+2x}\cdot$ Also find the value of k by comparing the result with (4.3.7).

8. According to the Maxwell-Boltzmann law of theoretical physics, the probability density of $\mathbf{v}$, the velocity of a gas molecule, is

$$f(v) = \begin{cases} kv^2e^{-\beta v^2} & \text{for } v > 0 \\ 0 & \text{for } v \leq 0 \end{cases}$$

where β depends on its mass and the absolute temperature, and k is an appropriate constant. Show that the probability density of the kinetic energy $\mathbf{E}$, whose values are related to those of v by means of the equation $E = \frac{1}{2}mv^2$, is a gamma distribution.

APPLIED EXERCISES

9. With reference to the illustration on page 126, find the values of $\mathbf{x}$ which correspond to the next four values of $\mathbf{y}$.

10. It can easily be shown by the method of this section that if a random variable $\mathbf{x}$ has the normal distribution (4.3.8), then the random variable $\mathbf{u}$, whose values are related to those of $\mathbf{x}$ by means of the equation $u = \dfrac{x - \alpha}{\beta}$, has the normal distribution (4.3.8) with $\alpha = 0$ and $\beta = 1$. Thus, we can simulate random variables having *any* normal distribution by first obtaining values of $\mathbf{u}$ with the use of random numbers and the graph of Figure 4.14, and then calculating the corresponding values of $\mathbf{x}$ by means of the equation $x = \alpha + \beta u$.

(a) Suppose that the length of the trout in a certain lake (in millimeters) is a random variable having a normal distribution with $\alpha = 88$ and $\beta = 12$. Use random numbers to simulate the length of 12 trout caught in this lake, and calculate their average length.

(b) Suppose that the length of time it takes an adult to memorize a list of twelve French verbs is a random variable having a normal distribution with $\alpha = 5.4$ and $\beta = 1.3$, where time is measured in minutes. Use random numbers to simulate the length of time it takes eight adults to memorize this list of French verbs.

11. Suppose that the durability of a paint (in years) is a random variable having an exponential density (4.3.2) with $\theta = 2$. Draw the graph of the distribution function of this random variable and use it to simulate (with random numbers) an experiment in which this paint is applied to 10 houses and it is observed in each case how long the paint lasts (that is, how long it takes until there are certain agreed-upon signs of deterioration). *On the average*, how long did the paint last in this "experiment"?

12. The simulation technique described on page 125 applies also to *discrete* random variables. Thus, use two-digit random numbers and Figure 3.2 on page 72 to simulate the number of points obtained in 12 rolls of a pair of balanced dice.

13. Draw the distribution function of a random variable having the binomial distribution (3.3.3) with $n = 2$ and $\theta = \frac{1}{2}$, and use it to simulate with two-digit random numbers the *number of heads* obtained in each of 20 flips of a pair of "honest" coins.

14. The simulation of Exercise 13 could have been performed just as well (in fact, more easily) *without the graph of the distribution function* by using two-digit random numbers, letting those from 00 through 24 represent *zero heads*, those from 25 through 74 represent *one head*, and those from 75 through 99 represent *two heads*.

(a) What three-digit random numbers might thus be associated with 0, 1, 2, and 3 heads, if we wanted to simulate the number of heads obtained in repeated flips of *three* "honest" coins?

(b) Use Table I at the end of the book to construct a table which shows what four-digit random numbers might be associated with 0, 1, 2, . . . , and 9 successes in 9 trials (for which the conditions underlying the binomial distribution are met) and $\theta = 0.35$.

(c) Use Table II at the end of the book to construct a table which shows what four-digit random numbers might be associated with 0, 1, 2, . . . , and 7 successes, if we wanted to simulate a random variable having the Poisson distribution (3.3.9) with $\lambda = 1$.

4.5 MULTIVARIATE PROBABILITY DENSITIES

Let us now generalize the work of Section 4.2 by considering n random variables $\mathbf{x}_1$, $\mathbf{x}_2$, ..., and $\mathbf{x}_n$. If the probability that $\mathbf{x}_1$ takes on a value less than or equal to x_1, $\mathbf{x}_2$ takes on a value less than or equal to x_2, ..., and $\mathbf{x}_n$ takes on a value less than or equal to x_n, is given by

$$F(x_1, x_2, \ldots, x_n) = \int_{-\infty}^{x_n} \cdots \int_{-\infty}^{x_2} \int_{-\infty}^{x_1} f(t_1, t_2, \ldots, t_n)\, dt_1\, dt_2 \cdots dt_n \tag{4.5.1}$$

for all real values of x_1, x_2, ..., and x_n, we refer to $\mathbf{F}$ as the *joint distribution function* of the n random variables, and to $\mathbf{f}$ as their *joint probability density*. As in Section 4.2, we shall limit our discussion to the case where the joint distribution function is continuous everywhere and partially differentiable with respect to the n variables for all but a finite set of values of each of the n variables. Analogous to the relationship $f(x) = F'(x)$ in the *univariate* case, partial differentation of (4.5.1) thus leads to

$$f(x_1, x_2, \ldots, x_n) = \frac{\partial^n}{\partial x_1\, \partial x_2 \cdots \partial x_n} F(x_1, x_2, \ldots, x_n) \tag{4.5.2}$$

wherever these partial derivatives exist. [Analogous to what we did in Section 4.2, we shall generally let the values of joint probability densities equal *zero* wherever they are not defined by (4.5.2).]

To illustrate, let us consider the joint distribution function of two random variables $\mathbf{x}_1$ and $\mathbf{x}_2$ which is given by

$$F(x_1, x_2) = \begin{cases} (1 - e^{-x_1})(1 - e^{-x_2}) & \text{for } x_1 > 0 \text{ and } x_2 > 0 \\ 0 & \text{elsewhere} \end{cases}$$

Note that $F(-\infty, -\infty) = 0$, $F(\infty, \infty) = 1$ and that the function is *non-decreasing* in both variables. Now, if we substitute $x_1 = 2$ and $x_2 = 3$, we find that the probability that $\mathbf{x}_1$ will take on a value *less than* 2 while $\mathbf{x}_2$ will take on a value *less than* 3 is given by

$$F(2, 3) = (1 - e^{-2})(1 - e^{-3}) = (0.865)(0.950) = 0.822$$

and it should be observed that (as in the one-variable case) it does not matter whether or not we include $x_1 = 2$ or $x_2 = 3$. Similar substitutions yield

$$F(2, 1) = (1 - e^{-2})(1 - e^{-1}) = (0.865)(0.632) = 0.547$$

$$F(1, 3) = (1 - e^{-1})(1 - e^{-3}) = (0.632)(0.950) = 0.600$$

$$F(1, 1) = (1 - e^{-1})(1 - e^{-1}) = (0.632)(0.632) = 0.399$$

and if we combine all these results we find that the probability that $\mathbf{x}_1$ will take on a value on the interval from 1 to 2 while $\mathbf{x}_2$ will take on a value on the interval from 1 to 3 is given by

$$F(2, 3) - F(2, 1) - F(1, 3) + F(1, 1) = 0.074$$

[To justify these additions and subtractions, we have only to observe that $F(2, 3) - F(2, 1)$ is the probability that $\mathbf{x}_1$ takes on a value less than 2 and $\mathbf{x}_2$ takes on a value on the interval from 1 to 3, while $F(1, 3) - F(1, 1)$ is the probability that $\mathbf{x}_1$ takes on a value less than 1 and $\mathbf{x}_2$ takes on a value on the interval from 1 to 3.]

We have done all these calculations mainly to demonstrate that in problems like this it can be much easier to work with the joint probability density instead of the joint distribution function. Performing the necessary partial differentiations, we get

$$\frac{\partial^2}{\partial x_1\, \partial x_2} F(x_1, x_2) = \begin{cases} e^{-(x_1+x_2)} & \text{for } x_1 > 0 \text{ and } x_2 > 0 \\ 0 & \text{for } x_1 < 0 \text{ or } x_2 < 0 \end{cases}$$

so that the joint probability density of the two random variables is given by

$$f(x_1, x_2) = \begin{cases} e^{-(x_1+x_2)} & \text{for } x_1 > 0 \text{ and } x_2 > 0 \\ 0 & \text{elsewhere} \end{cases}$$

Now, the probability that $\mathbf{x}_1$ will take on a value on the interval from 1 to 2 and $\mathbf{x}_2$ will take on a value on the interval from 1 to 3 is given by the double integral

$$\int_1^3 \int_1^2 e^{-(x_1+x_2)}\, dx_1\, dx_2 = e^{-2} - e^{-3} - e^{-4} + e^{-5} = 0.074$$

For two random variables, the probability density is, geometrically speaking, a *surface*, and the probability which we have calculated is given by the *volume* under this surface shown in Figure 4.15.

To illustrate the inverse problem of determining the joint distribution function which corresponds to a given joint probability density, let us use the *tri-variate* probability density given by

$$f(x_1, x_2, x_3) = \begin{cases} (x_1 + x_2)e^{-x_3} & \text{for } 0 < x_1 < 1,\ 0 < x_2 < 1,\ x_3 > 0 \\ 0 & \text{elsewhere} \end{cases}$$

No integration is required to show that $F(x_1, x_2, x_3) = 0$ so long as either x_1, x_2, or x_3 is negative, but for $0 < x_1 < 1$, $0 < x_2 < 1$, and $x_3 > 0$,

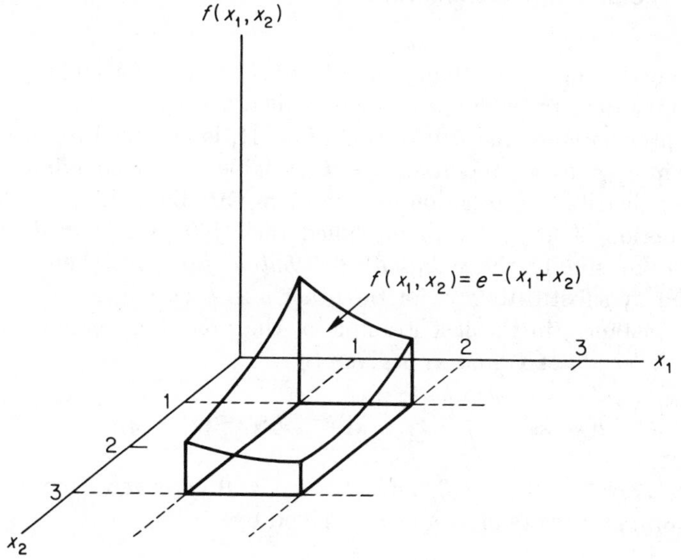

FIGURE 4.15 Joint density of two random variables.

we get

$$F(x_1, x_2, x_3) = \int_0^{x_3}\int_0^{x_2}\int_0^{x_1} (t_1 + t_2)e^{-t_3}\,dt_1\,dt_2\,dt_3$$
$$= \tfrac{1}{2}x_1x_2(x_1 + x_2)(1 - e^{-x_3})$$

Also, for $x_1 > 1$, $0 < x_2 < 1$, and $x_3 > 0$, we get

$$F(x_1, x_2, x_3) = \int_0^{x_3}\int_0^{x_2}\int_0^{1} (t_1 + t_2)e^{-t_3}\,dt_1\,dt_2\,dt_3 = \tfrac{1}{2}x_2(1 + x_2)(1 - e^{-x_3})$$

for $0 < x_1 < 1$, $x_2 > 1$, and $x_3 > 0$, we get

$$F(x_1, x_2, x_3) = \int_0^{x_3}\int_0^{1}\int_0^{x_1} (t_1 + t_2)e^{-t_3}\,dt_1\,dt_2\,dt_3 = \tfrac{1}{2}x_1(x_1 + 1)(1 - e^{-x_3})$$

and for $x_1 > 1$, $x_2 > 1$, and $x_3 > 0$, we get

$$F(x_1, x_2, x_3) = \int_0^{x_3}\int_0^{1}\int_0^{1} (t_1 + t_2)e^{-t_3}\,dt_1\,dt_2\,dt_3 = 1 - e^{-x_3}$$

To complete the picture, let us add that the boundary between any two of these regions can be included in either one; in other words, in the inequalities which specify the regions for which the various parts of the distribution function are defined, $\leq$ can be substituted for $<$, and $\geq$ for $>$.

4.5.1 Joint Marginal Distributions

Given the joint probability density of n random variables, the joint probability density of any k of them, called their *joint marginal density* (or simply the *marginal density* when $k = 1$), is obtained by integrating out, from $-\infty$ to ∞, the other $n - k$ variables. Correspondingly, given the joint distribution function of n random variables, the joint distribution function of any k of them, called their *joint marginal distribution function* (or simply the *marginal distribution function* when $k = 1$), is obtained by substituting ∞ for the other $n - k$ variables.

For instance, in the last example of the preceding section, the *joint marginal density* of $\mathbf{x}_1$ and $\mathbf{x}_3$ is given by

$$g(x_1, x_3) = \int_0^1 (x_1 + x_2)e^{-x_3}\,dx_2 = (x_1 + \tfrac{1}{2})e^{-x_3}$$

for $0 < x_1 < 1$ and $x_3 > 0$, and $g(x_1, x_3) = 0$ elsewhere. Similarly, the *joint marginal density* of $\mathbf{x}_1$ and $\mathbf{x}_2$ is given by

$$h(x_1, x_2) = \int_0^\infty (x_1 + x_2)e^{-x_3}\,dx_3 = x_1 + x_2$$

for $0 < x_1 < 1$ and $0 < x_2 < 1$, and $h(x_1, x_2) = 0$ elsewhere; and the *marginal density* of x_2 alone is given by

$$\begin{aligned} m(x_2) &= \int_0^1 \int_0^\infty f(x_1, x_2, x_3)\,dx_3\,dx_1 = \int_0^1 h(x_1, x_2)\,dx_1 \\ &= \int_0^1 (x_1 + x_2)\,dx_1 = \tfrac{1}{2} + x_2 \end{aligned}$$

for $0 < x_2 < 1$, and $m(x_2) = 0$ elsewhere.

Working with the corresponding joint distribution function, we find that the *joint marginal distribution function* of x_1 and x_3 is given by

$$G(x_1, x_3) = F(x_1, \infty, x_3) = \begin{cases} 0 & \text{for } x_1 \le 0 \text{ or } x_3 \le 0 \\ \tfrac{1}{2}x_1(x_1 + 1)(1 - e^{-x_3}) & \text{for } 0 < x_1 < 1,\ x_3 > 0 \\ 1 - e^{-x_3} & \text{for } x_1 \ge 1,\ x_3 > 0 \end{cases}$$

Similarly, the *joint marginal distribution function* of x_1 and x_2 is given by

$$H(x_1, x_2) = F(x_1, x_2, \infty) = \begin{cases} 0 & \text{for } x_1 \le 0 \text{ or } x_2 \le 0 \\ \tfrac{1}{2}x_1(x_1 + 1) & \text{for } 0 < x_1 < 1,\ x_2 \ge 1 \\ \tfrac{1}{2}x_2(1 + x_2) & \text{for } 0 < x_2 < 1,\ x_1 \ge 1 \\ \tfrac{1}{2}x_1x_2(x_1 + x_2) & \text{for } 0 < x_1 < 1,\ 0 < x_2 < 1 \\ 1 & \text{for } x_1 \ge 1,\ x_2 \ge 1 \end{cases}$$

and the *marginal distribution function* of x_2 is given by

$$M(x_2) = F(\infty, x_2, \infty) = \begin{cases} 0 & \text{for } x_2 \leq 0 \\ \frac{1}{2}x_2(1 + x_2) & \text{for } 0 < x_2 < 1 \\ 1 & \text{for } x_2 \geq 1 \end{cases}$$

The work of this section was not very difficult; we have to be careful, though, when we are dealing with functions which are defined separately for various different intervals or regions.

4.5.2 Conditional Probability Densities

Analogous to (3.4.3) we shall define the *conditional probability density* of the random variable $\mathbf{x}_1$ *given that the random variable* $\mathbf{x}_2$ *takes on the value* x_2 as

$$\varphi(x_1|x_2) = \frac{f(x_1, x_2)}{g(x_2)}, \qquad g(x_2) \neq 0 \tag{4.5.3}$$

where $f(x_1, x_2)$ and $g(x_2)$ are, respectively, values of the joint probability density of $\mathbf{x}_1$ and $\mathbf{x}_2$ and the marginal density of $\mathbf{x}_2$.

For instance, if two random variables $\mathbf{x}_1$ and $\mathbf{x}_2$ have the joint probability density

$$f(x_1, x_2) = \begin{cases} \frac{2}{3}(x_1 + 2x_2) & \text{for } 0 < x_1 < 1, 0 < x_2 < 1 \\ 0 & \text{elsewhere} \end{cases}$$

the marginal density of $\mathbf{x}_2$ is given by

$$g(x_2) = \int_0^1 \frac{2}{3}(x_1 + 2x_2)\,dx_1 = \frac{1}{3}(1 + 4x_2) \qquad \text{for } 0 < x_2 < 1$$

and $g(x_2) = 0$ elsewhere. Hence, the conditional probability density of $\mathbf{x}_1$ given that $\mathbf{x}_2$ takes on the value x_2 is given by

$$\varphi(x_1|x_2) = \frac{\frac{2}{3}(x_1 + 2x_2)}{\frac{1}{3}(1 + 4x_2)} = \frac{2x_1 + 4x_2}{1 + 4x_2} \qquad \text{for } 0 < x_1 < 1$$

and $\varphi(x_1|x_2) = 0$ elsewhere. Having obtained this result, we might use it, for example, to determine the probability that $\mathbf{x}_1$ will take on a value *less than* $\frac{1}{2}$ given that the value of $\mathbf{x}_2$ actually equals $\frac{1}{2}$. Substituting $x_2 = \frac{1}{2}$ and performing the required integration, we find that the desired probability is

$$\int_0^{\frac{1}{2}} \frac{2x_1 + 4(\frac{1}{2})}{1 + 4(\frac{1}{2})}\,dx_1 = \frac{5}{12}$$

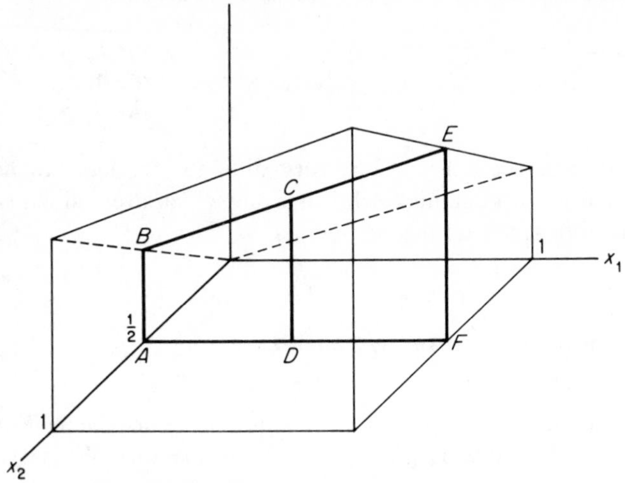

FIGURE 4.16 Conditional probability density.

and it is of interest to note that in Figure 4.16 it is given by the *ratio* of the area of trapezoid $ABCD$ to the area of trapezoid $ABEF$.

The concept of a conditional probability density can easily be extended so that it applies to situations involving more than two random variables, and, of course, we can also speak of *conditional distribution functions*. For instance, in the above example the conditional distribution function of $\mathbf{x}_1$ given that $\mathbf{x}_2$ takes on the value x_2 can be written as

$$\Phi(x_1|x_2) = 0 \qquad \text{for } x_1 \leq 0$$

$$\Phi(x_1|x_2) = \int_0^{x_1} \frac{2x_1 + 4x_2}{1 + 4x_2}\, dx_1 = \frac{x_1^2 + 4x_1x_2}{1 + 4x_2} \qquad \text{for } 0 < x_1 < 1$$

$$\Phi(x_1|x_2) = 1 \qquad \text{for } x_1 \geq 1$$

Note that the probability which we calculated above is, in fact, $\Phi(\frac{1}{2}|\frac{1}{2}) = \frac{5}{12}$.

In Section 3.4 we saw that in the discrete case we can define various kinds of conditional probability functions when we are dealing with more than two random variables—as might be expected, this is true also in the continuous case. For instance, for four random variables $\mathbf{x}_1$, $\mathbf{x}_2$, $\mathbf{x}_3$, and $\mathbf{x}_4$, whose joint density is given by $f(x_1, x_2, x_3, x_4)$, we can write

$$p(x_3|x_1, x_2, x_4) = \frac{f(x_1, x_2, x_3, x_4)}{g(x_1, x_2, x_4)}$$

for the conditional probability density of $\mathbf{x}_3$ *given the values of* $\mathbf{x}_1$, $\mathbf{x}_2$, *and* $\mathbf{x}_4$, provided that $g(x_1, x_2, x_4)$, the corresponding value of the joint marginal density of $\mathbf{x}_1$, $\mathbf{x}_2$, and $\mathbf{x}_4$, is not equal to zero. We can also write

$$q(x_2, x_4|x_1, x_3) = \frac{f(x_1, x_2, x_3, x_4)}{m(x_1, x_3)}$$

for the *joint* conditional probability density of $\mathbf{x}_2$ and $\mathbf{x}_4$ *given the values of* $\mathbf{x}_1$ and $\mathbf{x}_3$, provided that $m(x_1, x_3)$, the corresponding value of the joint marginal density of $\mathbf{x}_1$ and $\mathbf{x}_3$, is not equal to zero; and among numerous other possibilities we can write

$$r(x_2, x_3, x_4|x_1) = \frac{f(x_1, x_2, x_3, x_4)}{b(x_1)}$$

for the *joint* conditional probability density of $\mathbf{x}_2$, $\mathbf{x}_3$, and $\mathbf{x}_4$ given the value of $\mathbf{x}_1$, provided that $b(x_1)$, the corresponding value of the marginal density of $\mathbf{x}_1$, is not equal to zero.

As in the discrete case, the question of *independence* is of special importance whenever we deal with more than one random variable, and analogous to (3.4.9) on page 94 let us thus make the following definition: *If $f(x_1, x_2, \ldots, x_n)$ and $f_i(x_i)$ are, respectively, the values of the joint density and the marginal densities of n random variables $\mathbf{x}_1$, $\mathbf{x}_2$, . . . , and $\mathbf{x}_n$, these random variables are independent if and only if*

$$f(x_1, x_2, \ldots, x_n) = \prod_{i=1}^{n} f_i(x_i) \tag{4.5.4}$$

for all values within the range of these random variables for which $f(x_1, x_2, \ldots, x_n)$ is defined.

With this definition of independence, it can easily be verified that in the example on page 130 the three random variables whose joint density was given by

$$f(x_1, x_2, x_3) = \begin{cases} (x_1 + x_2)e^{-x_3} & \text{for } 0 < x_1 < 1,\ 0 < x_2 < 1,\ x_3 > 0 \\ 0 & \text{elsewhere} \end{cases}$$

are *not independent.* On the other hand, it can be shown that the two random variables $\mathbf{x}_1$ and $\mathbf{x}_3$ are *pairwise independent*, and so are the random variables $\mathbf{x}_2$ and $\mathbf{x}_3$ (see Exercise 7 on page 137). Also, in the example on page 130 where we had

$$f(x_1, x_2) = \begin{cases} e^{-(x_1+x_2)} & \text{for } x_1 > 0 \text{ and } x_2 > 0 \\ 0 & \text{elsewhere} \end{cases}$$

the two random variables *are independent*, since the two marginal densities are

$$f_1(x_1) = \begin{cases} e^{-x_1} & \text{for } x_1 > 0 \\ 0 & \text{for } x_1 \leq 0 \end{cases} \quad \text{and} \quad f_2(x_2) = \begin{cases} e^{-x_2} & \text{for } x_2 > 0 \\ 0 & \text{for } x_2 \leq 0 \end{cases}$$

so that $f(x_1, x_2) = f_1(x_1) \cdot f_2(x_2)$ for all real values of x_1 and x_2.

THEORETICAL EXERCISES

1. If two random variables $\mathbf{x}_1$ and $\mathbf{x}_2$ have the joint probability density

$$f(x_1, x_2) = \begin{cases} 2 & \text{for } x_1 > 0,\ x_2 > 0, \text{ and } x_1 + x_2 < 1 \\ 0 & \text{elsewhere} \end{cases}$$

which might be referred to as a *uniform density* defined over a triangle in the x_1x_2-plane, find the probabilities that
(a) both random variables will take on a value less than $\frac{1}{2}$;
(b) $\mathbf{x}_1$ will take on a value less than $\frac{1}{4}$ and $\mathbf{x}_2$ will take on a value greater than $\frac{1}{4}$;
(c) the *sum* of the values taken on by the two random variables will exceed $\frac{2}{3}$.

2. Find an expression for the values of the joint distribution function of the two random variables of Exercise 1 which applies when
(a) $x_1 \leq 0$ or $x_2 \leq 0$;
(b) $x_1 \geq 1$, $x_2 \geq 1$;
(c) $x_1 > 0$, $x_2 > 0$, $x_1 + x_2 < 1$, and use it to verify the result of part (a) of Exercise 1.
For how many *other* parts of the x_1x_2-plane will the values of the joint distribution function of the given random variables have to be defined separately?

3. If two random variables $\mathbf{x}_1$ and $\mathbf{x}_2$ have the joint distribution function

$$F(x_1, x_2) = \begin{cases} (1 - e^{-x_1^2})(1 - e^{-x_2}) & \text{for } x_1 > 0,\ x_2 > 0 \\ 0 & \text{elsewhere} \end{cases}$$

find the probabilities that
(a) both random variables will take on values less than 2;
(b) $\mathbf{x}_1$ will take on a value less than 3 and $\mathbf{x}_2$ will take on a value less than 4;
(c) $\mathbf{x}_1$ will take on a value less than 2 and $\mathbf{x}_2$ will take on a value less than 4;

(d) $\mathbf{x}_1$ will take on a value less than 3 and $\mathbf{x}_2$ will take on a value less than 2.

Also combine the results of parts (a), (b), (c), and (d) to find the probability that

(e) $\mathbf{x}_1$ will take on a value on the interval from 2 to 3 and $\mathbf{x}_2$ will take on a value on the interval from 2 to 4.

4. Find the joint probability density of the two random variables of Exercise 3 and use the result to verify part (e) of that exercise.

5. If three random variables $\mathbf{x}_1$, $\mathbf{x}_2$, and $\mathbf{x}_3$ have the joint probability density

$$f(x_1, x_2, x_3) = \begin{cases} \frac{3}{8}x_1x_2x_3^2 & \text{for } 0 < x_1 < 1,\ 0 < x_2 < 2,\ 0 < x_3 < 2 \\ 0 & \text{elsewhere} \end{cases}$$

find the probabilities that

(a) $\mathbf{x}_1$ will take on a value less than $\frac{1}{2}$ and $\mathbf{x}_2$ and $\mathbf{x}_3$ will both take on values less than 1;

(b) $\mathbf{x}_1$ will take on a value less than $\frac{1}{3}$, $\mathbf{x}_2$ will take on a value greater than $\frac{3}{2}$, and $\mathbf{x}_3$ will take on a value greater than $\frac{4}{3}$;

(c) $\mathbf{x}_1$ will take on a value on the interval from $\frac{1}{4}$ to $\frac{3}{4}$, $\mathbf{x}_2$ will take on a value on the interval from $\frac{3}{4}$ to 1, and $\mathbf{x}_3$ will take on a value on the interval from 1 to $\frac{5}{4}$.

6. Evidently, the joint distribution function of the three random variables of Exercise 5 is given by $F(x_1, x_2, x_3) = 0$ so long as either $x_1 \leq 0$, $x_2 \leq 0$, or $x_3 \leq 0$.

(a) For how many *other* regions of the $x_1x_2x_3$-space will this distribution function have to be determined separately?

(b) Find an expression for $F(x_1, x_2, x_3)$ when $0 < x_1 < 1, 0 < x_2 < 2$, and $0 < x_3 < 2$.

(c) Find an expression for $F(x_1, x_2, x_3)$ when $0 < x_1 < 1$, $x_2 \geq 2$, and $0 < x_3 < 2$.

(d) Find an expression for $F(x_1, x_2, x_3)$ when $0 < x_1 < 1$, $x_2 \geq 2$, and $x_3 \geq 2$.

(e) Find an expression for $F(x_1, x_2, x_3)$ when $x_1 \geq 1$, $x_2 \geq 2$, and $x_3 \geq 2$.

7. With reference to the illustration near the bottom of page 130, find

(a) the joint marginal density of $\mathbf{x}_2$ and $\mathbf{x}_3$;

(b) the marginal density of $\mathbf{x}_1$;

(c) the marginal density of $\mathbf{x}_3$.

Also use these results as well as those obtained in the text to verify that

(d) the three random variables are not independent;

(e) $\mathbf{x}_1$ and $\mathbf{x}_3$ are pairwise independent;

(f) $\mathbf{x}_2$ and $\mathbf{x}_3$ are pairwise independent;
(g) $\mathbf{x}_1$ and $\mathbf{x}_2$ are not pairwise independent.

8. If the two random variables $\mathbf{x}_1$ and $\mathbf{x}_2$ have the joint probability density

$$f(x_1, x_2) = \frac{1}{2\pi} e^{-\frac{1}{2}(x_1^2 - 4x_1x_2 + 5x_2^2)}$$

for $-\infty < x_1 < \infty$ and $-\infty < x_2 < \infty$, find
(a) the marginal density of $\mathbf{x}_2$;
(b) the conditional density of $\mathbf{x}_1$ given that $\mathbf{x}_2$ takes on the value x_2.
[*Hint:* To perform the integration required for part (a) write $x_1^2 - 4x_1x_2 + 5x_2^2$ as $(x_1^2 - 4x_1x_2 + 4x_2^2) + x_2^2 = (x_1 - 2x_2)^2 + x_2^2$.] The probability density of this exercise is a special form of that of the *bivariate normal distribution*, and it should be observed that the results obtained for parts (a) and (b) are both *normal densities*.

9. With reference to the illustration near the middle of page 133 find the conditional density of $\mathbf{x}_2$ given that $\mathbf{x}_1$ takes on the value x_1.

10. Given the joint probability density

$$f(x_1, x_2) = \begin{cases} \dfrac{2}{(1 + x_1 + x_2)^3} & \text{for } x_1 > 0 \text{ and } x_2 > 0 \\ 0 & \text{elsewhere} \end{cases}$$

find
(a) the marginal densities of $\mathbf{x}_1$ and $\mathbf{x}_2$;
(b) the conditional density of $\mathbf{x}_1$ given that $\mathbf{x}_2$ takes on the value x_2;
(c) the conditional density of $\mathbf{x}_2$ given that $\mathbf{x}_1$ takes on the value x_1.
Also check whether the two random variables are independent.

11. Check for each of the following probability densities whether the two random variables are independent:
(a) $f(x_1, x_2) = \frac{1}{81}x_1^2x_2^2$ for $0 < x_1 < 3$, $0 < x_2 < 3$, and $f(x_1, x_2) = 0$ elsewhere;
(b) $f(x_1, x_2) = \frac{2}{81}x_1^2x_2^2$ for $0 < x_1 < x_2 < 3$, and $f(x_1, x_2) = 0$ elsewhere.

12. Suppose that someone shoots at a target which is located at the origin of a system of rectangular coordinates and that the x- and y-coordinates of the point which he hits are values of *independent* random variables with the *same marginal density*, whose joint probability density depends *only* on the distance from the point to the origin. Thus, we can write $f(x, y) = g(x) \cdot g(y) = \varphi(r)$, where $\varphi(r)$ is the value of some function which depends only on $r = \sqrt{x^2 + y^2}$, the distance from the origin to the point (x, y).

(a) Show that if we introduce polar coordinates by letting $x = r \cdot \cos \theta$ and $y = r \cdot \sin \theta$, partial differentiation of $g(x) \cdot g(y) = \varphi(r)$ with respect to θ yields $\frac{g'(x)}{xg(x)} = \frac{g'(y)}{yg(y)}$, where in $g'(x)$ the prime means differentiation with respect to x, and in $g'(y)$ it means differentiation with respect to y.

(b) Show that since the result of part (a) must hold for any values of x and y, the original random variables must have *normal probability densities* given by the solution of the differential equation

$$\frac{g'(x)}{xg(x)} = c$$

APPLIED EXERCISES

13. If the joint probability density of the price **p** of a certain commodity (in dollars) and total sales **s** (in 10,000 units) is given by

$$f(s, p) = \begin{cases} 5pe^{-ps} & \text{for } 0.20 < p < 0.40,\ s > 0 \\ 0 & \text{elsewhere} \end{cases}$$

find

(a) the marginal density of **p**;

(b) the conditional density of **s** given that **p** takes on the value p;

(c) the probability that sales will exceed 20,000 units when $p = 0.25$.

14. If **x** is the proportion of persons who will respond to one kind of mail-order solicitation, **y** is the proportion of persons who will respond to another kind of mail-order solicitation, and the joint probability density of **x** and **y** is given by

$$f(x, y) = \begin{cases} \frac{2}{5}(x + 4y) & \text{for } 0 < x < 1,\ 0 < y < 1 \\ 0 & \text{elsewhere} \end{cases}$$

find

(a) the marginal densities of **x** and **y**;

(b) the conditional density of **y** given that **x** takes on the value x;

(c) the conditional density of **x** given that **y** takes on the value y.

Also find the probabilities that

(d) there will be at most a 40 percent response to the first kind of mail-order solicitation;

(e) there will be at least a 20 percent response to the second kind of mail-order solicitation;

(f) there will be at most a 40 percent response to the first kind of mail-order solicitation given that there has been only a 20 percent response to the second kind of mail-order solicitation;

(g) there will be at least a 20 percent response to the second kind of mail-order solicitation given that there has been an 80 percent response to the first kind of mail-order solicitation.

REFERENCES

The problem of changing variables becomes fairly complicated when we are dealing with more than one random variable; it requires the concept of a *Jacobian*, which is ordinarily not studied in a first course in calculus. Thus, we discussed only the one-variable case in this book, but more advanced treatments may be found, for example, in

Wilks, S. S., *Mathematical Statistics*. New York: John Wiley & Sons, Inc., 1962.

and in

Keeping, E. S., *Introduction to Statistical Inference*. New York: D. Van Nostrand Co., Inc., 1962.

5

MATHEMATICAL EXPECTATION

5.1 INTRODUCTION

Originally, the concept of a mathematical expectation was introduced in connection with games of chance, and in its simplest form *a mathematical expectation is the product of a player's potential gain and the probability that this gain will actually be realized.* Thus, if we are to get \$10.00 if and only if a balanced coin comes up heads, our mathematical expectation is $10 \cdot \frac{1}{2} = \$5.00$, and if we hold 1 of 1,000 raffle tickets for which the prize is \$250.00, our mathematical expectation is $250(0.001) = \$0.25$. Of course, nobody will actually get \$0.25, but 999 of the raffle tickets will not pay anything at all, one ticket will pay \$250.00, so that *on the average* the 1,000 tickets will pay \$0.25 per ticket.

To generalize the concept of a mathematical expectation, let us change the raffle of the preceding paragraph so that there is also a second prize of \$100.00 and a third prize of \$50.00. Now we can argue that 997 of the tickets will not pay anything at all, one ticket will pay \$250.00, one ticket will pay \$100.00, and one ticket will pay \$50.00, so that altogether they will pay \$400.00 or *on the average* \$0.40 per ticket. Note that this mathematical expectation of \$0.40 could also have been obtained by *adding* the products which are obtained by multiplying each amount by the probability that it will be won by the holder of any given ticket, namely, by writing

$$0(0.997) + 250(0.001) + 100(0.001) + 50(0.001) = \$0.40$$

The amount a person stands to win in a lottery is a random variable, and as we have just indicated, the mathematical expectation of such a random variable is the sum of the products which are obtained by multiplying each value of the random variable by its probability. More generally, *the mathematical expectation of a discrete random variable* $\mathbf{x}$ *is given by*

$$E(\mathbf{x}) = \Sigma\, x \cdot f(x) \tag{5.1.1a}$$

where $f(x)$ is the value of its probability function at x, and the summation extends over the entire range of $\mathbf{x}$. Correspondingly, *in the continuous case the mathematical expectation of a random variable* $\mathbf{x}$ *is given by*

$$E(\mathbf{x}) = \int_{-\infty}^{\infty} x \cdot f(x)\, dx \tag{5.1.1b}$$

where $f(x)$ is the value of its probability density at x. [In (5.1.1a) and (5.1.1b) it is assumed, of course, that the corresponding sum or integral exists; otherwise, the mathematical expectation does not exist (see, for example, Exercise 6 on page 145).] For instance, if $\mathbf{x}$ is the number of points rolled with a balanced die, then $f(x) = \frac{1}{6}$ for $x = 1, 2, 3, 4, 5, 6$, and its mathematical expectation is

$$E(\mathbf{x}) = 1 \cdot \tfrac{1}{6} + 2 \cdot \tfrac{1}{6} + 3 \cdot \tfrac{1}{6} + 4 \cdot \tfrac{1}{6} + 5 \cdot \tfrac{1}{6} + 6 \cdot \tfrac{1}{6} = 3\tfrac{1}{2}$$

Also, if $\mathbf{x}$ has the uniform density $f(x) = \frac{1}{2}$ for $2 < x < 4$ and $f(x) = 0$ elsewhere, then its mathematical expectation is

$$E(\mathbf{x}) = \int_2^4 x \cdot \tfrac{1}{2}\, dx = \tfrac{1}{4}x^2 \Big]_2^4 = 3$$

In many problems of statistics we are interested not only in the *expected value* (that is, the mathematical expectation) of a random variable $\mathbf{x}$, but also in the expected values of random variables related to $\mathbf{x}$. Thus, we might be interested in the random variable $\mathbf{y}$, whose values are related to those of $\mathbf{x}$ by means of the equation $y = g(x)$, and to simplify our notation we shall denote this random variable $g(\mathbf{x})$. To find the mathematical expectation of $g(\mathbf{x})$, we could use the change of variable technique of Section 4.4, but it is usually easier and more straightforward to substitute directly into the formula

$$E[g(\mathbf{x})] = \Sigma\, g(x) f(x) \tag{5.1.2a}$$

when $\mathbf{x}$ is *discrete*, and

$$E[g(\mathbf{x})] = \int_{-\infty}^{\infty} g(x) f(x)\, dx \tag{5.1.2b}$$

when $\mathbf{x}$ is *continuous*. As the reader will be asked to verify in Exercises 3 and 4 on page 144, the result will be the same either way. To illustrate, let us refer to the two examples which we used above, and let us determine in each case the mathematical expectation of the random variable $g(\mathbf{x}) = \mathbf{x}^2$. For the number of points rolled with a balanced die we get

$$E(\mathbf{x}^2) = \Sigma x^2 \cdot \tfrac{1}{6} = 1 \cdot \tfrac{1}{6} + 4 \cdot \tfrac{1}{6} + 9 \cdot \tfrac{1}{6} + 16 \cdot \tfrac{1}{6} + 25 \cdot \tfrac{1}{6} + 36 \cdot \tfrac{1}{6} = 15\tfrac{1}{6}$$

and for the random variable with the uniform density $f(x) = \frac{1}{2}$ for $2 < x < 4$ and $f(x) = 0$ elsewhere we get

$$E(\mathbf{x}^2) = \int_2^4 x^2 \cdot \tfrac{1}{2}\, dx = \tfrac{1}{6}x^3 \Big]_2^4 = 9\tfrac{1}{3}$$

The determination of mathematical expectations can often be simplified by means of the following theorems, which are immediate consequences of (5.1.2a) and (5.1.2b):

THEOREM 5.1*

If c is a constant, $E(\mathbf{c}) = c$.

THEOREM 5.2

If c is a constant, $E[c \cdot g(\mathbf{x})] = c \cdot E[g(\mathbf{x})]$.

THEOREM 5.3

$$E\left[\sum_{i=1}^{k} g_i(\mathbf{x})\right] = \sum_{i=1}^{k} E[g_i(\mathbf{x})]$$

The proofs of these theorems, which the reader will be asked to give in Exercise 1 on page 144, merely require substitution into (5.1.2a) or (5.1.2b) and the rules for summations given in the Appendix on page 419. The following is another useful theorem about mathematical expectations, which the reader will be asked to prove in Exercise 2 on page 144:

THEOREM 5.4

$$E[(a\mathbf{x} + b)^n] = \sum_{i=0}^{n} \binom{n}{i} a^{n-i} b^i E(\mathbf{x}^{n-i})$$

For instance, for $n = 1$ and $n = 2$ this theorem yields $E(a\mathbf{x} + b) = aE(\mathbf{x}) + b$ and $E[(a\mathbf{x} + b)^2] = a^2E(\mathbf{x}^2) + 2abE(\mathbf{x}) + b^2$.

The concept of a mathematical expectation can easily be extended to situations involving more than one random variable. For instance, if $\mathbf{z}$ is a random variable whose values are related to those of two random variables $\mathbf{x}$ and $\mathbf{y}$ by means of the equation $z = g(x, y)$, we write its mathematical expectation as

$$E[g(\mathbf{x}, \mathbf{y})] = \Sigma\, \Sigma\, g(x, y) f(x, y) \tag{5.1.3a}$$

when $\mathbf{x}$ and $\mathbf{y}$ are discrete random variables, and as

$$E[g(\mathbf{x}, \mathbf{y})] = \int_{-\infty}^{\infty} \int_{-\infty}^{\infty} g(x, y) f(x, y)\, dx\, dy \tag{5.1.3b}$$

when the two random variables are continuous. Of course, in (5.1.3a) $f(x, y)$ is the value of the joint probability function of $\mathbf{x}$ and $\mathbf{y}$ at (x, y) and

* Note that we have to use boldface type for the random variable which can take on only the value c.

the summation extends over the entire range of $\mathbf{x}$ and $\mathbf{y}$; in (5.1.3b) $f(x, y)$ is the corresponding value of the joint probability density.

To give a simple example, let us calculate $E(\mathbf{xy})$ for the bivariate distribution on page 91, where we are rolling a pair of balanced dice, $\mathbf{x}$ is the number of dice which come up 1, and $\mathbf{y}$ is the number of dice which come up either 5 or 6. Omitting all terms where the value of either random variable is 0, we get

$$E(\mathbf{xy}) = 1 \cdot 1 \cdot \tfrac{4}{36} = \tfrac{1}{9}$$

This means that if we repeatedly roll a pair of balanced dice and in each case multiply the number of dice which come up 1 by the number of dice which come up either 5 or 6, we should *on the average* get $\frac{1}{9}$, although most of the time we will of course get zero.

THEORETICAL EXERCISES

1. Prove Theorems 5.1, 5.2, and 5.3
(a) for the discrete case;
(b) for the continuous case.

2. Use Theorems 5.1, 5.2, and 5.3 to prove Theorem 5.4.

3. On page 119 we considered the random variable $\mathbf{x}$ which has a binomial distribution with $n = 5$ and $\theta = \frac{1}{4}$ and obtained the probability function of the random variable $\mathbf{y}$, whose values are related to those of $\mathbf{x}$ by means of the equation $y = \dfrac{1}{1 + x}$.
(a) Find $E(\mathbf{y})$ using (5.1.1a) and the probability function of $\mathbf{y}$ which we obtained on page 120.
(b) Find $E(\mathbf{y})$ by evaluating $E\left(\dfrac{1}{1 + \mathbf{x}}\right)$ in accordance with (5.1.2a).

Note that there is actually no difference between what we have to do in part (a) and in part (b).

4. On page 121 we considered the random variable $\mathbf{x}$ which has an exponential density with $\theta = 1$ and obtained the probability density of the random variable $\mathbf{y}$, whose values are related to those of $\mathbf{x}$ by means of the equation $y = \sqrt{x}$.
(a) Find $E(\mathbf{y})$ using (5.1.1b) and the probability density which we obtained for $\mathbf{y}$.
(b) Find $E(\mathbf{y})$ by evaluating $E(\sqrt{\mathbf{x}})$ in accordance with (5.1.2b).
[*Hint:* To evaluate the integral in part (a) make the substitution $u = y^2$ and then refer to (4.3.3) and the hint to Exercise 9 on page 117;

the last part of this suggestion applies to part (b), and it should be noted that the two methods of obtaining $E(\mathbf{y})$ are, in fact, equivalent.]

5. If the random variable $\mathbf{x}$ has the geometric distribution (3.3.6) with $\theta = \frac{1}{2}$, show that $E(2^{\mathbf{x}})$ *does not exist.* [This is the famous *Petersburg paradox,* according to which a player's mathematical expectation is *infinite* if he is to receive 2^n dollars when the *first* head (in a series of flips of a balanced coin) occurs on the nth trial.]

6. Show that if a random variable has the *Cauchy distribution* of Exercise 11 on page 117 its mathematical expectation *does not exist.*

APPLIED EXERCISES

7. If someone were to give us \$6.00 each time we roll a 2 with a balanced die, how much should we pay him when we roll a 1, 3, 4, 5, or 6 to make the game *equitable?* (A game is said to be equitable if the mathematical expectation of each player is zero. Note also that if the amount a player will receive is taken to be *positive,* the amount he will have to pay must be *negative.*)

8. In a winner-take-all tournament among three professional golfers the prize money is \$12,000. If Mr. Brown, one of these golfers, figures that the odds against his winning are 3 to 2, what is his mathematical expectation? Would he figure that it is more profitable for him to make a secret agreement with the other golfers to divide the prize money evenly regardless of who wins?

9. The probability that Mr. Jones will sell his house at a profit of \$2,000 is $\frac{3}{20}$, the probability that he will sell it at a profit of \$1,000 is $\frac{7}{20}$, the probability that he will break even is $\frac{9}{20}$, and the probability that he will lose \$1,000 is $\frac{1}{20}$. What is his *expected profit?*

10. **GAMBLER'S RUIN** Mr. Adams and Mr. Smith are betting on repeated flips of a balanced coin. At the start of the game Mr. Adams has a dollars, Mr. Smith has b dollars, at each flip the loser pays the winner one dollar, and the game is continued until either player has lost all the money with which he began. Making use of the fact that in an *equitable game* each player's mathematical expectation is zero, find the probability that Mr. Adams will win Mr. Smith's b dollars before he loses his a dollars.

11. The manager of a bakery knows that the number of chocolate cakes he can sell on any given day is a random variable with the probability function $f(x) = \frac{1}{6}$ for $x = 0, 1, 2, 3, 4,$ and 5. He also knows that there is a profit of \$1.00 on each cake which he sells and a loss (due to

spoilage) of \$0.40 on each cake which he does not sell. Assuming that each cake can be sold only on the day it is made, find the baker's expected profit for
(a) a day on which he bakes 5 of these cakes;
(b) a day on which he bakes 4 of these cakes;
(c) a day on which he bakes 3 of these cakes.

12. If a contractor's profit on a construction job can be looked upon as a continuous random variable having the probability density

$$f(x) = \begin{cases} \frac{1}{18}(x+1) & \text{for } -1 < x < 5 \\ 0 & \text{elsewhere} \end{cases}$$

where the units are one thousand dollars, what is his *expected profit?*

13. With reference to Exercise 8 on page 110, what mileage can a car owner *expect* to get (in the sense of a mathematical expectation) with one of these tires?

14. With reference to Exercise 19 on page 119, what is the city's *expected* power consumption for any given day? [*Hint:* Make use of (4.3.3) in the evaluation of the required integral.]

15. With reference to Exercise 21 on page 119, what proportion of its donut shops can the chain *expect* to make a profit during their first year of operation?

5.2 MOMENTS

If we let $g(\mathbf{x}) = \mathbf{x}^r$ in (5.1.2a) and (5.1.2b), we obtain what is called the *rth moment about the origin* of the distribution of the random variable $\mathbf{x}$. Symbolically, we shall write

$$\mu_r' = E(\mathbf{x}^r) = \Sigma x^r \cdot f(x) \tag{5.2.1a}$$

for $r = 0, 1, 2, 3, \ldots$, when $\mathbf{x}$ is a discrete random variable, and

$$\mu_r' = E(\mathbf{x}^r) = \int_{-\infty}^{\infty} x^r \cdot f(x)\, dx \tag{5.2.1b}$$

when $\mathbf{x}$ is a continuous random variable. It is of interest to note that the term "moment" is borrowed from the field of physics—if the quantities $f(x)$ in (5.2.1a) were point masses acting perpendicular to the x-axis at distances x from the origin, μ_1' would be the x-coordinate of the *center of gravity*, namely, the *first moment* divided by $\Sigma f(x) = 1$; similarly, μ_2' would give the *moment of inertia*. This also explains why the moments μ_r' are called moments *about the origin;* in the analogy to physics, the length of the lever arm is in each case the distance from the origin. (Incidentally, this analogy to physics applies also in the continuous case, where μ_1' and μ_2' might be the x-coordinate of the center of gravity and the moment of inertia of a rod of variable density like the one pictured in Figure 4.6.)

When $r = 0$ we get $\mu_0' = 1$, and this merely expresses the fact that a random variable must take on one of its values, namely, that for the entire sample space $P(S) = 1$. When $r = 1$ we get the *mathematical expectation* of the random variable $\mathbf{x}$, itself, and since this is of special importance in statistics we abbreviate its symbol to μ. Also, we refer to $\mu = \mu_1' = E(\mathbf{x})$ as the *mean* of the distribution of the random variable $\mathbf{x}$. Informally, the mean is what in everyday language is referred to as an "average": In the first example on page 142 we saw that the mean is $\mu = E(\mathbf{x}) = 3\frac{1}{2}$, which means that if we repeatedly roll a balanced die the number of points we get should *average* $3\frac{1}{2}$; and in the second example on that page we saw that $\mu = E(\mathbf{x}) = 3$, which means that if we repeatedly observe a random variable having the uniform density $f(x) = \frac{1}{2}$ for $2 < x < 4$ and $f(x) = 0$ elsewhere, the values we get should *average* 3.

When $g(\mathbf{x}) = (\mathbf{x} - \mu)^r$ in (5.1.2a) and (5.1.2b), we obtain what is called the *rth moment about the mean* of the distribution of the random variable $\mathbf{x}$. Symbolically, we shall write

$$\mu_r = E[(\mathbf{x} - \mu)^r] = \Sigma\, (x - \mu)^r \cdot f(x) \tag{5.2.2a}$$

for $r = 0, 1, 2, 3, \ldots$, when $\mathbf{x}$ is a discrete random variable, and

$$\mu_r = E[(\mathbf{x} - \mu)^r] = \int_{-\infty}^{\infty} (x - \mu)^r \cdot f(x)\, dx \tag{5.2.2b}$$

when $\mathbf{x}$ is a continuous random variable. Note that $\mu_0 = 1$ and $\mu_1 = 0$ for the distribution of *any* random variable $\mathbf{x}$ as can easily be verified (see Exercise 5 on page 153).

Moments about the mean are of special importance in statistics, as they can be used to describe the *shape* of the distribution of a random variable (namely, the shape of the graph of its probability function or its probability density). In particular, the second moment about the mean, called the *variance* and denoted σ^2, is indicative of the *spread or dispersion* of a distribution. This is illustrated in Figure 5.1, which shows the histograms of four probability functions with the same mean μ, but variances equalling 0.88, 1.66, 3.18, and 5.26, respectively. As we shall see in Section 5.2.1, σ^2 (or its positive square root σ, called the *standard deviation*) thus tells us how close one can expect a value of a random variable to be to the mean of its distribution. A brief discussion of how the *third*

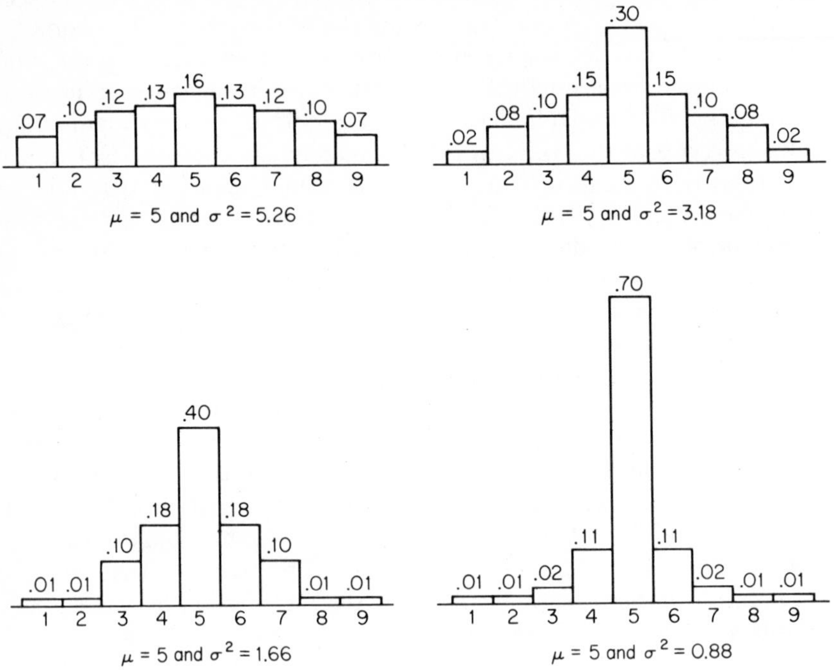

FIGURE 5.1 Distributions with different "spread" or "dispersion."

moment about the mean is used to describe the *symmetry* or lack of symmetry (*skewness*) of a distribution is given in Exercise 9 on page 154.

In many instances moments about the mean are obtained by first calculating moments about the origin and then expressing the μ_r in terms of the μ_r'. To this end we use

THEOREM 5.5

$$\mu_r = \mu_r' - \binom{r}{1}\mu_{r-1}'\mu + \cdots + (-1)^i\binom{r}{i}\mu_{r-i}'\mu^i + \cdots + (-1)^{r-1}(r-1)\mu^r$$

for $r = 1, 2, 3, \ldots$.

To prove this theorem, we have only to let $a = 1$, $b = -\mu$, and $n = r$ in Theorem 5.4, and combine the last two terms by writing

$$\begin{aligned}(-1)^{r-1}\binom{r}{r-1}\mu_1'\mu^{r-1} + (-1)^r\binom{r}{r}\mu_0'\mu^r &= (-1)^{r-1}r\mu^r - (-1)^{r-1}\mu^r \\ &= (-1)^{r-1}(r-1)\mu^r\end{aligned}$$

since $\mu_1' = \mu$ and $\mu_0' = 1$. Theorem 5.5 is of special importance when $r = 2$, in which case it yields the formula

$$\sigma^2 = \mu_2' - \mu^2 \tag{5.2.3}$$

for the *variance* of the distribution of a random variable. Sometimes we write the variance of the distribution of a random variable $\mathbf{x}$ as $V(\mathbf{x})$ or as $\text{var}(\mathbf{x})$ instead of σ^2; this is done especially when we are dealing with the distributions of several random variables in one and the same problem.

5.2.1 Chebyshev's Theorem

In actual practice we often use the *standard deviation* σ instead of the variance σ^2 *because* σ *is in the same units as the values of the random variable, itself*. To demonstrate *how* σ is indicative of the "spread" or "dispersion" of the distribution of a random variable, let us now prove the following theorem, called *Chebyshev's theorem* after the nineteenth century Russian mathematician P. L. Chebyshev:

THEOREM 5.6

*If μ and σ are, respectively, the mean and the standard deviation of the distribution of the random variable $\mathbf{x}$, then for any positive constant k the probability that $\mathbf{x}$ will take on a value which is at most $\mu - k\sigma$ or at least $\mu + k\sigma$ is less than or equal to $1/k^2$.**

Before we prove this theorem, let us illustrate its significance with reference to an experiment which consists of 100 flips of a balanced coin. As the reader will be asked to show in Exercise 23 on page 172, the mean and the standard deviation of the distribution of the total number of heads are, respectively, $\mu = 50$ and $\sigma = 5$, so that for $k = 3$ we can assert that the probability of getting at least 65 heads or at most 35 is less than or equal to $\frac{1}{9}$. Similarly, for $k = 5$ we can assert that the probability of getting at least 75 heads or at most 25 is less than or equal to 0.04.

Theorem 5.6 holds regardless of whether $\mathbf{x}$ is a continuous random variable or whether it is discrete; to prove it for the discrete case, let us begin with

$$\sigma^2 = \Sigma\ (x - \mu)^2 f(x)$$

* Many statisticians write the results of this theorem as $P(|\mathbf{x} - \mu| \geq k\sigma) \leq 1/k^2$, a notation which we have been avoiding as it fails to distinguish between the random variable, a function, and the *value* of a random variable, the corresponding dependent variable.

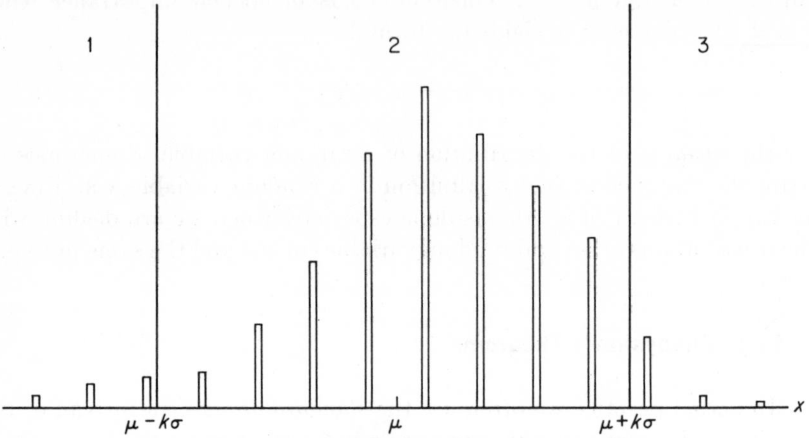

FIGURE 5.2 Proof of Chebyshev's Theorem.

and divide the sum into three parts as indicated in Figure 5.2. To be more specific, let us write

$$\sigma^2 = \sum_1 (x - \mu)^2 f(x) + \sum_2 (x - \mu)^2 f(x) + \sum_3 (x - \mu)^2 f(x)$$

where $\sum_1$ is summed over all x less than or equal to $\mu - k\sigma$, $\sum_2$ is summed over all x greater than $\mu - k\sigma$ but less than $\mu + k\sigma$, and $\sum_3$ is summed over all x greater than or equal to $\mu + k\sigma$. Since the quantities $(x - \mu)^2 f(x)$ are all non-negative, we have

$$\sigma^2 \geq \sum_1 (x - \mu)^2 f(x) + \sum_3 (x - \mu)^2 f(x)$$

and since, furthermore, the absolute value of the difference between x and μ is at least $k\sigma$ for all terms in $\sum_1$ as well as $\sum_3$, it follows that

$$\sigma^2 \geq \sum_1 k^2\sigma^2 f(x) + \sum_3 k^2\sigma^2 f(x)$$

and, hence, that

$$\frac{1}{k^2} \geq \sum_1 f(x) + \sum_3 f(x)$$

This assumes, of course, that $\sigma \neq 0$. Since the expression on the right-hand side of the last inequality represents the probability that **x** will take on a value less than or equal to $\mu - k\sigma$ or greater than or equal to $\mu + k\sigma$, this completes the proof of Theorem 5.6.

As an immediate consequence of Theorem 5.6 we find that values of $\mathbf{x}$ falling *between* $\mu - k\sigma$ and $\mu + k\sigma$ account for at least $1 - 1/k^2$ of the total probability of 1. In other words, the probability that a random variable $\mathbf{x}$ takes on a value within two standard deviations of the mean of its distribution is *at least* $\frac{3}{4}$, the probability that it takes on a value within four standard deviations of the mean of its distribution is *at least* $\frac{15}{16}$, and so forth. It is in this sense that σ "controls" the spread or dispersion of the distribution of a random variable $\mathbf{x}$.

5.2.2 Moment Generating Functions

Although the moments of most distributions can be determined directly by evaluating the necessary integrals or sums, there exists an alternate technique which often provides considerable simplifications. This technique is based on so-called moment generating functions, which are defined as follows: *If it exists, the moment generating function of the distribution of a random variable* $\mathbf{x}$ *is given by the expected value of* $e^{t\mathbf{x}}$, *namely, by*

$$M_{\mathbf{x}}(t) = E(e^{t\mathbf{x}}) = \Sigma\, e^{tx} f(x) \tag{5.2.4a}$$

when $\mathbf{x}$ *is discrete, and by*

$$M_{\mathbf{x}}(t) = E(e^{t\mathbf{x}}) = \int_{-\infty}^{\infty} e^{tx} f(x)\, dx \tag{5.2.4b}$$

when $\mathbf{x}$ *is continuous.* Substituting for e^{tx} its *Maclaurin's series*

$$1 + tx + \frac{t^2x^2}{2!} + \cdots + \frac{t^rx^r}{r!} + \cdots$$

we obtain for the discrete case

$$\begin{aligned} M_{\mathbf{x}}(t) &= \sum \left[1 + tx + \frac{t^2x^2}{2!} + \cdots + \frac{t^rx^r}{r!} + \cdots\right] f(x) \\ &= \sum f(x) + t\cdot\sum xf(x) + \frac{t^2}{2!}\cdot\sum x^2 f(x) \\ &\qquad + \cdots + \frac{t^r}{r!}\cdot\sum x^r f(x) + \cdots \\ &= 1 + \mu t + \mu_2'\cdot\frac{t^2}{2!} + \cdots + \mu_r'\cdot\frac{t^r}{r!} + \cdots \end{aligned}$$

and this explains the term "moment generating function"—*if we expand $M_{\mathbf{x}}(t)$ as a power series in t, the coefficient of $\frac{t^r}{r!}$ is μ_r', the rth moment about the origin of the distribution of* **x**. Of course, as in the case of the moments, themselves, a moment generating function may or may not exist.

The main difficulty in using the series expansion of a moment generating function to read off the moments of the distribution of a random variable is usually *not* that of determining $M_{\mathbf{x}}(t)$, but that of expanding it as a power series in t. If we are interested only in the first few moments of a distribution (say, μ_1' and μ_2'), their determination can usually be simplified by making use of the fact that in a Maclaurin's series of a function the coefficient of $\frac{t^r}{r!}$ is the rth derivative of the function with respect to t at $t = 0$. Thus,

$$\left[\frac{d^r M_{\mathbf{x}}(t)}{dt}\right]_{t=0} = \mu_r' \tag{5.2.5}$$

provides a third way of determining the moments of a distribution. All these methods will be illustrated in the sections which follow, that is, in the sections where we shall study the moments of the special distributions introduced in Chapters 3 and 4. After that, in Section 5.4, we shall use moment generating functions to simplify proofs concerning limiting distributions; for instance, we shall use moment generating functions to simplify the proof of Section 3.3.7, where we showed that under certain conditions the Poisson distribution is a limiting form of the binomial distribution.

Work with moment generating functions can often be simplified by making use of the following theorem:

THEOREM 5.7

If a and b are constants, then

(a) $M_{\mathbf{x}+a}(t) = E[e^{(\mathbf{x}+a)t}] = e^{at} \cdot M_{\mathbf{x}}(t)$;

(b) $M_{b\mathbf{x}}(t) = E(e^{b\mathbf{x}t}) = M_{\mathbf{x}}(bt)$;

(c) $M_{\frac{\mathbf{x}+a}{b}}(t) = E\left[e^{\left(\frac{\mathbf{x}+a}{b}\right)t}\right] = e^{\frac{a}{b}t} \cdot M_{\mathbf{x}}\left(\frac{t}{b}\right)$.

The proof of this theorem, which merely requires substitution into (5.2.4a) or (5.2.4b), will be left to the reader in Exercise 15 on page 155; as we shall see later on, part (a) of the theorem is of special importance when

$a = -\mu$, and part (c) is of special importance when $a = -\mu$ and $b = \sigma$ (where μ and σ are the mean and the standard deviation of the distribution of a given random variable).

THEORETICAL EXERCISES

1. Find μ, μ_2', and then σ^2 by means of (5.2.3) for the distribution of a random variable having the probability function $f(x) = \frac{1}{2}$ for $x = -2$ and $x = 2$.
2. Find μ, μ_2', and then σ by means of (5.2.3) for the distribution of a random variable having the probability density $f(x) = \frac{1}{2}x$ for $0 < x < 2$ and $f(x) = 0$ elsewhere.
3. Use the results obtained for $E(\mathbf{x})$ and $E(\mathbf{x}^2)$ on page 142 to find the variance of the distribution of the number of points rolled with a balanced die.
4. Use the results obtained for $E(\mathbf{x})$ and $E(\mathbf{x}^2)$ on page 142 to find the variance of the probability density $f(x) = \frac{1}{2}$ for $2 < x < 4$ and $f(x) = 0$ elsewhere.
5. Show that $\mu_0 = 1$ and $\mu_1 = 0$ for the distribution of any discrete or continuous random variable.
6. Express var($\mathbf{y}$) in terms of var($\mathbf{x}$) if
   (a) the values of $\mathbf{y}$ are related to those of $\mathbf{x}$ by means of the equation $y = x + c$;
   (b) the values of $\mathbf{y}$ are related to those of $\mathbf{x}$ by means of the equation $y = cx$.
7. If the distribution of the random variable $\mathbf{x}$ has the mean μ and the variance σ^2, show that the mean and the variance of the random variable $\mathbf{y}$, whose values are related to those of $\mathbf{x}$ by means of the equation $y = \dfrac{x - \mu}{\sigma}$, are $E(\mathbf{y}) = 0$ and var($\mathbf{y}$) $= 1$. (A distribution which has the mean 0 and the variance 1 is said to be in *standard form*, and when we perform the above change of variable $y = \dfrac{x - \mu}{\sigma}$ we are said to be *standardizing* the distribution of $\mathbf{x}$.)
8. Check whether the mean and the variance of the following distributions exist.
   (a)
   $$f(x) = \begin{cases} 2x^{-3} & \text{for } x > 1 \\ 0 & \text{for } x \leq 1 \end{cases}$$
   (b) The Cauchy distribution of Exercise 11 on page 117.

9. **SYMMETRY AND SKEWNESS** The symmetry or lack thereof (skewness) of a probability distribution is often measured by means of the quantity

$$\alpha_3 = \frac{\mu_3}{\sigma^3} \tag{5.2.6}$$

Find α_3 for each of the following distributions, which (as can easily be verified) have equal means and standard deviations:
(a) $f(1) = 0.05$, $f(2) = 0.15$, $f(3) = 0.30$, $f(4) = 0.30$, $f(5) = 0.15$, and $f(6) = 0.05$;
(b) $f(1) = 0.05$, $f(2) = 0.20$, $f(3) = 0.15$, $f(4) = 0.45$, $f(5) = 0.10$, and $f(6) = 0.05$.
Also draw histograms of these two distributions and note that whereas the first is *symmetrical*, the second has a "tail" on the left-hand side and is said to be *negatively skewed.*

10. **KURTOSIS** The extent to which a distribution is peaked or flat is often measured by means of the quantity

$$\alpha_4 = \frac{\mu_4}{\sigma^4} \tag{5.2.7}$$

Find α_4 for each of the following distributions, of which the first is more peaked (narrow humped) than the second:
(a) $f(-3) = 0.06$, $f(-2) = 0.09$, $f(-1) = 0.10$, $f(0) = 0.50$, $f(1) = 0.10$, $f(2) = 0.09$, and $f(3) = 0.06$;
(b) $f(-3) = 0.04$, $f(-2) = 0.11$, $f(-1) = 0.20$, $f(0) = 0.30$, $f(1) = 0.20$, $f(2) = 0.11$, and $f(3) = 0.04$.
Also draw the histograms of these two distributions.

11. Duplicating the steps used in the text, prove Chebyshev's theorem for continuous random variables.

12. What is the least value of k in Chebyshev's theorem for which the probability that a random variable takes on a value which is at least $\mu + k\sigma$ or at most $\mu - k\sigma$ is (a) 0.05 or less, and (b) 0.01 or less.

13. If we let $k\sigma = c$ in Chebyshev's theorem, what does this theorem assert about the probability that a random variable will take on a value which is at least $\mu + c$ or at most $\mu - c$?

14. Given the probability density $f(x) = \frac{1}{2}e^{-|x|}$ for $-\infty < x < \infty$, show that the corresponding moment generating function is $M_x(t) = \dfrac{1}{1 - t^2}$. Also find the variance of this distribution (a) by expanding $M_x(t)$ as

an infinite series and reading off the necessary coefficients, and (b) by finding its first and second derivatives at $t = 0$.

15. Prove all three parts of Theorem 5.7.

16. Explain why there can be no random variable for which $M_{\mathbf{x}}(t) = \frac{t}{1-t}$.

APPLIED EXERCISES

17. In Exercise 3 the reader was asked to find the variance of the distribution of the number of points rolled with a balanced die. Using this result, what can we assert about the values of this random variable according to Chebyshev's theorem with $k = 1.25$? What is the corresponding *exact* probability?

18. In Exercise 4 the reader was asked to find the variance of the probability density $f(x) = \frac{1}{2}$ for $2 < x < 4$ and $f(x) = 0$ elsewhere. Using this result, what can we assert about a random variable having this density according to Chebyshev's theorem with $k = 1.5$? What is the corresponding *exact* probability?

5.3 MOMENTS OF SPECIAL PROBABILITY DISTRIBUTIONS

In this section we shall study some of the *lower moments*, mainly μ and σ^2, of the many special distributions (probability functions and probability densities) introduced in Chapters 3 and 4. To hold down the size of this book, only the most important ones will be discussed in the text, while the others will be left as exercises for the reader.

As we have indicated earlier, there are several ways of obtaining the moments of a distribution. We can evaluate the corresponding integrals or sums, we can expand the moment generating function as an infinite series and read off the coefficients of the appropriate terms, or we can differentiate the moment generating function and make use of (5.2.5). Although it would seem logical to use in each case whichever method is simplest, or requires the least effort, we shall treat some distributions in more ways than one. This will be done partly because the results will be needed later on, and partly because it will provide the reader with useful experience.

5.3.1 Moments of the Binomial Distribution

To determine the *mean* of the binomial distribution (3.3.3), let us directly evaluate the sum

$$\mu = \sum_{x=0}^{n} x \cdot \binom{n}{x} \theta^x (1 - \theta)^{n-x}$$

$$= \sum_{x=1}^{n} \frac{n!}{(x-1)!(n-x)!} \theta^x (1 - \theta)^{n-x}$$

where we omitted the term for which $x = 0$, which is 0, and cancelled the x against the first factor of $x! = x(x - 1)!$ in the denominator of $\binom{n}{x}$. Then, factoring out the factor n in $n! = n(n - 1)!$ and one factor θ, we get

$$\mu = n\theta \cdot \sum_{x=1}^{n} \binom{n-1}{x-1} \theta^{x-1} (1 - \theta)^{n-x}$$

and, letting $y = x - 1$ and $m = n - 1$, we finally obtain

$$\mu = n\theta \cdot \sum_{y=0}^{m} \binom{m}{y} \theta^y (1 - \theta)^{m-y} = n\theta$$

since the last summation is the sum of *all* the values of a binomial probability function with the parameters θ and m, and hence equal to 1.

To obtain expressions for the *second moment about the origin* and the *variance* of the binomial distribution, let us make use of the fact that $E(\mathbf{x}^2) = E[\mathbf{x}(\mathbf{x} - 1)] + E(\mathbf{x})$, and begin by evaluating $E[\mathbf{x}(\mathbf{x} - 1)]$, duplicating for all practical purposes the steps used above. We thus get

$$E[\mathbf{x}(\mathbf{x} - 1)] = \sum_{x=0}^{n} x(x - 1) \binom{n}{x} \theta^x (1 - \theta)^{n-x}$$

$$= \sum_{x=2}^{n} \frac{n!}{(x-2)!(n-x)!} \theta^x (1 - \theta)^{n-x}$$

$$= n(n - 1)\theta^2 \cdot \sum_{x=2}^{n} \binom{n-2}{x-2} \theta^{x-2} (1 - \theta)^{n-x}$$

and, letting $y = x - 2$ and $m = n - 2$, we finally obtain

$$E[\mathbf{x}(\mathbf{x} - 1)] = n(n - 1)\theta^2 \cdot \sum_{y=0}^{m} \binom{m}{y} \theta^y (1 - \theta)^{m-y}$$
$$= n(n - 1)\theta^2$$

Thus, $\mu_2' = E[\mathbf{x}(\mathbf{x} - 1)] + E(\mathbf{x}) = n(n - 1)\theta^2 + n\theta$, and (5.2.3) yields

$$\sigma^2 = n(n - 1)\theta^2 + n\theta - n^2\theta^2$$
$$= n\theta(1 - \theta)$$

To summarize our results, we have shown that

THEOREM 5.8

The mean and the variance of the binomial distribution (3.3.3) *are* $\mu = n\theta$ and $\sigma^2 = n\theta(1 - \theta)$.

The fact that the mean of the binomial distribution is given by the formula $\mu = n\theta$ should really have been expected. After all, if a balanced coin is flipped 200 times, we expect (in the sense of a mathematical expectation, namely, in the sense of an *average*) $200 \cdot \frac{1}{2} = 100$ heads and 100 tails; similarly, if a balanced die is rolled 240 times we expect $240 \cdot \frac{1}{6} = 40$ sixes, and if the probability that any one person will recover from a given disease is 0.80, we would expect $50(0.80) = 40$ recoveries among 50 persons having this disease.

The formula for the variance of the binomial distribution does not have any immediate (intuitive) significance, but being a measure of variation or dispersion, it has many important applications. Consider the random variable $\mathbf{y}$, for example, whose values are related to those of a random variable $\mathbf{x}$ having the binomial distribution (3.3.3) by means of the equation $y = \frac{x}{n}$; in other words, the values of $\mathbf{y}$ are the *proportion of successes* in n trials (meeting the conditions underlying the binomial distribution). Using the results just obtained together with Theorem 5.2 and the result of part (b) of Exercise 6 on page 153, we have

$$E(\mathbf{y}) = \theta \quad \text{and} \quad \text{var}(\mathbf{y}) = \frac{\theta(1 - \theta)}{n} \tag{5.3.1}$$

and if we apply Chebyshev's theorem in the form in which it is given in Exercise 13 on page 154, we can assert that *for any positive constant c, the probability that the proportion of successes is at least* $\theta + c$ *or at most* $\theta - c$

is less than or equal to $\frac{\theta(1-\theta)}{nc^2}\cdot$ Hence, when $n \to \infty$, the probability that the proportion of successes will differ from θ by more than any arbitrary constant c approaches 0. In other words, when $n \to \infty$, the probability that the proportion of successes will take on a value arbitrarily close to θ approaches 1. This result is often referred to as the *law of large numbers,* and it should be observed that it applies to the proportion of successes, not to their actual number. It is a fallacy to assume that when n is large the *number of successes* must necessarily be close to $n\theta$.

Since the moment generating function of the binomial distribution is easy to obtain, let us use it to illustrate also the third method of determining moments which we listed in Section 5.3, namely, the one based on (5.2.5). Substituting into (5.2.4a) the expression for the binomial probabilities $b(x; n, \theta)$ given by (3.3.3), we obtain

$$M_{\mathbf{x}}(t) = \sum_{x=0}^{n} e^{xt}\binom{n}{x}\theta^x(1-\theta)^{n-x}$$

$$= \sum_{x=0}^{n}\binom{n}{x}(\theta e^t)^x(1-\theta)^{n-x}$$

and recognizing this last summation as the binomial expansion of $[\theta e^t + (1-\theta)]^n$, we have thus shown that

THEOREM 5.9

The moment generating function of the binomial distribution is given by $M_{\mathbf{x}}(t) = [1 + \theta(e^t - 1)]^n$.

Differentiating twice with respect to t, we obtain

$$M'_{\mathbf{x}}(t) = n\theta e^t[1 + \theta(e^t - 1)]^{n-1}$$

$$M''_{\mathbf{x}}(t) = n\theta e^t[1 + \theta(e^t - 1)]^{n-1} + n(n-1)\theta^2 e^{2t}[1 + \theta(e^t - 1)]^{n-2}$$

$$= n\theta e^t(1 - \theta + n\theta e^t)[1 + \theta(e^t - 1)]^{n-2}$$

and, upon substituting $t = 0$, we get $\mu'_1 = n\theta$ and $\mu'_2 = n\theta(1 - \theta + n\theta)$. Thus, $\mu = \mu'_1 = n\theta$ and $\sigma^2 = \mu'_2 - \mu^2 = n\theta(1 - \theta + n\theta) - (n\theta)^2 = n\theta(1-\theta)$, which agrees (as it should) with the results stated in Theorem 5.8.

From the work of this section it may seem easier to find the moments of the binomial distribution with the moment generating function than to evaluate them directly, although it must be apparent that the differentiation would become fairly involved if we wanted to determine, say, μ_3' or μ_4'. Actually, there exists yet an easier way of determining the moments of the binomial distribution; it is based on the so-called *factorial moment generating function*, which is explained in Exercise 5 on page 169.

5.3.2 Moments of the Hypergeometric Distribution

Using methods which are very similar to the ones used in the first part of the preceding section, let us now prove the following theorem:

THEOREM 5.10

The mean and the variance of the hypergeometric distribution (3.3.5) *are*

$$\mu = \frac{na}{a+b} \quad \textit{and} \quad \sigma^2 = \frac{nab(a+b-n)}{(a+b)^2(a+b-1)}$$

To obtain the formula for the mean, let us directly evaluate the sum

$$\mu = \sum_{x=0}^{n} x \cdot \frac{\binom{a}{x}\binom{b}{n-x}}{\binom{a+b}{n}} = \sum_{x=1}^{n} \frac{a!}{(x-1)!(a-x)!} \cdot \frac{\binom{b}{n-x}}{\binom{a+b}{n}}$$

where we omitted the term for which $x = 0$, which is 0, and cancelled the x against the first factor of $x! = x(x-1)!$ in the denominator of $\binom{a}{x}$. Then, factoring out $a/\binom{a+b}{n}$, we obtain

$$\mu = \frac{a}{\binom{a+b}{n}} \cdot \sum_{x=1}^{n} \binom{a-1}{x-1}\binom{b}{n-x}$$

and, letting $y = x - 1$ and $m = n - 1$, we get

$$\mu = \frac{a}{\binom{a+b}{n}} \cdot \sum_{y=0}^{m} \binom{a-1}{y}\binom{b}{m-y}$$

Finally, using Theorem 1.8 on page 28 we can write

$$\mu = \frac{a}{\binom{a+b}{n}} \cdot \binom{a+b-1}{m} = \frac{a}{\binom{a+b}{n}} \cdot \binom{a+b-1}{n-1} = \frac{na}{a+b}$$

To obtain the formula for σ^2, we proceed as on page 156, namely, by first evaluating $E[\mathbf{x}(\mathbf{x}-1)]$ and then making use of the fact that $E(\mathbf{x}^2) = E[\mathbf{x}(\mathbf{x}-1)] + E(\mathbf{x})$ and, hence, that $\sigma^2 = E[\mathbf{x}(\mathbf{x}-1)] + E(\mathbf{x}) - \mu^2$. Leaving it to the reader to show that

$$E[\mathbf{x}(\mathbf{x}-1)] = \frac{a(a-1)n(n-1)}{(a+b)(a+b-1)}$$

we thus obtain

$$\begin{aligned}\sigma^2 &= \frac{a(a-1)n(n-1)}{(a+b)(a+b-1)} + \frac{na}{a+b} - \left[\frac{na}{a+b}\right]^2 \\ &= \frac{nab(a+b-n)}{(a+b)^2(a+b-1)}\end{aligned}$$

Since the moment generating function of the hypergeometric distribution is fairly complicated, we shall not treat it in this book. It may be found, though, in the book by M. G. Kendall and A. Stuart listed among the references on page 181.

5.3.3 Moments of the Poisson Distribution

Having derived the Poisson distribution in Chapter 3 as a limiting case of the binomial distribution when $n \to \infty$, $\theta \to 0$, while $n\theta = \lambda$ remains constant, we can now obtain the mean and the variance of the Poisson distribution by applying the same limiting process to the mean and the variance of the binomial distribution. Since $\mu = n\theta = \lambda$ and $\sigma^2 = n\theta(1-\theta) = \lambda(1-\theta) \to \lambda$ when $\theta \to 0$, we have

THEOREM 5.11

The mean and the variance of the Poisson distribution (3.3.9) *are* $\mu = \lambda$ *and* $\sigma^2 = \lambda$.

These formulas can also be derived directly, that is, without reference to the binomial distribution. For instance, for the mean of the Poisson distribution we can write

$$\mu = \sum_{x=0}^{\infty} x \cdot \frac{\lambda^x e^{-\lambda}}{x!} = \lambda \cdot \sum_{x=1}^{\infty} \frac{\lambda^{x-1} e^{-\lambda}}{(x-1)!}$$

where we again omitted the term for which $x = 0$, which is 0, cancelled the x against the first factor of $x(x - 1)!$ in the denominator, and factored out λ. Then, letting $y = x - 1$, we get

$$\mu = \lambda \cdot \sum_{y=0}^{\infty} \frac{\lambda^y e^{-\lambda}}{y!} = \lambda$$

since the summation which we finally obtained is the sum of *all* the values of the Poisson probability function (3.3.9).

The formula for σ^2 may, similarly, be obtained by first determining $E[\mathbf{x}(\mathbf{x} - 1)]$ as in the case of the binomial and hypergeometric distributions. Easier, though, is the use of the *recursion formula*, which the reader will be asked to prove in Exercise 9 on page 170.

To obtain the moment generating function of the Poisson distribution, we substitute (3.3.9) into (5.2.4a), getting

$$M_{\mathbf{x}}(t) = \sum_{x=0}^{\infty} e^{xt} \cdot \frac{\lambda^x e^{-\lambda}}{x!} = e^{-\lambda} \cdot \sum_{x=0}^{\infty} \frac{(\lambda e^t)^x}{x!}$$

Recognizing the last summation as the Maclaurin's series for e^z with $z = \lambda e^t$, we conclude that

THEOREM 5.12

The moment generating function of the Poisson distribution is given by $M_{\mathbf{x}}(t) = e^{\lambda(e^t - 1)}$.

Differentiating twice with respect to t, we get

$$M'_{\mathbf{x}}(t) = \lambda e^t e^{\lambda(e^t-1)}$$

$$M''_{\mathbf{x}}(t) = \lambda e^t e^{\lambda(e^t-1)} + \lambda^2 e^{2t} e^{\lambda(e^t-1)}$$

so that $\mu'_1 = M'_{\mathbf{x}}(0) = \lambda$ and $\mu'_2 = M''_{\mathbf{x}}(0) = \lambda + \lambda^2$. Thus, $\mu = \lambda$ and $\sigma^2 = (\lambda + \lambda^2) - \lambda^2 = \lambda$, which agrees (as it should) with the results given in Theorem 5.11.

5.3.4 Moments of the Gamma Distribution

Substituting (4.3.6) into (5.2.1b) and letting $y = \frac{x}{\beta}$, we find that for the gamma distribution the rth moment about the origin is given by

$$\mu_r' = \int_0^\infty x^r \cdot \frac{1}{\beta^\alpha \Gamma(\alpha)} x^{\alpha-1} e^{-\frac{x}{\beta}} \, dx = \frac{\beta^r}{\Gamma(\alpha)} \cdot \int_0^\infty y^{\alpha+r-1} e^{-y} \, dy$$

Since the last integral is $\Gamma(r + \alpha)$ according to (4.3.3), we have thus shown that

THEOREM 5.13

The rth moment about the origin of the gamma distribution is $\mu_r' = \frac{\beta^r \Gamma(\alpha + r)}{\Gamma(\alpha)}$.

In particular, $\mu_1' = \frac{\beta \Gamma(\alpha + 1)}{\Gamma(\alpha)} = \beta\alpha$, $\mu_2' = \frac{\beta^2 \Gamma(\alpha + 2)}{\Gamma(\alpha)} = \beta^2(\alpha + 1)\alpha$, so that the mean of the gamma distribution is $\mu = \beta\alpha$ and its variance is

$$\sigma^2 = \beta^2(\alpha + 1)\alpha - (\beta\alpha)^2 = \beta^2\alpha \tag{5.3.2}$$

Since the exponential distribution (4.3.2) is a special case of the gamma distribution with $\alpha = 1$ and $\beta = \theta$, let us also state the result that

THEOREM 5.14

The mean and the variance of the exponential distribution (4.3.2) are $\mu = \theta$ *and* $\sigma^2 = \theta^2$.

To obtain the moment generating function of the gamma distribution, we substitute (4.3.6) into (5.2.4b), getting

$$M_x(t) = \int_0^\infty e^{xt} \cdot \frac{1}{\beta^\alpha \Gamma(\alpha)} x^{\alpha-1} e^{-\frac{x}{\beta}} \, dx = \frac{1}{\beta^\alpha \Gamma(\alpha)} \cdot \int_0^\infty x^{\alpha-1} e^{-x\left(\frac{1}{\beta}-t\right)} \, dx$$

Then, if we let $y = x\left(\frac{1}{\beta} - t\right)$, we get

$$M_x(t) = \frac{1}{\beta^\alpha \Gamma(\alpha) \left(\frac{1}{\beta} - t\right)^\alpha} \cdot \int_0^\infty y^{\alpha-1} e^{-y} \, dy$$

and, identifying the last integral as $\Gamma(\alpha)$ in accordance with (4.3.3), we conclude that

Theorem 5.15

The moment generating function of the gamma distribution is given by $M_x(t) = (1 - \beta t)^{-\alpha}$.

Using this moment generating function, we *could* obtain the moments of the gamma distribution by differentiation, but it is easier in this case to use the second method mentioned in Section 5.3. Writing the moment generating function as the *binomial series*

$$M_x(t) = 1 + \alpha\beta t + \alpha(\alpha + 1)\beta^2 \cdot \frac{t^2}{2!} + \alpha(\alpha + 1)(\alpha + 2)\beta^3 \cdot \frac{t^3}{3!} + \cdots \tag{5.3.3}$$

we can immediately *read off* the first three moments about the origin as $\mu_1' = \alpha\beta$, $\mu_2' = \alpha(\alpha + 1)\beta^2$, and $\mu_3' = \alpha(\alpha + 1)(\alpha + 2)\beta^3$—the corresponding coefficients of t, $\frac{t^2}{2!}$, and $\frac{t^3}{3!}$. Note that the Maclaurin's series (5.3.3) of the moment generating function of the gamma distribution converges for $|\beta t| < 1$.

5.3.5 Moments of the Beta Distribution

Before we derive the formulas for the mean and the variance of the beta distribution, let us point out that since (4.3.7) is a probability density, its integral from $-\infty$ to ∞ must equal 1, and hence

$$\int_0^1 x^{\alpha-1}(1 - x)^{\beta-1}\,dx = \frac{\Gamma(\alpha)\cdot\Gamma(\beta)}{\Gamma(\alpha + \beta)} \tag{5.3.4}$$

In fact, this integral *defines* the so-called *beta function*, whose values are denoted $B(\alpha, \beta)$; in other words, $B(\alpha, \beta) = \frac{\Gamma(\alpha)\cdot\Gamma(\beta)}{\Gamma(\alpha + \beta)}$. Using this function, let us now prove the following theorem:

Theorem 5.16

The mean and the variance of the beta distribution (4.3.7) are $\mu = \frac{\alpha}{\alpha + \beta}$ *and* $\sigma^2 = \frac{\alpha\beta}{(\alpha + \beta)^2(\alpha + \beta + 1)}$.

Substituting (4.3.7) into (5.2.1b) with $r = 1$ and recognizing the fact that the value of the integral is $B(\alpha + 1, \beta)$, we have

$$\begin{aligned}\mu &= \frac{\Gamma(\alpha+\beta)}{\Gamma(\alpha)\cdot\Gamma(\beta)}\cdot\int_0^1 x\cdot x^{\alpha-1}(1-x)^{\beta-1}\,dx \\ &= \frac{\Gamma(\alpha+\beta)}{\Gamma(\alpha)\cdot\Gamma(\beta)}\cdot\frac{\Gamma(\alpha+1)\cdot\Gamma(\beta)}{\Gamma(\alpha+\beta+1)} \\ &= \frac{\alpha}{\alpha+\beta}\end{aligned}$$

since $\Gamma(\alpha + 1) = \alpha\cdot\Gamma(\alpha)$ and $\Gamma(\alpha + \beta + 1) = (\alpha + \beta)\cdot\Gamma(\alpha + \beta)$ according to (4.3.3).

Similarly, substituting (4.3.7) into (5.2.1b) with $r = 2$ and recognizing the fact that the value of the integral is $B(\alpha + 2, \beta)$, we get

$$\begin{aligned}\mu_2' &= \frac{\Gamma(\alpha+\beta)}{\Gamma(\alpha)\cdot\Gamma(\beta)}\cdot\int_0^1 x^2\cdot x^{\alpha-1}(1-x)^{\beta-1}\,dx \\ &= \frac{\Gamma(\alpha+\beta)}{\Gamma(\alpha)\cdot\Gamma(\beta)}\cdot\frac{\Gamma(\alpha+2)\cdot\Gamma(\beta)}{\Gamma(\alpha+\beta+2)} \\ &= \frac{(\alpha+1)\alpha}{(\alpha+\beta+1)(\alpha+\beta)}\end{aligned}$$

since $\Gamma(\alpha + 2) = (\alpha + 1)\alpha\cdot\Gamma(\alpha)$ and $\Gamma(\alpha + \beta + 2) = (\alpha + \beta + 1)\cdot(\alpha + \beta)\cdot\Gamma(\alpha + \beta)$ according to (4.3.3). Finally, substitution into (5.2.3) yields

$$\begin{aligned}\sigma^2 &= \frac{(\alpha+1)\alpha}{(\alpha+\beta+1)(\alpha+\beta)} - \left(\frac{\alpha}{\alpha+\beta}\right)^2 \\ &= \frac{\alpha\beta}{(\alpha+\beta)^2(\alpha+\beta+1)}\end{aligned}$$

5.3.6 Moments of the Normal Distribution

Let us now demonstrate that the parameters α and β of the normal distribution (4.3.8) are, in fact, its mean and its standard deviation. Making the substitution $y = \dfrac{x - \alpha}{\beta}$ in the integral for the mean, we get

$$\mu = \int_{-\infty}^{\infty} x \cdot \frac{1}{\beta\sqrt{2\pi}} e^{-\frac{1}{2}\left(\frac{x-\alpha}{\beta}\right)^2} dx$$

$$= \frac{1}{\sqrt{2\pi}} \cdot \int_{-\infty}^{\infty} (\beta y + \alpha) e^{-\frac{1}{2}y^2}\, dy$$

$$= \frac{\beta}{\sqrt{2\pi}} \cdot \int_{-\infty}^{\infty} y \cdot e^{-\frac{1}{2}y^2}\, dy + \alpha \left[\frac{1}{\sqrt{2\pi}} \cdot \int_{-\infty}^{\infty} e^{-\frac{1}{2}y^2}\, dy\right]$$

Since the first of these two integrals is 0 because the integrand is an *odd function* (that is, only its sign changes when we substitute $-y$ for y) and the quantity in brackets equals 1 because it is the integral from $-\infty$ to ∞ of a normal density with $\alpha = 0$ and $\beta = 1$, we get $\mu = 0 + \alpha \cdot 1 = \alpha$.

Using this result, we can now write the variance of the normal distribution with the parameters $\alpha = \mu$ and β as

$$\sigma^2 = \frac{1}{\beta\sqrt{2\pi}} \cdot \int_{-\infty}^{\infty} (x - \mu)^2 e^{-\frac{1}{2}\left(\frac{x-\mu}{\beta}\right)^2} dx$$

$$= \frac{\beta^2}{\sqrt{2\pi}} \cdot \int_{-\infty}^{\infty} y^2 e^{-\frac{1}{2}y^2}\, dy$$

$$= \frac{2\beta^2}{\sqrt{2\pi}} \cdot \int_{0}^{\infty} y^2 e^{-\frac{1}{2}y^2}\, dy$$

where we first made the substitution $y = \dfrac{x - \mu}{\beta}$ and then recognized the fact that the integrand has become an *even function* (that is, its values remain unchanged when we substitute $-y$ for y). Then, if we let $z = \frac{1}{2}y^2$, we get

$$\sigma^2 = \frac{2\beta^2}{\sqrt{\pi}} \cdot \int_{0}^{\infty} z^{\frac{1}{2}} e^{-z}\, dz = \frac{2\beta^2}{\sqrt{\pi}} \cdot \Gamma(\tfrac{3}{2}) = \frac{2\beta^2}{\sqrt{\pi}} \cdot \frac{\sqrt{\pi}}{2} = \beta^2$$

where we made use of the fact that $\Gamma(\frac{3}{2}) = \frac{1}{2} \cdot \Gamma(\frac{1}{2})$ and the result of Exercise 4 on page 116. The density of the normal distribution can thus be written as

$$N(x; \mu, \sigma) = \frac{1}{\sigma\sqrt{2\pi}} e^{-\frac{1}{2}\left(\frac{x-\mu}{\sigma}\right)^2} \quad \text{for } -\infty < x < \infty \qquad (5.3.5)$$

where we used a notation similar to that of (3.3.3), (3.3.5), etc., to indicate explicitly that the two parameters of the normal distribution are its mean μ and its standard deviation σ.

To conclude this discussion of the moments and the moment generating functions of special distribution, let us prove the following theorem:

THEOREM 5.17

The moment generating function of the normal distribution is given by $M_x(t) = e^{\mu t + \frac{1}{2}t^2\sigma^2}$.

Substituting (5.3.5) into (5.2.4b), we get

$$M_x(t) = \int_{-\infty}^{\infty} e^{xt} \cdot \frac{1}{\sigma\sqrt{2\pi}} e^{-\frac{1}{2}\left(\frac{x-\mu}{\sigma}\right)^2} dx$$

$$= \frac{1}{\sigma\sqrt{2\pi}} \cdot \int_{-\infty}^{\infty} e^{-\frac{1}{2\sigma^2}[-2xt\sigma^2 + (x-\mu)^2]} dx$$

and if we then *complete the square* in the exponent, that is, use the identity

$$-2xt\sigma^2 + (x-\mu)^2 = [x - (\mu + t\sigma^2)]^2 - 2\mu t\sigma^2 - t^2\sigma^4$$

we obtain

$$M_x(t) = e^{\mu t + \frac{1}{2}t^2\sigma^2} \left\{ \frac{1}{\sigma\sqrt{2\pi}} \cdot \int_{-\infty}^{\infty} e^{-\frac{1}{2}\left[\frac{x-(\mu+t\sigma^2)}{\sigma}\right]^2} dx \right\}$$

Since the quantity inside the braces is the integral from $-\infty$ to ∞ of a normal density with the mean $\mu + t\sigma^2$ and the standard deviation σ, and hence equal to 1, we finally get

$$M_x(t) = e^{\mu t + \frac{1}{2}t^2\sigma^2}$$

In particular, the moment generating function of the *standard normal distribution*, namely, the normal distribution which has $\mu = 0$ and $\sigma = 1$ (see Exercise 7 on page 153), is given by

$$M_z(t) = e^{\frac{1}{2}t^2} \tag{5.3.6}$$

where the values of **z** (a random variable having the *standard* normal distribution) are related to those of **x** (a random variable having a normal distribution with the mean μ and the standard deviation σ) by means of the equation

$$z = \frac{x - \mu}{\sigma} \tag{5.3.7}$$

Since the normal distribution plays a central role in statistics and its density cannot be integrated directly, areas under the standard normal curve have been tabulated and they are given in Table III. More specifically, the entries of Table III, represented by the shaded area of Figure 5.3, are the values of

$$\int_0^z \frac{1}{\sqrt{2\pi}} e^{-\frac{1}{2}x^2}\, dx$$

(namely, the probability that a random variable having the standard normal distribution takes on a value on the interval from 0 to z) for $z = 0.00, 0.01, 0.02, \ldots, 3.08$, and 3.09. By virtue of the *symmetry* of the normal distribution about its mean, it is unnecessary to extend Table III to negative values of z.

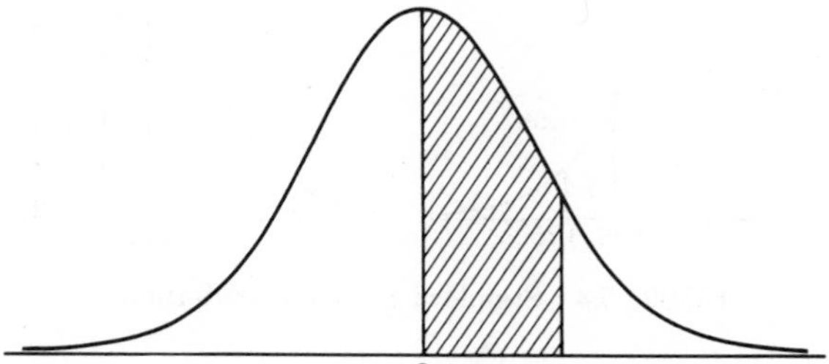

FIGURE 5.3 Area under the standard normal distribution.

To illustrate the use of Table III, let us find the probabilities that a random variable having the standard normal distribution will (1) take on a value less than 1.72, (2) take on a value less than −0.88, (3) take on a value between 1.30 and 1.75, and (4) take on a value between −0.25 and 0.45. To find the answer to (1) we look up the entry which corresponds to $z = 1.72$, add 0.5000 (see Figure 5.4), and get $0.4573 + 0.5000 = 0.9573$; to find the answer to (2) we look up the entry which corresponds to $z = 0.88$, subtract it from 0.5000 (see Figure 5.4), and get $0.5000 - 0.3106 = 0.1894$; to find the answer to (3) we look up the entries which correspond to $z = 1.75$ and $z = 1.30$, take their difference (see Figure 5.4), and get $0.4599 - 0.4032 = 0.0567$; and to find the answer to (4) we look up the entries which correspond to $z = 0.25$ and $z = 0.45$, add them (see Figure 5.4), and get $0.0987 + 0.1736 = 0.2723$.

If we want to use Table III to find probabilities relating to random variables having normal distributions which are not in standard form, we have only to use (5.3.7) and look up the entries for the corresponding z's. For instance, to find the probability that a random variable having the normal distribution with $\mu = 53$ and $\sigma = 0.8$ will take on a value greater than 55, we look up the entry which corresponds to $z = \frac{55 - 53}{0.8} = 2.50$, subtract from 0.5000 (see Figure 5.5), and get $0.5000 - 0.4938 = 0.0062$.

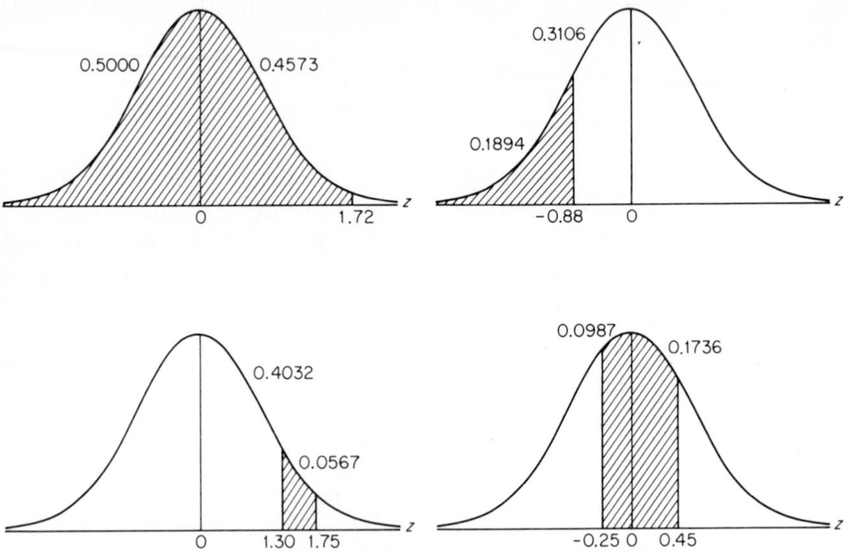

FIGURE 5.4 Standard normal distributions.

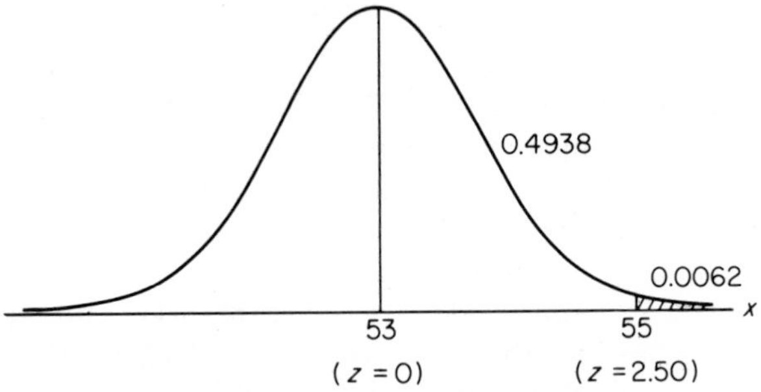

FIGURE 5.5 Normal distribution.

THEORETICAL EXERCISES

1. **THE DISCRETE UNIFORM DISTRIBUTION** Use the formulas for the sum of the first n positive integers and the sum of their squares (see Appendix) to show that the discrete uniform distribution (3.3.1) has
   (a) the mean $\mu = \dfrac{k+1}{2}$;
   (b) the variance $\sigma^2 = \dfrac{k^2-1}{12}$.

Also use these formulas to verify the results obtained in Exercise 3 on page 153.

2. **THE DISCRETE UNIFORM DISTRIBUTION** Show that if $\mathbf{x}$ is a random variable having the discrete uniform distribution (3.3.1), its moment generating function is given by

$$M_{\mathbf{x}}(t) = \frac{e^t(1 - e^{kt})}{k(1 - e^t)}$$

Also find the mean of this distribution by evaluating $M'_{\mathbf{x}}(t)$ as $t \to 0$, and compare the result with that of part (a) of Exercise 1. (*Hint:* Use l'Hospital's rule.)

3. **THE BERNOULLI DISTRIBUTION** This distribution was not studied separately in this chapter as it can be looked upon as a binomial distribution with $n = 1$. Show that $\mu'_r = \theta$ for the Bernoulli distribution for $r = 1, 2, 3, \ldots$
   (a) by substituting (3.3.2) into (5.2.1a);
   (b) by letting $n = 1$ in Theorem 5.9 and examining the Maclaurin's series of the moment generating function.

   Also show that for the Bernoulli distribution
   (c) $\alpha_3 = \dfrac{1 - 2\theta}{\sqrt{\theta(1 - \theta)}}$, where α_3 is the measure of skewness defined in Exercise 9 on page 154;
   (d) $\alpha_4 = \dfrac{1 - 3\theta(1 - \theta)}{\theta(1 - \theta)}$, where α_4 is the measure of peakedness defined in Exercise 10 on page 154.

4. In Sections (5.3.1) and (5.3.2) we found the variance by first evaluating $E[\mathbf{x}(\mathbf{x} - 1)]$, and if we had wanted to find μ_3 or μ_4 we would first have evaluated $E[\mathbf{x}(\mathbf{x} - 1)(\mathbf{x} - 2)]$ and $E[\mathbf{x}(\mathbf{x} - 1)(\mathbf{x} - 2)(\mathbf{x} - 3)]$. These quantities are referred to as *factorial moments,* and, in general, the *rth factorial moment* of the distribution of $\mathbf{x}$ is given by

$$\mu'_{(r)} = E[\mathbf{x}(\mathbf{x} - 1)(\mathbf{x} - 2)\ldots(\mathbf{x} - r + 1)]$$

   Express μ'_2, μ'_3, and μ'_4 in terms of factorial moments.

5. The *factorial moment generating function* of a discrete random variable $\mathbf{x}$ is given by the expected value of $t^{\mathbf{x}}$, namely, by

$$F_{\mathbf{x}}(t) = E(t^{\mathbf{x}}) = \Sigma\, t^x \cdot f(x)$$

   (a) Show that the rth derivative of $F_{\mathbf{x}}(t)$ with respect to t at $t = 1$ is $\mu'_{(r)}$, the rth factorial moment defined in Exercise 4.
   (b) Show that for the Bernoulli distribution (3.3.2) the factorial moment generating function is given by $F_{\mathbf{x}}(t) = 1 - \theta + \theta t$, and hence $\mu'_{(1)} = \theta$ and $\mu'_{(r)} = 0$ for $r > 1$.

(c) Find the factorial moment generating function of the binomial distribution and use it to find μ and σ^2.

(d) Find the factorial moment generating function of the Poisson distribution and use it to find μ and σ^2.

6. Use either the moment generating function of Theorem 5.9 or the factorial moment generating function of part (c) of Exercise 5 to show that for the binomial distribution $\alpha_3 = \frac{1 - 2\theta}{\sqrt{n\theta(1-\theta)}}$, where α_3 is as defined in Exercise 9 on page 154.

7. Use either the moment generating function of Theorem 5.12 or the factorial moment generating function of part (d) of Exercise 5 to show that for the Poisson distribution $\alpha_3 = \frac{1}{\sqrt{\lambda}}$, where α_3 is as defined in Exercise 9 on page 154.

8. Show that if we let $\theta = \frac{a}{a+b}$ and $N = a + b$, the mean and the variance of the hypergeometric distribution can be written as $\mu = n\theta$ and $\sigma^2 = n\theta(1-\theta)\cdot\frac{N-n}{N-1}$. How do these results tie in with the discussion on page 79?

9. Differentiating with respect to λ the expressions on both sides of the equation

$$\mu_r = \sum_{x=0}^{\infty} (x - \lambda)^r \cdot \frac{\lambda^x e^{-\lambda}}{x!}$$

derive the following recursion formula for the *moments about the mean* of the Poisson distribution:

$$\mu_{r+1} = \lambda\left[r\mu_{r-1} + \frac{d\mu_r}{d\lambda}\right]$$

for $r = 1, 2, 3, \ldots$. Use this recursion formula and the fact that $\mu_0 = 1$ and $\mu_1 = 0$ to find μ_2, μ_3, and μ_4. Also verify the result given for α_3 in Exercise 7.

10. In part (a) of Theorem 5.7 we expressed the moment generating function of $\mathbf{y}$ in terms of that of $\mathbf{x}$, if the values of $\mathbf{y}$ are related to those of $\mathbf{x}$ by means of the equation $y = x + a$. If $a = -\mu$, where μ is the mean of the distribution of $\mathbf{x}$, the result can be written as

$$M_{\mathbf{y}}(t) = M_{\mathbf{x}-\mu}(t) = e^{-\mu t}\cdot M_{\mathbf{x}}(t)$$

(a) Show that the rth derivative with respect to t of $M_{x-\mu}(t)$ at $t = 0$ gives the rth moment about the mean of the distribution of $\mathbf{x}$.

(b) Find such a generating function for moments about the mean of the Poisson distribution, and verify that the second derivative at $t = 0$ equals λ.

(c) Find such a generating function for moments about the mean of the binomial distribution, and verify that the second derivative at $t = 0$ equals $n\theta(1 - \theta)$.

11. **THE GEOMETRIC DISTRIBUTION** Differentiating with respect to θ the expressions on both sides of the equation

$$\sum_{x=1}^{\infty} \theta(1 - \theta)^{x-1} = 1$$

show that the mean of the geometric distribution (3.3.6) is given by $\mu = \frac{1}{\theta}\cdot$ Also, differentiating the expressions on both sides of the equation *twice* with respect to θ show that $\mu_2' = \frac{2 - \theta}{\theta^2}$, and hence $\sigma^2 = \frac{1 - \theta}{\theta^2}\cdot$

12. Show that the moment generating function of the geometric distribution (3.3.6) is given by

$$M_{\mathbf{x}}(t) = \frac{\theta e^t}{1 - e^t(1 - \theta)}$$

and use it to verify that $\mu = \frac{1}{\theta}$ and $\sigma^2 = \frac{1 - \theta}{\theta^2}\cdot$

13. **THE NEGATIVE BINOMIAL DISTRIBUTION** Using steps similar to those employed in the beginning of Section (5.3.1), show that the mean of the negative binomial distribution (3.3.7) is $\mu = \frac{k}{\theta}\cdot$ Also show that the variance of the negative binomial distribution is $\sigma^2 = \frac{k}{\theta}\left(\frac{1}{\theta} - 1\right)$ by first evaluating $E[\mathbf{x}(\mathbf{x} + 1)]$.

14. Find the mean and the variance of the negative binomial distribution by the method of Exercise 11.

15. **THE UNIFORM DISTRIBUTION** Show that for the uniform distribution (4.3.1) the mean is $\mu = \frac{\alpha + \beta}{2}$ and the variance is

$$\sigma^2 = \tfrac{1}{12}(\beta - \alpha)^2$$

16. A random variable is said to have the *Rayleigh distribution* if its probability density is given by

$$f(x) = \begin{cases} 2\alpha x e^{-\alpha x^2} & \text{for } x > 0 \\ 0 & \text{for } x \le 0 \end{cases}$$

where $\alpha > 0$.

(a) Show that $\mu = \frac{1}{2}\sqrt{\frac{\pi}{\alpha}}$ by evaluating the required integral, making the substitution $u = \alpha x^2$.

(b) Show that $\sigma^2 = \frac{1}{\alpha}\left(1 - \frac{\pi}{4}\right)$.

17. Verify that $\mu = \beta\alpha$ and $\sigma^2 = \beta^2\alpha$ for the gamma distribution (4.3.6) by differentiation of its moment generating function given in Theorem 5.15.

18. Use Theorem 5.16 to show that for the beta distribution (4.3.7) the parameters α and β can be expressed as follows in terms of the mean and the variance of the distribution:

(a) $\alpha = \mu\left[\frac{\mu(1-\mu)}{\sigma^2} - 1\right]$;

(b) $\beta = (1-\mu)\left[\frac{\mu(1-\mu)}{\sigma^2} - 1\right]$.

19. Twice differentiating the moment generating function of Theorem 5.17, verify that the parameters μ and σ^2 are, indeed, the mean and the variance of the normal distribution.

20. Explain why α_3 and α_4 (defined in Exercises 9 and 10 on page 154) of the distribution of a random variable equal the third and fourth moments about the origin of the corresponding *standardized* random variable (see Exercise 7 on page 153). Then use (5.3.6) to show that for a normal distribution $\alpha_3 = 0$ and $\alpha_4 = 3$.

21. Show that for the normal distribution $M_{\mathbf{x}-\mu}(t) = e^{\frac{1}{2}t^2\sigma^2}$, where $M_{\mathbf{x}-\mu}(t)$ is as defined in Exercise 10.

22. If we let $K_{\mathbf{x}}(t) = \ln M_{\mathbf{x}-\mu}(t)$, where $M_{\mathbf{x}-\mu}(t)$ is as defined in Exercise 10, the coefficient of $\frac{t^r}{r!}$ in the Maclaurin's series of $K_{\mathbf{x}}(t)$ is called the *rth cumulant* and it is denoted κ_r. Equating coefficients of like powers, show that

(a) $\kappa_2 = \mu_2$;

(b) $\kappa_3 = \mu_3$;

(c) $\kappa_4 = \mu_4 - 3\mu_2^2$;

(d) $\kappa_5 = \mu_5 - 10\mu_3\mu_2$.

Also, use the generating function for moments about the mean of Exercise 21 to show that for a normal distribution $\kappa_2 = \sigma^2$ while all other cumulants are zero.

APPLIED EXERCISES

23. Verify the statement made on page 149 that according to Chebyshev's theorem the probability of getting at least 65 heads or at most 35 heads in 100 flips of a balanced coin is less than or equal to $\frac{1}{9}$.

24. Use Chebyshev's theorem to verify that the probability is at least $\frac{35}{36}$ that

(a) in 400 flips of a balanced coin the proportion of heads will differ from 0.50 by less than 0.15;

(b) in 10,000 flips of a balanced coin the proportion of heads will differ from 0.50 by less than 0.03;

(c) in 1,000,000 flips of a balanced coin the proportion of heads will differ from 0.50 by less than 0.003.

25. If the number of policies an insurance salesman sells during a month is a random variable having the Poisson distribution with $\lambda = 16$, with what probability can we assert, according to Chebyshev's theorem, that he will sell at least 26 or at most 6 policies during any given month?

26. If $\mathbf{z}$ is a random variable having the standard normal distribution, use Table III to find

(a) the probability that $\mathbf{z}$ takes on a value greater than 1.04;

(b) the probability that $\mathbf{z}$ takes on a value less than 1.38;

(c) the probability that $\mathbf{z}$ takes on a value greater than -1.55;

(d) the probability that $\mathbf{z}$ takes on a value less than -0.05;

(e) the probability that $\mathbf{z}$ takes on a value between 0.65 and 1.65;

(f) the probability that $\mathbf{z}$ takes on a value between -0.48 and -1.22;

(g) the probability that $\mathbf{z}$ takes on a value between -2.40 and 1.95.

27. If $\mathbf{z}$ is a random variable having the standard normal distribution, use Table III to find its value z for which

(a) the probability that $\mathbf{z}$ will take on a value less than z is 0.9881;

(b) the probability that $\mathbf{z}$ will take on a value greater than z is 0.9099;

(c) the probability that $\mathbf{z}$ will take on a value less than z is 0.2676;

(d) the probability that $\mathbf{z}$ will take on a value greater than z is 0.1056.

28. If $\mathbf{z}$ is a random variable having the standard normal distribution, use Table III to find its value z for which

(a) the probability that $\mathbf{z}$ will take on a value greater than z is 0.05;

(b) the probability that $\mathbf{z}$ will take on a value greater than z is 0.01;

(c) the probability that $\mathbf{z}$ will take on a value between $-z$ and z is 0.95;

(d) the probability that $\mathbf{z}$ will take on a value between $-z$ and z is 0.99.

29. If the amount of instant coffee which a filling machine puts into a 6-ounce can is a random variable having a normal distribution with $\mu = 6.00$ ounces and $\sigma = 0.08$ ounces, find the probabilities that

(a) one of these cans will contain at least 6.10 ounces;

(b) one of these cans will contain less than 5.82 ounces;

(c) one of these cans will contain anywhere from 5.96 to 6.04 ounces.

30. The burning time of an experimental rocket is a random variable having a normal distribution with $\mu = 5.24$ seconds and $\sigma = 0.04$ seconds. Find the probabilities that such a rocket will (a) burn less than 5.18 seconds, (b) burn at least 5.22 seconds, and (c) burn anywhere from 5.23 to 5.25 seconds.

31. The time required to perform a certain job is a random variable having a normal distribution with a mean of 72 minutes and a standard deviation of 12 minutes. What is the probability that
(a) the job will take more than 93 minutes;
(b) the job will take less than 80 minutes;
(c) the job will take anywhere from 63 to 78 minutes?

32. The lifetime of a certain kind of energy cell is a random variable having a normal distribution with a standard deviation of 25 hours. Find the mean of this distribution if the probability that any one of these energy cells will last over 400 hours is 0.10.

5.4 MOMENT GENERATING FUNCTIONS AND LIMITING DISTRIBUTIONS

In Section (3.3.7) we showed that when $n \to \infty$, $\theta \to 0$, and $n\theta = \lambda$ remains constant, the binomial distribution approaches the Poisson distribution. In this section we shall demonstrate that under the same limiting conditions the moment generating function of the binomial distribution approaches that of the Poisson distribution. Actually, this requires only one step, for if we substitute $\frac{\lambda}{n}$ for θ and make use of the fact that $\lim_{n\to\infty}\left(1 + \frac{x}{n}\right)^n = e^x$, with which we are familiar from elementary calculus, we get

$$M_{\mathbf{x}}(t) = [1 + \theta(e^t - 1)]^n = \left[1 + \frac{\lambda(e^t - 1)}{n}\right]^n$$

and, hence,

$$\lim_{n\to\infty} M_{\mathbf{x}}(t) = e^{\lambda(e^t-1)}$$

The fact that the moment generating function of the binomial distribution approaches that of the Poisson distribution under the given limiting condition does not by itself constitute proof that the binomial distribution approaches the Poisson distribution under these conditions. To this end we need two theorems which we shall state without proof:

(1) *There is a one-to-one correspondence between moment generating functions and probability distributions when the former exist.*

(2) *If the moment generating function of the distribution of one random variable approaches that of the distribution of another random variable, then the distribution of the first random variable approaches that of the second under the same limiting conditions.*

Although the proof of Section (3.3.7) and the one which we gave here applies when $n \to \infty$ and $\theta \to 0$ while $n\theta$ remains constant, we often approximate binomial probabilities with Poisson probabilities so long as n is "fairly large" and θ is "fairly small." An acceptable rule of thumb is to use this approximation if $n \geq 20$ and $\theta \leq 0.05$; if $n \geq 100$, the approximation is generally excellent so long as $n\theta \leq 10$.

When n is large and θ is not necessarily small, we often use the normal distribution with the mean $\mu = n\theta$ and the variance $\sigma^2 = n\theta(1 - \theta)$ to approximate binomial probabilities. This is justified by the following theorem:

THEOREM 5.18

If $\mathbf{x}$ *is a random variable having the binomial distribution (3.3.3), then the moment generating function of the random variable* $\mathbf{z}$, *whose values are related to those of* $\mathbf{x}$ *by means of the equation* $z = \dfrac{x - n\theta}{\sqrt{n\theta(1 - \theta)}}$, *approaches that of the standard normal distribution when* $n \to \infty$.

Making use of Theorem 5.9 and part (c) of Theorem 5.7 on page 152, we have

$$M_{\mathbf{z}}(t) = M_{\frac{\mathbf{x}-\mu}{\sigma}}(t) = e^{-\frac{\mu t}{\sigma}} \cdot [1 + \theta(e^{\frac{t}{\sigma}} - 1)]^n$$

where $\mu = n\theta$ and $\sigma = \sqrt{n\theta(1 - \theta)}$. Then, if we take the logarithm of the expressions on both sides of this equation and substitute its Maclaurin's series for $e^{\frac{t}{\sigma}}$, we get

$$\begin{aligned} \ln M_{\frac{\mathbf{x}-\mu}{\sigma}}(t) &= -\frac{\mu t}{\sigma} + n \cdot \ln\,[1 + \theta(e^{\frac{t}{\sigma}} - 1)] \\ &= -\frac{\mu t}{\sigma} + n \cdot \ln\left[1 + \theta\left\{\frac{t}{\sigma} + \frac{1}{2}\left(\frac{t}{\sigma}\right)^2 + \frac{1}{6}\left(\frac{t}{\sigma}\right)^3 + \cdots\right\}\right] \end{aligned}$$

Using the infinite series $\ln(1 + x) = x - \frac{1}{2}x^2 + \frac{1}{3}x^3 - \cdots$, which converges for $|x| < 1$, to expand this logarithm, we get

$$\ln M_{\frac{X-\mu}{\sigma}}(t) = -\frac{\mu t}{\sigma} + n\theta\left[\frac{t}{\sigma} + \frac{1}{2}\left(\frac{t}{\sigma}\right)^2 + \frac{1}{6}\left(\frac{t}{\sigma}\right)^3 + \cdots\right]$$
$$- \frac{n\theta^2}{2}\left[\frac{t}{\sigma} + \frac{1}{2}\left(\frac{t}{\sigma}\right)^2 + \frac{1}{6}\left(\frac{t}{\sigma}\right)^3 + \cdots\right]^2$$
$$+ \frac{n\theta^3}{3}\left[\frac{t}{\sigma} + \frac{1}{2}\left(\frac{t}{\sigma}\right)^2 + \frac{1}{6}\left(\frac{t}{\sigma}\right)^3 + \cdots\right]^3 - \cdots$$

and collecting powers of t, we obtain

$$\ln M_{\frac{X-\mu}{\sigma}}(t) = \left(-\frac{\mu}{\sigma} + \frac{n\theta}{\sigma}\right)t + \left(\frac{n\theta}{2\sigma^2} - \frac{n\theta^2}{2\sigma^2}\right)t^2$$
$$+ \left(\frac{n\theta}{6\sigma^3} - \frac{n\theta^2}{2\sigma^3} + \frac{n\theta^3}{3\sigma^3}\right)t^3 + \cdots$$
$$= \frac{1}{\sigma^2}\left(\frac{n\theta - n\theta^2}{2}\right)t^2 + \frac{n}{\sigma^3}\left(\frac{\theta - 3\theta^2 + 2\theta^3}{6}\right)t^3 + \cdots$$

since $\mu = n\theta$. Finally, substituting $\sigma = \sqrt{n\theta(1-\theta)}$, we get

$$\ln M_{\frac{X-\mu}{\sigma}}(t) = \frac{1}{2}t^2 + \frac{n}{\sigma^3}\left(\frac{\theta - 3\theta^2 + 2\theta^3}{6}\right)t^3 + \cdots$$

where for $r > 2$ the coefficient of t^r is a constant times $\frac{n}{\sigma^r}$, which approaches 0 when $n \to \infty$. Thus, we obtain

$$\lim_{n\to\infty} \ln M_{\frac{X-\mu}{\sigma}}(t) = \tfrac{1}{2}t^2 \quad \text{and hence} \quad \lim_{n\to\infty} M_{\frac{X-\mu}{\sigma}}(t) = e^{\frac{1}{2}t^2}$$

since the limit of a logarithm equals the logarithm of the limit (provided these limits exist). This completes the proof of Theorem 5.18, and if we refer to the two theorems mentioned on page 175, we can conclude that when $n \to \infty$ the *standardized* binomial distribution approaches the standard normal distribution. This result can also be obtained directly, without the use of moment generating functions, and a reference to such a proof is given at the end of this chapter. A third proof, based on the *central limit theorem*, will be given in Chapter 6.

Although, strictly speaking, Theorem 5.18 applies when $n \to \infty$, normal distributions are often used to approximate binomial distribution even when n is relatively small. To be more specific, a good rule of thumb is to use this approximation only when $n\theta$ and $n(1 - \theta)$ are *both* greater than 5. Also, this approximation requires that 0 is represented by the interval from $-\frac{1}{2}$ to $\frac{1}{2}$, 1 is represented by the interval from $\frac{1}{2}$ to $1\frac{1}{2}, \ldots,$ and in general the positive integer k is represented by the interval from $k - \frac{1}{2}$ to $k + \frac{1}{2}$. (In this way the values of a *discrete* binomial random variable are "spread" over a continuous scale.) Thus, if we wanted to approximate the binomial probability of getting eight heads in 20 flips of a balanced coin, we would have to find the probability that a random variable having a normal distribution with $\mu = 20 \cdot \frac{1}{2} = 10$ and $\sigma = \sqrt{20 \cdot \frac{1}{2} \cdot \frac{1}{2}} = 2.236$ takes on a value between 7.5 and 8.5. Similarly, if we wanted to approximate the probability of getting fewer than 35 sixes in 240 rolls of a balanced die, we would have to find the probability that a random variable having a normal distribution with $\mu = 240 \cdot \frac{1}{6} = 40$ and $\sigma = \sqrt{240 \cdot \frac{1}{6} \cdot \frac{5}{6}} = 5.77$ takes on a value less than 34.5.

THEORETICAL EXERCISES

1. Use Theorem 5.12 and part (c) of Theorem 5.7 on page 152 to show that when $\lambda \to \infty$ the moment generating function of the *standardized* Poisson distribution approaches that of the standard normal distribution.
2. Use Theorem 5.14 and part (c) of Theorem 5.7 on page 152 to show that when $\alpha \to \infty$ and β remains constant the moment generating function of the *standardized* gamma distribution approaches that of the standard normal distribution.

APPLIED EXERCISES

3. Work the two problems suggested in the last paragraph of this section, that is, use the normal distribution to approximate
   (a) the probability of obtaining 8 heads in 20 flips of a balanced coin;
   (b) the probability of getting fewer than 35 sixes in 240 rolls of a balanced die.

   Also, compare the result of part (a) with the corresponding exact probability.
4. If 60 percent of the customers of a large department store charge all of their purchases, what is the probability that among 200 customers (randomly chosen) at least 125 charge all of their purchases?

5. A true-false test contains 80 questions. What is the probability that a student who answers each question by flipping a coin (heads is "true" and tails is "false") will get anywhere from 40 to 50 correct answers?
6. A manufacturer knows that on the average 2 percent of his products are defective. Use the normal curve approximation to find the probabilities that in a lot of 400 there will be
   (a) exactly 10 defectives;
   (b) at least 10 defectives.
7. A television station claims that its Monday night movie regularly has 35 percent of the total viewing audience. If this claim is correct, what is the probability that among 100 viewers reached by phone on a Monday night fewer than 25 were watching the movie?
8. Rework Exercise 25 on page 173 using the normal approximation of the Poisson distribution (see Exercise 1) instead of Chebyshev's theorem.

5.5 PRODUCT MOMENTS

Given a pair of random variables $\mathbf{x}$ and $\mathbf{y}$, the *rth and sth product moment about the origin* of their joint distribution is given by

$$\mu'_{r,s} = E(\mathbf{x}^r\mathbf{y}^s) = \Sigma\,\Sigma\, x^r y^s f(x, y) \tag{5.5.1a}$$

when the two random variables are discrete, and by

$$\mu'_{r,s} = \int_{-\infty}^{\infty}\int_{-\infty}^{\infty} x^r y^s f(x, y)\, dx\, dy \tag{5.5.1b}$$

when they are continuous. [The double summation in (5.5.1a) extends over the entire range of the two random variables.] Analogous to (5.2.2a) and (5.2.2b) we also define the *rth and sth product moment about the respective means* as $\mu_{r,s} = E[(\mathbf{x} - \mu'_{1,0})^r(\mathbf{y} - \mu'_{0,1})^s]$, where $\mu'_{1,0}$ and $\mu'_{0,1}$ are, respectively, the means of the marginal distributions of $\mathbf{x}$ and $\mathbf{y}$. In statistics, $\mu_{1,1}$ is of special interest, and it is called the *covariance* of the joint distribution of $\mathbf{x}$ and $\mathbf{y}$; symbolically, it is also denoted $\sigma_{\mathbf{xy}}$, $\text{cov}(\mathbf{x}, \mathbf{y})$, or $C(\mathbf{x}, \mathbf{y})$.

As in the case of "ordinary" moments, product moments about the respective means are usually expressed (and evaluated) in terms of product moments about the origin. For instance, for the covariance we have

$$\begin{aligned}\sigma_{\mathbf{xy}} &= \mu'_{1,1} - (\mu'_{1,0})(\mu'_{0,1}) \\ &= E(\mathbf{xy}) - E(\mathbf{x})\cdot E(\mathbf{y})\end{aligned} \tag{5.5.2}$$

which the reader will be asked to verify in Exercise 1 on page 180. To illustrate, let us determine the covariance of the joint distribution of Exercise 1 on page 136, where we had

$$f(x, y) = \begin{cases} 2 & \text{for } x > 0,\ y > 0, \text{ and } x + y < 1 \\ 0 & \text{elsewhere} \end{cases}$$

Evaluating the necessary integrals, we get

$$E(\mathbf{xy}) = \int_0^1 \int_0^{1-x} 2xy\,dy\,dx = \tfrac{1}{12}, \qquad E(\mathbf{x}) = \int_0^1 \int_0^{1-x} 2x\,dy\,dx = \tfrac{1}{3}$$

and

$$E(\mathbf{y}) = \int_0^1 \int_0^{1-x} 2y\,dy\,dx = \tfrac{1}{3}$$

so that (5.5.2) yields $\sigma_{\mathbf{xy}} = \frac{1}{12} - \frac{1}{3}\cdot\frac{1}{3} = -\frac{1}{36}$.

In a sense, the covariance is a measure of the relationship, or association, between the values of **x** and **y**. If there is a high probability that large values of **x** will go with large values of **y** and small values of **x** with small values of **y**, the covariance will be *positive;* if there is a high probability that large values of **x** will go with small values of **y** and vice versa, the covariance will be *negative* (as in our example). If **x** and **y** are independent, the covariance is zero, for in that case

$$E(\mathbf{xy}) = \int_{-\infty}^{\infty} \int_{-\infty}^{\infty} xyf(x, y)\,dx\,dy = \int_{-\infty}^{\infty} xg(x)\,dx \cdot \int_{-\infty}^{\infty} yh(y)\,dy = E(\mathbf{x})\cdot E(\mathbf{y})$$

and substitution into (5.5.2) yields $\sigma_{\mathbf{xy}} = E(\mathbf{x})\cdot E(\mathbf{y}) - E(\mathbf{x})\cdot E(\mathbf{y}) = 0$. Note that we made use of the fact that for independent random variables the values $f(x, y)$ of their joint density equal the product of the corresponding values $g(x)$ and $h(y)$ of their respective marginal densities.

It is of interest to note that the independence of two random variables implies a zero covariance, but that *a zero covariance does not necessarily imply their independence.* This is illustrated by the following example, where the probabilities in the table pertain to the corresponding pairs of values of the discrete random variables **x** and **y**:

		Values of **x**		
		−1	0	+1
	−1	$\frac{1}{6}$	$\frac{1}{3}$	$\frac{1}{6}$
Values of **y**	0	0	0	0
	+1	$\frac{1}{6}$	0	$\frac{1}{6}$

Evidently, $E(\mathbf{xy}) = (-1)(-1)\frac{1}{6} + (0)(-1)\frac{1}{3} + (+1)(-1)\frac{1}{6} + (-1)(+1)\frac{1}{6} + (+1)(+1)\frac{1}{6} = 0$, $E(\mathbf{x}) = (-1)\frac{1}{3} + (0)\frac{1}{3} + (+1)\frac{1}{3} = 0$, $E(\mathbf{y}) = (-1)\frac{2}{3} + (0)0 + (+1)\frac{1}{3} = -\frac{1}{3}$, so that

$$\sigma_{\mathbf{xy}} = 0 - 0(-\tfrac{1}{3}) = 0$$

yet the random variables are *not independent*, for the conditional distribution of $\mathbf{x}$ given that $\mathbf{y}$ takes on the value -1 is not the same as the conditional distribution of $\mathbf{x}$ given that $\mathbf{y}$ takes on the value $+1$.

THEORETICAL EXERCISES

1. Prove (5.5.2) for the case where $\mathbf{x}$ and $\mathbf{y}$ are continuous random variables.
2. Find the covariance of the joint probability distribution of Exercise 1 on page 96.
3. Use the marginal densities on page 132 to find
   (a) $\text{cov}(\mathbf{x}_1, \mathbf{x}_2)$;
   (b) $\text{cov}(\mathbf{x}_1, \mathbf{x}_3)$.
4. Find the covariance of the joint distribution of Exercise 14 on page 139.
5. Given the joint probability function of two random variables $\mathbf{x}$ and $\mathbf{y}$, whose values $f(x, y)$ are $f(-1, 0) = 0$, $f(-1, 1) = \frac{1}{4}$, $f(0, 0) = \frac{1}{6}$, $f(0, 1) = 0$, $f(1, 0) = \frac{1}{12}$, and $f(1, 1) = \frac{1}{2}$,
   (a) check whether the two random variables are independent;
   (b) find $\text{cov}(\mathbf{x}, \mathbf{y})$.
6. For k random variables $\mathbf{x}_1, \mathbf{x}_2, \ldots, \mathbf{x}_k$, the values of the *joint moment generating function* of their joint distribution are given by

$$E(e^{t_1\mathbf{x}_1 + t_2\mathbf{x}_2 + \cdots + t_k\mathbf{x}_k})$$

   (a) Show for either the discrete case or the continuous case that the partial derivative of the joint moment generating function with respect to t_i at $t_1 = t_2 = \cdots = t_k = 0$ is $E(\mathbf{x}_i)$.
   (b) Show for either the discrete case or the continuous case that the second partial derivative of the joint moment generating function with respect to t_i and t_j at $t_1 = t_2 = \cdots = t_k = 0$ is $E(\mathbf{x}_i\mathbf{x}_j)$.
   (c) Show that the joint moment generating function of the multinomial distribution (3.4.12) is

$$(\theta_1 e^{t_1} + \theta_2 e^{t_2} + \cdots + \theta_k e^{t_k})^n$$

   and, hence, that for $i \neq j$

$$\text{cov}(\mathbf{x}_i, \mathbf{x}_j) = -n\theta_i\theta_j$$

REFERENCES

A discussion of the moment generating function of the beta distribution may be found in

Kendall, M. G., and Stuart, A., *The Advanced Theory of Statistics*. London: Charles Griffin & Co., 1958 and 1961.

A direct proof that the standardized binomial distribution approaches the standard normal distribution when $n \to \infty$ is given in

Keeping, E. S., *Introduction to Statistical Inference*. Princeton, N.J.: D. Van Nostrand Co., Inc., 1962.

6

SUMS OF RANDOM VARIABLES

6.1 SUMS OF RANDOM VARIABLES

In this chapter we shall be concerned with *sums* and *linear combinations* of random variables; that is, given a set of random variables $\mathbf{x}_1$, $\mathbf{x}_2$, ..., and $\mathbf{x}_n$, we shall be concerned with random variables such as $\mathbf{y}$ and $\mathbf{z}$, whose values are related to those of the $\mathbf{x}$'s by means of the equations $y = \sum_{i=1}^{n} x_i$ and $z = \sum_{i=1}^{n} a_i x_i$, where the a's are constants. This is what we mean when we write $\mathbf{y} = \mathbf{x}_1 + \mathbf{x}_2 + \cdots + \mathbf{x}_n$ and $\mathbf{z} = a_1\mathbf{x}_1 + a_2\mathbf{x}_2 + \cdots + a_n\mathbf{x}_n$.

Leaving applications to later chapters, we shall be concerned here

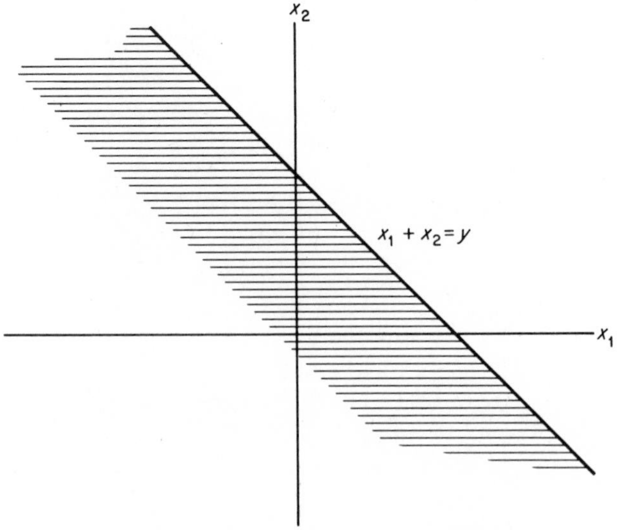

FIGURE 6.1 Sum of two random variables.

mainly with the problem of finding the distributions of sums of random variables given their joint distribution, and the problem of expressing some of the moments of linear combinations of random variables in terms of the moments and product moments of their joint distribution.

To illustrate a straightforward method of obtaining the probability density of the sum of two random variables, let us refer to Figures 6.1 and 6.2. In the case of *two* random variables, the probability $F(y)$ that

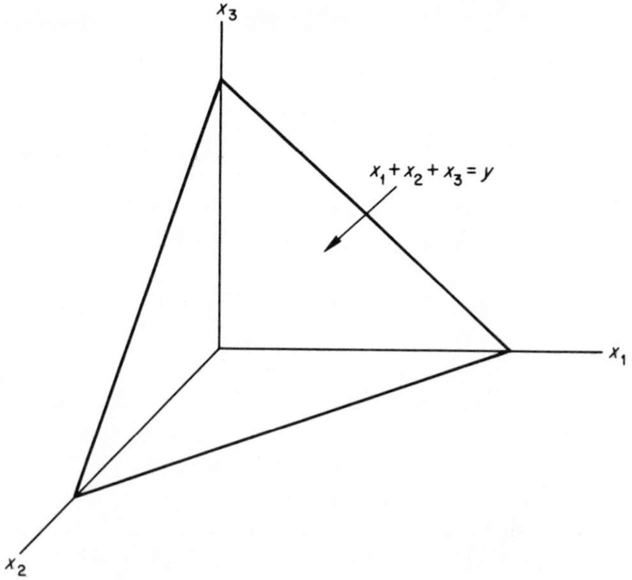

FIGURE 6.2 Sum of three random variables.

$\mathbf{y} = \mathbf{x}_1 + \mathbf{x}_2$ will take on a value less than or equal to y can be obtained directly by integrating the joint probability density of $\mathbf{x}_1$ and $\mathbf{x}_2$ over the shaded region of Figure 6.1. Then, the probability density of $\mathbf{y}$ is obtained by making use of the fact that $f(y) = F'(y)$. In the case of *three* random variables, the probability $F(y)$ that $\mathbf{y} = \mathbf{x}_1 + \mathbf{x}_2 + \mathbf{x}_3$ will take on a value less than or equal to y can be obtained directly by integrating the joint density of $\mathbf{x}_1$, $\mathbf{x}_2$, and $\mathbf{x}_3$ over the three-dimensional region which lies below the plane indicated in Figure 6.2. As before, the probability density of $\mathbf{y}$ is then obtained by differentiation. In the case of n random variables, the general procedure is the same, but the region over which we have to integrate the joint probability density cannot be pictured when $n > 3$.

To give an example, let us consider the random variables $\mathbf{x}_1$ and $\mathbf{x}_2$

whose joint density is

$$f(x_1, x_2) = \begin{cases} 6e^{-3x_1-2x_2} & \text{for } x_1 > 0 \text{ and } x_2 > 0 \\ 0 & \text{elsewhere} \end{cases}$$

If we integrate this joint density over the shaded region of Figure 6.1, we get

$$F(y) = \int_0^y \int_0^{y-x_2} 6e^{-3x_1-2x_2}\,dx_1\,dx_2 = 1 + 2e^{-3y} - 3e^{-2y}$$

for $y > 0$ and $F(y) = 0$ for $y \leq 0$, and if we then differentiate with respect to x, we obtain

$$f(y) = \begin{cases} 6(e^{-2y} - e^{-3y}) & \text{for } y > 0 \\ 0 & \text{for } y \leq 0 \end{cases}$$

The situation would have been more complicated if we had considered the random variables $\mathbf{x}_1$ and $\mathbf{x}_2$ whose joint density is

$$f(x_1, x_2) = \begin{cases} 1 & \text{for } 0 < x_1 < 1 \text{ and } 0 < x_2 < 1 \\ 0 & \text{elsewhere} \end{cases}$$

As can be seen from Figure 6.3, we now have to consider separately the cases where $y \geq 2$, $1 < y < 2$, $0 < y < 1$, and $y \leq 0$, and the details of this will be left to the reader in Exercise 3 on page 190.

In the next two sections we shall introduce two general methods of finding the distributions of sums of random variables, of which the first has the same theoretical foundation as the method which we have just discussed. The method of Section 6.1.2, based on moment generating functions, has many important applications, but it is limited to the case where the random variables $\mathbf{x}_1, \mathbf{x}_2, \ldots,$ and $\mathbf{x}_n$ are independent.

6.1.1 Sums of Random Variables: Convolutions

The method of this section is based, essentially, on the work of Section 4.4, namely, a change of variable. Beginning with two random variables $\mathbf{x}_1$ and $\mathbf{x}_2$, whose joint density is given by $f(x_1, x_2)$, we can write the values of the joint density of $\mathbf{y} = \mathbf{x}_1 + \mathbf{x}_2$ and $\mathbf{x}_2$ as

$$g(y, x_2) = f(x_1, x_2) \cdot \left| \frac{dx_1}{dy} \right| \tag{6.1.1}$$

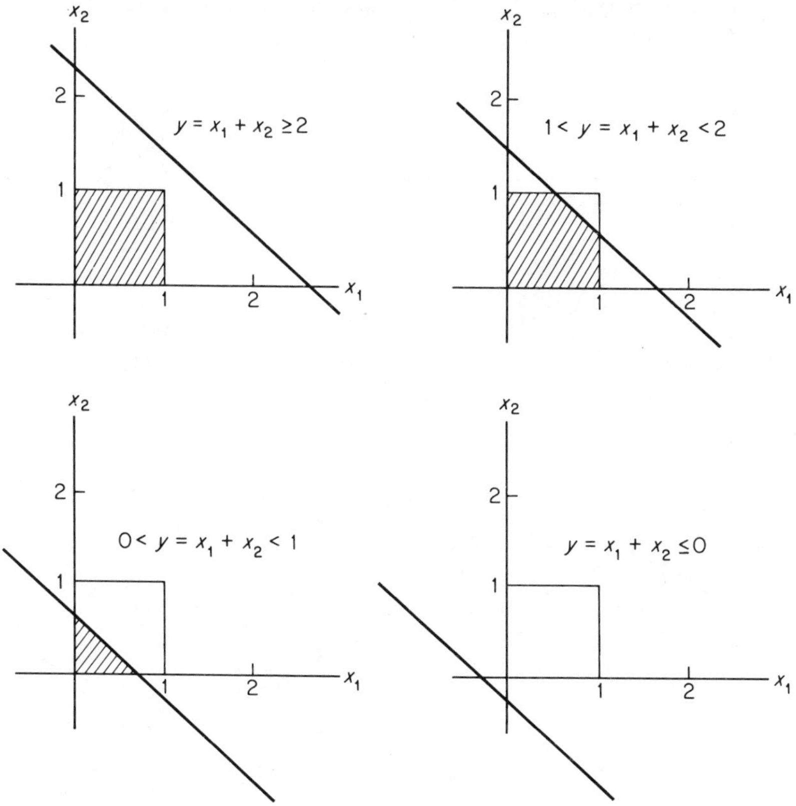

FIGURE 6.3 Diagrams for Exercise 3 on page 190.

in accordance with (4.4.1). Note that we are, in fact, changing variable from $\mathbf{x}_1$ to $\mathbf{y}$ treating x_2 as a constant in $y = x_1 + x_2$, so that $\left|\frac{dx_1}{dy}\right| = 1$ and (6.1.1) becomes

$$g(y, x_2) = f(x_1, x_2) = f(y - x_2, x_2)$$

Finally, integrating out x_2 from $-\infty$ to ∞, we find that the marginal density of $\mathbf{y} = \mathbf{x}_1 + \mathbf{x}_2$ is given by

$$h(y) = \int_{-\infty}^{\infty} f(y - x_2, x_2)\, dx_2 \tag{6.1.2}$$

In particular, when $\mathbf{x}_1$ and $\mathbf{x}_2$ are *independent* and $f(x_1, x_2) = f_1(x_1)\cdot f_2(x_2)$, (6.1.2) becomes

$$h(y) = \int_{-\infty}^{\infty} f_1(y - x_2) f_2(x_2)\, dx_2 \tag{6.1.3}$$

The integrals in (6.1.2) and (6.1.3) are called *convolution integrals* or simply *convolutions*, and it should be noted that if we reverse the roles played by $\mathbf{x}_1$ and $\mathbf{x}_2$, we also have

$$h(y) = \int_{-\infty}^{\infty} f(x_1, y - x_1)\, dx_1 \tag{6.1.4}$$

analogous to (6.1.2), and

$$h(y) = \int_{-\infty}^{\infty} f_1(x_1) f_2(y - x_1)\, dx_1 \tag{6.1.5}$$

analogous to (6.1.3) when $\mathbf{x}_1$ and $\mathbf{x}_2$ are independent.

To illustrate the use of a convolution integral in determining the probability density of the sum of two random variables, suppose that

$$f(x_1, x_2) = \frac{1}{\pi^2} \cdot \frac{1}{1 + x_1^2} \cdot \frac{1}{1 + x_2^2} \qquad \text{for } -\infty < x_1 < \infty \text{ and } -\infty < x_2 < \infty$$

namely, that $\mathbf{x}_1$ and $\mathbf{x}_2$ are independent random variables having the same *Cauchy distribution* of Exercise 11 on page 117. Substituting into (6.1.2) and integrating by the method of *partial fractions*, we get

$$\begin{aligned} f(y) &= \frac{1}{\pi^2} \cdot \int_{-\infty}^{\infty} \frac{1}{1 + (y - x_2)^2} \cdot \frac{1}{1 + x_2^2}\, dx_2 \\ &= \frac{1}{\pi} \cdot \frac{2}{4 + y^2} \qquad \text{for } -\infty < y < \infty \end{aligned}$$

which is interesting, as it shows that the distribution of the sum of two independent random variables having the same Cauchy distribution is *also* a Cauchy distribution.

To consider another example, suppose that $\mathbf{x}_1$ and $\mathbf{x}_2$ have the joint density given in the middle of page 184. Since

$$f(y - x_2, x_2) = \begin{cases} 1 & \text{for } 0 < x_2 < 1,\ y = x_1 + x_2 > x_2, \text{ and} \\ & x_1 = y - x_2 < 1 \text{ (which is shaded in Figure 6.4)} \\ 0 & \text{elsewhere} \end{cases}$$

substitution into (6.1.2) yields

$$h(y) = \begin{cases} 0 & \text{for } y \le 0 \\ \int_0^y 1\, dx_2 = y & \text{for } 0 < y < 1 \\ \int_{y-1}^1 1\, dx_2 = 2 - y & \text{for } 1 < y < 2 \\ 0 & \text{for } y \ge 2 \end{cases}$$

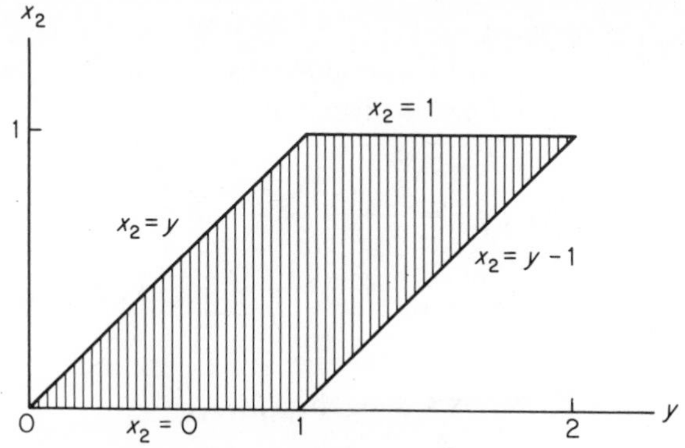

FIGURE 6.4 Sample space for x_2 and y.

and to make the function continuous, we shall let $h(1) = 1$. We have thus shown that the sum of the given random variables [which are, in fact, independent random variables having the uniform density (4.3.1) with $\alpha = 0$ and $\beta = 1$] has the *triangular probability density*, whose graph is shown in Figure 6.5.

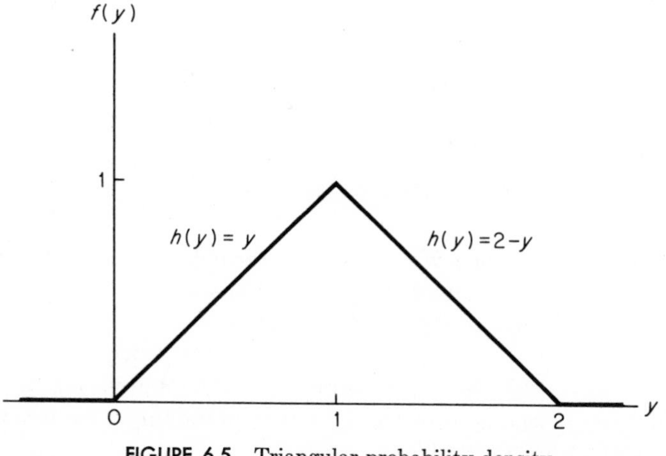

FIGURE 6.5 Triangular probability density.

Although we have presented convolutions with reference to continuous random variables, they can be used also in the discrete case. Replacing the integral in (6.1.2) with a sum, we get

$$h(y) = \Sigma\, f(y - x_2, x_2) \tag{6.1.6}$$

where the summation extends over the entire range of $\mathbf{x}_2$; evidently, similar sums can be defined analogous to (6.1.3), (6.1.4), and (6.1.5).

To illustrate the use of (6.1.6), suppose that $\mathbf{x}_1$ and $\mathbf{x}_2$ are independent random variables having Poisson distributions (3.3.9) with the parameters λ_1 and λ_2, so that

$$f(x_1, x_2) = \frac{e^{-\lambda_1}(\lambda_1)^{x_1}}{x_1!} \cdot \frac{e^{-\lambda_2}(\lambda_2)^{x_2}}{x_2!}$$

for $x_1 = 0, 1, 2, \ldots$, and $x_2 = 0, 1, 2, \ldots$. Substituting into (6.1.6), we get

$$h(y) = \sum_{x_2=0}^{y} \frac{e^{-(\lambda_1+\lambda_2)}(\lambda_2)^{x_2}(\lambda_1)^{y-x_2}}{x_2!(y-x_2)!}$$

where the summation goes from $x_2 = 0$ to $x_2 = y$, since x_2 cannot exceed $y = x_1 + x_2$. Factoring out $e^{-(\lambda_1+\lambda_2)}$ and multiplying and dividing by $y!$, we obtain

$$h(y) = \frac{e^{-(\lambda_1+\lambda_2)}}{y!} \cdot \sum_{x_2=0}^{y} \frac{y!}{x_2!(y-x_2)!} (\lambda_2)^{x_2}(\lambda_1)^{y-x_2}$$

and, identifying this sum as the binomial expansion of $(\lambda_1 + \lambda_2)^y$, we finally get

$$h(y) = \frac{e^{-(\lambda_1+\lambda_2)}(\lambda_1 + \lambda_2)^y}{y!} \qquad \text{for } y = 0, 1, 2, \ldots$$

We have thus shown that

THEOREM 6.1

The sum of two independent random variables having Poisson distributions with the parameters λ_1 and λ_2 has a Poisson distribution with the parameter $\lambda = \lambda_1 + \lambda_2$.

So far we have considered only sums of two random variables; to find the distribution of the sum of three or more random variables, we can generalize (6.1.2), for example, and write

$$h(y) = \int_{-\infty}^{\infty} \int_{-\infty}^{\infty} f(y - x_2 - x_3, x_2, x_3)\, dx_2\, dx_3 \tag{6.1.7}$$

for the probability density of $\mathbf{y} = \mathbf{x}_1 + \mathbf{x}_2 + \mathbf{x}_3$, and

$$h(y) = \int_{-\infty}^{\infty} \int_{-\infty}^{\infty} \int_{-\infty}^{\infty} f(y - x_2 - x_3 - x_4, x_2, x_3, x_4)\, dx_2\, dx_3\, dx_4 \tag{6.1.8}$$

for the probability density of $\mathbf{y} = \mathbf{x}_1 + \mathbf{x}_2 + \mathbf{x}_3 + \mathbf{x}_4$. For instance, if the three random variables $\mathbf{x}_1$, $\mathbf{x}_2$, and $\mathbf{x}_3$ have the joint probability density

$$f(x_1, x_2, x_3) = \begin{cases} e^{-(x_1+x_2+x_3)} & \text{for } x_1 > 0,\ x_2 > 0, \text{ and } x_3 > 0 \\ 0 & \text{elsewhere} \end{cases}$$

substitution into (6.1.7) yields

$$h(y) = \int_0^y \int_0^{y-x_3} e^{-y}\, dx_2\, dx_3 = \tfrac{1}{2}y^2e^{-y} \qquad \text{for } y > 0$$

and $h(y) = 0$ for $y \leq 0$. Note that we had to be careful with the limits of integration, since the integrand is *zero* when x_2 exceeds $y - x_3$ (which would make x_1 negative) or when x_3 exceeds y (which would make $x_1 + x_2$ negative).

An alternate approach to finding the distribution of the sum of more than two random variables is to begin by finding the distribution of $\mathbf{y} = \mathbf{x}_1 + \mathbf{x}_2$, and then using the same method again to find the distribution of $\mathbf{u} = \mathbf{y} + \mathbf{x}_3$, $\mathbf{v} = \mathbf{u} + \mathbf{x}_4$, and so forth. This is illustrated in Exercise 7 below.

Although the method of this section may seem different from that of Section 6.1, their theoretical basis is the same. In the case of two random variables the method of Section 6.1 yields

$$F(y) = \int_{-\infty}^{\infty} \left[\int_{-\infty}^{y-x_2} f(x_1, x_2)\, dx_1 \right] dx_2$$

and, differentiating with respect to y, we get

$$F'(y) = \int_{-\infty}^{\infty} f(y - x_2, x_2)\, dx_2$$

namely, the convolution integral (6.1.2).

THEORETICAL EXERCISES

1. Given that $\mathbf{x}_1$ and $\mathbf{x}_2$ are independent random variables having exponential densities (4.3.2) with the parameters θ_1 and θ_2, use the method of Section 6.1 to find the probability density of $\mathbf{y} = \mathbf{x}_1 + \mathbf{x}_2$
   (a) when $\theta_1 \neq \theta_2$;
   (b) when $\theta_1 = \theta_2$.
   Note that the illustration on page 184 is a special case of part (a) with $\theta_1 = \frac{1}{3}$ and $\theta_2 = \frac{1}{2}$.
2. Given that $\mathbf{x}_1$ and $\mathbf{x}_2$ are independent random variables having the

normal distribution (5.3.5) with $\mu = 0$ and $\sigma = 1$, show that $\mathbf{y} = \mathbf{x}_1 + \mathbf{x}_2$ also has a normal distribution. (*Hint:* Complete the square in the exponent.) What is the mean and the variance of this distribution?

3. On page 184 we gave the joint probability density of two random variables, which (as we pointed out on page 187) may be looked upon as independent random variables having the uniform density (4.3.1) with $\alpha = 0$ and $\beta = 1$. Referring to Figure 6.3, find expressions for the values of the distribution function of $\mathbf{y} = \mathbf{x}_1 + \mathbf{x}_2$ which apply
(a) when $y \leq 0$;
(b) when $0 < y < 1$;
(c) when $1 < y < 2$;
(d) when $y \geq 2$.
Also find the probability density of $\mathbf{y}$ and compare with the result obtained on page 186.

4. Given that $\mathbf{x}_1$ and $\mathbf{x}_2$ are independent random variables having binomial distributions (3.3.3) with the respective parameters n_1 and θ *and* n_2 and θ, show that $\mathbf{y} = \mathbf{x}_1 + \mathbf{x}_2$ has the binomial distribution with the parameters $n_1 + n_2$ and θ. (*Hint:* Make use of Theorem 1.8.)

5. Given that $\mathbf{x}_1$ and $\mathbf{x}_2$ are independent random variables having the geometric distribution (3.3.6), use (6.1.6) to find the probability distribution of $\mathbf{y} = \mathbf{x}_1 + \mathbf{x}_2$ and verify that it is the negative binomial distribution (3.3.7) with $k = 2$. [*Hint:* Note that in (6.1.6) the limits of summation are $x_2 = 1$ and $x_2 = y - 1$.]

6. Use (6.1.7) to find the distribution of the sum of three independent random variables having the exponential density (4.3.2) with the same parameter θ. Identify the resulting distribution and the values of its parameters.

7. On page 186 we obtained the probability density of the sum of two independent random variables having the uniform density (4.3.1) with $\alpha = 0$ and $\beta = 1$. Given a third random variable $\mathbf{x}_3$ which has the same uniform density and is independent of $\mathbf{x}_1$ and $\mathbf{x}_2$, show that if

$$\mathbf{u} = \mathbf{y} + \mathbf{x}_3 = \mathbf{x}_1 + \mathbf{x}_2 + \mathbf{x}_3$$

(a) the joint density of $\mathbf{u}$ and $\mathbf{y}$ is given by $g(u, y) = y$ for regions I and II of Figure 6.6, $g(u, y) = 2 - y$ for regions III and IV of Figure 6.6, and $g(u, y) = 0$ elsewhere;
(b) the probability density of u is given by

$$h(u) = \begin{cases} 0 & \text{for } u \leq 0 \\ \frac{1}{2}u^2 & \text{for } 0 < u < 1 \\ \frac{1}{2}u^2 - \frac{3}{2}(u-1)^2 & \text{for } 1 < u < 2 \\ \frac{1}{2}u^2 - \frac{3}{2}(u-1)^2 + \frac{3}{2}(u-2)^2 & \text{for } 2 < u < 3 \\ 0 & \text{for } u \geq 3 \end{cases}$$

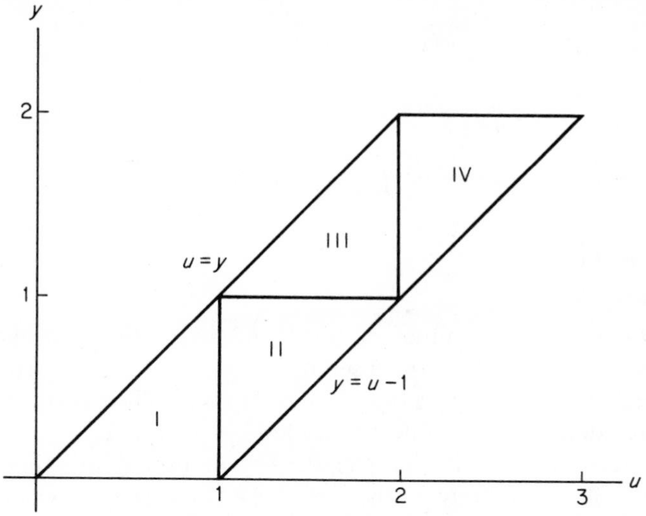

FIGURE 6.6 Diagram for Exercise 7.

and if we let $h(1) = h(2) = \frac{1}{2}$ this will make the function continuous. Note that on the interval from 0 to 3 the graph of the function consists of parts of three different parabolas, and that its shape is very similar to that of the normal distribution.

6.1.2 Sums of Random Variables: Moment Generating Functions

The method of this section is based on the theorem that the moment generating function of the sum of n independent random variables equals the product of their moment generating functions, namely,

THEOREM 6.2

If $\mathbf{x}_1, \mathbf{x}_2, \ldots,$ *and* $\mathbf{x}_n$ *are independent random variables with the moment generating functions* $M_{\mathbf{x}_1}(t), M_{\mathbf{x}_2}(t), \ldots, M_{\mathbf{x}_n}(t)$, *and* $\mathbf{y} = \mathbf{x}_1 + \mathbf{x}_2 + \cdots + \mathbf{x}_n$, *then*

$$M_{\mathbf{y}}(t) = \prod_{i=1}^{n} M_{\mathbf{x}_i}(t)$$

Making use of the fact that the random variables are independent and, hence, that $f(x_1, x_2, \ldots, x_n) = f_1(x_1)\cdot f_2(x_2) \cdots f_n(x_n)$ according to (4.5.4), we have

$$\begin{aligned}
M_{\mathbf{y}}(t) &= E(e^{\mathbf{y}t}) = E[e^{(\mathbf{x}_1+\mathbf{x}_2+\cdots+\mathbf{x}_n)t}] \\
&= \int_{-\infty}^{\infty} \cdots \int_{-\infty}^{\infty} e^{(x_1+x_2+\cdots+x_n)t} f(x_1, x_2, \ldots, x_n)\, dx_1\, dx_2 \ldots dx_n \\
&= \int_{-\infty}^{\infty} e^{x_1t} f_1(x_1)\, dx_1 \cdot \int_{-\infty}^{\infty} e^{x_2t} f_2(x_2)\, dx_2 \cdots \int_{-\infty}^{\infty} e^{x_nt} f_n(x_n)\, dx_n \\
&= \prod_{i=1}^{n} M_{\mathbf{x}_i}(t)
\end{aligned}$$

and this completes the proof of the theorem for the continuous case. (To prove Theorem 6.2 for the discrete case we have only to replace all of the integrals by sums.) Note that if we want to use Theorem 6.2 to find the probability density of the random variable $\mathbf{y} = \mathbf{x}_1 + \mathbf{x}_2 + \cdots + \mathbf{x}_n$, we must be able to identify whatever density or probability function corresponds to $M_{\mathbf{y}}(t)$ and rely on the first of the two theorems which we gave on page 175.

To illustrate this technique, suppose that $\mathbf{x}_1, \mathbf{x}_2, \ldots,$ and $\mathbf{x}_n$ are independent random variables having exponential distributions (4.3.2). Since the exponential distribution is a gamma distribution with $\alpha = 1$ and $\beta = 0$ (see page 113), we have

$$M_{\mathbf{x}_i}(t) = (1 - \theta t)^{-1}$$

according to Theorem 5.15, and hence

$$M_{\mathbf{y}}(t) = \prod_{i=1}^{n} (1 - \theta t)^{-1} = (1 - \theta t)^{-n}$$

according to the second of the special rules for products given on page 419. Identifying the resulting moment generating function as that of the gamma distribution with $\alpha = n$ and $\beta = \theta$, we conclude that *the distribution of the sum of n independent random variables having exponential distributions with the same parameter θ is a gamma distribution* (4.3.6) *with the parameters $\alpha = n$ and $\beta = \theta$.* Note that this agrees with the result of Exercise 6 on page 190, where we had $n = 3$ and the distribution of the sum of the three random variables was a gamma distribution with $\alpha = 3$ and $\beta = \theta$.

To show how this method works in the discrete case, suppose that $\mathbf{x}_1, \mathbf{x}_2, \ldots,$ and $\mathbf{x}_n$ are independent random variables having Poisson distributions with the parameters $\lambda_1, \lambda_2, \ldots, \lambda_n$. Making use of Theorem 5.12 we have

$$M_{\mathbf{x}_i}(t) = e^{\lambda_i(e^t-1)}$$

and, hence,

$$M_{\mathbf{y}}(t) = \prod_{i=1}^{n} e^{\lambda_i(e^t-1)} = e^{(\lambda_1+\lambda_2+\cdots+\lambda_n)(e^t-1)}$$

which can readily be identified as the moment generating function of the Poisson distribution (3.3.9) with the parameter $\lambda = \lambda_1 + \lambda_2 + \cdots + \lambda_n$. Referring again to the *uniqueness theorem* (that is, the first of the two theorems on page 175), we have thus shown that *the distribution of the sum of n independent random variables having Poisson distributions with the parameters λ_i is a Poisson distribution with the parameter*

$$\lambda = \lambda_1 + \lambda_2 + \cdots + \lambda_n$$

Note that on page 188 we already proved this for $n = 2$.

Theorem 6.2 also provides an easy and elegant way of deriving the moment generating function of the binomial distribution. Suppose that $\mathbf{x}_1$, $\mathbf{x}_2$, . . . , and $\mathbf{x}_n$ are independent random variables having the Bernoulli distribution (3.3.2) with the same parameter θ. Substituting (3.3.2) into (5.2.4a) we get

$$M_{\mathbf{x}_i}(t) = e^{0\cdot t}(1-\theta) + e^{1\cdot t}\theta = 1 + \theta(e^t - 1)$$

so that Theorem 6.2 yields

$$M_{\mathbf{y}}(t) = \prod_{i=1}^{n} [1 + \theta(e^t - 1)] = [1 + \theta(e^t - 1)]^n$$

which is readily identified as the moment generating function of the binomial distribution with the parameters n and θ. Of course, $\mathbf{y} = \mathbf{x}_1 + \mathbf{x}_2 + \cdots + \mathbf{x}_n$ *is* here the total number of successes in n trials, since $\mathbf{x}_1$ is the number of successes on the first trial, $\mathbf{x}_2$ is the number of successes on the second trial, . . . , and $\mathbf{x}_n$ is the number of successes on the nth trial. As we shall see in Chapter 6, this is a fruitful way of looking at the binomial distribution.

THEORETICAL EXERCISES

1. Repeat Exercise 4 on page 190 using the method of Section 6.1.2. How can the result be generalized?
2. Find the moment generating function of the negative binomial distribution (3.3.7) by making use of the fact that *if k independent ran-*

dom variables have the geometric distribution with the same parameter θ, their sum has the negative binomial distribution with the parameters θ and k (see also Exercise 5 on page 190). (*Hint:* Make use of the result of Exercise 12 on page 171.)

3. If n independent random variables have gamma distributions with the same parameters α and β, find the moment generating function of their sum and, if possible, identify the corresponding distribution.

4. If n independent random variables $\mathbf{x}_i$ have normal distributions with the means μ_i and the standard deviations σ_i, find the moment generating function of their sum and identify the corresponding distribution, its mean, and its variance.

5. Prove the following generalization of Theorem 6.2: *If* $\mathbf{x}_1, \mathbf{x}_2, \ldots, \mathbf{x}_n$ *are independent random variables with the moment generating functions* $M_{\mathbf{x}_i}(t)$, *then* $\mathbf{y} = a_1\mathbf{x}_1 + a_2\mathbf{x}_2 + \cdots + a_n\mathbf{x}_n$ *has the moment generating function*

$$M_{\mathbf{y}}(t) = \prod_{i=1}^{n} M_{\mathbf{x}_i}(a_i t)$$

6. Use the result of Exercise 5 to show that if n independent random variables $\mathbf{x}_i$ have normal distributions with the means μ_i and the standard deviations σ_i, then $\mathbf{y} = a_1\mathbf{x}_1 + a_2\mathbf{x}_2 + \cdots + a_n\mathbf{x}_n$ has also a normal distribution. What are the mean and the variance of this distribution?

APPLIED EXERCISES

7. If the number of fish a person catches per hour at Woods Canyon Lake is a random variable having the Poisson distribution with $\lambda = 1.8$, find the probability that
(a) a person fishing there will catch 3 fish in 2 hours;
(b) a person fishing there will catch at least 5 fish in 3 hours;
(c) a person fishing there will catch only 2 fish in 5 hours.
(*Hint:* Use Table II at the end of the book.)

8. If the number of incoming calls reaching the switchboard of a large company per minute is a random variable having the Poisson distribution with $\lambda = 1.2$, find the probability that there will be
(a) exactly 10 incoming calls during a period of 5 minutes;
(b) fewer than 10 incoming calls during a period of 8 minutes;
(c) at least 15 incoming calls during a period of 10 minutes.

9. If the number of minutes a doctor spends with a patient is a random variable having an exponential distribution with the parameter $\theta = 8$, what is the probability that it will take the doctor at least 20 minutes to treat
   (a) two patients;
   (b) three patients.
10. If the number of minutes it takes a secretary to fill in a certain form is a random variable having an exponential distribution with the parameter $\theta = 4$, what is the probability that
   (a) it will take her at least 10 minutes to fill in 3 of the forms;
   (b) it will take her less than 6 minutes to fill in 2 of the forms.

6.2 MOMENTS OF LINEAR COMBINATIONS OF RANDOM VARIABLES

In this section we shall derive expressions for the mean and the variance of a linear combination of n random variables and the covariance of two linear combinations of n random variables; applications of these results will be treated in subsections 6.2.1, 6.2.2, and 6.2.3.

THEOREM 6.3

If $\mathbf{x}_1, \mathbf{x}_2, \ldots, \mathbf{x}_n$ *are random variables,* $a_1, a_2, \ldots, a_n$ *are constants, and* $\mathbf{y} = a_1\mathbf{x}_1 + a_2\mathbf{x}_2 + \cdots + a_n\mathbf{x}_n$, *then*

$$E(\mathbf{y}) = \sum_{i=1}^{n} a_i E(\mathbf{x}_i) \tag{6.2.1}$$

and

$$\operatorname{var}(\mathbf{y}) = \sum_{i=1}^{n} a_i^2 \cdot \operatorname{var}(\mathbf{x}_i) + 2 \cdot \sum_{i<j}\sum a_i a_j \cdot \operatorname{cov}(\mathbf{x}_i, \mathbf{x}_j) \tag{6.2.2}$$

where $\sum_{i<j}\sum$ *means that the summation extends over all values of* i *and* j, *from* 1 *to* n, *for which* $i < j$.

Using first Theorem 5.3 and then Theorem 5.2, we have

$$E(\mathbf{y}) = E\left(\sum_{i=1}^{n} a_i\mathbf{x}_i\right) = \sum_{i=1}^{n} E(a_i\mathbf{x}_i) = \sum_{i=1}^{n} a_i E(\mathbf{x}_i)$$

and this proves the first part of Theorem 6.3. To derive the expression for the variance of $\mathbf{y}$, let us write μ_i for $E(\mathbf{x}_i)$, so that we get

$$\text{var}(\mathbf{y}) = E([\mathbf{y} - E(\mathbf{y})]^2) = E\left(\left[\sum_{i=1}^{n} a_i\mathbf{x}_i - \sum_{i=1}^{n} a_iE(\mathbf{x}_i)\right]^2\right)$$

$$= E\left(\left[\sum_{i=1}^{n} a_i(\mathbf{x}_i - \mu_i)\right]^2\right)$$

Finally, expanding this expression by means of the *multinomial theorem* according to which $(a + b + c + d)^2$, for example, equals $a^2 + b^2 + c^2 + d^2 + 2ab + 2ac + 2ad + 2bc + 2bd + 2cd$, we obtain

$$\text{var}(\mathbf{y}) = \sum_{i=1}^{n} a_i^2E[(\mathbf{x}_i - \mu_i)^2] + 2\cdot\sum_{i<j}\sum a_ia_jE[(\mathbf{x}_i - \mu_i)(\mathbf{x}_j - \mu_j)]$$

$$= \sum_{i=1}^{n} a_i^2\cdot\text{var}(\mathbf{x}_i) + 2\cdot\sum_{i<j}\sum a_ia_j\cdot\text{cov}(\mathbf{x}_i, \mathbf{x}_j)$$

Note that we have tacitly made use of the fact that $\text{cov}(\mathbf{x}_i, \mathbf{x}_j) = \text{cov}(\mathbf{x}_j, \mathbf{x}_i)$. An immediate collorary of Theorem 6.3 is that if the random variables $\mathbf{x}_1, \mathbf{x}_2, \ldots$, and $\mathbf{x}_n$ are *independent*, then

$$\text{var}(\mathbf{y}) = \sum_{i=1}^{n} a_i^2\cdot\text{var}(\mathbf{x}_i) \tag{6.2.3}$$

since the covariances are all zero according to the proof on page 179.

Several applications of (6.2.3) will be discussed in Sections 6.2.1 and 6.2.2; an application of the more general formula (6.2.2), where the random variables $\mathbf{x}_i$ are not assumed to be independent, will be given in Section 6.2.3.

The following is another important theorem about linear combinations of random variables; it concerns the covariance of two linear combinations of a set of n random variables:

THEOREM 6.4

If $\mathbf{x}_1, \mathbf{x}_2, \ldots, \mathbf{x}_n$ are random variables, $a_1, a_2, \ldots, a_n$ and $b_1, b_2, \ldots, b_n$ are constants, $\mathbf{y}_1 = a_1\mathbf{x}_1 + a_2\mathbf{x}_2 + \cdots + a_n\mathbf{x}_n$, and $\mathbf{y}_2 = b_1\mathbf{x}_1 + b_2\mathbf{x}_2 + \cdots + b_n\mathbf{x}_n$, then

$$\text{cov}(\mathbf{y}_1, \mathbf{y}_2) = \sum_{i=1}^{n} a_ib_i\cdot\text{var}(\mathbf{x}_i) + \sum_{i<j}\sum (a_ib_j + a_jb_i)\cdot\text{cov}(\mathbf{x}_i, \mathbf{x}_j) \tag{6.2.4}$$

The proof of this theorem, which is very similar to that of Theorem 6.3, will be left to the reader in Exercise 4 on page 201. Analogous to (6.2.3), we have the corollary that if the random variables $\mathbf{x}_i$ are independent,

$$\operatorname{cov}(\mathbf{y}_1, \mathbf{y}_2) = \sum_{i=1}^{n} a_i b_i \cdot \operatorname{var}(\mathbf{x}_i) \tag{6.2.5}$$

since the covariances are again all equal to zero. A very important application of (6.2.5) will be given in Section 6.2.1.

6.2.1 The Distribution of the Mean

Given a set of random variables $\mathbf{x}_1, \mathbf{x}_2, \ldots,$ and $\mathbf{x}_n$, we refer to the random variable $\bar{\mathbf{x}}$ (called x-bar), whose values are related to those of the $\mathbf{x}$'s by means of the equation

$$\bar{x} = \frac{x_1 + x_2 + \cdots + x_n}{n} \tag{6.2.6}$$

as the *mean* of the n random variables. The mean of a set of random variables is of special importance when these random variables are *independent and identically distributed,* for we can then use values of $\bar{\mathbf{x}}$ as estimates of the mean μ of their common distribution. In fact, *under these conditions we refer to the n random variables as a random sample, and we refer to their common distribution as the "population from which we are sampling."* (In the past it was common practice to apply the term "random sample" to the *values* of the random variables instead of the random variables, themselves. Intuitively, this may make more sense, but it does not conform with current usage.)

Since the distribution of $\bar{\mathbf{x}}$ plays an important role in problems of statistical inference, let us now prove the following theorem:

THEOREM 6.5

If $\mathbf{x}_1, \mathbf{x}_2, \ldots, \mathbf{x}_n$ are independent random variables having the same distribution with the mean μ and the variance σ^2, then

$$E(\bar{\mathbf{x}}) = \mu \quad \textit{and} \quad \operatorname{var}(\bar{\mathbf{x}}) = \frac{\sigma^2}{n}$$

To prove this theorem we have only to substitute $E(\mathbf{x}_i) = \mu$, $\operatorname{var}(\mathbf{x}_i) = \sigma^2$, and $a_i = \frac{1}{n}$ into (6.2.1) and (6.2.3), getting

$$E(\bar{\mathbf{x}}) = \sum_{i=1}^{n} \frac{1}{n} \cdot \mu = n\left(\frac{1}{n} \cdot \mu\right) = \mu$$

and

$$\operatorname{var}(\bar{\mathbf{x}}) = \sum_{i=1}^{n} \frac{1}{n^2}\cdot\sigma^2 = n\left(\frac{1}{n^2}\cdot\sigma^2\right) = \frac{\sigma^2}{n}$$

It is customary to write $E(\bar{\mathbf{x}})$ as $\mu_{\bar{x}}$, $\operatorname{var}(\bar{\mathbf{x}})$ as $\sigma_{\bar{x}}^2$, and refer to $\sigma_{\bar{x}}$ as the *standard error of the mean.*

The formula $\sigma_{\bar{x}} = \frac{\sigma}{\sqrt{n}}$ shows that the standard deviation of the distribution of $\bar{\mathbf{x}}$ decreases when n, the *sample size*, is increased. This means that when n becomes larger and we actually have more information (the values of more random variables), values of $\bar{\mathbf{x}}$ can be expected to be closer to μ, the quantity they are intended to estimate. Referring to Chebyshev's theorem as it is formulated in Exercise 13 on page 154, we can express this more precisely as follows: *For any positive constant c, the probability that $\bar{\mathbf{x}}$ will take on a value between $\mu - c$ and $\mu + c$ is at least $1 - \frac{\sigma^2}{nc^2}$; when $n \to \infty$, this probability approaches* 1.

In Section 5.2 we defined the variance of a probability distribution, a measure of its "spread" or "dispersion," in terms of the squared deviations from the mean of the distribution. Since we shall later, in Chapter 7, want to define a corresponding measure of variability based on values of the random variables $\mathbf{x}_1 - \bar{\mathbf{x}}$, $\mathbf{x}_2 - \bar{\mathbf{x}}$, $\ldots$, and $\mathbf{x}_n - \bar{\mathbf{x}}$, let us now prove the following theorem for future reference:

THEOREM 6.6

If $\mathbf{x}_1, \mathbf{x}_2, \ldots, \mathbf{x}_n$ are independent random variables having the same distribution with the variance σ^2 and $\bar{\mathbf{x}}$ is their mean, then

$$\operatorname{cov}(\mathbf{x}_r - \bar{\mathbf{x}}, \bar{\mathbf{x}}) = 0$$

for $r = 1, 2, \ldots,$ or n.

To prove this theorem, we have only to substitute appropriate values for the a's and b's of (6.2.5). Since

$$x_r - \bar{x} = \frac{n-1}{n}\cdot x_r - \frac{1}{n}\left[\left(\sum_{i=1}^{n} x_i\right) - x_r\right]$$

for the values of the corresponding random variables, it follows that $a_r = \frac{n-1}{n}$ while all the other a's are equal to $-\frac{1}{n}$. Thus, if we substitute these values together with $b_i = \frac{1}{n}$ and $\text{var}(\mathbf{x}_i) = \sigma^2$ into (6.2.5), we get

$$\begin{aligned}\text{cov}(\mathbf{x}_r - \bar{\mathbf{x}}, \bar{\mathbf{x}}) &= \sum_{i=1}^{n} a_i b_i \cdot \text{var}(\mathbf{x}_i) \\ &= \left[\frac{n-1}{n} + \left(-\frac{1}{n}\right) + \cdots + \left(-\frac{1}{n}\right)\right] \cdot \frac{1}{n} \cdot \sigma^2 \\ &= 0\end{aligned}$$

6.2.2 Differences between Means and Differences between Proportions

In many problems of statistics we must decide whether differences between the observed values of random variables can be attributed to chance, or whether they indicate that the distributions of the random variables differ in some respect. For instance, if a sample of 50 light bulbs of one kind lasted on the average 442 hours (of continuous use) while a sample of 50 light bulbs of another kind lasted on the average 435 hours, we may want to decide whether the difference between 442 and 435 can be attributed to chance, or whether there is actually a difference in quality. Similarly, if 284 of 400 randomly selected housewives in Dallas preferred Detergent A to Detergent B while 375 of 500 randomly selected housewives in Atlanta preferred Detergent A to Detergent B, we may want to decide whether the difference between the corresponding proportions, $\frac{284}{400} = 0.71$ and $\frac{375}{500} = 0.75$, can be attributed to chance, or whether it is indicative of an actual difference between the preferences of all the housewives in these cities.

To develop theory pertinent to the analysis of problems of this kind, let us consider $n_1 + n_2$ random variables, where the first n_1 are denoted $\mathbf{x}_{11}, \mathbf{x}_{12}, \ldots, \mathbf{x}_{1n_1}$ and the other n_2 are denoted $\mathbf{x}_{21}, \mathbf{x}_{22}, \ldots, \mathbf{x}_{2n_2}$. If these random variables are all independent, with the first n_1 having identical distributions, and the other n_2 also having identical distributions, we refer to them as a pair of *independent random samples.* Using this notation, let us now prove the following theorem:

THEOREM 6.7

If $\mathbf{x}_{11}, \mathbf{x}_{12}, \ldots, \mathbf{x}_{1n_1}, \mathbf{x}_{21}, \mathbf{x}_{22}, \ldots, \mathbf{x}_{2n_2}$ are independent random variables, the first n_1 have identical distributions with the mean μ_1 and the variance σ_1^2, and the other n_2 have identical distributions *with the mean* μ_2 *and the variance* σ_2^2, *then*

$$E(\bar{\mathbf{x}}_1 - \bar{\mathbf{x}}_2) = \mu_1 - \mu_2$$

and

$$\text{var}(\bar{\mathbf{x}}_1 - \bar{\mathbf{x}}_2) = \frac{\sigma_1^2}{n_1} + \frac{\sigma_2^2}{n_2}$$

where

$$\bar{\mathbf{x}}_1 = \frac{1}{n_1}(\mathbf{x}_{11} + \mathbf{x}_{12} + \cdots + \mathbf{x}_{1n_1})$$

and

$$\bar{\mathbf{x}}_2 = \frac{1}{n_2}(\mathbf{x}_{21} + \mathbf{x}_{22} + \cdots + \mathbf{x}_{2n_2})$$

To prove this theorem we have only to substitute $\frac{1}{n_1}$ for the first n_1 a's and $-\frac{1}{n_2}$ for the other n_2 a's in (6.2.1) and (6.2.3), getting

$$E(\bar{\mathbf{x}}_1 - \bar{\mathbf{x}}_2) = \sum_{i=1}^{n_1} \frac{1}{n_1}\cdot\mu_1 + \sum_{i=1}^{n_2}\left(-\frac{1}{n_2}\right)\cdot\mu_2$$

$$= \mu_1 - \mu_2$$

and

$$\text{var}(\bar{\mathbf{x}}_1 - \bar{\mathbf{x}}_2) = \sum_{i=1}^{n_1}\left(\frac{1}{n_1}\right)^2\cdot\sigma_1^2 + \sum_{i=1}^{n_2}\left(-\frac{1}{n_2}\right)^2\cdot\sigma_2^2$$

$$= \frac{\sigma_1^2}{n_1} + \frac{\sigma_2^2}{n_2}$$

It is customary to refer to the standard deviation of the distribution of $\bar{\mathbf{x}}_1 - \bar{\mathbf{x}}_2$ as the *standard error of the difference between two means.*

An important special case arises when the first n_1 random variables have Bernoulli distributions (3.3.2) with the parameter $\theta = \theta_1$ while the other n_2 random variables have Bernoulli distributions with the parameter $\theta = \theta_2$. In that case $\bar{\mathbf{x}}_1$ is the *proportion of successes* in the first n_1 trials, which we shall denote $\mathbf{p}_1$, $\bar{\mathbf{x}}_2$ is the *proportion of successes* in the other n_2 trials, which we shall denote $\mathbf{p}_2$, and substitution of $\mu_1 = \theta_1$, $\mu_2 = \theta_2$, $\sigma_1^2 = \theta_1(1 - \theta_1)$, and $\sigma_2^2 = \theta_2(1 - \theta_2)$ into the two formulas of Theorem 6.7 yields

$$E(\mathbf{p}_1 - \mathbf{p}_2) = \theta_1 - \theta_2 \tag{6.2.7}$$

and

$$\text{var}(\mathbf{p}_1 - \mathbf{p}_2) = \frac{\theta_1(1 - \theta_1)}{n_1} + \frac{\theta_2(1 - \theta_2)}{n_2} \tag{6.2.8}$$

It is customary to refer to the standard deviation of the distribution of $\mathbf{p}_1 - \mathbf{p}_2$ as the *standard error of the difference between two proportions.*

We have taken up the material of this section at this time because it follows immediately from the work of Section 6.2. Various applications will be given later in Chapter 11.

THEORETICAL EXERCISES

1. Given the independent random variables $\mathbf{x}_1$, $\mathbf{x}_2$, and $\mathbf{x}_3$, whose distributions have the means 3, 5, and 2, and the variances 8, 12, and 18, find the mean and the variance of the distribution of
   (a) $\mathbf{x}_1 + 4\mathbf{x}_2 + 2\mathbf{x}_3$;
   (b) $3\mathbf{x}_1 - \mathbf{x}_2 - \mathbf{x}_3$.
2. Repeat both parts of Exercise 1, dropping the assumption of independence and adding the information that $\text{cov}(\mathbf{x}_1, \mathbf{x}_2) = 1$, $\text{cov}(\mathbf{x}_2, \mathbf{x}_3) = 2$, and $\text{cov}(\mathbf{x}_1, \mathbf{x}_3) = -3$.
3. Given the random variables $\mathbf{x}$ and $\mathbf{y}$, express $\text{var}(\mathbf{x} + \mathbf{y})$, $\text{var}(\mathbf{x} - \mathbf{y})$, and $\text{cov}(\mathbf{x} + \mathbf{y}, \mathbf{x} - \mathbf{y})$ in terms of the variances and the covariance of $\mathbf{x}$ and $\mathbf{y}$.
4. Prove Theorem 6.4 on page 196.
5. If $\mathbf{x}_1, \mathbf{x}_2, \ldots, \mathbf{x}_n$ are independent random variables having the Bernoulli distribution (3.3.2) with the respective parameters $\theta_1, \theta_2, \ldots, \theta_n$, and $\mathbf{y} = \mathbf{x}_1 + \mathbf{x}_2 + \cdots + \mathbf{x}_n$, show that

   (a) $E(\mathbf{y}) = \sum_{i=1}^{n} \theta_i$;

   (b) $\text{var}(\mathbf{y}) = n\theta(1 - \theta) - \sum_{i=1}^{n} (\theta_i - \theta)^2$, where $\theta = \frac{1}{n} \cdot \sum_{i=1}^{n} \theta_i$.
6. If $\mathbf{x}_1, \mathbf{x}_2, \ldots$ and $\mathbf{x}_n$ have the multinomial distribution (3.4.12), find the mean and the variance of $\mathbf{x}_r + \mathbf{x}_s$, the sum of any two of these random variables (assuming that $n > 2$). [*Hint:* Use the fact that the marginal distribution of $\mathbf{x}_i$ is a binomial distribution with the parameters n and θ_i, and the result of part (c) of Exercise 6 on page 180.]
7. Show that under the conditions of Theorem 6.6

   (a) $\text{cov}(\mathbf{x}_s, \mathbf{x}_r - \bar{\mathbf{x}}) = -\frac{1}{n}\sigma^2$ for $r \neq s$;

   (b) $\text{cov}(\mathbf{x}_r, \mathbf{x}_r - \bar{\mathbf{x}}) = \left(1 - \frac{1}{n}\right)\sigma^2$.

APPLIED EXERCISES

8. If the inside diameter of a cylindrical tube is a random variable with a mean of 3 inches and a standard deviation of 0.02 inches, the thickness of the tube is a random variable with a mean of 0.3 inches and a standard deviation of 0.005 inches, and these two random variables are independent, what is the mean and the standard deviation of the distribution of the outside diameter of the tube?

9. If the length of certain bricks is a random variable with a mean of 8, inches and a standard deviation of 0.1 inches, the thickness of the mortar between two bricks is a random variable with a mean of 0.5 inches and a standard deviation of 0.03 inches, what is the mean and the standard deviation of the length of walls made of 50 of these bricks laid side by side, assuming that all random variables involved are independent?

10. If *heads* is a success when flipping a coin, getting a *six* is a success when rolling a die, and getting an *ace* is a success when drawing a card from an ordinary deck of 52 playing cards, find the mean and the standard deviation of the *total number of successes* when we

(a) flip a balanced coin, roll a balanced die, and then draw a card from a well-shuffled deck;

(b) flip a balanced coin three times, roll a balanced die twice, and then draw a card from a well-shuffled deck.

11. If we alternately flip a balanced coin and a coin which is loaded so that the probability for heads is 0.45, what is the mean and the variance of the distribution of the number of heads which we obtain in a total of 10 flips of these coins?

12. A random sample of size $n = 100$ is taken from a population whose mean is $\mu = 75$ and whose variance is $\sigma^2 = 225$. Using Chebyshev's theorem, with what probability can we assert that the value which we obtain for $\bar{\mathbf{x}}$ will fall between 69 and 81?

13. A random sample of size $n = 81$ is taken from a population whose mean is $\mu = 145$ and whose standard deviation is $\sigma = 6.3$. Using Chebyshev's theorem, with what probability can we assert that the value which we obtain for $\bar{\mathbf{x}}$ will be at least 146.4 or at most 143.6?

14. Random samples of size 400 are taken from each of two populations having the same mean μ and the standard deviations $\sigma_1 = 40$ and $\sigma_2 = 50$. Using Chebyshev's theorem, what can we assert with a probability of at least 0.99 about the value which we will get for the random variable $\bar{\mathbf{x}}_1 - \bar{\mathbf{x}}_2$?

15. The actual proportion of men who liked a certain television program is 0.30, while the corresponding proportion for women is 0.25. If 500 men and 500 women are interviewed about this program and their individual responses are looked upon as the values of independent random variables having Bernoulli distributions with the respective parameters $\theta = 0.30$ and $\theta = 0.25$, what can we assert (according to Chebyshev's theorem) with a probability of at least 0.9975 about the value which we will get for $\mathbf{p}_1 - \mathbf{p}_2$, the difference between the sample proportions of favorable responses?

6.2.3 The Distribution of the Mean: Finite Populations

If an experiment consists of selecting one or more values from a finite set of numbers $\{c_1\ c_2, \ldots, c_N\}$, this set is referred to as a *finite population of size* N. If the selection is *without replacement*, $\mathbf{x}_1$ is the first number selected, $\mathbf{x}_2$ is the second number selected, ..., and $\mathbf{x}_n$ is the nth number selected, these random variables are said to constitute a *random sample of size* n from this finite population provided their joint probability function is given by

$$f(x_1, x_2, \ldots, x_n) = \frac{1}{N(N-1)\cdots(N-n+1)} \tag{6.2.9}$$

for each possible *ordered* n-tuple of values selected from the set $\{c_1, c_2, \ldots, c_N\}$. It immediately follows from this that the probability for each possible n-tuple *regardless of the order in which the values are obtained* is

$$\frac{n!}{N(N-1)\cdots(N-n+1)} = \frac{1}{\binom{N}{n}} \tag{6.2.10}$$

and this is often given as the requirement for the selection of a random sample of size n from a finite population of size N.

It also follows from (6.2.9) with $n = 1$ and the fact that the order of the $\mathbf{x}$'s is immaterial, that the marginal distribution of any one of the n random variables $\mathbf{x}_1, \mathbf{x}_2, \ldots, \mathbf{x}_n$ is given by

$$f(x_r) = \frac{1}{N} \qquad \text{for } x_r = c_1, c_2, \ldots, c_N \tag{6.2.11}$$

and it is customary to refer to

$$E(\mathbf{x}_r) = \sum_{i=1}^{N} \frac{1}{N} \cdot c_i = \mu \tag{6.2.12}$$

and

$$\mathrm{var}(\mathbf{x}_r) = \sum_{i=1}^{N} \frac{1}{N}(c_i - \mu)^2 = \sigma^2 \tag{6.2.13}$$

as *the mean and the variance of the finite population.*

Similarly, the joint marginal distribution of any two of the random variables $\mathbf{x}_1, \mathbf{x}_2, \ldots, \mathbf{x}_n$ is given by

$$g(x_r, x_s) = \frac{1}{N(N-1)} \tag{6.2.14}$$

for each ordered pair of values of the finite population. Then, making use of fact that

$$\sum_{i=1}^{N} c_i = \mu N \quad \text{and} \quad \sum_{i=1}^{N} c_i^2 = N(\sigma^2 + \mu^2) \tag{6.2.15}$$

which the reader will be asked to verify in Exercise 1 on page 205, we find that the covariance of any two of the random variables can be written as

$$\mathrm{cov}(\mathbf{x}_r, \mathbf{x}_s) = \sum_{\substack{i=1 \\ i \neq j}}^{N} \sum_{j=1}^{N} \frac{1}{N(N-1)}(c_i - \mu)(c_j - \mu)$$

$$= \frac{\left[\sum_{i=1}^{N} c_i\right]^2 - \sum_{i=1}^{N} c_i^2}{N(N-1)} - \mu^2$$

$$= -\frac{\sigma^2}{N-1} \tag{6.2.16}$$

Making use of all these results, let us now prove the following theorem:

THEOREM 6.8

If $\bar{\mathbf{x}}$ is the mean of a random sample of size n selected from a finite population with the mean μ and the variance σ^2, then

$$E(\bar{\mathbf{x}}) = \mu \quad \textit{and} \quad \mathrm{var}(\bar{\mathbf{x}}) = \frac{\sigma^2}{n} \cdot \frac{N-n}{N-1}$$

Substituting $a_i = \frac{1}{n}$, $\mathrm{var}(\mathbf{x}_i) = \sigma^2$, and $\mathrm{cov}(\mathbf{x}_i, \mathbf{x}_j) = -\frac{\sigma^2}{N-1}$ into (6.2.1) and (6.2.2), we get

$$E(\bar{\mathbf{x}}) = \sum_{i=1}^{n} \frac{1}{n} \cdot \mu = \mu$$

and

$$\text{var}(\bar{\mathbf{x}}) = \sum_{i=1}^{n} \frac{1}{n^2}\cdot\sigma^2 + 2\cdot \sum_{i<j}\sum \frac{1}{n^2}\left(-\frac{\sigma^2}{N-1}\right)$$

$$= \frac{\sigma^2}{n} + 2\cdot\frac{n(n-1)}{2}\cdot\frac{1}{n^2}\left(-\frac{\sigma^2}{N-1}\right)$$

$$= \frac{\sigma^2}{n}\cdot\frac{N-n}{N-1}$$

which completes the proof.

It is of interest to note that the two formulas which we obtained for $\text{var}(\bar{\mathbf{x}})$, the one of Theorem 6.5 and the one of Theorem 6.8, differ only by the *finite population correction factor* $\frac{N-n}{N-1}\cdot$ Indeed, when N, the size of the population, is large compared to n, the size of the sample, the difference between the two formulas for $\text{var}(\bar{\mathbf{x}})$ is generally negligible, and the formula $\sigma_{\bar{x}} = \frac{\sigma}{\sqrt{n}}$ is often used as an approximation when we are dealing with relatively small samples from very large finite populations.

THEORETICAL EXERCISES

1. Using (6.2.12) and (6.2.13), prove both parts of (6.2.15).
2. Explain why the results of Theorem 6.5 apply instead of those of Theorem 6.8 when we sample *with replacement* from a finite population.
3. Explain the results of Exercise 8 on page 170 in the light of Theorem 6.8.
4. If a random sample of size n is selected from the finite population which consists of the first N positive integers, show that
   (a) the mean of the distribution of $\bar{\mathbf{x}}$ is $\frac{N+1}{2}$;
   (b) the variance of the distribution of $\bar{\mathbf{x}}$ is $\frac{(N+1)(N-n)}{12n}$;
   (c) the mean and the variance of the distribution of $\mathbf{y} = n\cdot\bar{\mathbf{x}}$ are

$$E(\mathbf{y}) = \frac{n(N+1)}{2} \quad\text{and}\quad \text{var}(\mathbf{y}) = \frac{n(N+1)(N-n)}{12}$$

   (*Hint:* Use the results of Exercise 1 on page 168 or the formulas for the sum of the first N positive integers and the sum of their squares, given in the Appendix at the end of the book.)

6.3 THE CENTRAL LIMIT THEOREM

The theorem which we shall study in this section is one of the most important theorems of statistics. Essentially, the *central limit theorem*, as it is called, concerns the limiting distribution of the *standardized mean* of n random variables when $n \to \infty$. We shall prove it here for the case where the random variables $\mathbf{x}_1$, $\mathbf{x}_2$, . . . , and $\mathbf{x}_n$ are independent and have the same distribution, whose moment generating function exists. More general conditions under which the theorem holds are given in Exercises 1 and 2 on page 208, and the most general conditions under which it holds are referred to at the end of this chapter.

THEOREM 6.9

If $\mathbf{x}_1, \mathbf{x}_2, \ldots,$ *and* $\mathbf{x}_n$ *are independent random variables having the same distribution with the mean* μ, *the variance* σ^2, *and the moment generating function* $M_{\mathbf{x}}(t)$, *then if* $n \to \infty$ *the limiting distribution of the random variable* $\mathbf{z} = \dfrac{\bar{\mathbf{x}} - \mu}{\sigma/\sqrt{n}}$ *is the standard normal distribution.*

To be able to use results which have already been proved, let us write $\mathbf{z}$ as $\dfrac{n\bar{\mathbf{x}} - n\mu}{\sigma\sqrt{n}} = \dfrac{\mathbf{y} - n\mu}{\sigma\sqrt{n}}$, where $\mathbf{y} = \mathbf{x}_1 + \mathbf{x}_2 + \cdots + \mathbf{x}_n$. Then, using part (c) of Theorem 5.7 and Theorem 6.2, we get

$$M_{\mathbf{z}}(t) = M_{\frac{\mathbf{y}-n\mu}{\sigma\sqrt{n}}}(t) = e^{-\frac{n\mu}{\sigma\sqrt{n}}t} \cdot M_{\mathbf{y}}\left(\frac{t}{\sigma\sqrt{n}}\right) = e^{-\frac{\sqrt{n}\mu}{\sigma}t} \cdot \left[M_{\mathbf{x}}\left(\frac{t}{\sigma\sqrt{n}}\right)\right]^n$$

and, hence,

$$\ln M_{\mathbf{z}}(t) = -\frac{\sqrt{n}\,\mu}{\sigma}t + n \cdot \ln M_{\mathbf{x}}\left(\frac{t}{\sigma\sqrt{n}}\right)$$

Expanding $M_{\mathbf{x}}\left(\dfrac{t}{\sigma\sqrt{n}}\right)$ as a power series in t, we obtain

$$\ln M_{\mathbf{z}}(t) = -\frac{\sqrt{n}\,\mu}{\sigma}t + n \cdot \ln\left[1 + \mu_1' \frac{t}{\sigma\sqrt{n}} + \mu_2' \frac{t^2}{2\sigma^2 n} + \mu_3' \frac{t^3}{6\sigma^3 n\sqrt{n}} + \cdots\right]$$

where $\mu_1', \mu_2', \mu_3', \ldots$, are the moments of the distribution of the original random variables $\mathbf{x}_i$. If n is sufficiently large, we can then expand $\ln(1 + x)$ as a power series in x (as on page 176), getting

$$\begin{aligned}\ln M_{\mathbf{z}}(t) = -\frac{\sqrt{n}\,\mu}{\sigma}t + n\Bigg\{&\left[\mu_1'\frac{t}{\sigma\sqrt{n}} + \mu_2'\frac{t^2}{2\sigma^2 n} + \mu_3'\frac{t^3}{6\sigma^3 n\sqrt{n}} + \cdots\right]\\ &-\frac{1}{2}\left[\mu_1'\frac{t}{\sigma\sqrt{n}} + \mu_2'\frac{t^2}{2\sigma^2 n} + \mu_3'\frac{t^3}{6\sigma^3 n\sqrt{n}} + \cdots\right]^2\\ &+\frac{1}{3}\left[\mu_1'\frac{t}{\sigma\sqrt{n}} + \mu_2'\frac{t^2}{2\sigma^2 n} + \mu_3'\frac{t^3}{6\sigma^3 n\sqrt{n}} + \cdots\right]^3 - \cdots\Bigg\}\end{aligned}$$

Collecting powers of t, we then obtain

$$\begin{aligned}\ln M_{\mathbf{z}}(t) = &\left(-\frac{\sqrt{n}\,\mu}{\sigma} + \frac{\sqrt{n}\,\mu_1'}{\sigma}\right)t + \left(\frac{\mu_2'}{2\sigma^2} - \frac{\mu_1'^2}{2\sigma^2}\right)t^2\\ &+\left(\frac{\mu_3'}{6\sigma^3\sqrt{n}} - \frac{\mu_1'\cdot\mu_2'}{2\sigma^3\sqrt{n}} + \frac{\mu_1'^3}{3\sigma^3\sqrt{n}}\right)t^3 + \cdots\end{aligned}$$

and since $\mu_1' = \mu$ and $\mu_2' - \mu_1'^2 = \sigma^2$, this reduces to

$$\ln M_{\mathbf{z}}(t) = \frac{1}{2}t^2 + \left(\frac{\mu_3'}{6} - \frac{\mu_1'\mu_2'}{2} + \frac{\mu_1'^3}{3}\right)\frac{t^3}{\sigma^3\sqrt{n}} + \cdots$$

Finally, observing that the coefficient of t^3 is a constant times $\frac{1}{\sqrt{n}}$ and in general the coefficient of t^r is a constant times $\frac{1}{\sqrt{n^{r-2}}}$, we get

$$\lim_{n\to\infty} \ln M_{\mathbf{z}}(t) = \tfrac{1}{2}t^2$$

and hence

$$\lim_{n\to\infty} M_{\mathbf{z}}(t) = e^{\frac{1}{2}t^2}$$

since the limit of a logarithm equals the logarithm of the limit (provided these limits exist). Identifying this limiting moment generating function as that of the standard normal distribution, we need only the two theorems stated on page 175 to complete the proof of Theorem 6.9.

Sometimes the central limit theorem is interpreted incorrectly as implying that the distribution of $\bar{\mathbf{x}}$ approaches a normal distribution when $n \to \infty$. This is incorrect because $\mathrm{var}(\bar{\mathbf{x}}) \to 0$ when $n \to \infty$; on the other hand, the central limit theorem justifies our *approximating* the distribution of $\bar{\mathbf{x}}$ with a normal distribution having the mean μ and the variance $\frac{\sigma^2}{n}$ when n is fairly large. It is of interest to note that if the distribution of the original random variables $\mathbf{x}_i$ actually *is* a normal distribution with the mean μ and the variance σ^2, then the distribution of $\bar{\mathbf{x}}$ *is* a normal distribution with the mean μ and the variance $\frac{\sigma^2}{n}$ regardless of the size of n. This will be proved in Section 7.2.1.

THEORETICAL EXERCISES

1. The following is a *sufficient* condition for the central limit theorem: *If the random variables $\mathbf{x}_1, \mathbf{x}_2, \ldots,$ and $\mathbf{x}_n$ are independent and uniformly bounded (that is, there is a positive constant k such that the probability that any one of the $\mathbf{x}_i$ takes on a value greater than k or less than $-k$ is 0), then if the variance of $\mathbf{y}_n = \mathbf{x}_1 + \mathbf{x}_2 + \cdots + \mathbf{x}_n$ becomes infinite when $n \to \infty$, the distribution of the standardized mean of the $\mathbf{x}_i$ approaches the standard normal distribution.* Show that this sufficient condition holds for a sequence of independent random variables $\mathbf{x}_i$ having the probability functions

$$f_i(x_i) = \begin{cases} \frac{1}{2} & \text{for } x_i = 1 - (\frac{1}{2})^i \\ \frac{1}{2} & \text{for } x_i = (\frac{1}{2})^i - 1 \end{cases}$$

2. The following is a *sufficient* condition, the *Laplace-Liapounoff* condition, for the central limit theorem: *If $\mathbf{x}_1, \mathbf{x}_2, \mathbf{x}_3, \ldots,$ is a sequence of independent random variables, each having an absolute third moment*

$$c_i = E(|\mathbf{x}_i - \mu_i|^3)$$

and if

$$\lim_{n\to\infty} [\text{var}(\mathbf{y}_n)]^{-\frac{3}{2}} \sum_{i=1}^{n} c_i = 0$$

where $\mathbf{y}_n = \mathbf{x}_1 + \mathbf{x}_2 + \cdots + \mathbf{x}_n$, then the distribution of the standardized mean of the $\mathbf{x}_i$ approaches the standard normal distribution when $n \to \infty$. Use this condition to show that the central limit theorem holds for the sequence of random variables of Exercise 1.

3. If $\mathbf{x}_1, \mathbf{x}_2, \mathbf{x}_3, \ldots,$ is a sequence of independent random variables having the uniform densities

$$f_i(x_i) = \begin{cases} \dfrac{1}{2 - \dfrac{1}{i}} & \text{for } 0 < x_i < 2 - \dfrac{1}{i} \\ 0 & \text{elsewhere} \end{cases}$$

show that the central limit theorem holds using

(a) the conditions of Exercise 1;
(b) the conditions of Exercise 2.

4. Looking at the binomial distribution as we did on page 193, use the central limit theorem to prove Theorem 5.18.

APPLIED EXERCISES

5. Repeat Exercise 12 on page 202, using the central limit theorem instead of Chebyshev's theorem.
6. Repeat Exercise 13 on page 202, using the central limit theorem instead of Chebyshev's theorem.

REFERENCES

Necessary and sufficient conditions for the strongest form of the central limit theorem for independent random variables, the so-called *Lindeberg-Feller* conditions, are given in

Feller, W., *An Introduction to Probability Theory and its Applications*, Vol. I, 2nd ed. New York: John Wiley & Sons, Inc., 1957

and, of course, in many advanced texts on probability theory.

7

SAMPLING DISTRIBUTIONS

7.1 INTRODUCTION

The mean $\mathbf{\bar{x}}$ is typical of what we mean by a *statistic*—it is a random variable whose values are determined by sample data, namely, by the values of a set of random variables $\mathbf{x}_1, \mathbf{x}_2, \ldots, \mathbf{x}_n$. Other examples of statistics are the *sample range* $\mathbf{r}$, whose values are always given by the largest sample observation minus the smallest, and the *sample variance* $\mathbf{s}^2$, whose values are given by

$$s^2 = \frac{1}{n-1} \cdot \sum_{i=1}^{n} (x_i - \bar{x})^2 \qquad (7.1.1)$$

and, hence, are indicative of the variability of the x's which are actually observed. [Note that (7.1.1) differs from (6.2.13), which defined the variance of a finite population, only inasmuch as we are dividing by $n - 1$ instead of n. The reason for this will be explained in Section 9.3.1, where we shall use values of $\mathbf{s}^2$ to *estimate* σ^2, the variance of the "population" from which the data are obtained.]

Since the values of statistics will vary from sample to sample, it is customary to refer to their distributions (that is, their probability densities or their probability functions) as *sampling distributions*. Thus, we shall devote most of this chapter to the sampling distributions of various statistics which play important roles in applied statistics.

7.2 SAMPLING FROM NORMAL POPULATIONS

To distinguish between the two kinds of random samples defined in Sections 6.2.3 and 6.2.1, we refer to them, respectively, as samples from *finite* and *infinite* populations, where "finite" does not require an explanation, but "infinite" does. To justify this terminology, suppose that we are sampling *with replacement* from a finite population of size N and that for each drawing the probability of each element is $1/N$. Then, there is no limit to the size of the sample, the population is for all practical purposes infinite, and the observations are values of random variables satisfying the conditions for a random sample of Section 6.2.1.

In Sections 7.2.1 through 7.2.4 we shall concern ourselves only with random samples from infinite populations having normal distributions; or to put it more briefly, we shall limit our discussion to *random samples from normal populations.* In particular, we shall investigate the sampling distributions of the statistics $\bar{\mathbf{x}}$, $\mathbf{s}^2$, $\frac{(\bar{\mathbf{x}} - \mu)\sqrt{n}}{\mathbf{s}}$, and $\mathbf{s}_1^2/\mathbf{s}_2^2$, where μ is the mean of the population from which the sample is obtained and $\mathbf{s}_1^2$ and $\mathbf{s}_2^2$ are the variances of independent random samples from two normal populations.

7.2.1 The Distribution of $\bar{\mathbf{x}}$

Having seen in Section 6.3 that *for large n* the sampling distribution of $\bar{\mathbf{x}}$ can (under very general conditions) be approximated with a normal distribution, let us now show that for random samples from normal populations the sampling distribution of $\bar{\mathbf{x}}$ is a normal distribution *regardless of the size of the sample.* To prove this we have only to substitute into

$$M_{\bar{\mathbf{x}}}(t) = \left[M_{\mathbf{x}}\left(\frac{t}{n}\right)\right]^n \tag{7.2.1}$$

which follows immediately from Theorems 5.7 and 6.2 (as the reader will be asked to verify in Exercise 1 on page 216) or as a special case of Exercise 5 on page 194.

Since the moment generating function of the normal distribution with the mean μ and the variance σ^2 is given by

$$M_{\mathbf{x}}(t) = e^{\mu t + \frac{1}{2}t^2\sigma^2}$$

according to Theorem 5.17, substitution into (7.2.1) yields

$$\begin{aligned} M_{\bar{\mathbf{x}}}(t) &= \left[e^{\mu\cdot\frac{t}{n} + \frac{1}{2}\left(\frac{t}{n}\right)^2\sigma^2}\right]^n \\ &= e^{\mu t + \frac{1}{2}t^2\left(\frac{\sigma^2}{n}\right)} \end{aligned}$$

namely, the moment generating function of the normal distribution with the mean μ and the variance $\frac{\sigma^2}{n}$. Referring again to the one-to-one correspondence between moment generating functions and probability distributions mentioned on page 175, we have thus shown that

THEOREM 7.1

If $\bar{\mathbf{x}}$ is the mean of a random sample of size n from a normal population with the mean μ and the variance σ^2, then the sampling distribution of $\bar{\mathbf{x}}$ is a normal distribution with the mean μ and the variance $\frac{\sigma^2}{n}$.

Of course, the fact that the mean and the variance of the sampling distribution of $\bar{\mathbf{x}}$ are, respectively, μ and $\frac{\sigma^2}{n}$ was already shown in Theorem 6.5; the only new part of Theorem 7.1 is that under the given conditions the sampling distribution of $\bar{\mathbf{x}}$ *is* a normal distribution.

7.2.2 The Chi-Square Distribution and the Distribution of s^2

In Section 4.4 we showed that if $\mathbf{x}$ has the standard normal distribution, then $\mathbf{x}^2$ has a gamma distribution with $\alpha = \frac{1}{2}$ and $\beta = 2$. In view of this relationship, the gamma distribution plays an important role in theory connected with sampling from normal populations; especially, the gamma distribution (4.3.6) with the parameters $\alpha = \nu/2$ and $\beta = 2$, which is given by

$$f(x) = \begin{cases} \dfrac{1}{2^{\nu/2}\Gamma(\nu/2)} x^{\frac{\nu-2}{2}} e^{-\frac{x}{2}} & \text{for } x > 0 \\ 0 & \text{for } x \leq 0 \end{cases} \tag{7.2.2}$$

It is customary to refer to (7.2.2) as the density of the *chi-square distribution* (also written χ^2 *distribution*) with ν degrees of freedom; its parameter, ν (*nu*), is called the *number of degrees of freedom*. Since the chi-square distribution is a special gamma distribution and, hence, not really new, we know from Theorem 5.15 that its moment generating function is given by

$$M_{\mathbf{x}}(t) = (1 - 2t)^{-\nu/2} \tag{7.2.3}$$

and as a result of Theorem 5.13 that its mean and variance are, respectively, ν and 2ν.

The chi-square distribution has several important mathematical properties, which are given in Theorems 7.2 through 7.5. First, let us restate the result obtained on page 124 as

THEOREM 7.2

If $\mathbf{x}$ has the standard normal distribution, then $\mathbf{x}^2$ has the chi-square distribution with 1 degree of freedom.

More generally, let us show that

THEOREM 7.3

If $\mathbf{x}_1, \mathbf{x}_2, \ldots, \mathbf{x}_n$ are independent random variables having standard normal distributions, then

$$\mathbf{y} = \sum_{i=1}^{n} \mathbf{x}_i^2$$

has the chi-square distribution with n degrees of freedom.

Using Theorems 6.2 and 7.2 [and, hence, (7.2.3) with $\nu = 1$], we get

$$M_{\mathbf{y}}(t) = \prod_{i=1}^{n} (1 - 2t)^{-\frac{1}{2}} = (1 - 2t)^{-n/2}$$

and this moment generating function can be identified by inspection as that of the chi-square distribution with n degrees of freedom.

Two further properties of the chi-square distribution are given in the following two theorems, which the reader will be asked to prove in Exercises 5 and 6 on page 216:

THEOREM 7.4

If $\mathbf{x}_1, \mathbf{x}_2, \ldots, \mathbf{x}_n$ are independent random variables having chi-square distributions with $\nu_1, \nu_2, \ldots, \nu_n$ degrees of freedom, then

$$\mathbf{y} = \sum_{i=1}^{n} \mathbf{x}_i$$

has the chi-square distribution with $\nu_1 + \nu_2 + \cdots + \nu_n$ degrees of freedom.

THEOREM 7.5

If $\mathbf{x}_1$ and $\mathbf{x}_2$ are independent random variables, $\mathbf{x}_1$ has a chi-square distribution with ν_1 degrees of freedom, and $\mathbf{x}_1 + \mathbf{x}_2$ has a chi-square distribution with $\nu > \nu_1$ degrees of freedom, then $\mathbf{x}_2$ has a chi-square distribution with $\nu - \nu_1$ degrees of freedom.

The chi-square distribution has many important applications, some of which will be discussed in Chapters 9 through 11. Foremost, there are those based, directly or indirectly, on the following theorem:

THEOREM 7.6

If $\mathbf{s}^2$ *is the variance of a random sample of size n from a normal population with the mean* μ *and the variance* σ^2, *then* $\frac{(n-1)\mathbf{s}^2}{\sigma^2}$ *has a chi-square distribution with* $n - 1$ *degrees of freedom.*

In spite of the importance of this theorem, we shall merely outline the major steps of its proof; some of the more advanced work that is required to fill in the details is referred to on page 230.

First, let us observe that for a random sample of size n from a normal population with the mean μ and the variance σ^2

$$\sum_{i=1}^{n}\left(\frac{\mathbf{x}_i-\mu}{\sigma}\right)^2 = \frac{1}{\sigma^2}\cdot\sum_{i=1}^{n}(\mathbf{x}_i-\mu)^2 \qquad (7.2.4)$$

is a random variable having the chi-square distribution with n degrees of freedom; this follows from Theorem 7.3. Now, if we write the random variable with which we are concerned in Theorem 7.6 as

$$\frac{(n-1)\mathbf{s}^2}{\sigma^2} = \frac{1}{\sigma^2}\cdot\sum_{i=1}^{n}(\mathbf{x}_i-\bar{\mathbf{x}})^2 \qquad (7.2.5)$$

it can be seen that (7.2.4) and (7.2.5) differ only insofar as $\bar{\mathbf{x}}$ has taken the place of μ; in other words, in (7.2.4) we are dealing with the squared deviations from μ and in (7.2.5) we are dealing with the squared deviations from $\bar{\mathbf{x}}$. As the reader will be asked to verify in Exercise 7 on page 216, the exact relationship between these two random variables is given by

$$\frac{1}{\sigma^2}\cdot\sum_{i=1}^{n}(\mathbf{x}_i-\mu)^2 = \left(\frac{\bar{\mathbf{x}}-\mu}{\sigma/\sqrt{n}}\right)^2 + \frac{(n-1)\mathbf{s}^2}{\sigma^2} \qquad (7.2.6)$$

so that Theorem 7.5 would lead to the result that $\frac{(n-1)\mathbf{s}^2}{\sigma^2}$ has a chi-square distribution with $n - 1$ degrees of freedom *provided the random variables* $\left(\frac{\bar{\mathbf{x}}-\mu}{\sigma/\sqrt{n}}\right)^2$ *and* $\frac{(n-1)\mathbf{s}^2}{\sigma^2}$ *are independent.* After all, $\left(\frac{\bar{\mathbf{x}}-\mu}{\sigma/\sqrt{n}}\right)^2$ has a chi-square distribution with 1 degree of freedom according to

Theorems 7.1 and 7.2. Proofs of this independence are referred to on page 230, and in Exercise 9 on page 217 the reader will be asked to prove it for the special case where $n = 2$. It should also be noted that $\mathbf{s}^2$ is expressed in terms of the deviations from the mean $\mathbf{x}_i - \bar{\mathbf{x}}$ and that we showed in Theorem 6.6 that $\text{cov}(\mathbf{x}_i - \bar{\mathbf{x}}, \bar{\mathbf{x}}) = 0$ under the given conditions. Of course, we saw on page 179 that a zero covariance does *not* imply independence, but it does serve as some indication of how the corresponding random variables "vary together" (see discussion on page 179).

Since the chi-square distribution arises in many important applications, integrals of its density have been extensively tabulated. Table V at the end of the book contains the values of $\chi^2_{\alpha,\nu}$, where

$$\int_{\chi^2_{\alpha,\nu}}^{\infty} f(x)\, dx = \alpha$$

and $f(x)$ is given by (7.2.2), for $\alpha = 0.995, 0.99, 0.975, 0.95, 0.05, 0.025, 0.01, 0.005$, and $\nu = 1, 2, \ldots, 30$ (see Figure 7.1). When ν is greater

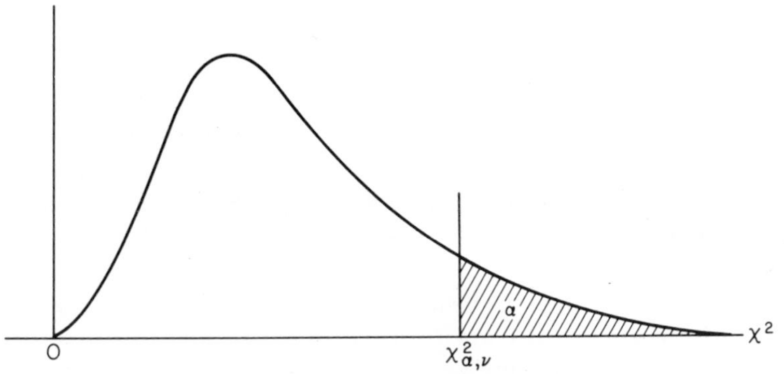

FIGURE 7.1 The Chi-square distribution.

than 30, probabilities related to chi-square distributions are usually approximated with the use of normal distributions (see Exercises 10 and 11 on page 217).

To give a numerical application of Theorem 7.6, suppose that the thickness of a part used in a semiconductor is its critical dimension, and that the process of manufacturing these parts is considered to be under control if the "true" variation among the thicknesses of the parts is given by a standard deviation not greater than $\sigma = 0.60$ thousandth of an inch. To keep a check on the process, random samples of size $n = 20$ are taken periodically, and it is regarded to be "out of control" if the probability that $\mathbf{s}^2$ will take on a value greater than or equal to the observed sample value is *at most 0.01* (even though $\sigma = 0.60$). What should they conclude about the process if they obtained a standard deviation of $s = 0.84$ thousandth of an inch (where the standard deviation is, as before, the positive square root of the variance)?

Translating the criterion into the language of Theorem 7.6, the process will be declared "out of control" if $\frac{(n-1)s^2}{\sigma^2}$ with $n = 20$ and $\sigma = 0.60$ exceeds $\chi^2_{.01,19} = 36.191$, and since

$$\frac{(n-1)s^2}{\sigma^2} = \frac{19(0.84)^2}{(0.60)^2} = 37.24$$

this *is* the case in our example. (Of course, it is assumed in this analysis that the sample may be regarded as a random sample from a normal population.)

THEORETICAL EXERCISES

1. Verify (7.2.1) using (a) Theorems 5.7 and 6.2, and (b) Exercise 5 on page 194.
2. Differentiating the function given by (7.2.1) twice with respect to t and evaluating these derivatives at $t = 0$, prove Theorem 6.5.
3. If $\mathbf{x}_1$ and $\mathbf{x}_2$ are independent random variables having normal distributions with the means μ_1 and μ_2 and the variances σ_1^2 and σ_2^2, show that $\mathbf{x}_1 - \mathbf{x}_2$ has a normal distribution with the mean $\mu_1 - \mu_2$ and the variance $\sigma_1^2 + \sigma_2^2$.
4. Use Theorem 7.1 and the result of Exercise 3 to find the distribution of $\bar{\mathbf{x}}_1 - \bar{\mathbf{x}}_2$, where $\bar{\mathbf{x}}_1$ and $\bar{\mathbf{x}}_2$ are the means of independent random samples of size n_1 and n_2 from normal populations having the means μ_1 and μ_2 and the variances σ_1^2 and σ_2^2. What are the mean and the variance of this distribution?
5. Prove Theorem 7.4.
6. Prove Theorem 7.5.
7. Verify the identity (7.2.6).
8. Use Theorem 7.6 and the expressions for the mean and the variance of the chi-square distribution with ν degrees of freedom (see page 212), to show that for random samples of size n from a normal population with the variance σ^2 the sampling distribution of $\mathbf{s}^2$ has the mean σ^2 and the variance $\frac{2\sigma^4}{n-1}\cdot$ (A general formula for the variance of the sampling distribution of $\mathbf{s}^2$ for random samples from any population

having a finite second and fourth moment may be found in the book by H. Cramér listed on page 230.)

9. Given the independent random variables $\mathbf{x}_1$ and $\mathbf{x}_2$ having standard normal distributions, show that
   (a) the joint density of $\mathbf{x}_1$ and $\bar{\mathbf{x}}$ is given by

$$g(x_1, \bar{x}) = \frac{1}{\pi} \cdot e^{-\bar{x}^2} e^{-(x_1-\bar{x})^2}$$

   for $-\infty < x_1 < \infty$ and $-\infty < \bar{x} < \infty$;
   (b) $\mathbf{s}^2 = 2(\mathbf{x}_1 - \bar{\mathbf{x}})^2$;
   (c) the joint density of $\bar{\mathbf{x}}$ and $\mathbf{s}^2$ is given by

$$h(s^2, \bar{x}) = \frac{1}{\sqrt{\pi}} e^{-\bar{x}^2} \cdot \frac{1}{\sqrt{2\pi}} (s^2)^{-\frac{1}{2}} e^{-\frac{1}{2}s^2}$$

   for $s^2 > 0$ and $-\infty < \bar{x} < \infty$, and hence $\mathbf{s}^2$ and $\bar{\mathbf{x}}$ are *independent.*

10. Given a random variable $\mathbf{x}$ whose range is the positive real numbers, show that
   (a) the probability that the random variable $\sqrt{2\mathbf{x}} - \sqrt{2n}$ takes on a value less than k equals the probability that the random variable $\frac{\mathbf{x} - n}{\sqrt{2n}}$ takes on a value less than $k + \frac{k^2}{2\sqrt{2n}}$;
   (b) if $\mathbf{x}$ has a chi-square distribution with n degrees of freedom, then for large n the distribution of $\sqrt{2\mathbf{x}} - \sqrt{2n}$ can be approximated with the standard normal distribution.

   Also use the result of part (b) and Theorem 7.6 to show that for large n the variance of the sampling distribution of $\mathbf{s}$ is approximately $\frac{\sigma^2}{2(n-1)}$.

11. Find approximate values for the probability that a random variable $\mathbf{x}$ having the chi-square distribution with 50 degrees of freedom takes on a value greater than 68.0
   (a) by treating $\frac{\mathbf{x} - \nu}{\sqrt{2\nu}}$ with $\nu = 50$ as a random variable having the standard normal distribution;
   (b) by treating $\sqrt{2\mathbf{x}} - \sqrt{2\nu}$ with $\nu = 50$ as a random variable having the standard normal distribution.

   (Note that the actual value of this probability is 0.04596.)

12. Derive the following computing formula for the sample variance (7.1.1):

$$s^2 = \frac{n \cdot \sum_{i=1}^{n} x_i^2 - \left(\sum_{i=1}^{n} x_i\right)^2}{n(n-1)}$$

APPLIED EXERCISES

13. A random sample of size 64 is to be taken from a normal population with $\mu = 51.4$ and $\sigma = 6.8$. Find the probabilities that the mean of the sample will
(a) exceed 53.1;
(b) fall on the interval from 50.4 to 52.4;
(c) be less than 50.8.

14. A random sample of size 100 is taken from a normal population with the variance 625. What is the probability that the mean of the sample will differ from the mean of the population by 4 or more *either way?*

15. Independent random samples of size 40 are taken from two normal populations having equal means and the variances $\sigma_1^2 = 20$ and $\sigma_2^2 = 60$. What is the probability that the difference between the two sample means will be numerically less than 2? (*Hint:* Use the results of Exercise 4.)

16. Independent random samples of size $n_1 = 30$ and $n_2 = 50$ are taken from two normal populations having the means $\mu_1 = 78$ and $\mu_2 = 75$ and the variances $\sigma_1^2 = 150$ and $\sigma_2^2 = 200$. What is the probability that the mean of the first sample will exceed that of the second sample by at least 4.5?

17. Perform the necessary integration of an appropriate chi-square density to find the probability that the variance of a random sample of size 5 from a normal population with $\sigma^2 = 25$ will fall between 20 and 30.

18. The claim that the variance of a normal population is $\sigma^2 = 25$ is to be rejected if the variance of a random sample of size 16 exceeds 50.963 or is less than 8.715. What is the probability that this claim will be rejected even though $\sigma^2 = 25$?

19. The claim that the variance of a normal population is $\sigma^2 = 4$ is to be rejected if the variance of a random sample of size 9 exceeds 8.7675. What is the probability this claim will be rejected even though $\sigma^2 = 4$?

7.2.3 The F Distribution

Another distribution which plays an important role in connection with sampling from normal populations is the F *distribution*, which we shall derive by means of a transformation from the beta distribution (4.3.7). Letting the parameters of this distribution be $\alpha = \nu_1/2$ and

$\beta = \nu_2/2$, we have

$$f(x) = \begin{cases} k \cdot x^{\frac{\nu_1}{2}-1}(1-x)^{\frac{\nu_2}{2}-1} & \text{for } 0 < x < 1 \\ 0 & \text{elsewhere} \end{cases} \tag{7.2.7}$$

where

$$k = \frac{\Gamma\left(\frac{\nu_1+\nu_2}{2}\right)}{\Gamma\left(\frac{\nu_1}{2}\right)\Gamma\left(\frac{\nu_2}{2}\right)}$$

and it will be our goal to find the probability of the random variable **y**, whose values are related to those of the random variable **x** [which has the density (7.2.7)] by means of the equation

$$y = \frac{\nu_2 x}{\nu_1(1-x)} \tag{7.2.8}$$

Solved for x, this equation becomes $x = \dfrac{y\nu_1}{\nu_2 + y\nu_1}$, so that $\dfrac{dx}{dy} = \dfrac{\nu_1\nu_2}{(\nu_2 + y\nu_1)^2}$ and substitution into $g(y) = f(x) \cdot \left|\dfrac{dx}{dy}\right|$ in accordance with Theorem 4.1 yields

$$g(y) = \begin{cases} \dfrac{\Gamma\left(\frac{\nu_1+\nu_2}{2}\right)}{\Gamma\left(\frac{\nu_1}{2}\right)\Gamma\left(\frac{\nu_2}{2}\right)}\left(\dfrac{\nu_1}{\nu_2}\right)^{\frac{\nu_1}{2}} \cdot y^{\frac{\nu_1}{2}-1}\left(1 + \dfrac{\nu_1}{\nu_2}\cdot y\right)^{-\frac{1}{2}(\nu_1+\nu_2)} & \text{for } y > 0 \\ 0 & \text{elsewhere} \end{cases} \tag{7.2.9}$$

The distribution whose density is given by (7.2.9) is called the *F distribution with ν_1 and ν_2 degrees of freedom*, and in view of its importance it has been tabulated extensively. Table VI, for instance, contains the values of F_{α,ν_1,ν_2}, where

$$\int_{F_{\alpha,\nu_1,\nu_2}}^{\infty} g(y)\,dy = \alpha$$

and $g(y)$ is given by (7.2.9), for $\alpha = 0.05$ and 0.01, and various values of ν_1 and ν_2 (see Figure 7.2).

The importance of the F distribution in applied statistics is due mainly to the following theorem:

THEOREM 7.7

If $\mathbf{x}_1$ and $\mathbf{x}_2$ are independent random variables having chi-square distributions with ν_1 and ν_2 degrees of freedom, then

$$\mathbf{y} = \frac{\mathbf{x}_1/\nu_1}{\mathbf{x}_2/\nu_2}$$

has the F distribution with ν_1 and ν_2 degrees of freedom.

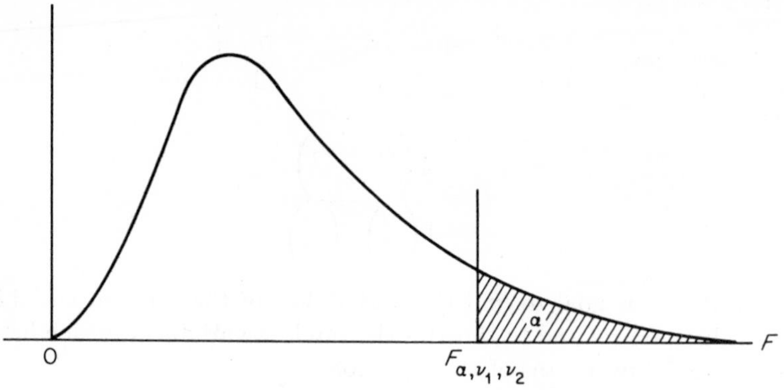

FIGURE 7.2 The F distribution.

The details of this proof will be left to the reader in Exercise 1 on page 223; essentially, it consists of beginning with the joint density of $\mathbf{x}_1$ and $\mathbf{x}_2$, performing a change of scale (by the method of Section 4.4) to get the joint density of $\mathbf{y}$ and $\mathbf{x}_2$, and then integrating out x_2 to obtain the marginal density of $\mathbf{y}$.

One application of Theorem 7.7 arises in problems in which we are interested in comparing the variances σ_1^2 and σ_2^2 of two normal populations. For example, when we want to test whether $\sigma_1^2 = \sigma_2^2$ on the basis of independent random samples, for which we know according to Theorem 7.6 that

$$\frac{(n_1 - 1)\mathbf{s}_1^2}{\sigma_1^2} \quad \text{and} \quad \frac{(n_2 - 1)\mathbf{s}_2^2}{\sigma_2^2}$$

are (independent) random variables having respective chi-square distributions with $n_1 - 1$ and $n_2 - 1$ degrees of freedom. Under the assumption that σ_1^2 equals σ_2^2, we thus find that $\mathbf{s}_1^2/\mathbf{s}_2^2$ has an F distribution with $n_1 - 1$ and $n_2 - 1$ degrees of freedom, and we can base our decision (about the hypothesis $\sigma_1^2 = \sigma_2^2$) on the value which we obtain for the *ratio* of the two sample variances. This method will be illustrated in Section 11.3, and for the time being let us merely point out that it is for this reason that the F distribution is also known as the *variance-ratio distribution.*

7.2.4 The t Distribution

Earlier in this chapter we showed that for random samples from a normal population with the mean μ and the variance σ^2, $\bar{\mathbf{x}}$ has a normal distribution with the mean μ and the variance $\frac{\sigma^2}{n}$; that is, $\frac{\bar{\mathbf{x}} - \mu}{\sigma/\sqrt{n}}$ has the standard normal distribution. The major difficulty in applying this result is that in actual practice σ is usually unknown, which makes it necessary to replace it with a value of the sample standard deviation $\mathbf{s}$ (or some other estimate). The theory which follows leads to the *exact* sampling distribution of $\frac{\bar{\mathbf{x}} - \mu}{\mathbf{s}/\sqrt{n}}$ for random samples from normal populations.

To derive this sampling distribution, let us first study the more general problem of obtaining the probability density of the random variable

$$\mathbf{t} = \frac{\mathbf{x}}{\sqrt{\mathbf{y}/\nu}} \tag{7.2.10}$$

where $\mathbf{x}$ and $\mathbf{y}$ are independent, $\mathbf{x}$ has the standard normal distribution, and $\mathbf{y}$ has the chi-square distribution with ν degrees of freedom. The joint density of $\mathbf{x}$ and $\mathbf{y}$ is thus given by

$$f(x, y) = \frac{1}{\sqrt{2\pi}} e^{-\frac{1}{2}x^2} \cdot \frac{1}{\Gamma\left(\frac{\nu}{2}\right) 2^{\frac{\nu}{2}}} y^{\frac{\nu}{2}-1} e^{-\frac{y}{2}}$$

for $-\infty < x < \infty$ and $y > 0$, and $f(x, y) = 0$ elsewhere, and the method of Section 4.4 leads to $g(t, y) = f(x, y) \cdot \left|\frac{dx}{dt}\right|$, namely,

$$g(t, y) = \begin{cases} \dfrac{1}{\sqrt{2\pi}\,\Gamma\left(\frac{\nu}{2}\right) 2^{\frac{\nu}{2}}} y^{\frac{\nu-1}{2}} e^{-\frac{y}{2}\left(1+\frac{t^2}{\nu}\right)} & \text{for } -\infty < t < \infty \text{ and } y > 0 \\ 0 & \text{elsewhere} \end{cases}$$

since $t = \frac{x}{\sqrt{y/\nu}}$ and, hence, $\frac{dx}{dt} = \sqrt{y/\nu}$ where y is held fixed. Finally, integrating out y with the aid of the substitution $z = \frac{y}{2}\left(1 + \frac{t^2}{\nu}\right)$, we get

$$f(t) = \frac{\Gamma\left(\frac{\nu+1}{2}\right)}{\sqrt{\pi\nu}\,\Gamma\left(\frac{\nu}{2}\right)} \cdot \left(1 + \frac{t^2}{\nu}\right)^{-\frac{\nu+1}{2}} \qquad \text{for } -\infty < t < \infty \tag{7.2.11}$$

This is the density of the *t distribution with ν degrees of freedom.* It was first obtained by W. S. Gosset, who published his research under the pen name "Student"; hence, the distribution is also known as the *Student-t distribution.*

In view of its importance, the t distribution has been tabulated extensively; Table IV, for example, contains the values of $t_{\alpha,\nu}$, where

$$\int_{t_{\alpha,\nu}}^{\infty} f(t)\, dt = \alpha$$

and $f(t)$ is given by (7.2.11), for $\alpha = 0.10, 0.05, 0.025, 0.01, 0.005$, and $\nu = 1, 2, \ldots, 29$ (see Figure 7.3). The table does not contain values of

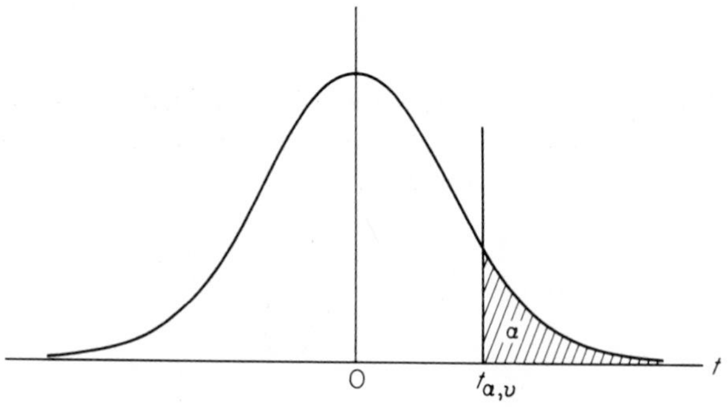

FIGURE 7.3 The t distribution.

$t_{\alpha,\nu}$ for $\alpha > 0.50$, since the density is *symmetrical* about $t = 0$ and, hence, $t_{1-\alpha,\nu} = -t_{\alpha,\nu}$. When ν is 30 or more, probabilities related to the t distribution are usually approximated with the use of normal distributions (see Exercises 7 and 9 on pages 223 and 224).

Although the t distribution has many other applications, some of which will be treated in Chapter 11, its major application (for which it was originally developed) is based on the following theorem:

THEOREM 7.8

If $\bar{\mathbf{x}}$ *and* $\mathbf{s}^2$ *are the mean and the variance of a random sample of size n from a normal population with the mean* μ *and the variance* σ^2, *then* $\dfrac{\bar{\mathbf{x}} - \mu}{\mathbf{s}/\sqrt{n}}$ *has the t distribution with* $n - 1$ *degrees of freedom.*

To prove this theorem, we have only to let $\mathbf{x}$ in (7.2.10) be the random variable $\dfrac{\bar{\mathbf{x}} - \mu}{\sigma/\sqrt{n}}$, and let $\mathbf{y}$ in (7.2.10) be the random variable $\dfrac{(n-1)\mathbf{s}^2}{\sigma^2}$. For random samples from a normal population with the mean μ and the variance σ^2, these two random variables are known to have, respectively, the standard normal distribution and the chi-square distribution with $n - 1$ degrees of freedom.

To illustrate the use of Theorem 7.8, suppose that we want to test the claim that the average gasoline consumption of a certain type of engine is 12.0 gallons per hour, and that we have at our disposal the results of 16 one-hour test runs for which the gasoline consumption averaged 16.4 gallons with a standard deviation of 2.1 gallons. Substituting $n = 16$, $\mu = 12.0$, $\bar{x} = 16.4$, and $s = 2.1$, we get

$$t = \frac{\bar{x} - \mu}{s/\sqrt{n}} = \frac{16.4 - 12.0}{2.1/\sqrt{16}} = 8.38$$

Since Table IV shows that the probability of getting a value of t greater than 2.947 is 0.005 for 15 degrees of freedom, the probability of getting a value greater than 8 must be *negligible*, and it would seem reasonable to conclude that the "true" average hourly gasoline consumption of the engine must be greater than 12.0 gallons.

THEORETICAL EXERCISES

1. Following the suggestions on page 220, give a detailed proof of Theorem 7.7.
2. If the random variable $\mathbf{x}$ has an F distribution with ν_1 and ν_2 degrees of freedom, show that the random variable $\mathbf{y} = \frac{1}{\mathbf{x}}$ has an F distribution with ν_2 and ν_1 degrees of freedom.
3. Using the result of Exercise 2, show that $F_{1-\alpha,\nu_1,\nu_2} = 1/F_{\alpha,\nu_2,\nu_1}$.
4. Show that for $\nu_2 > 2$ the mean of the F distribution is $\frac{\nu_2}{\nu_2 - 2}$, and investigate what happens when ν_2 equals 1 and 2.
5. If $\mathbf{x}$ has an F distribution with ν_1 and ν_2 degrees of freedom, show that when $\nu_2 \to \infty$ the distribution of $\nu_1\mathbf{x}$ approaches the chi-square distribution with ν_1 degrees of freedom.
6. Show that if $\mathbf{x}$ has the t distribution with ν degrees of freedom, then $\mathbf{x}^2$ has the F distribution with 1 and ν degrees of freedom.
7. Use Stirling's formula of Exercise 10 on page 30 to show that when $\nu \to \infty$ the t distribution approaches the standard normal distribution.
8. Show that the t distribution with 1 degree of freedom is a Cauchy distribution.

9. Show that for $\nu > 2$ the variance of the t distribution with ν degrees of freedom is $\frac{\nu}{\nu - 2}$.

10. Show that the F distribution with 4 and 4 degrees of freedom is given by

$$g(F) = \begin{cases} 6F(1+F)^{-4} & \text{for } F > 0 \\ 0 & \text{for } F \leq 0 \end{cases}$$

and use this density to determine the probability that for independent random samples of size 5 from two normal populations having the same variance, $\mathbf{s}_1^2/\mathbf{s}_2^2$ will take on a value less than $\frac{1}{2}$ or greater than 2.

APPLIED EXERCISES

11. If independent random samples of size 10 from two normal populations have the sample variances $s_1^2 = 12.8$ and $s_2^2 = 3.2$, what can we conclude from this about the claim that the two populations have the same variance?

12. A random sample of size 25 from a normal population has the mean $\bar{x} = 47$ and the standard deviation $s = 6$. Basing our decision on the statistic of Theorem 7.8, can we say that the given information supports the hypothesis that the mean of the population is $\mu = 42$?

13. A random sample of size 9 from a normal population has the mean $\bar{x} = 27.7$ and the variance $s^2 = 3.24$. Basing our decision on the statistic of Theorem 7.8, can we say that the given information supports the claim that the mean of the population is $\mu = 28.5$?

7.3 SAMPLING DISTRIBUTIONS OF ORDER STATISTICS

Given a random sample of size n from an infinite population with a continuous density, suppose that we arrange the values taken on by $\mathbf{x}_1, \mathbf{x}_2, \ldots, \mathbf{x}_n$ according to size and look upon the smallest of the x's as a value of the random variable $\mathbf{y}_1$, the next largest as a value of the random variable $\mathbf{y}_2$, the next largest after that as a value of the random variable $\mathbf{y}_3, \ldots$, and the largest as a value of the random variable $\mathbf{y}_n$. The random variables thus defined are referred to as *order statistics;* in particular, $\mathbf{y}_1$ is the *first order statistic*, $\mathbf{y}_2$ is the *second order statistic*, $\mathbf{y}_3$ is the *third order statistic*, and so on. (Note that we have limited this discussion to infinite populations with continuous densities so that the probability will be zero that any two of the x's will be alike.)

To be more specific, consider the case where $n = 2$ and the relationship between the values of $\mathbf{x}_1$ and $\mathbf{x}_2$ *and* $\mathbf{y}_1$ and $\mathbf{y}_2$ is

$$y_1 = x_1 \text{ and } y_2 = x_2 \text{ when } x_1 < x_2$$

$$y_1 = x_2 \text{ and } y_2 = x_1 \text{ when } x_2 < x_1$$

Similarly, for $n = 3$ the relationship between the values of the respective random variables is

$$y_1 = x_1,\ y_2 = x_2, \text{ and } y_3 = x_3, \text{ when } x_1 < x_2 < x_3$$

$$y_1 = x_1,\ y_2 = x_3, \text{ and } y_3 = x_2, \text{ when } x_1 < x_3 < x_2$$

$$\cdots$$

$$y_1 = x_3,\ y_2 = x_2, \text{ and } y_3 = x_1, \text{ when } x_3 < x_2 < x_1$$

To determine the probability density of the rth order statistic, let us consider the following argument: Suppose that the real axis is divided into three intervals, one from $-\infty$ to y_r, a second from y_r to $y_r + h$ (where h is a positive constant), and the third from $y_r + h$ to ∞. Then, if the density of the population from which we are sampling is given by $f(x)$, the probability that $r - 1$ of the sample values fall into the first interval, one falls into the second interval, and $n - r$ fall into the third interval is

$$\frac{n!}{(r-1)!1!(n-r)!}\left[\int_{-\infty}^{y_r} f(x)\,dx\right]^{r-1}\left[\int_{y_r}^{y_r+h} f(x)\,dx\right]\left[\int_{y_r+h}^{\infty} f(x)\,dx\right]^{n-r}$$

according to the formula for the multinomial distribution (3.4.12). Using the *law of the mean* from calculus, we have

$$\int_{y_r}^{y_r+h} f(x)\,dx = f(\xi)\cdot h \qquad \text{where } y_r \le \xi \le y_r + h$$

and if we let $h \to 0$, we finally get

$$g_r(y_r) = \frac{n!}{(r-1)!(n-r)!}\left[\int_{-\infty}^{y_r} f(x)\,dx\right]^{r-1} f(y_r)\left[\int_{y_r}^{\infty} f(x)\,dx\right]^{n-r} \quad (7.3.1)$$

for $-\infty < y_r < \infty$ for the *probability density of the rth order statistic* $\mathbf{y}_r$.

In particular, the distribution of $\mathbf{y}_1$, the *smallest* value in a random sample of size n, is given by

$$g_1(y_1) = n\cdot f(y_1)\left[\int_{y_1}^{\infty} f(x)\,dx\right]^{n-1} \qquad \text{for } -\infty < y_1 < \infty \quad (7.3.2)$$

while the distribution of $\mathbf{y}_n$, the *largest* value in a random sample of size n, is given by

$$g_n(y_n) = n \cdot f(y_n) \left[\int_{-\infty}^{y_n} f(x)\, dx\right]^{n-1} \quad \text{for } -\infty < y_n < \infty \qquad (7.3.3)$$

Also, in a random sample of size $n = 2m + 1$ the *sample median* $\tilde{\mathbf{x}}$ is $\mathbf{y}_{m+1}$, so that its sampling distribution is given by

$$h(\tilde{x}) = \frac{(2m+1)!}{m!m!}\left[\int_{-\infty}^{\tilde{x}} f(x)\, dx\right]^m f(\tilde{x}) \left[\int_{\tilde{x}}^{\infty} f(x)\, dx\right]^m \quad \text{for } -\infty < \tilde{x} < \infty \qquad (7.3.4)$$

according to (7.3.1). [For random samples of size $n = 2m$ the median is defined as $\frac{1}{2}(\mathbf{y}_m + \mathbf{y}_{m+1})$.]

In some instances it is possible to find expressions for the integrals of (7.3.2), (7.3.3), and (7.3.4); for other populations there may be no choice but to approximate these integrals using numerical methods. The first is the case, for example, for random samples from the exponential population (4.3.2). As the reader will be asked to show in Exercise 1 on page 227, the distributions of $\mathbf{y}_1$ and $\mathbf{y}_n$ for random samples of size n from the exponential population (4.3.2) are given by

$$g_1(y_1) = \begin{cases} \dfrac{n}{\theta} \cdot e^{-\frac{ny_1}{\theta}} & \text{for } y_1 > 0 \\ 0 & \text{for } y_1 \leq 0 \end{cases}$$

and

$$g_n(y_n) = \begin{cases} \dfrac{n}{\theta} \cdot e^{-\frac{y_n}{\theta}}[1 - e^{-\frac{y_n}{\theta}}]^{n-1} & \text{for } y_n > 0 \\ 0 & \text{for } y_n \leq 0 \end{cases}$$

Also, for random samples of size $n = 2m + 1$ from this kind of population, the sampling distribution of the median is given by

$$h(\tilde{x}) = \begin{cases} \dfrac{(2m+1)!}{m!m!\theta} \cdot e^{-\frac{\tilde{x}(m+1)}{\theta}}[1 - e^{-\frac{\tilde{x}}{\theta}}]^m & \text{for } \tilde{x} > 0 \\ 0 & \text{for } \tilde{x} \leq 0 \end{cases}$$

The following is an interesting result about the sampling distribution of the median which holds when the population density $f(x)$ is continuous and non-zero at the *population median* $\tilde{\mu}$ (which is such that $\int_{-\infty}^{\tilde{\mu}} f(x)\, dx = \frac{1}{2}$):

THEOREM 7.9

For large n, the sampling distribution of the median for random samples of size $2n + 1$ *is approximately normal with the mean* $\tilde{\mu}$ *and the variance* $\dfrac{1}{8[f(\tilde{\mu})]^2 n}$.

A proof of this theorem is referred to on page 230. Note that for random samples of size $2n + 1$ from a normal population we have $\mu = \tilde{\mu}$, so that $f(\tilde{\mu}) = f(\mu) = \dfrac{1}{\sigma\sqrt{2\pi}}$ and the variance of the median is approximately $\dfrac{\pi\sigma^2}{4n}$. If we compare this with the variance of the mean, which for random samples of size $2n + 1$ is $\dfrac{\sigma^2}{2n+1}$, we find that for large samples from normal populations *the mean is more reliable than the median*, that is, the mean is subject to smaller chance fluctuations than the median.

THEORETICAL EXERCISES

1. Verify the results given in the text for the sampling distributions of $\mathbf{y}_1$, $\mathbf{y}_n$, and $\tilde{\mathbf{x}}$ for random samples from exponential populations.
2. Find the sampling distributions of $\mathbf{y}_1$ and $\mathbf{y}_n$ for random samples of size n from the *uniform population* (4.3.1) with $\alpha = 0$ and $\beta = 1$. Also find the sampling distribution of $\tilde{\mathbf{x}}$ for random samples of size $2m + 1$ from this population.
3. Find the mean and the variance of the sampling distribution of $\mathbf{y}_1$ for random samples of size n from the uniform population of Exercise 2.
4. Find the sampling distributions of $\mathbf{y}_1$ and $\mathbf{y}_n$ for random samples of size n from a population having the beta distribution (4.3.7) with $\alpha = 3$ and $\beta = 2$. Also find the sampling distribution of $\tilde{\mathbf{x}}$ for random samples of size $2m + 1$ from this population.
5. Find the sampling distribution of $\mathbf{y}_1$ for random samples of size $n = 2$ taken
   (a) without replacement from the first five positive integers;
   (b) with replacement from the first five positive integers.
   (*Hint:* Enumerate all possibilities.)
6. Duplicating the method on page 225, it can be shown that for a random sample of size n the joint density of $\mathbf{y}_1$ and $\mathbf{y}_n$ is given by

$$g(y_1, y_n) = n(n-1)f(y_1)f(y_n)\left[\int_{y_1}^{y_n} f(x)\,dx\right]^{n-2}$$

   for $-\infty < y_1 < y_n < \infty$.

   (a) Use this result to find the joint density of $\mathbf{y}_1$ and $\mathbf{y}_n$ for random samples of size n from a population having the exponential distribution (4.3.2).
   (b) Use this result to find the joint density of $\mathbf{y}_1$ and $\mathbf{y}_n$ for random

samples of size n from a population having the uniform distribution of Exercise 2. Also find the covariance of $\mathbf{y}_1$ and $\mathbf{y}_n$.

7. Use the joint density of $\mathbf{y}_1$ and $\mathbf{y}_n$ given in Exercise 6 and the method of Section 4.4 to find an expression for the joint density of $\mathbf{y}_1$ and the *sample range* $\mathbf{R} = \mathbf{y}_n - \mathbf{y}_1$.

8. Use the result of Exercise 7 to find the sampling distribution of $\mathbf{R}$ for random samples of size n from a population having the exponential distribution (4.3.2).

9. Use the result of Exercise 7 to find the sampling distribution of $\mathbf{R}$ for random samples of size n from a population having the uniform distribution of Exercise 2. Also find the mean and the variance of this sampling distribution of $\mathbf{R}$.

10. **TOLERANCE LIMITS** There are many problems, particularly in industrial applications, in which we are interested in the proportion of a population which lies between certain limits. The following steps lead to the sampling distribution of the statistic $\mathbf{p}$, which is the proportion of a population (having a continuous density) that lies between the smallest and the largest values of a random sample of size n.

    (a) Use the joint density of Exercise 6 and the method of Section 4.4 to show that the joint density of $\mathbf{y}_1$ and $\mathbf{p}$, whose values are given by

    $$p = \int_{y_1}^{y_n} f(x)\,dx$$

    is

    $$h(y_1, p) = n(n-1)f(y_1)p^{n-2}$$

    (b) Use the result of part (a) and the method of Section 4.4 to show that the joint density of $\mathbf{p}$ and $\mathbf{w}$, whose values are given by

    $$w = \int_{-\infty}^{y_1} f(x)\,dx$$

    is

    $$\varphi(w, p) = n(n-1)p^{n-2}$$

    for $w > 0$, $p > 0$, $w + p < 1$, and $\varphi(w, p) = 0$ elsewhere.

    (c) Use the result of part (b) to show that the marginal density of $\mathbf{p}$ is given by

    $$g(p) = \begin{cases} n(n-1)p^{n-2}(1-p) & \text{for } 0 < p < 1 \\ 0 & \text{elsewhere} \end{cases}$$

    This is the desired density of the proportion of the population that lies between the smallest and the largest values of a random sample of size n, and it is of interest to note that it does not depend on the distribution of the population, itself.

11. TOLERANCE LIMITS, CONTINUED Use part (c) of Exercise 10 to show that $E(\mathbf{p}) = \dfrac{n-1}{n+1}$ and $\text{var}(\mathbf{p}) = \dfrac{2(n-1)}{(n+1)^2(n+2)}$. What can we conclude from this about the distribution of p when n is *very large*.

APPLIED EXERCISES

12. Use the result of Exercise 2 to find the probability that in a random sample of size 4 from the given uniform population the smallest value is at least 0.20.

13. Use the result of Exercise 4 to find the probability that in a random sample of size 3 from the given beta population the largest value is less than 0.90.

14. Use the result of Exercise 9 to find the probability that the range of a random sample of size 5 from the given uniform population is at least 0.75.

15. Use the result of part (c) of Exercise 10 to find the probability that in a random sample of size 10 at least 80 percent of the population will fall between the smallest and the largest values.

16. Use the result of part (c) of Exercise 10 to set up an equation in n, whose solution would give the sample size that is required to be able to assert with the probability $1 - \alpha$ that the proportion of the population contained between the smallest and largest sample values is at least P. Show that for $P = 0.90$ and $\alpha = 0.05$ this equation can be written as

$$(0.90)^{n-1} = \frac{1}{2n+18}$$

This kind of equation is difficult to solve, but it can be shown that an approximate solution for n is given by $\dfrac{1}{2} + \dfrac{1}{4} \cdot \dfrac{1+P}{1-P} \cdot \chi^2_{\alpha,4}$ where $\chi^2_{\alpha,4}$ must be looked up in Table V. Use this method to find an approximate solution of the given equation.

REFERENCES

Extensive tables of the normal, chi-square, F, and t distributions may be found in

Pearson, E. S., and Hartley, H. O., *Biometrika Tables for Statisticians*, Vol. I. New York: Cambridge University Press, 1966.

A general formula for the variance of the sampling distribution of the second sample moment $\mathbf{m}_2$ (which differs from $\mathbf{s}^2$ only insofar as we divide by n instead of $n - 1$) is derived in

Cramér, H., *Mathematical Methods of Statistics*. Princeton, N.J.: Princeton University Press, 1950

and a proof of Theorem 7.9 is given in

Wilks, S. S., *Mathematical Statistics*. New York: John Wiley & Sons, Inc., 1962.

Proofs of the independence of $\bar{\mathbf{x}}$ and $\mathbf{s}^2$ for random samples from normal populations are given in many more advanced texts on mathematical statistics. For instance, a proof based on moment generating functions may be found in the above-mentioned book by S. S. Wilks, and a somewhat more elementary proof, illustrated for $n = 3$, may be found in

Keeping, E. S., *Introduction to Statistical Inference*. Princeton, N.J: D. Van Nostrand Co., Inc., 1962.

8

DECISION THEORY

8.1 INTRODUCTION

In Chapter 5 we introduced mathematical expectations to study expected values of random variables; in particular, the moments of their distributions. In applied situations, mathematical expectations are often used as a guide in choosing between alternatives, that is, in making decisions, for it is generally considered "rational" to select alternatives with the "most promising" mathematical expectations—the ones which *maximize expected profits*, *minimize expected losses*, *maximize expected sales*, *minimize expected costs*, and so on. For instance, in Exercise 11 on page 145 it would have been "rational" for the baker to bake 4 chocolate cakes each day (rather than 3 or 5), since this would maximize his expected profit.

Although this approach to decision making has great intuitive appeal and sounds very logical, it is not without complications—there are many problems in which it is very difficult, if not impossible, to assign numerical values to all possible "payoffs" and their respective probabilities. To illustrate some of these difficulties, suppose that the manager of an oil company must decide whether to continue drilling for oil at a certain location. If he decides to continue drilling and there is oil, this will be worth $8,000,000 to his company; if he decides to continue drilling and there is no oil, this will entail a loss of $4,800,000; if he decides to stop drilling and there is oil (for a competitor to use), this will entail a loss of $3,200,000; and if he decides to stop drilling and there is no oil, this will be worth $800,000 to his company (because funds allocated to the project will remain unspent). Schematically, all this information can be represented as in the table on the next page, where the entries are the *losses* which correspond to the various possibilities, and, hence, *gains* are represented by negative numbers.*

* The only reason why we are representing the company's gains by negative numbers in this example (and its losses by positive numbers) is to make this illustration fit the general scheme which we shall introduce in Sections 8.2 and 8.3. (Otherwise, it would, of course, be more logical to represent gains by positive numbers and losses by negative numbers.)

	Manager's Decision Continue drilling	*Manager's Decision* Stop drilling
There is oil	−8,000,000	3,200,000
There is no oil	4,800,000	−800,00

Clearly, it is advantageous to continue drilling if and only if there is oil, and the manager's decision will, therefore, have to depend on the probability that this is the case. Suppose, for example, that (on the basis of many years of experience) the manager feels that the *odds* against there being any oil at this location are 3 to 2, namely, that the probability for oil is 0.40. He can then argue that *if he decides to continue drilling*, the expected losses are

$$-8{,}000{,}000(0.40) + 4{,}800{,}000(0.60) = -320{,}000$$

and *if he decides to stop drilling*, the expected losses are

$$3{,}200{,}000(0.40) + (-800{,}000)(0.60) = 800{,}000$$

Since an expected gain (negative loss) of \$320,000 is obviously preferable to an expected loss of \$800,000, it stands to reason that the manager should continue drilling for oil at the given location. Of course, *this assumed that his assessment of the odds is correct*, and in Exercise 2 on page 240 the reader will be asked to rework this example when the odds against there being any oil are 4 to 1 and when they are 2 to 1. [The analysis is also based on the assumption that the figures given for the various "payoffs" (losses or gains) are correct, and it is easy to see that this may give rise to all sorts of criticisms.]

Let us also examine briefly what the manager of the oil company might have done in our example if he had no idea about the probability of there being any oil. To suggest one possibility, suppose that he is a *confirmed pessimist;* that is, he is the kind of person who always expects the worst. If this were the case, he might argue that continued drilling could lead to a loss of \$4,800,000, stopping could lead to a loss of \$3,200,000, and, hence, he can protect his company against the worst that can happen by stopping the drilling. The criterion on which this decision is based is called the *minimax criterion*, for by stopping the drilling he actually *min*imizes the *max*imum loss to which his company is exposed.

There are numerous other criteria on which the manager of the oil company could base his decision in the absence of any knowledge about the probability that there is oil. One such criterion based on *optimism* is referred to in Exercise 3 on page 240, and another based on the *fear of losing out on a "good deal"* is referred to in Exercise 6 on page 241.

8.2 THE THEORY OF GAMES

The example of the preceding section may well have given the impression that the manager of the oil company is playing a game—a game between the manager and *Nature* (or call it *fate* or whatever "controls" whether there is any oil at the given location) as his opponent. Each of the "players" has the choice of two moves: The manager of the oil company has the choice between actions a_1 and a_2 (to continue or to stop drilling), Nature controls the choice between θ_1 and θ_2 (whether or not there is to be any oil), and depending on the choice of their moves, there are the "payoffs" shown in the following table:

		Player A (The Manager) a_1	a_2
Player B (Nature)	θ_1	$L(a_1, \theta_1)$	$L(a_2, \theta_1)$
	θ_2	$L(a_1, \theta_2)$	$L(a_2, \theta_2)$

In statistics, we refer to the amounts $L(a_1, \theta_1)$, $L(a_2, \theta_1)$, . . . , as values of the *loss function* which characterizes the particular "game." In other words, $L(a_i, \theta_j)$ is the *loss* of Player A (the amount he has to pay to Player B) when he chooses alternative a_i and Player B chooses alternative θ_j. Although it does not really matter, we shall assume throughout our discussion that these amounts are in dollars; in actual practice, they could be expressed in terms of any goods or services, in units of utility (desirability, or satisfaction), or even in terms of life or death (as in Russian roulette or the conduct of a war).

This whole analogy is not at all far-fetched; the problem of Section 8.1 is typical of the kind of situation treated in the *theory of games*, a relatively new branch of mathematics which has stimulated considerable interest in recent years. This theory is not limited to parlor games, as its name might suggest, but it applies to any kind of competitive situation and, as we shall see, it has led to a unified approach to the problems of statistics.

To introduce some of the basic concepts of the theory of games, let us begin by explaining what we mean by a *zero-sum two-person game*. In this term, "two-person" means that there are two players (or, more generally, two parties with conflicting interests), and "zero-sum" means that whatever one player loses the other player wins. Thus, in a zero-sum two-person game there is no "cut for the house" as in professional gambling, and no capital is created or destroyed during the course of play. Of course, the theory of games also includes games which are neither "zero-sum" nor limited to two players; as can well be imagined, however, in such games everything becomes much more complicated, and we shall not go into it in this text.

Games are also classified according to the number of *strategies* (moves, choices, or alternatives) each player has at his disposal. For instance, if each player has to choose one of two alternatives (as in the illustration of Section 8.1), we say that they are playing a 2×2 game; if one player has 3 possible moves while the other has 4, we say that they are playing a 3×4 game, and so on. In this section we shall consider only *finite* games, that is, games in which each player has only a finite, or fixed, number of possible moves, but later we shall consider also games in which the choice among moves is infinite.

It is customary in the Theory of Games to refer to the two players as Player A and Player B as we did in the table on page 233, but the moves (choices, or alternatives) of Player A are usually labeled I, II, III, ..., instead of a_1, a_2, a_3, ..., and those of Player B are usually labeled 1, 2, 3, ..., instead of θ_1, θ_2, θ_3, ... The amounts of money which change hands when the players choose their respective strategies are usually shown in a table like that on page 233, which is referred to as a *payoff matrix* in the theory of games. (As before, positive payoffs represent losses of Player A while negative payoffs represent losses of Player B.) Let us also add that it is always assumed in the theory of games that *each player must choose his strategy without knowledge of what his opponent has done or is planning to do, and that once a player has made his choice, it cannot be changed.*

The objectives of the theory of games are to determine *optimum strategies* (namely, strategies which are most profitable to the respective players), and the corresponding payoff, which is called the *value* of the game. To illustrate, let us consider the 2×2 zero-sum two-person game given by the following payoff matrix

		Player A I	*Player A* II
Player B	1	7	−4
Player B	2	8	10

As can be seen by inspection, it would be foolish for Player B to choose Strategy 1, since Strategy 2 will yield more than Strategy 1 *regardless of the choice made by Player A.* In a situation like this we say that Strategy 1 *is dominated by* Strategy 2 (or that Strategy 2 *dominates* Strategy 1), and it stands to reason that any strategy which is dominated by another should be disregarded. If we do this in our example, we find that Player B's *optimum strategy* is Strategy 2, the only one left, and that Player A's *optimum strategy* is Strategy I, since a loss of \$8.00 is obviously preferable to a loss of \$10.00. Also, the value of the game is the payoff corresponding to this pair of strategies, namely, \$8.00.

To illustrate this further, let us consider the following 3×2 zero-sum two-person game

		Player A I	II	III
Player B	1	−4	1	7
	2	4	3	5

Here Player A has the choice of three different moves, Player B has the choice of two, and all but one of the payoff amounts are advantageous to Player B. In this case neither strategy of Player B dominates the other, but the third strategy of Player A is dominated by each of the other two—clearly, a gain of \$4.00 or a loss of \$1.00 is preferable to a loss of \$7.00, and a loss of \$4.00 or a loss of \$3.00 is preferable to a loss of \$5.00. Thus, we can disregard the third column and study the *equivalent* 2×2 game

		Player A I	II
Player B	1	−4	1
	2	4	3

in which Strategy 2 of Player B *now* dominates Strategy 1. Thus, the *optimum choice* for Player B is Strategy 2, the optimum choice for Player A is Strategy II (since a loss of \$3.00 is preferable to a loss of \$4.00), and the *value* of the game is \$3.00.

The process of disregarding dominated strategies can be of great help in the *solution* of a game (namely, in finding optimum strategies and the value of the game), but it is the exception rather than the rule that this process will lead to the *complete* solution of a game. The fact that dominances need not even exist is illustrated in the following 3×3 zero-sum two-person game:

		Player A I	II	III
Player B	1	−1	6	−2
	2	2	4	6
	3	−2	−6	12

Evidently, there are no dominances among the strategies of either player, but if we look at the game from the point of view of Player A, we might argue as follows: if he chooses Strategy I the worst that can happen is that he loses \$2.00; if he chooses Strategy II the worst that can happen is that he loses \$6.00; and if he chooses Strategy III the worst that can happen is that he loses \$12.00. Thus, he could *minimize his maximum losses* by choosing Strategy I.

If we apply the same kind of argument to select a strategy for Player B, we find that if he chooses Strategy 1 the worst that can happen is that he loses \$2.00; if he chooses Strategy 2 the worst that can happen is that he wins \$2.00; and if he chooses Strategy 3 the worst that can happen is that he loses \$6.00. Thus, he could *minimize his maximum losses* (or *maximize his minimum gain*, which is the same) by choosing Strategy 2.

The selection of Strategies I and 2, appropriately called *minimax strategies* (or strategies based on the *minimax criterion*), is really quite reasonable. By choosing Strategy I, Player A makes sure that his opponent can win at most \$2.00, and by choosing Strategy 2, Player B makes sure that he will actually win this amount. Thus, the value of the game is \$2.00, which means that it favors Player B, but we could make it *equitable* by charging Player B \$2.00 for playing the game, while letting Player A play for free.

A very important aspect of the minimax strategies I and 2 of this example is that they are completely "spyproof" in the sense that neither player can profit from any knowledge about the other's choice. Even if Player A announced publicly that he will choose Strategy I, it would still be best for Player B to choose Strategy 2, and if Player B announced publicly that he will choose Strategy 2, it would still be best for Player A to choose Strategy I. Unfortunately, this is not always the case, as is illustrated by the following game:

		Player A I	II
Player B	1	8	−5
	2	2	6

Here, Player A can minimize his maximum losses by choosing Strategy II, Player B can minimize his maximum losses by choosing Strategy 2, and the corresponding payoff is \$6.00. However, if Player A knew that Player B was going to base his choice on the minimax criterion, he could switch to Strategy I and thus *reduce* his loss from \$6.00 to \$2.00. Of course, if Player B knew that Player A would try to outsmart him in this way, he could in turn switch to Strategy 1 and *increase* his winnings to \$8.00. In any case, the minimax strategies of the two players are *not spyproof* in this example.

Fortunately, there exists a fairly easy way of determining for any given game whether minimax strategies are spyproof. What we have to look for are *saddle points*, namely, pairs of strategies for which the corresponding entry in the payoff matrix is *the smallest value of its row and the greatest value of its column.* In the last example there is no saddle point because the smallest value of each row is also the smallest value of its column. On the other hand, in the game on page 234 there *is* a saddle point corresponding to Strategies I and 2 (since 8, the smallest value of the second row, is the greatest value of the first column). Also, the 3×2 game on page 235 has a saddle point corresponding to Strategies II and 2 (since 3, the smallest value of the second row, is the greatest value of the second column), and the 3×3 game on page 235 has a saddle point corresponding to Strategies I and 2 (since 2, the smallest value of the second row, is the greatest value of the first column). In general, if a game has a saddle point it is said to be *strictly determined*, and the strategies which correspond to the saddle point provide spyproof (and, hence, optimum) minimax strategies for the two players. (The fact that there can be more than one saddle point is illustrated in Exercise 10 on page 241, but it also follows from that exercise that it does not matter in that case which of the saddle points is used to determine the optimum strategies of the two players.)

If a game does not have a saddle point, minimax strategies are not spyproof, and each player can outsmart the other if he knows how his opponent will react in a given situation. To avoid this possibility, it suggests itself that each player should somehow mix up his behavior patterns intentionally, and the best way of doing this is by introducing an element of chance into the selection of his strategy. Suppose, for instance, that in the last example Player B uses a gambling device (dice, cards, numbered slips of paper, or random numbers) which leads to the choice of Strategy 1 with the probability x, and to the choice of Strategy 2 with the probability $1 - x$. He can then argue that *if Player A chooses Strategy I*, he (Player B) can *expect* to win

$$E = 8x + 2(1 - x)$$

dollars, and *if Player A chooses Strategy II*, he can expect to win

$$E = -5x + 6(1 - x)$$

dollars. Graphically, this situation is described in Figure 8.1, where we have plotted the lines whose equations are $E = 8x + 2(1 - x)$ and $E = -5x + 6(1 - x)$ for values of x from 0 to 1.

Now suppose that we apply the minimax criterion to the expected winnings of Player B, namely, *maximize his minimum expected gains.* As is apparent from Figure 8.1, the smaller of the two values of E for any

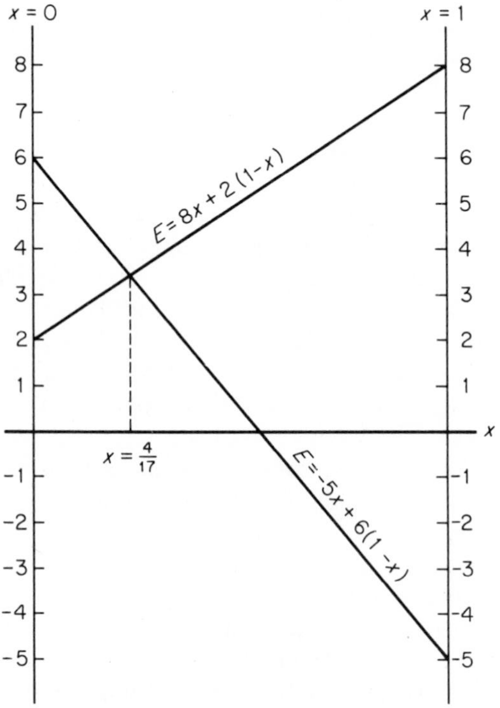

FIGURE 8.1 The expected winnings of Player B.

given value of x is *greatest* where the two lines intersect, and to find the corresponding value of x we have only to solve the equation

$$8x + 2(1 - x) = -5x + 6(1 - x)$$

which yields $x = \frac{4}{17}$. This means that if Player B labels four slips of paper "Strategy 1," 13 slips of paper "Strategy 2," shuffles them thoroughly, and then acts according to which kind he randomly draws, he will be applying the minimax criterion to his expected winnings, and he will be assuring for himself expected winnings of $8(\frac{4}{17}) + 2(\frac{13}{17}) = 3\frac{7}{17}$ or \$3.41 to the nearest cent.

So far as Player A is concerned, the analysis is very much the same, and in Exercise 11 on page 242 the reader will be asked to show that Player A can minimize his maximum expected losses by choosing between Strategies I and II with respective probabilities of $\frac{11}{17}$ and $\frac{6}{17}$, and that he can thus hold his expected losses down to $3\frac{7}{17}$ or \$3.41 to the nearest cent. Incidentally, the \$3.41 to which Player B can raise his expected winnings and Player A can hold down his expected losses is again called the *value* of the game. Also, if a player's ultimate choice is thus left to chance, his over-all strategy is referred to as *randomized* or *mixed*, whereas the original Strategies I and II of Player A and the original Strategies 1 and 2 of Player B are referred to as *pure.*

The examples of this section were all given without any "physical" interpretation because we were interested only in introducing some of the basic concepts of the theory of games. If we now apply these methods to the problem of Section 8.1, we would find that the "game" does not have a saddle point and that the manager of the oil company can hold his company's expected losses down to \$533,333 by *randomizing* his decision whether to continue or stop drilling with respective probabilities of $\frac{5}{21}$ and $\frac{16}{21}$ (see Exercise 13 on page 242). Evidently, this is preferable to the potential losses of \$4,800,000 or \$3,200,000 to which his company is exposed by his two *pure strategies* of continuing or stopping to drill. It must be understood, of course, that this game-theoretical analysis of the problem is based on the assumption that Nature (which controls whether or not there is going to be any oil) is a *malevolent opponent,* who is trying to make things as difficult as possible for the manager and his company. In fact, it would seem that in a problem like this the manager *should* have some idea about the chances of success, and, hence, that the problem should be solved by the method which we first discussed in Section 8.1. Like the discussion in the text, the exercises which follow deal only with zero-sum two-person games; hence, it may be well to remind the reader that such special games constitute only a small part of the subject matter studied in the theory of games.

THEORETICAL EXERCISE

1. If a zero-sum two-person game has a saddle point corresponding to Strategies II and 4 and another corresponding to Strategies IV and 1, explain why there must also be a saddle point corresponding to Strategies II and 1 and another corresponding to Strategies IV and 4, and that all four of these saddle points must have the same payoff.

APPLIED EXERCISE

2. With reference to the illustration on page 232, which decision would minimize the company's expected losses if the manager felt that
   (a) the odds against there being any oil are 4 to 1;
   (b) the odds against there being any oil are 2 to 1?
3. With reference to the illustration on page 232, what should we expect the manager of the oil company to do if we knew that he is a confirmed optimist?
4. Mr. Smith is planning to attend a sales meeting in San Diego, and he must send in his room reservation immediately. The sales meeting is so large that the activities are held partly in Hotel A and partly in Hotel B, and Mr. Smith does not know whether the particular session he wants to attend will be held at Hotel A or Hotel B. He is planning to stay only one day, which would cost him \$23.00 at Hotel A and \$20.00 at Hotel B, but it will cost him an extra \$5.00 for cab fare if he stays at the wrong hotel.
   (a) Present all this information in a table like that on page 232 with Mr. Smith being Player A.
   (b) Where should he make his reservation if he wants to minimize his expected expenses and feels that the odds are 5 to 1 that the session will be held at Hotel A?
   (c) Where should he make his reservation if the odds quoted in part (b) should have been 2 to 1 instead of 5 to 1?
   (d) Where should be make his reservation if the odds quoted in part (b) should have been 4 to 1 instead of 5 to 1?
5. A dinner guest wants to show his appreciation to his hostess by sending her either a pound of candy or a bottle of wine. He remembers, though, that she is either on a strict reducing diet or a teetotaler, but he can't remember which. In any case, he feels that her reaction to his gift will be as shown in the following table, where the numbers are in units of "appreciation":

	Hostess is Dieting	*Hostess is Teetotaler*
Send Candy	−2	6
Send Wine	3	−9

   (a) What should he send so as to maximize the expected appreciation of his gift, if he feels that the odds are 2 to 1 that his hostess is a teetotaler rather than dieting?

(b) What should he send so as to maximize the expected appreciation of his gift, if he feels that the odds are 7 to 1 that his hostess is dieting rather than a teetotaler?
(c) What should he send so as to minimize the maximum lack of appreciation?

6. Suppose that the manager of the oil company referred to in the text is the kind of person who always worries about "what might have been." Looking at the table on page 232, he finds that if he continues to drill and there is no oil, his company will lose out by \$5,600,000 (the difference between the amount they would lose and the amount they would have gained if he had decided to stop). Referring to this difference as the corresponding *regret*, he also finds that the regret associated with his decision to stop drilling when there is oil is \$11,200,000, and that in the two other cases the regret is 0. What should he decide to do if he wanted to *minimize the maximum regret?*

7. Referring to the definition of *regret* in Exercise 6, find
(a) what Mr. Smith of Exercise 4 should decide to do if he wanted to minimize the maximum regret;
(b) what the dinner guest of Exercise 5 should do if he wanted to minimize his maximum regret?

8. Each of the following is the payoff matrix (the payments Player A makes to Player B) for a zero-sum two-person game. Eliminate all dominated strategies and determine the best strategy for each player as well as the value of the game:

(a)

3	−2
5	7

(b)

14	11
16	−2

(c)

−5	0	3
−6	−3	−3
−12	−1	1

(d)

7	10	8
8	8	11
7	5	9

9. Find the saddle point of each of the games of Exercise 8.

10. Each of the following is the payoff matrix of a zero-sum two-person game. Find the saddle point (or saddle points) and the value of each game:

(a)

3	−4	−3
1	−1	−2
−6	3	−4

(b)

9	3	2	4
3	4	4	4
6	5	6	5
9	5	7	5

11. Verify the probabilities of $\frac{11}{17}$ and $\frac{6}{17}$ given for the mixed strategy of Player *A* on page 239.

12. The following is the payoff matrix of a 2×2 zero-sum two-person game:

3	−4
−3	1

(a) What randomized strategy should Player *A* use so as to minimize his maximum expected losses?
(b) What randomized strategy should Player *B* use so as to maximize his minimum expected gain?
(c) What is the value of the game?

13. With reference to the example of Section 8.1, verify the randomized optimum strategy of the manager of the oil company given on page 239. Also find the corresponding randomized optimum strategy of Nature.

14. With reference to Exercise 4, which randomized strategy minimizes Mr. Smith's maximum expected expenses?

15. A country has two airfields with installations worth \$2,000,000 and \$10,000,000, respectively, of which it can defend only one against an attack by its enemy. The enemy, on the other hand, can attack only one of these airfields and take it successfully only if it is left undefended. Considering the "payoff" to the country to be the total value of the installations it holds after the attack, find the optimum strategy of the country as well as that of its enemy, and the value of the "game."

16. Two persons agree to play the following game: The first writes either 9 or 4 on a slip of paper, while the other writes either 5 or 6 on another slip of paper. If S, the sum of the two numbers is *even*, the first player wins $\frac{S}{2} - 5$ dollars from his opponent, and if S is *odd*, the second player wins $8 - \frac{S}{3}$ dollars from his opponent.

(a) Find the optimum strategy of each player.

(b) How much should the second player pay the first for the "privilege" of playing the game, so as to make the game equitable?

8.3 STATISTICAL GAMES

In *statistical inference* we base decisions about populations on sample data, and it is by no means far-fetched to look upon such an inference as a game between Nature, which controls the relevant feature (or features) of the population, and the person (scientist, or statistician) who must arrive at some decision about Nature's choice. For instance, if we want to estimate the mean μ of a normal population on the basis of a random sample of size n, we could say that Nature has control over the "true" value of μ, we might estimate μ in terms of the value of the sample mean or the sample median, and presumably, there is some penalty or reward which depends mostly on how far we are off.

In spite of the obvious similarity between this problem and the ones of the preceding section, there are essentially two features in which *statistical games* differ from those studied in the theory of games. First, there is the question which we already met when we tried to apply the theory of games to the decision problem of Section 8.1, namely, the question *whether it is reasonable to treat Nature as a malevolent opponent.* Obviously no, but this hurts rather than helps. If we could treat Nature as a "rational" opponent, we would know what to expect—otherwise, anything goes.

The other distinction is that in the theory of games each player must choose his strategy without any knowledge of what his opponent has done or is planning to do, whereas *in a statistical game the statistician is supplied with sample data which provide him with some information about Nature's "move."* This also complicates matters, but what it really amounts to is that it leads to a more complex kind of game. To illustrate, suppose that a coin is either two-headed or that it is a balanced coin with heads on one side and tails on the other. We cannot inspect the coin, but we can flip it once, observe whether it comes up heads or tails, and then we must decide whether it is two-headed or not. Furthermore, there is a penalty of \$1.00 if our decision is wrong, and no penalty (or reward) if our decision is right. If we ignored the fact that we can observe one flip of the coin, we could treat the problem as the following game

		Player A (The Statistician) a_1	a_2
Player B (Nature)	θ_1	$L(a_1, \theta_1) = 0$	$L(a_2, \theta_1) = 1$
	θ_2	$L(a_1, \theta_2) = 1$	$L(a_2, \theta_2) = 0$

which should remind the reader of the scheme on page 233. Now, θ_1 is the "state of Nature" that the coin is two headed, θ_2 is the "state of Nature" that the coin is balanced with heads on one side and tails on the other, a_1 is the statistician's decision that the coin is two-headed, a_2 is the statistician's decision that the coin is balanced with heads on one side and tails on the other, and the entries in the table are (as on page 233) the corresponding values of the loss function.

Now let us consider the fact that we (the *statistician*, or *Player A*) know about the flip of the coin, namely, that we know whether a random variable $\mathbf{x}$ has taken on the value $x = 0$ (*heads*) or $x = 1$ (*tails*). Since we shall want to make use of this information in choosing between a_1 and a_2, we need a function, a *decision function*, which tells us what action to take when $x = 0$ and what action to take when $x = 1$. One possibility is to choose a_1 when $x = 0$ and a_2 when $x = 1$, and we can express this symbolically by writing

$$d_1(x) = \begin{cases} a_1 & \text{when } x = 0 \\ a_2 & \text{when } x = 1 \end{cases}$$

or more simply $d_1(0) = a_1$ and $d_1(1) = a_2$. The purpose of the subscript is to distinguish this decision function from others, for instance, from

$$d_2(0) = a_1 \quad \text{and} \quad d_2(1) = a_1$$

which tells us to choose a_1 regardless of the outcome of the experiment,

$$d_3(0) = a_2 \quad \text{and} \quad d_3(1) = a_2$$

which tells us to choose a_2 regardless of the experiment, and

$$d_4(0) = a_2 \quad \text{and} \quad d_4(1) = a_1$$

which tells us to choose a_2 when $x = 0$ and a_1 when $x = 1$.

To compare the merits of all these decision functions, we shall have to determine the *expected* losses to which they lead for the various strategies of Nature, namely, the values of the *risk function*

$$R(\mathbf{d}_i, \theta_j) = E\{L[d_i(\mathbf{x}), \theta_j]\} \tag{8.3.1}$$

where the expectation is taken with respect to the random variable $\mathbf{x}$. Since the probabilities for $x = 0$ and $x = 1$ are, respectively, 1 and 0 for θ_1 and $\frac{1}{2}$ and $\frac{1}{2}$ for θ_2, we find that

$$R(\mathbf{d}_1, \theta_1) = 1 \cdot L(a_1, \theta_1) + 0 \cdot L(a_2, \theta_1) = 1 \cdot 0 + 0 \cdot 1 = 0$$

$$R(\mathbf{d}_1, \theta_2) = \tfrac{1}{2} \cdot L(a_1, \theta_2) + \tfrac{1}{2} \cdot L(a_2, \theta_2) = \tfrac{1}{2} \cdot 1 + \tfrac{1}{2} \cdot 0 = \tfrac{1}{2}$$

$$R(\mathbf{d}_2, \theta_1) = 1 \cdot L(a_1, \theta_1) + 0 \cdot L(a_1, \theta_1) = 1 \cdot 0 + 0 \cdot 0 = 0$$

$$R(\mathbf{d}_2, \theta_2) = \tfrac{1}{2} \cdot L(a_1, \theta_2) + \tfrac{1}{2} \cdot L(a_1, \theta_2) = \tfrac{1}{2} \cdot 1 + \tfrac{1}{2} \cdot 1 = 1$$

$$R(\mathbf{d}_3, \theta_1) = 1 \cdot L(a_2, \theta_1) + 0 \cdot L(a_2, \theta_1) = 1 \cdot 1 + 0 \cdot 1 = 1$$

$$R(\mathbf{d}_3, \theta_2) = \tfrac{1}{2} \cdot L(a_2, \theta_2) + \tfrac{1}{2} \cdot L(a_2, \theta_2) = \tfrac{1}{2} \cdot 0 + \tfrac{1}{2} \cdot 0 = 0$$

$$R(\mathbf{d}_4, \theta_1) = 1 \cdot L(a_2, \theta_1) + 0 \cdot L(a_1, \theta_1) = 1 \cdot 1 + 0 \cdot 0 = 1$$

$$R(\mathbf{d}_4, \theta_2) = \tfrac{1}{2} \cdot L(a_2, \theta_2) + \tfrac{1}{2} \cdot L(a_1, \theta_2) = \tfrac{1}{2} \cdot 0 + \tfrac{1}{2} \cdot 1 = \tfrac{1}{2}$$

where the values of the loss function were obtained from the table on page 244.

We have thus arrived at the following 4×2 zero-sum two-person game, in which the payoffs are the corresponding values of the risk function:

		Player A (The Statistician) $\mathbf{d}_1$	$\mathbf{d}_2$	$\mathbf{d}_3$	$\mathbf{d}_4$
Player B (Nature)	θ_1	0	0	1	1
	θ_2	$\frac{1}{2}$	1	0	$\frac{1}{2}$

As can be seen by inspection, $\mathbf{d}_2$ is *dominated* by $\mathbf{d}_1$ and $\mathbf{d}_4$ is *dominated* by $\mathbf{d}_3$, so that $\mathbf{d}_2$ and $\mathbf{d}_4$ can be disregarded—in decision theory we say that they are *inadmissible*. [Actually, this should not come as a surprise, since in $\mathbf{d}_2$ as well as $\mathbf{d}_4$ we accept the alternative a_1 (that the coin is two-headed) even though it came up tails.]

This leaves us with the 2×2 zero-sum two-person game in which Player A has to choose between $\mathbf{d}_1$ and $\mathbf{d}_3$. If he looks upon Nature as a malevolent opponent, his optimum strategy would be to randomize between $\mathbf{d}_1$ and $\mathbf{d}_3$ with respective probabilities of $\frac{2}{3}$ and $\frac{1}{3}$, so that his expected risk will be $\frac{1}{3}$ (of a dollar); otherwise, he will have to use some other *decision rule* for choosing between $\mathbf{d}_1$ and $\mathbf{d}_3$, and this will be discussed in Section 8.4. Incidentally, we introduced this problem with reference to a two-headed coin and an ordinary coin, but we could just as well have formulated it "more statistically" as a decision problem in which *we must decide on the basis of one observation whether a random variable has the Bernoulli distribution as formulated on page 74 with the parameter* $\theta = 0$ *or the parameter* $\theta = \frac{1}{2}$.

The two important concepts which we have introduced in this section are those of a loss function and a risk function, and to illustrate them further let us consider the following example, where Nature as well as the statistician has a *continuum* of strategies. Suppose that a random variable has the uniform density

$$f(x) = \begin{cases} \dfrac{1}{\theta} & \text{for } 0 < x < \theta \\ 0 & \text{elsewhere} \end{cases}$$

and that we want to *estimate* the parameter θ (the "move" of Nature) on the basis of a single observation. The decision function which we shall use is given by $d(x) = kx$, where $k \geq 1$, and it will be assumed that *the losses are proportional to the absolute value of the error*, namely, that

$$L(kx, \theta) = c|kx - \theta|$$

where c is a positive constant. (The choice of these particular decision and loss functions is at least in part a matter of *mathematical convenience;* for other decision or loss functions there may well be no solution or insurmountable mathematical difficulties.) According to (8.3.1) the risk function is given by

$$\begin{aligned} R(\mathbf{d}, \theta) &= \int_0^{\theta/k} c(\theta - kx) \cdot \frac{1}{\theta}\, dx + \int_{\theta/k}^{\theta} c(kx - \theta) \cdot \frac{1}{\theta}\, dx \\ &= c\theta \left(\frac{k}{2} - 1 + \frac{1}{k} \right) \end{aligned}$$

so that we can go even one step further and ask for the value of k which will minimize the statistician's risk. Obviously, there is nothing he can do about the factor θ, but he can minimize $\frac{k}{2} - 1 + \frac{1}{k}$ by letting $k = \sqrt{2}$. Thus, if the statistician actually took the observation and got $x = 5$, his *best* estimate of θ under the given conditions would be $5\sqrt{2}$ or approximately 7.07.

8.4 DECISION CRITERIA

In the last example of the preceding section we were able to find a decision function which minimized the statistician's risk *regardless of the true state of Nature* (that is, regardless of the true value of the parameter θ), but this is the exception rather than the rule. Had we not limited ourselves to decision functions of the form $d(x) = kx$, then the decision function given by $d(x) = \theta_1$ would be *best* when θ happens to equal θ_1, the one given by $d(x) = \theta_2$ would be *best* when θ happens to equal θ_2, . . . , and it is obvious that there can be no decision function which is *best* for all values of θ.

Generally, we thus have to be satisfied with decision functions that are *best*, at least, with respect to some criterion, and the two criteria which we shall study in this chapter are: (1) the *minimax criterion*, according to which we choose the decision function for which $R(\mathbf{d}, \theta)$, maximized with respect to θ, is a minimum; and (2) the *Bayes criterion*, according to which we choose the decision function for which the *Bayes risk* $E[R(\mathbf{d}, \boldsymbol{\theta})]$ is a minimum, where the expectation is taken with respect to $\boldsymbol{\theta}$). (Of course, this requires that we look upon $\boldsymbol{\theta}$ as a random variable which has a given distribution.) It is of interest to note that in the solution of the decision problem of Section 8.1 we actually used these two criteria. When we quoted odds about there being any oil, we were really assigning probabilities to the two states of Nature, θ_1 and θ_2, and when we suggested that the manager of the oil company should minimize the corresponding expected losses, we were, in fact, suggesting that he use the *Bayes criterion*. Also, when we asked on page 232 what the manager of the oil company should do if he were a confirmed pessimist, we suggested that he protect himself against the worst that can happen by using the *minimax criterion*.

8.4.1 The Minimax Criterion

If we apply the minimax principle to the first decision problem of Section 8.3, the one dealing with the two-headed coin, we find that for $\mathbf{d}_1$ the maximum risk is $\frac{1}{2}$, that for $\mathbf{d}_3$ the maximum risk is 1, and, hence, that the minimax decision function is $\mathbf{d}_1$.

To consider another example, suppose we want to *estimate* the parameter θ of a binomial distribution on the basis of the observed number of "successes" in n trials, and that the decision function is to be of the form

$$d(x) = \frac{x + a}{n + b}$$

where $\mathbf{x}$ is the number of "successes" and a and b are constants. Note that this decision function is, in fact, *linear* in x. Furthermore, suppose that the loss function is to be

$$L\left(\frac{x+a}{n+b}, \theta\right) = c\left(\frac{x+a}{n+b} - \theta\right)^2$$

where c is a positive constant, so that *our problem is to find the values of a and b which will minimize the corresponding risk function after it has been maximized with respect to θ.* After all, we have control over the choice of a and b, while Nature, our (presumed) opponent, has control over the choice of θ.

Since $E(\mathbf{x}) = n\theta$ and $E(\mathbf{x}^2) = n\theta(1 - \theta + n\theta)$, which we proved on page 158, we find that

$$R(\mathbf{d}, \theta) = E\left[c\left(\frac{\mathbf{x}+a}{n+b} - \theta\right)^2\right]$$

$$= \frac{c}{(n+b)^2}\left[\theta^2(b^2 - n) + \theta(n - 2ab) + a^2\right] \qquad (8.4.1)$$

and we could proceed to determine with the use of calculus the value of θ which maximizes (8.4.1), and then minimize $R(\mathbf{d}, \theta)$ for this value of θ with respect to a and b. This would not be very difficult, but it would involve quite a bit of algebraic detail, which we shall leave to the reader in Exercise 6 on page 252.

Instead, we shall make use of the so-called *equalizer principle*, according to which (under fairly general conditions) *the risk function of a minimax decision rule is a constant;* that is, in this case it would not depend on the parameter θ.* To justify this principle, at least intuitively, let us point out that for Player B's minimax strategy on page 238 his expected winnings were $3\frac{7}{17}$ dollars *regardless of whether Player A chose Strategy I or Strategy II*, and that for the oil company manager's minimax strategy (which we gave on page 239) the expected losses were $533,333 *regardless of whether or not there was any oil.*

To make (8.4.1) independent of θ, the coefficients of θ and θ^2 will both have to equal 0, so that $b^2 - n = 0$ and $n - 2ab = 0$, and, hence, $b = \sqrt{n}$ and $a = \frac{1}{2}\sqrt{n}$. Thus, the required minimax decision function is given by

$$d(x) = \frac{x + \frac{1}{2}\sqrt{n}}{n + \sqrt{n}}$$

and if we actually obtained 39 successes in 100 trials, we would estimate the parameter of this binomial distribution as $\frac{39+5}{100+10} = 0.40$.

* The exact conditions under which the equalizer principle holds are given in the book by T. S. Ferguson listed on page 253.

8.4.2 The Bayes Criterion

To solve the first decision problem of Section 8.3 by means of the Bayes criterion, we will have to assign probabilities to the two strategies of Nature, θ_1 and θ_2. If we assign θ_1 and θ_2 respective probabilities of p and $1 - p$, it can be seen from the table on page 245 that for $\mathbf{d}_1$ the *Bayes risk* is

$$0 \cdot p + \tfrac{1}{2} \cdot (1 - p) = \tfrac{1}{2} \cdot (1 - p)$$

and that for $\mathbf{d}_3$ the *Bayes risk* is

$$1 \cdot p + 0 \cdot (1 - p) = p$$

It follows that the Bayes risk of $\mathbf{d}_1$ is less than that of $\mathbf{d}_3$ (and $\mathbf{d}_1$ is to be preferred to $\mathbf{d}_3$) when $p > \frac{1}{3}$, and that the Bayes risk of $\mathbf{d}_3$ is less than that of $\mathbf{d}_1$ (and $\mathbf{d}_3$ is to be preferred to $\mathbf{d}_1$) when $p < \frac{1}{3}$. When $p = \frac{1}{3}$, the two Bayes risks are equal, and we can use either $\mathbf{d}_1$ or $\mathbf{d}_3$.

To consider another example, let us return to the problem on page 246, where we wanted to estimate the parameter θ of the given uniform distribution on the basis of a single observation. This time, however, let us not restrict the form of the decision function, but use the (quadratic) loss function whose values are given by

$$L[d(x), \theta] = c[d(x) - \theta]^2$$

and assume that the parameter is a random variable having the density

$$h(\theta) = \begin{cases} \theta e^{-\theta} & \text{for } \theta > 0 \\ 0 & \text{for } \theta \leq 0 \end{cases}$$

Since $\boldsymbol{\theta}$ is now a random variable, we look upon the original density as the *conditional density*

$$f(x|\theta) = \begin{cases} \dfrac{1}{\theta} & \text{for } 0 < x < \theta \\ 0 & \text{elsewhere} \end{cases}$$

and in Exercise 8 on page 252 the reader will be asked to show by the methods of Section 4.5.2 that the joint density of $\mathbf{x}$ and $\boldsymbol{\theta}$ is given by

$$f(x, \theta) = \begin{cases} e^{-\theta} & \text{for } 0 < x < \theta \\ 0 & \text{elsewhere} \end{cases}$$

and, hence, that the marginal density of $\mathbf{x}$ and the conditional density of $\boldsymbol{\theta}$ given x are given by

$$g(x) = \begin{cases} e^{-x} & \text{for } x > 0 \\ 0 & \text{for } x < 0 \end{cases} \quad \text{and} \quad \varphi(\theta|x) = \begin{cases} e^{x-\theta} & \text{for } x < \theta \\ 0 & \text{for } x > \theta \end{cases}$$

Now, the Bayes risk $E[R(\mathbf{d}, \boldsymbol{\theta})]$ which we shall want to minimize is given by the double integral

$$\int_0^\infty \left\{ \int_0^\theta c[d(x) - \theta]^2 f(x|\theta)\, dx \right\} h(\theta)\, d\theta$$

which, upon making use of the fact that $f(x|\theta) \cdot h(\theta) = \varphi(\theta|x) \cdot g(x)$ and changing the order of integration, can be written as

$$\int_0^\infty \left\{ \int_x^\infty c[d(x) - \theta]^2 \varphi(\theta|x)\, d\theta \right\} g(x)\, dx$$

To minimize this double integral, we must choose $d(x)$ for each x so that the integral

$$\int_x^\infty c[d(x) - \theta]^2 \varphi(\theta|x)\, d\theta = \int_x^\infty c[d(x) - \theta]^2 e^{x-\theta}\, d\theta$$

is as small as possible; so, differentiating with respect to $d(x)$ and putting the derivative equal to 0, we get

$$2ce^x \cdot \int_x^\infty [d(x) - \theta] e^{-\theta}\, d\theta = 0$$

and this yields

$$d(x) \cdot \int_x^\infty e^{-\theta}\, d\theta - \int_x^\infty \theta e^{-\theta}\, d\theta = 0$$

and, finally,

$$d(x) = \frac{\int_x^\infty \theta e^{-\theta}\, d\theta}{\int_x^\infty e^{-\theta}\, d\theta} = \frac{(x + 1)e^{-x}}{e^{-x}} = x + 1$$

Thus, if the value obtained for $\mathbf{x}$ is 5 (as on page 246), this decision function gives the Bayesian estimate $5 + 1 = 6$ for the parameter θ of the original distribution.

THEORETICAL EXERCISES

1. A statistician has to decide on the basis of one observation whether the parameter θ of a Bernoulli distribution is 0, $\frac{1}{2}$, or 1; his loss in dollars (a penalty which is deducted from his fee) is 100 times the absolute value of his error.
   (a) Construct a table showing the nine possible values of the loss function.

(b) List the nine possible decision functions and construct a table showing all the values of the corresponding risk function.*

(c) Show that five of the decision functions are not admissible, and that according to the minimax criterion the remaining decision functions are all equally good.

(d) Which decision function is best according to the Bayes criterion, if the three possible values of the parameter θ are regarded as equally likely?

2. A statistician has to decide on the basis of two observations whether the parameter θ of a binomial distribution is $\frac{1}{4}$ or $\frac{1}{2}$; his loss (a penalty which is deducted from his fee) is \$160.00 if he is wrong.

(a) Construct a table showing the four possible values of the loss function.

(b) List the eight possible decision functions and construct a table showing all the values of the corresponding risk function.

(c) Show that three of the decision functions are not admissible.

(d) Find the decision function which is best according to the minimax criterion.

(e) Find the decision function which is best according to the Bayes criterion, if the probabilities assigned to $\theta = \frac{1}{4}$ and $\theta = \frac{1}{2}$ are, respectively, $\frac{2}{3}$ and $\frac{1}{3}$.

3. With reference to the example on page 244, show that even if the coin were flipped n times, there would only be two admissible decision functions. Also construct a table showing the values of the risk function for these two admissible decision functions.

4. With reference to the example on page 246, show that if the losses are proportional to the *squared error* instead of the absolute value of the error, the risk function becomes

$$R(\mathbf{d}, \theta) = \frac{c\theta^2}{3}(k^2 - 3k + 3)$$

and that its minimum is at $k = \frac{3}{2}$.

5. A statistician has to decide on the basis of a single observation whether the parameter θ of the density

$$f(x) = \begin{cases} \dfrac{2x}{\theta^2} & \text{for } 0 < x < \theta \\ 0 & \text{elsewhere} \end{cases}$$

equals θ_1 or θ_2, where $\theta_1 < \theta_2$. If he decides on θ_1 when the observed value is less than the constant k, on θ_2 when the observed value is

* As in the text, it will be assumed in all of the exercises that the decision functions are *non-randomized;* namely, that each outcome must lead to the acceptance of one of the specified actions.

greater than or equal to the constant k, and he is fined C dollars for making the wrong decision, which value of k will minimize his maximum risk?

6. Find θ', the value of θ which maximizes (8.4.1), and then find the values of a and b which minimize $R(\mathbf{d}, \theta')$. Compare these values with the results obtained on page 248.
7. If we assume in the example on page 248 that $\boldsymbol{\theta}$ is a random variable having the uniform density (4.3.1) with $\alpha = 0$ and $\beta = 1$, show that the *Bayes risk* corresponding to (8.4.1) is given by

$$\frac{c}{(n+b)^2}\left[\tfrac{1}{3}(b^2 - n) + \tfrac{1}{2}(n - 2ab) + a^2\right]$$

Then, show that it is a minimum when $a = 1$ and $b = 2$, so that the corresponding Bayes decision rule is given by $d(x) = \dfrac{x+1}{n+2}$.
8. Verify the results obtained on page 249 for the joint density of $\mathbf{x}$ and $\boldsymbol{\theta}$, the marginal density of $\mathbf{x}$, and the conditional density of $\boldsymbol{\theta}$ given x.
9. Suppose we want to estimate the parameter θ of the geometric distribution (3.3.6) on the basis of a single observation. If the loss function is given by

$$L[d(x), \theta] = c[d(x) - \theta]^2$$

and $\boldsymbol{\theta}$ is looked upon as a random variable having the uniform density $h(\theta) = 1$ for $0 < \theta < 1$ and $h(\theta) = 0$ elsewhere, duplicate the steps of Section 8.4.2 to show that

(a) the conditional density of $\boldsymbol{\theta}$ given x is

$$\varphi(\theta|x) = \begin{cases} x(x+1)\theta(1-\theta)^{x-1} & \text{for } 0 < \theta < 1 \\ 0 & \text{elsewhere} \end{cases}$$

(b) the Bayes risk is minimized by the decision function

$$d(x) = \frac{2}{x+2}$$

[*Hint:* Make use of the fact that the integral of the beta density (4.3.7) is equal to 1.]

APPLIED EXERCISE

10. A manufacturer produces an item consisting of two components, which must both work for the item to function properly. The cost of returning one of the items to the manufacturer for repairs is α dollars,

the cost of inspecting one of the components is β dollars, and the cost of repairing a faulty component is φ dollars. He can ship each item without inspection with the guarantee that it will be put into perfect working condition at his factory in case it does not work; he can inspect both components and repair them if necessary; or he can randomly select one of the components and ship the item with the original guarantee if it works, or repair it and also check the other component.

(a) Construct a table showing the manufacturer's expected losses corresponding to his three "strategies" and the three "states" of Nature that 0, 1, or 2 of the components do not work.

(b) What should the manufacturer do if α = \$25.00, φ = 10.00, and he wants to minimize his maximum expected losses?

(c) What should the manufacturer do to minimize his Bayes risk if α = \$10.00, β = \$12.00, φ = \$30.00, and he felt that the probabilities for 0, 1, and 2 defective components are, respectively, 0.70, 0.20, and 0.10.

REFERENCES

Some fairly elementary material on the theory of games and decision theory can be found in

Chernoff, H., and Moses, L. E., *Elementary Decision Theory*. New York: John Wiley & Sons, Inc., 1959.

Dresher, M., *Games of Strategy: Theory and Applications*. Englewood Cliffs, N.J.: Prentice-Hall, Inc., 1961.

McKinsey, J. C. C., *Introduction to the Theory of Games*. New York: McGraw-Hill Book Company, 1952.

Williams, J. D., *The Compleat Strategyst*. New York: McGraw-Hill Book Company, 1954.

The *equalizer rule* mentioned on page 248 is treated in detail in the advanced text on mathematical statistics

Ferguson, T. S., *Mathematical Statistics, A Decision Theoretic Approach*. New York: Academic Press, Inc., 1967.

9

ESTIMATION

9.1 INTRODUCTION

Traditionally, statistical inference has been divided into *problems of estimation* and the *testing of hypotheses*, and we shall continue to make this distinction mainly to facilitate the organization of the material which we want to present in this book. Let us also point out, however, that all problems of statistical inference (namely, all problems in which we make generalizations on the basis of sample data) are essentially decision problems, and, hence, can be handled by a unified approach like that presented in Chapter 8. The main distinction is that in a problem of estimation we must choose a value of a parameter (that is, we must choose one particular strategy of Nature) from a possible continuum of alternatives, while in the testing of hypotheses we must decide whether to accept or reject one *specified* value (or set of values) of a parameter.

As we already pointed out on page 247, *perfect* decision functions do not exist, and this is another reason for distinguishing between problems of estimation and the testing of hypotheses—the methods which we are willing to accept as "second best" or the restrictions which we must impose to obtain optimum decision functions differ somewhat for the two kinds of problems.

9.2 POINT ESTIMATION

If we use a sample mean to estimate the mean of a population, a sample proportion to estimate the parameter θ of a binomial population, or a sample variance to estimate the variance of a population, we are in each case using a *point estimate* of the parameter in question. These estimates are called *point estimates* because they are single numbers, single points on the real axis, used, respectively, to estimate μ, θ, and σ^2.

Since we can hardly expect that a point estimate, which is a value of a random variable, will actually *equal* the parameter it is supposed to estimate, it is often desirable to give ourselves some leeway by replacing it with an interval. For instance, if we are asked to estimate the average I.Q. of a very large group of students on the basis of a sample, we might arrive at a point estimate of 113, or we might arrive at the *interval estimate* 109–117. Such interval estimates can be interpreted in two different ways, which we shall discuss in Sections 9.5 and 9.6 (and in another way, as *fiducial intervals*, which we shall not discuss). If we look upon the end-points of the interval as values of random variables, so that we can assert with a given probability whether or not such an interval will do its job (of containing the parameter it is supposed to estimate) we refer to it as a *confidence interval*. On the other hand, if we look upon the parameter which we want to estimate as a random variable (as in Section 8.4.2), so that we can assert with a certain probability that the "true" value of the parameter will fall between the given limits (109 and 117 in our example), we refer to the interval as a *Bayesian interval estimate*. Confidence intervals will be introduced in Section 9.5, and Bayesian interval estimates will be discussed in Section 9.6.

In the next few sections we shall study some basic problems of point estimation; first we shall analyze some of the properties (the disadvantages as well as the merits) of various methods of point estimation, and then we shall discuss methods of obtaining point estimates having certain desirable properties.

9.3 PROPERTIES OF ESTIMATORS

If we use the mean of a sample to estimate the mean of a population, we refer to the random variable $\bar{\mathbf{x}}$ as an *estimator* of μ and to the actual value which we obtain, say, $\bar{x} = 16.2$, as an *estimate* of μ. More generally, we call the statistic (the random variable, whose value we use to estimate a parameter) an *estimator*, and the value, itself, an *estimate*.

Defining estimators, thus, as random variables, we find that one of the key problems of point estimation is to study their distributions. For instance, when we estimate the variance of a population on the basis of a random sample, we know that the value of $\mathbf{s}^2$ which we get need not (and probably does not) equal σ^2, but it would be reassuring, at least, to know whether we can expect it to be close. Also, if we must decide whether to use a sample mean or a sample median to estimate the parameter of a population, it would be important to know (among other things) whether $\bar{\mathbf{x}}$ or $\tilde{\mathbf{x}}$ is more likely to yield a value which is actually close to this parameter.

Various statistical properties of estimators can, thus, be used to decide which estimator is most appropriate in a given situation, which will expose us to the smallest risk, which will give us the most information at the lowest cost, and so forth. The particular properties of statistics which we shall discuss in Sections 9.3.1 through 9.3.4 are *unbiasedness, minimum variance, consistency, relative efficiency,* and *sufficiency.*

9.3.1 Unbiased Estimators

Since there can be no "perfect" estimator which always gives the right answer, it would seem desirable that an estimator should do so at least *on the average.* In other words, it would seem desirable that *the expected value of an estimator equal the parameter which it is supposed to estimate.* If this is the case, the estimator is said to be *unbiased;* otherwise, it is said to be *biased.*

According to this definition, it follows from Theorems 6.5 and 6.8 that $\bar{\mathbf{x}}$ is an unbiased estimator of the mean of any population (whose mean exists). Similarly, it is easy to show that if $\mathbf{x}$ is a random variable having the binomial distribution (3.3.3), then the sample proportion $\frac{\mathbf{x}}{n}$ is an unbiased estimator of the parameter θ—since the mean of the binomial distribution is $E(\mathbf{x}) = n\theta$, we have $E\left(\frac{\mathbf{x}}{n}\right) = \frac{1}{n} \cdot E(\mathbf{x}) = \frac{1}{n} \cdot n\theta = \theta.$

It is of interest to note that the minimax estimator of the parameter θ of the binomial distribution which we gave on page 248, and the one based on the Bayes criterion of Exercise 7 on page 252, are both *biased.* For the one on page 248 we get

$$E\left(\frac{\mathbf{x} + \frac{1}{2}\sqrt{n}}{n + \sqrt{n}}\right) = \frac{E(\mathbf{x} + \frac{1}{2}\sqrt{n})}{n + \sqrt{n}} = \frac{n\theta + \frac{1}{2}\sqrt{n}}{n + \sqrt{n}} \neq \theta$$

and for the one of Exercise 7 on page 252 we get

$$E\left(\frac{\mathbf{x} + 1}{n + 2}\right) = \frac{E(\mathbf{x} + 1)}{n + 2} = \frac{n\theta + 1}{n + 2} \neq \theta$$

This illustrates the very important point that *the advantages of unbiasedness can be outweighed by other considerations.* Under the conditions given, respectively, on pages 247 and 248 and in Exercise 7 on page 252, these estimators are *better* than the sample proportion $\frac{\mathbf{x}}{n}$ even though they are biased.

Using the concept of unbiasedness, it is easy to explain why we divided by $n - 1$ and not by n when we defined the sample variance on page 210—

it makes $\mathbf{s}^2$ *an unbiased estimator of* σ^2 *for random samples from infinite populations.* To prove this for random samples from normal populations, we could refer to Theorem 7.6 and argue that since $\frac{(n-1)\mathbf{s}^2}{\sigma^2}$ is a random variable having a chi-square distribution with $n - 1$ degrees of freedom (whose mean is $n - 1$), the expected value of $\mathbf{s}^2$ is $\frac{\sigma^2}{n-1}\cdot(n-1) = \sigma^2$. To prove it in general, let us write

$$E(\mathbf{s}^2) = E\left[\frac{1}{n-1}\cdot\sum_{i=1}^{n}(\mathbf{x}_i - \bar{\mathbf{x}})^2\right]$$

$$= \frac{1}{n-1}\cdot E\left[\sum_{i=1}^{n}\{(\mathbf{x}_i - \mu) - (\bar{\mathbf{x}} - \mu)\}^2\right]$$

$$= \frac{1}{n-1}\left[\sum_{i=1}^{n} E\{(\mathbf{x}_i - \mu)^2\} - n\cdot E\{(\bar{\mathbf{x}} - \mu)^2\}\right]$$

Then, since by definition $E[(\mathbf{x}_i - \mu)^2] = \sigma^2$ and $E[(\bar{\mathbf{x}} - \mu)^2] = \frac{\sigma^2}{n}$ according to Theorem 6.5, it follows that

$$E(\mathbf{s}^2) = \frac{1}{n-1}\left[\sum_{i=1}^{n}\sigma^2 - n\cdot\frac{\sigma^2}{n}\right] = \sigma^2$$

and this completes the proof. Note, however, that $\mathbf{s}^2$ is *not* an unbiased estimator of the variance of a *finite population* as defined by (6.2.13), and that in either case $\mathbf{s}$ is not an unbiased estimator of σ. (The bias of $\mathbf{s}$ is discussed on page 209 of the book by E. S. Keeping referred to on page 287.)

If we have to choose between two or more unbiased estimators of one and the same parameter, we usually take the one which has the *smallest variance*. We already indicated this on page 227, although we did not mention at the time that the sample median $\tilde{\mathbf{x}}$ is also an unbiased estimator of the mean μ of a normal population. (Actually, it would seem that this does not require a formal proof since the normal distribution is symmetrical about its mean.) In general, when there is a choice between unbiased estimators we look for the one which has the smallest possible variance, and if such an estimator exists we refer to it as a *minimum variance unbiased estimator*. To check whether a given unbiased estimator has the smallest possible variance, we make use of the fact that if $\hat{\boldsymbol{\theta}}$ is an unbiased estimator of θ, it can be shown that under very general conditions (referred to on page 288) the variance of $\hat{\boldsymbol{\theta}}$ must satisfy the inequality

$$\text{var}(\hat{\boldsymbol{\theta}}) \geq \frac{1}{n \cdot E\left[\left(\frac{\partial \ln f(\mathbf{x})}{\partial \theta}\right)^2\right]} \tag{9.3.1}$$

where $f(x)$ is the population distribution from which the random sample of size n is obtained. This inequality is referred to as the *Cramér-Rao inequality*, and it leads to the following result:

THEOREM 9.1

If $\hat{\boldsymbol{\theta}}$ *is an unbiased estimator of* θ *and*

$$\text{var}(\hat{\boldsymbol{\theta}}) = \frac{1}{n \cdot E\left[\left(\frac{\partial \ln f(\mathbf{x})}{\partial \theta}\right)^2\right]}$$

then $\hat{\boldsymbol{\theta}}$ *is a minimum variance unbiased estimator of* θ.

To illustrate the use of this theorem, let us show that $\bar{\mathbf{x}}$ is, in fact, a minimum variance unbiased estimator of the mean of a normal population. Since

$$f(x) = \frac{1}{\sigma\sqrt{2\pi}} \cdot e^{-\frac{1}{2}\left(\frac{x-\mu}{\sigma}\right)^2} \quad \text{for } -\infty < x < \infty$$

it follows that

$$\ln f(x) = -\ln \sigma\sqrt{2\pi} - \frac{1}{2}\left(\frac{x-\mu}{\sigma}\right)^2$$

so that

$$\frac{\partial \ln f(x)}{\partial \mu} = \frac{1}{\sigma}\left(\frac{x-\mu}{\sigma}\right)$$

and, hence,

$$E\left[\left(\frac{\partial \ln f(\mathbf{x})}{\partial \mu}\right)^2\right] = \frac{1}{\sigma^2} \cdot E\left[\left(\frac{\mathbf{x}-\mu}{\sigma}\right)^2\right] = \frac{1}{\sigma^2} \cdot 1 = \frac{1}{\sigma^2}$$

Thus,

$$\frac{1}{n \cdot E\left[\left(\frac{\partial \ln f(\mathbf{x})}{\partial \mu}\right)^2\right]} = \frac{1}{n \cdot \frac{1}{\sigma^2}} = \frac{\sigma^2}{n}$$

and since $\bar{\mathbf{x}}$ is unbiased and $\text{var}(\bar{\mathbf{x}}) = \frac{\sigma^2}{n}$ according to Theorem 6.5, it follows that $\bar{\mathbf{x}}$ *is* a minimum variance unbiased estimator of μ.

It would be erroneous to conclude from this that $\bar{x}$ is necessarily a minimum variance unbiased estimator of the means of other kinds of populations. For instance, in Exercise 12 on page 265 the reader will be asked to show that for random samples of size *three* from the uniform population

$$f(x) = \begin{cases} 1 & \text{for } \theta - \frac{1}{2} < x < \theta + \frac{1}{2} \\ 0 & \text{elsewhere} \end{cases}$$

the variance of $\bar{x}$ is $\frac{1}{36}$ while the variance of the *mid-range* (namely, the mean of the largest value and the smallest) is $\frac{1}{40}$. Thus, $\bar{x}$ cannot be a minimum variance unbiased estimator in this particular situation.

9.3.2 Relative Efficiency

As we have indicated in the preceding section and also earlier in Chapter 7, we usually compare unbiased estimators in terms of their variances. If $\hat{\theta}_1$ and $\hat{\theta}_2$ are both unbiased estimators of a parameter θ and the variance of $\hat{\theta}_1$ is less than that of $\hat{\theta}_2$, we say that $\hat{\theta}_1$ is *relatively more efficient*. In fact, we use the ratio

$$\frac{\text{var}(\hat{\theta}_2)}{\text{var}(\hat{\theta}_1)} \tag{9.3.2}$$

as a measure of the *relative efficiency* of $\hat{\theta}_1$ with respect to $\hat{\theta}_2$. For instance, with reference to the last example of the preceding section we can say that for random samples of size 3 from the given population, the relative efficiency of the mean with respect to the mid-range is $\frac{\frac{1}{40}}{\frac{1}{36}} = 0.90$ or 90 percent.

To consider another example, suppose that we want to estimate the mean of a normal population on the basis of a random sample of size $2n + 1$, and that we can use either the median or the mean. So far as $\bar{x}$ is concerned, we know from Theorem 6.5 that it is unbiased and that it has the variance $\frac{\sigma^2}{2n+1}$; so far as $\tilde{x}$ is concerned we know that it is unbiased by virtue of the symmetry of the normal distribution, and the discussion following Theorem 7.9 tells us that for large samples its variance is approximately $\frac{\pi\sigma^2}{4n}$. Thus, for large samples, the relative efficiency of the median with respect to the mean is approximately

$$\frac{\frac{\sigma^2}{2n+1}}{\frac{\pi\sigma^2}{4n}} = \frac{4n}{\pi(2n+1)}$$

and the *asymptotic efficiency* of the median with respect to the mean (namely, the relative efficiency when $n \to \infty$) is $\frac{2}{\pi}$ or about 64 percent. The following is a somewhat more intuitive way of stating this result: *The mean requires only 64 percent as many observations as the median to assure the same "reliability," namely, the same size chance fluctuations.*

It is important to note that we restricted our definition of relative efficiency to unbiased estimators. If we included biased estimators, we could always assure ourselves of an estimator with *zero variance* by letting its values equal the same constant k regardless of the data which we may obtain. Thus, if $\hat{\boldsymbol{\theta}}$ is a biased estimator of the parameter θ, it is preferable to judge its merits and make efficiency comparisons on the basis of the *mean square error* $E[(\hat{\boldsymbol{\theta}} - \theta)^2]$ instead of its variance.

9.3.3 Consistent Estimators

The idea of choosing a minimum variance unbiased estimator is closely related to that of minimizing the risk function when the loss function is quadratic (as in the example on page 248). Of course, there are other kinds of loss functions and other ways of measuring the chance fluctuations of a statistic. The fact that the variance may not even provide a good criterion for this purpose is illustrated by the following example: Suppose we want to estimate on the basis of one observation the parameter θ of the population

$$f(x) = w \cdot \frac{1}{\sigma\sqrt{2\pi}} \cdot e^{-\frac{1}{2}\left(\frac{x-\theta}{\sigma}\right)^2} + (1 - w) \cdot \frac{1}{\pi} \cdot \frac{1}{1 + (x - \theta)^2}$$

for $-\infty < x < \infty$ and $0 < w < 1$. Evidently, this density is a *weighted average* of a normal density with the mean θ and the variance σ^2, and a Cauchy density (see Exercise 11 on page 117) which has been translated (shifted) so that it is symmetrical about $x = \theta$, and θ could thus be *defined* as its mean. Now, if w is very close to 1, say, $w = 1 - 10^{-100}$, and σ is very small, say, $\sigma = 10^{-100}$, the probability that a random variable having this distribution will take on a value which is extremely close to θ (and, hence, is an *excellent* estimate of θ) is *practically* 1. Yet, *the variance of this estimator is infinite,* which follows from the fact that the variance of the Cauchy distribution is infinite (see Exercise 8 on page 153).

The preceding example is somewhat out of the ordinary, of course, but it suggests that we pay more attention to the probabilities that the values of estimators will be close to the parameters which they are supposed to estimate. Perhaps, the reader will recall that we already touched upon questions concerning the "closeness" of estimates in Sections 5.3.1 and 6.2.1. Basing our argument on Chebyshev's theorem, we showed in Section 5.3.1 that *when* $n \to \infty$ *the probability that the sample proportion* $\frac{\mathbf{x}}{n}$ *will take on a value which differs numerically from* θ *by more than any arbitrary positive constant c approaches* 0. Also using Chebyshev's theorem, we showed in Section 6.2.1 that *when* $n \to \infty$ *the probability that* $\bar{\mathbf{x}}$ *will take on a value which differs numerically from* μ *by more than any arbitrary positive constant c approaches* 0. Note that in both of these examples we were practically assured that, at least for large n, the estimators will take on values which are very close to the respective parameters.

In general, we say that $\hat{\boldsymbol{\theta}}$ is a *consistent* estimator of the parameter θ of a given population, *if the probability that* $\hat{\boldsymbol{\theta}}$ *will take on a value which differs numerically from* θ *by more than any arbitrary constant c approaches* 0 *when* $n \to \infty$. Thus, we showed in Sections 5.3.1 and 6.2.1 that $\frac{\mathbf{x}}{n}$ is a consistent estimator of the parameter θ of a binomial population, and that $\bar{\mathbf{x}}$ is a consistent estimator of the mean μ of a population which has the finite variance σ^2. Note that consistency is an *asymptotic property* (namely, a *limiting property*) of an estimator; informally, it says that when n is sufficiently large, we can be practically certain that the *error* made with a consistent estimator will be numerically less than any small preassigned positive constant.

In actual practice, we can often judge whether an estimator is consistent by using the following *sufficient conditions* (though not *necessary conditions*), which are an immediate consequence of Chebyshev's theorem:

THEOREM 9.2

$\hat{\boldsymbol{\theta}}$ is a consistent estimator of the parameter θ of a given population if (1) $\hat{\boldsymbol{\theta}}$ *is unbiased, and* (2) *var*$(\hat{\boldsymbol{\theta}}) \to 0$ *when* $n \to \infty$.

To prove that these are not necessary conditions, we have only to show that an estimator can be consistent without being unbiased. An example of such an estimator is the one of Exercise 7 on page 252, and in Exercise 13 on page 265 the reader will be asked to verify that $\frac{\mathbf{x}+1}{n+2}$ is, indeed, a consistent estimator of the parameter θ of a binomial population. (We might add, though, that a biased estimator can be consistent only if it is *asymptotically unbiased*, namely, that it becomes unbiased when $n \to \infty$. Thus, the results on page 256 show that the minimax estimator on page 248 as well as the estimator mentioned above are both asymptotically unbiased.)

To illustrate the use of Theorem 9.2, let us show that $\mathbf{s}^2$ is a consistent estimator of σ^2 for random samples of size n from a normal population. Having already shown on page 257 that $\mathbf{s}^2$ is unbiased, it remains to be seen whether the variance of $\mathbf{s}^2$ approaches 0 when $n \to \infty$. Referring to the result of Exercise 8 on page 216 (or Theorem 7.6 on which this exercise was based), we find that

$$\text{var}(\mathbf{s}^2) = \frac{2\sigma^4}{n-1}$$

and, hence, that $\text{var}(\mathbf{s}^2) \to 0$ when $n \to \infty$, so long as σ is finite. This completes the proof.

9.3.4 Sufficient Estimators

An estimator $\hat{\boldsymbol{\theta}}$ is said to be *sufficient* if it utilizes all the information (relevant to the estimation of θ) that is contained in a sample. For instance, we will be able to show that the mean of a random sample is a sufficient estimator of the mean of a normal population with a known variance, and this means that there is *nothing to be gained for this purpose* by actually specifying the individual values of the sample or the order in which they were obtained. Similarly, we will be able to show that a sample proportion is a sufficient estimator of the parameter θ of a binomial population, and this means that there is nothing to be gained for this purpose by specifying the order in which the various "successes" and "failures" were actually obtained.

There are several ways in which the concept of sufficiency can be formulated more rigorously; let us state it as follows: *$\hat{\boldsymbol{\theta}}$ is a sufficient estimator of the parameter θ if $f(x_1, x_2, \ldots, x_n|\hat{\theta})$ does not depend on θ.* Here $f(x_1, x_2 \ldots, x_n|\hat{\theta})$ is the conditional joint density or probability function of the random sample $\mathbf{x}_1, \mathbf{x}_2, \ldots$, and $\mathbf{x}_n$, given that $\hat{\boldsymbol{\theta}}$ takes on the value $\hat{\theta}$. Thus, for the estimation of the parameter θ of a binomial population, the random variables $\mathbf{x}_1, \mathbf{x}_2, \ldots$, and $\mathbf{x}_n$ are independent Bernoulli random variables with the same parameter θ, $\hat{\theta} = \frac{x}{n}$, where x is the total number of "successes," and

$$f(x_1, x_2, \ldots, x_n|\hat{\theta}) = \frac{1}{\binom{n}{x}} = \frac{1}{\binom{n}{n\hat{\theta}}}$$

which evidently does not depend on θ. Actually, this merely expresses the fact that each possible arrangement of "successes" and "failures" is equally likely. For instance, if we have observed two "successes" in four trials, each of the six possibilities SSFF, SFSF, SFFS, FSSF, FSFS, and FFSS has the probability $\frac{1}{\binom{4}{2}} = \frac{1}{6}$.

Since the above definition can make it quite tedious to check whether a statistic is actually a sufficient estimator of a given parameter θ, it often helps to use the following factorization theorem:

THEOREM 9.3

$\hat{\boldsymbol{\theta}}$ is a sufficient estimator of θ if and only if

$$f(x_1, x_2, \ldots, x_n; \theta) = g(\hat{\theta}; \theta) \cdot h(x_1, x_2, \ldots, x_n)$$

where $f(x_1, x_2, \ldots, x_n; \theta)$ is the joint density or probability function of the random sample, $g(\hat{\theta}; \theta)$ is the density or probability function of the estimator $\hat{\boldsymbol{\theta}}$, and $h(x_1, x_2, \ldots, x_n)$ does not depend on θ.

To illustrate the use of this theorem (a proof of which is given in the book by S. S. Wilks referred to on page 287), let us show that $\bar{\mathbf{x}}$ is a sufficient estimator of the mean μ of a normal population with the known variance σ^2. Making use of the fact that

$$f(x_1, x_2, \ldots, x_n; \mu) = \left(\frac{1}{\sigma\sqrt{2\pi}}\right)^n \cdot e^{-\frac{1}{2} \cdot \sum_{i=1}^{n} \left(\frac{x_i - \mu}{\sigma}\right)^2}$$

and that

$$\sum_{i=1}^{n} (x_i - \mu)^2 = \sum_{i=1}^{n} [(x_i - \bar{x}) - (\mu - \bar{x})]^2$$

$$= \sum_{i=1}^{n} (x_i - \bar{x})^2 + \sum_{i=1}^{n} (\bar{x} - \mu)^2$$

$$= \sum_{i=1}^{n} (x_i - \bar{x})^2 + n(\bar{x} - \mu)^2$$

we get

$$f(x_1, x_2, \ldots, x_n; \mu) = \frac{\sqrt{n}}{\sigma\sqrt{2\pi}} \cdot e^{-\frac{1}{2}\left(\frac{\bar{x}-\mu}{\sigma/\sqrt{n}}\right)^2} \frac{1}{\sqrt{n}} \left(\frac{1}{\sigma\sqrt{2\pi}}\right)^{n-1} \cdot e^{-\frac{1}{2} \cdot \sum_{i=1}^{n} \left(\frac{x_i - \bar{x}}{\sigma}\right)^2}$$

where the first factor on the right-hand side is the density of $\bar{x}$ (namely, a normal density with the mean μ and the variance $\frac{\sigma^2}{n}$), and the second factor does not involve μ. This proves, according to Theorem 9.3, that $\bar{x}$ is a sufficient estimator of the mean of a normal population with the known variance σ^2.

THEORETICAL EXERCISES

1. Use (7.3.4) to show that for random samples of size 3 the median is an unbiased estimator of the parameter θ of the uniform distribution given on page 259.
2. Use the result obtained on page 226 to show that for $n = 3$ the median is a biased estimator of the parameter θ of the exponential distribution (4.3.2).
3. Given a random sample of size n from a population which has the known mean μ and the finite variance σ^2, show that $\frac{1}{n} \cdot \sum_{i=1}^{n} (x_i - \mu)^2$ is an unbiased estimator of σ^2.
4. Show that if $\hat{\theta}$ is an unbiased estimator of θ and $\text{var}(\hat{\theta})$ does not equal 0, then $\hat{\theta}^2$ is not an unbiased estimator of θ^2.
5. Use the results of Theorem 6.5 to show that $\bar{x}^2$ is an *asymptotically unbiased* estimator of μ^2.
6. Show that the sample proportion $\frac{x}{n}$ is a minimum variance unbiased estimator of the parameter θ of the binomial distribution. [*Hint:* The sample proportion is the mean of a random sample of size n from the Bernoulli distribution (3.3.2).]
7. Show that for a random sample of size n from a Poisson population (3.3.9), $\bar{x}$ is a minimum variance unbiased estimator of the parameter λ.
8. If x_1, x_2, and x_3 are a random sample from a normal population with the mean μ and the variance σ^2, what is the relative efficiency of the estimator $\hat{\mu} = \frac{x_1 + 2x_2 + x_3}{4}$ with respect to $\bar{x}$?
9. If $\bar{x}_1$ is the mean of a random sample of size n from a normal population with the mean μ and the variance σ_1^2, and $\bar{x}_2$ is the mean of a random sample of size n from a normal population with the mean μ and the variance σ_2^2, show that
   (a) $w \cdot \bar{x}_1 + (1 - w) \cdot \bar{x}_2$, where $0 \leq w \leq 1$, is an unbiased estimator of μ;

(b) the variance of this estimator is a minimum when

$$w = \frac{\sigma_2^2}{\sigma_1^2 + \sigma_2^2}$$

10. If $\bar{\mathbf{x}}_1$ and $\bar{\mathbf{x}}_2$ are the means of independent random samples of size n_1 and n_2 from a normal population with the mean μ and the variance σ^2, show that the variance of the unbiased estimator $w \cdot \bar{\mathbf{x}}_1 + (1 - w) \cdot \bar{\mathbf{x}}_2$ is a minimum when $w = \dfrac{n_1}{n_1 + n_2}$.

11. If $\mathbf{x}$ is a random variable having the binomial distribution (3.3.3), $\hat{\boldsymbol{\theta}}_1 = \dfrac{\mathbf{x}}{n}$ and $\hat{\boldsymbol{\theta}}_2 = \dfrac{\mathbf{x} + 1}{n + 2}$, for what values of θ is $E[(\hat{\boldsymbol{\theta}}_2 - \theta)^2]$ less than $E[(\hat{\boldsymbol{\theta}}_1 - \theta)^2]$?

12. Since the variance of the mean and the mid-range are not affected if the same constant is added to each observation, we can determine these variances for random samples of size 3 from the uniform population on page 259 by referring instead to the uniform population

$$f(x) = \begin{cases} 1 & \text{for } 0 < x < 1 \\ 0 & \text{elsewhere} \end{cases}$$

(a) Show that $E(\mathbf{x}) = \frac{1}{2}$, $E(\mathbf{x}^2) = \frac{1}{3}$, and $\text{var}(\mathbf{x}) = \frac{1}{12}$ for this distribution, so that for a random sample of size 3, $\text{var}(\bar{\mathbf{x}}) = \frac{1}{36}$.

(b) Use the results of Exercise 2 on page 227 and part (b) of Exercise 6 on page 227 (or derive the necessary densities and joint density) to show that for a random sample of size three from this distribution the order statistics $\mathbf{y}_1$ and $\mathbf{y}_3$ have $E(\mathbf{y}_1) = \frac{1}{4}$, $E(\mathbf{y}_1^2) = \frac{1}{10}$, $E(\mathbf{y}_3) = \frac{3}{4}$, $E(\mathbf{y}_3^2) = \frac{3}{5}$, and $E(\mathbf{y}_1\mathbf{y}_3) = \frac{1}{5}$, so that $\text{var}(\mathbf{y}_1) = \frac{3}{80}$, $\text{var}(\mathbf{y}_3) = \frac{3}{80}$, and $\text{cov}(\mathbf{y}_1, \mathbf{y}_3) = \frac{1}{80}$.

(c) Use the results of part (b) and Theorem 6.3 to show that $E\left(\dfrac{\mathbf{y}_1 + \mathbf{y}_3}{2}\right) = \dfrac{1}{2}$ and $\text{var}\left(\dfrac{\mathbf{y}_1 + \mathbf{y}_3}{2}\right) = \dfrac{1}{40}$.

13. Show that the estimator of Exercise 7 on page 252 is a consistent estimator of the parameter θ of the binomial distribution.

14. Show that the estimator of Exercise 9 is consistent.

15. Suppose we use the largest value of a random sample of size n (namely, the order statistic $\mathbf{y}_n$) to estimate the parameter θ of the population

$$f(x) = \begin{cases} \dfrac{1}{\theta} & \text{for } 0 < x < \theta \\ 0 & \text{elsewhere} \end{cases}$$

Check whether this estimator is (a) unbiased, and (b) consistent.

16. If $\mathbf{x}_1$ and $\mathbf{x}_2$ are independent random variables having, respectively, binomial distributions with the parameters θ and n, and θ and m, show that $\dfrac{\mathbf{x}_1 + \mathbf{x}_2}{n + m}$ is a sufficient estimator of θ.

17. If $\mathbf{x}_1$ and $\mathbf{x}_2$ are independent random variables having Poisson distributions with the same parameter λ, show that their mean is a sufficient estimator of λ.

18. Show that the estimator of Exercise 3 is a sufficient estimator of the variance of a normal population having the known mean μ.

9.4 METHODS OF POINT ESTIMATION

As we have already seen in this chapter and in Chapter 8, there can be many different ways of estimating one and the same parameter of a population. In decision theory, we will generally get different estimators depending on the form of the loss function and the criterion which we apply to the risk function in order to choose a decision function (that is, an estimator) which is thus considered "best." In actual practice, where we seldom have enough information to calculate a Bayes risk (and where it may be difficult even to decide on the functional form of the loss function), it is desirable to have some general methods which yield estimators with as many as possible of the properties discussed in the preceding sections. In Sections 9.4.1 and 9.4.2 we shall treat two such methods, the *method of moments* and the method of *maximum likelihood;* some further discussion of *Bayesian estimation* will be given in Section 9.6, and another method, the *method of least squares,* will be taken up in Chapter 13.

9.4.1 The Method of Moments

Historically, one of the oldest methods of estimation, called the *method of moments,* consists of equating the first few moments of a population with the corresponding moments of a sample, thus getting as many equations as are needed to solve for the unknown parameters of the population distribution. The kth *sample moment* of a set of observations $x_1, x_2, \ldots,$ and x_n is defined as

$$m'_k = \frac{\sum_{i=1}^{n} x_i^k}{n} \tag{9.4.1}$$

namely, as the mean of the kth power of the observations, just as the kth moment of a distribution is the expected value of the kth power of a corresponding random variable $\mathbf{x}$. In the method of moments, we use as many of the equations

$$m'_k = \mu'_k \tag{9.4.2}$$

for $k = 1, 2, \ldots$, as are needed to solve for the unknown parameters of the population distribution.

To illustrate this method, let us find estimates of the two parameters α and β of the gamma distribution (4.3.6) on the basis of n observations, where $n > 1$. The two equations which we shall have to solve are

$$m'_1 = \mu'_1 \quad \text{and} \quad m'_2 = \mu'_2$$

and since $\mu'_1 = \alpha\beta$ and $\mu'_2 = \beta^2\alpha(\alpha + 1)$ according to Theorem 5.13, we get

$$m'_1 = \alpha\beta \quad \text{and} \quad m'_2 = \beta^2\alpha(\alpha + 1)$$

Thus, solving these two equations for α and β, we find that the estimates of the two parameters of the gamma distribution are

$$\alpha = \frac{(m'_1)^2}{m'_2 - (m'_1)^2} \quad \text{and} \quad \beta = \frac{m'_2 - (m'_1)^2}{m'_1}$$

(and if we wanted to speak of the corresponding *estimators* we would simply give everything in boldface type).

In the above example we estimated the parameters of a *specific* population. It is important to note, however, that when the parameters to be estimated happen to be themselves, the moments of the population distribution, the method of moments can be used even without knowledge of the exact functional form of the population.

9.4.2 The Method of Maximum Likelihood

In two papers published in the early 1920's, R. A. Fisher, one of the foremost statisticians of our time, proposed a general method of estimation, called the *method of maximum likelihood*. At the time, he also demonstrated the advantages of this method by showing that (1) it yields sufficient estimators whenever they exist, and (2) it yields estimators which are *asymptotically* (when $n \to \infty$) minimum variance unbiased estimators.

In principle, *the method of maximum likelihood consists of selecting that value of the parameter* θ *under consideration for which* $f(x_1, x_2, \ldots, x_n; \theta)$, *the probability (or the value of the joint density) of obtaining the sample values, is a maximum.* Looking at $f(x_1, x_2, \ldots, x_n; \theta)$, thus, as a value of a function of θ, we refer to it as a *likelihood*, and to the corresponding function as the *likelihood function;* hence, the name "method of maximum likelihood."

To illustrate the method of maximum likelihood, let us use it to estimate the parameter θ of the binomial distribution on the basis of an "experiment" in which we obtained x "successes" in n trials. To find the value of θ which maximizes

$$L(\theta) = \binom{n}{x} \theta^x (1 - \theta)^{n-x}$$

it will be convenient to make use of the fact that the value of θ which maximizes $L(\theta)$ will also maximize

$$\ln L(\theta) = \ln \binom{n}{x} + x \cdot \ln \theta + (n - x) \cdot \ln (1 - \theta)$$

Thus, we get

$$\frac{d[\ln L(\theta)]}{d\theta} = \frac{x}{\theta} - \frac{n - x}{1 - \theta}$$

and, equating this derivative to 0 and solving for θ, we find that the likelihood function has a maximum at $\theta = \frac{x}{n}$. Hence, the maximum likelihood *estimator* of the parameter θ of the binomial distribution is the sample proportion $\frac{\mathbf{x}}{n}$.

To consider another example, suppose that $x_1, x_2, \ldots,$ and x_n are the values of a random sample from an exponential population, and that we want to use the method of maximum likelihood to estimate its parameter θ. Since the likelihood function is given by

$$L(\theta) = \left(\frac{1}{\theta}\right)^n \cdot e^{-\frac{1}{\theta}\left(\sum_{i=1}^{n} x_i\right)}$$

differentiation of $\ln L(\theta)$ with respect to θ yields

$$\frac{d[\ln L(\theta)]}{d\theta} = -\frac{n}{\theta} + \frac{1}{\theta^2} \cdot \sum_{i=1}^{n} x_i$$

Equating this derivative to 0 and solving for θ, we get $\theta = \frac{1}{n} \cdot \sum_{i=1}^{n} x_i$, and we have thus shown that the maximum likelihood *estimator* of the parameter θ of an exponential population is the sample mean $\bar{\mathbf{x}}$.

The method of maximum likelihood can also be used for the simultaneous estimation of several parameters of a given population. To illustrate, let us find simultaneous maximum likelihood estimators of the mean and the variance of a normal population, given a random sample of size n. Since the likelihood function is given by

$$L(\mu, \sigma^2) = \left(\frac{1}{\sigma\sqrt{2\pi}}\right)^n \cdot e^{-\frac{1}{2\sigma^2} \cdot \sum_{i=1}^{n} (x_i - \mu)^2}$$

partial differentiation of $\ln L(\mu, \sigma^2)$ with respect to μ and σ^2 yields

$$\frac{\partial[\ln L(\mu, \sigma^2)]}{\partial \mu} = \frac{1}{\sigma^2} \cdot \sum_{i=1}^{n} (x_i - \mu)$$

and

$$\frac{\partial[\ln L(\mu, \sigma^2)]}{\partial \sigma^2} = -\frac{n}{2\sigma^2} + \frac{1}{2\sigma^4} \cdot \sum_{i=1}^{n} (x_i - \mu)^2$$

Equating the first of these partial derivatives to 0 and solving for μ, we get $\mu = \frac{1}{n} \cdot \sum_{i=1}^{n} x_i = \bar{x}$, and equating the second partial derivative to 0 and solving for σ^2 after substituting $\mu = \bar{x}$, we get

$$\sigma^2 = \frac{1}{n} \cdot \sum_{i=1}^{n} (x_i - \bar{x})^2$$

Thus, the maximum likelihood *estimator* of μ is the sample mean $\bar{\mathbf{x}}$, and the maximum likelihood estimator of σ^2 is the *second sample moment about the mean*, namely, $\frac{n-1}{n} \cdot \mathbf{s}^2$. This demonstrates, incidentally, that *maximum likelihood estimators need not be unbiased.* (Although we maximized the logarithm of the likelihood function in all of our examples instead of the likelihood function, itself, this is by no means necessary; it so happened that it was convenient in each case.)

With regard to the preceding example, it should also be observed that we did not prove that $\sqrt{\frac{n-1}{n}}\cdot\mathbf{s}$ is necessarily a maximum likelihood estimator of σ. However, it can be shown (see reference on page 288) that maximum likelihood estimators have the *invariance property* that if $\hat{\boldsymbol{\theta}}$ is a maximum likelihood estimator of θ and the function given by $g(\theta)$ is continuous, then $g(\hat{\boldsymbol{\theta}})$ is also a maximum likelihood estimator of $g(\theta)$. From this we can conclude that $\sqrt{\frac{n-1}{n}}\cdot\mathbf{s}$ is also a maximum likelihood estimator of σ.

Finally, let us consider an example in which the methods of elementary calculus cannot be used to find the maximum value of the likelihood function. Suppose that $x_1, x_2, \ldots,$ and x_n are the values of a random sample from a uniform population having the density

$$f(x;\theta) = \begin{cases} \frac{1}{\theta} & \text{for } 0 < x < \theta \\ 0 & \text{elsewhere} \end{cases}$$

and that we want to use the method of maximum likelihood to estimate the parameter θ. Since the likelihood function is given by

$$L(\theta) = \left(\frac{1}{\theta}\right)^n$$

it can be seen that the values of the function *increase* when θ *decreases,* so that we should make θ as small as possible. Evidently, θ cannot be less than any of the sample values which are actually observed, so that the likelihood function attains its maximum when we set θ equal to the largest value of the sample. Thus, the maximum likelihood estimator of θ is $\mathbf{y}_n$, the *nth order statistics.*

THEORETICAL EXERCISES

1. Use the method of moments to find an estimator for the parameter θ of the uniform density given in the preceding paragraph.
2. If $x_1, x_2, \ldots,$ and x_n are the values of a random sample of size n from a Poisson population with the parameter λ, find an estimate of λ using
   (a) the method of moments;
   (b) the method of maximum likelihood.
3. If $x_1, x_2, \ldots,$ and x_n are the values of a random sample of size n from a beta population (4.3.7) with $\beta = 1$, find an estimate of the parameter α using

(a) the method of moments;
(b) the method of maximum likelihood.

4. If a random variable has the binomial distribution with the parameters θ and $n = 2$, and in N experiments it took on the values 0, 1, and 2 with respective frequencies of n_0, n_1, and n_2, find an estimator of θ using
(a) the method of moments;
(b) the method of maximum likelihood.

5. Give a random sample of size n from a normal population with the *known* mean μ, find the maximum likelihood estimator of σ.

6. If $x_1, x_2, \ldots,$ and x_n are the values of a random sample of size n from a population having the geometric distribution (3.3.6), find an estimate of its parameter θ using
(a) the method of moments;
(b) the method of maximum likelihood.

7. Given a random sample of size n from a population having the density

$$f(x;\theta) = \begin{cases} e^{-(x-\theta)} & \text{for } x > \theta \\ 0 & \text{elsewhere} \end{cases}$$

find an estimator of the parameter θ by the method of maximum likelihood.

8. Given a random sample of size n from a population having the uniform distribution (4.3.1), find simultaneous maximum likelihood estimators of the parameters α and β.

9. Given a random sample of size n from a population having the gamma distribution (4.3.6) with the *known* parameter α, find the maximum likelihood estimator of β.

10. Given independent random samples $\mathbf{x}_1, \mathbf{x}_2, \ldots, \mathbf{x}_n$, and $\mathbf{u}_1, \mathbf{u}_2, \ldots, \mathbf{u}_n$ from two normal populations having the means $\mu_1 = \alpha + \beta$ and $\mu_2 = \alpha - \beta$ and the common variance $\sigma^2 = 1$, find simultaneous maximum likelihood estimators for α and β.

9.5 CONFIDENCE INTERVALS

To introduce the idea of a *confidence interval* by means of an example, let us refer again to the sampling distribution of $\bar{\mathbf{x}}$ for random samples of size n from a normal population with the mean μ and the known variance σ^2. Now, if we define $z_{\alpha/2}$ such that the integral of the standard normal density from $z_{\alpha/2}$ to ∞ equals $\frac{\alpha}{2}$, we can assert with the probability $1 - \alpha$ that the random variable $\frac{\mathbf{\bar{x}} - \mu}{\sigma/\sqrt{n}}$ will take on a value between $-z_{\alpha/2}$ and $z_{\alpha/2}$; this follows from Theorem 7.1 and the symmetry of any normal distribution about its mean. Suppose now that we write

$$-z_{\alpha/2} < \frac{\bar{x} - \mu}{\sigma/\sqrt{n}} < z_{\alpha/2} \tag{9.5.1}$$

or equivalently

$$\bar{x} - z_{\alpha/2} \cdot \frac{\sigma}{\sqrt{n}} < \mu < \bar{x} + z_{\alpha/2} \cdot \frac{\sigma}{\sqrt{n}} \tag{9.5.2}$$

where $\bar{x}$ is a value of $\mathbf{\bar{x}}$ which we actually obtained in a sample. Then, since σ, n, $z_{\alpha/2}$, and $\bar{x}$ are all constants for a given sample, the double inequality given by (9.5.2) *must be either true or false*—either μ is contained between $\bar{x} - z_{\alpha/2} \cdot \frac{\sigma}{\sqrt{n}}$ and $\bar{x} + z_{\alpha/2} \cdot \frac{\sigma}{\sqrt{n}}$ or it is not.

For instance, if we want to estimate the mean of a normal population having the variance $\sigma^2 = 225$ and a random sample of size $n = 20$ yields $\bar{x} = 64.3$, then for $\alpha = 0.05$ (and, hence, $z_{.025} = 1.96$) the double inequality (9.5.2) becomes

$$64.3 - 1.96 \cdot \frac{15}{\sqrt{20}} < \mu < 64.3 + 1.96 \cdot \frac{15}{\sqrt{20}}$$

$$57.7 < \mu < 70.9$$

Although this double inequality must be either true or false, it is difficult to shake the feeling that somehow it is more apt to be true than false. After all, did we not let $\alpha = 0.05$ so that there is a probability of 0.95 that $\frac{\mathbf{\bar{x}} - \mu}{\sigma/\sqrt{n}}$ will take on a value between -1.96 and 1.96, or that $\mathbf{\bar{x}}$ will take on a value between $\mu - 1.96 \cdot \frac{\sigma}{\sqrt{n}}$ and $\mu + 1.96 \cdot \frac{\sigma}{\sqrt{n}}$?

Even though we cannot make probability statements about (9.5.2) or the double inequality $57.7 < \mu < 70.9$, which involve *nothing but constants*, we can assert with a probability of $1 - \alpha$ that the *random variables* $\mathbf{\bar{x}} - z_{\alpha/2} \cdot \frac{\sigma}{\sqrt{n}}$ and $\mathbf{\bar{x}} + z_{\alpha/2} \cdot \frac{\sigma}{\sqrt{n}}$ will take on values satisfying (9.5.2), *and it is in this sense that we assign* (9.5.2) *a "degree of confidence" equal to this probability*. In fact, we refer to (9.5.2) as a *confidence interval* for μ, and to $1 - \alpha$ as the corresponding *degree of confidence*. The interval from 57.7

to 70.9, which we obtained in our numerical example, is a 0.95 confidence interval for the mean of the population from which the sample was obtained. Of course, the double inequality $57.5 < \mu < 70.9$ must be either true or false, but if we had to bet, 95 to 5 (or 19 to 1) would be *fair odds* that it is true. These odds are *fair* because the method by which we obtained the interval works, so to speak, 95 percent of the time.

In general, to obtain a $1 - \alpha$ confidence interval for the parameter θ of a given population, we must find two random variables $\boldsymbol{\theta}_1$ and $\boldsymbol{\theta}_2$ for which $\boldsymbol{\theta}_1$ cannot take on a value greater than that of $\boldsymbol{\theta}_2$, and for which we can assert with the probability $1 - \alpha$ that their values will satisfy the double inequality $\theta_1 < \theta < \theta_2$. It is customary to refer to θ_1 and θ_2, the values taken on by the random variables $\boldsymbol{\theta}_1$ and $\boldsymbol{\theta}_2$, as the *lower and upper confidence limits for* θ, and to the interval from θ_1 to θ_2 as a *$1 - \alpha$ confidence interval for* θ; as before, $1 - \alpha$ is called the degree of confidence.

The method by which we shall construct confidence intervals in the next few sections consists, essentially, of finding suitable random variables *whose values are determined by the sample data as well as the parameters, but whose distributions do not involve the parameters in question.* This was the case in our example, where we used the random variable $\dfrac{\bar{\mathbf{x}} - \mu}{\sigma/\sqrt{n}}$, whose values cannot be calculated without knowledge of μ, but whose distribution, the standard normal distribution, does not involve μ. Although this method of obtaining confidence intervals, sometimes called the *pivotal method,* is very widely used, there exist more general methods, for instance, the one discussed in the book by A. M. Mood and F. A. Graybill referred to on page 287.

The fact that confidence intervals for a given parameter are not unique is illustrated in Exercises 2 and 3 on page 277. Note also that instead of (9.5.2) we could have given the $1 - \alpha$ confidence interval

$$\bar{x} - z_{2\alpha/3}\cdot\frac{\sigma}{\sqrt{n}} < \mu < \bar{x} + z_{\alpha/3}\cdot\frac{\sigma}{\sqrt{n}} \tag{9.5.3}$$

or the *one-sided* $1 - \alpha$ confidence interval

$$\mu < \bar{x} + z_{\alpha}\cdot\frac{\sigma}{\sqrt{n}} \tag{9.5.4}$$

and we could also have based the confidence interval, say, on the sample median instead of the sample mean. As in the case of point estimation, methods of obtaining confidence intervals must, thus, be judged by their various statistical properties. For instance, one desirable property is to have the length of a $1 - \alpha$ confidence interval *as short as possible,* and this is why (9.5.2) is preferable to (9.5.3) as well as (9.5.4), as the reader will be asked to verify in Exercise 4 on page 277. Another desirable property is to have the *expected length*, that is, $E(\boldsymbol{\theta}_2 - \boldsymbol{\theta}_1)$, as small as possible.

9.5.1 Confidence Intervals for Means

The confidence interval given by (9.5.2) was designed to estimate the mean of a normal population whose variance is known. However, when we deal with samples which are large enough to justify use of the central limit theorem, (9.5.2) can also be used to estimate the means of other kinds of populations with known variances. In connection with this, it is customary to regard a sample as "sufficiently large" if $n \geq 30$.

In order to construct a $1 - \alpha$ confidence interval for μ when σ^2 is unknown, let us make use of the fact that for random samples of size n from normal populations $\frac{\bar{\mathbf{x}} - \mu}{\mathbf{s}/\sqrt{n}}$ is a random variable having the t distribution with $n - 1$ degrees of freedom (see Theorem 7.8). Hence, we can assert with the probability $1 - \alpha$ that this random variable will take on a value between $-t_{\alpha/2, n-1}$ and $t_{\alpha/2, n-1}$, and for a given sample with the mean $\bar{x}$ and the standard deviation s we can assign the degree of confidence $1 - \alpha$ to

$$-t_{\alpha/2, n-1} < \frac{\bar{x} - \mu}{s/\sqrt{n}} < t_{\alpha/2, n-1} \tag{9.5.5}$$

or to the equivalent inequality

$$\bar{x} - t_{\alpha/2, n-1} \cdot \frac{s}{\sqrt{n}} < \mu < \bar{x} + t_{\alpha/2, n-1} \cdot \frac{s}{\sqrt{n}} \tag{9.5.6}$$

This $1 - \alpha$ confidence interval for μ applies only when we deal with random samples from normal populations; for other populations, an *approximate* large-sample confidence interval for μ may be obtained by substituting s for σ in (9.5.2).

To illustrate the use of (9.5.6), suppose that a paint manufacturer wants to determine the "true" average drying time of a new interior wall paint, and that for 12 test areas of equal size he obtained a mean drying time of 66.3 minutes and a standard deviation of 8.4 minutes. Substituting these values together with $t_{.025, 11} = 2.201$ into (9.5.6), we thus get

$$66.3 - 2.201 \cdot \frac{8.4}{\sqrt{12}} < \mu < 66.3 + 2.201 \cdot \frac{8.4}{\sqrt{12}}$$

$$61.0 < \mu < 71.6$$

which means that we can assert with the degree of confidence 0.95 that the interval from 61.0 minutes to 71.6 minutes contains the "true" average drying time of the paint.

9.5.2 Confidence Intervals for Proportions

In many problems in which we must estimate proportions, probabilities, percentages, or rates (say, the *proportion* of defectives in a large shipment of transistors, the *probability* that a car stopped at a road block will have faulty lights, the *percentage* of school children with I.Q.'s over 115, or the mortality *rate* of a disease) it is reasonable to assume that we are sampling from a binomial population, and, hence, that our problem is to estimate its parameter θ. Making use of the fact that for large n the binomial distribution can be approximated with a normal distribution, namely, that the random variable $\dfrac{\mathbf{x} - n\theta}{\sqrt{n\theta(1-\theta)}}$ can be treated as if it had the standard normal distribution, we can obtain a $1 - \alpha$ confidence interval for θ by solving the double inequality

$$-z_{\alpha/2} < \frac{x - n\theta}{\sqrt{n\theta(1-\theta)}} < z_{\alpha/2} \tag{9.5.7}$$

Leaving the details to the reader in Exercise 6 on page 277, let us merely state the result that the corresponding confidence limits are

$$\frac{x + \frac{1}{2}\cdot z_{\alpha/2}^2 \pm z_{\alpha/2}\sqrt{\dfrac{x(n-x)}{n} + \dfrac{1}{4}\cdot z_{\alpha/2}^2}}{n + z_{\alpha/2}^2} \tag{9.5.8}$$

A somewhat simpler large-sample approximation can be obtained by writing (9.5.7) as

$$\frac{x}{n} - z_{\alpha/2}\sqrt{\frac{\theta(1-\theta)}{n}} < \theta < \frac{x}{n} + z_{\alpha/2}\sqrt{\frac{\theta(1-\theta)}{n}} \tag{9.5.9}$$

and then substituting $\dfrac{x}{n}$ for θ inside the radicals.

Thus, if a sample yielded 140 "successes" in 400 trials and the assumptions underlying the binomial distribution are met, substitution into (9.5.8) with $z_{.025} = 1.96$ yields the following 0.95 confidence interval for the parameter θ:

$$0.305 < \theta < 0.398$$

If we substitute these same values into (9.5.9) with θ replaced by $\frac{140}{400} = 0.35$ inside the radicals, we obtain

$$0.303 < \theta < 0.397$$

and it should be observed that the two confidence intervals are nearly the same.

In actual practice, confidence limits for the parameter θ of the binomial distribution are generally obtained by means of the specially-constructed tables referred to on page 288. This not only saves a good deal of arithmetic, but it makes it possible to obtain confidence intervals for this parameter when n is small.

9.5.3 Confidence Intervals for Variances

Given a random sample of size n from a normal population, we can obtain a $1 - \alpha$ confidence interval for σ^2 by making use of Theorem 7.6, according to which $\frac{(n-1)\mathbf{s}^2}{\sigma^2}$ is a random variable having the chi-square distribution with $n - 1$ degrees of freedom. We can thus assert with a probability of $1 - \alpha$ that this random variable will take on a value between $\chi^2_{1-\alpha/2,n-1}$ and $\chi^2_{\alpha/2,n-1}$, or with the degree of confidence $1 - \alpha$ that for a given sample

$$\chi^2_{1-\alpha/2,n-1} < \frac{(n-1)s^2}{\sigma^2} < \chi^2_{\alpha/2,n-1} \tag{9.5.10}$$

or

$$\frac{(n-1)s^2}{\chi^2_{\alpha/2,n-1}} < \sigma^2 < \frac{(n-1)s^2}{\chi^2_{1-\alpha/2,n-1}} \tag{9.5.11}$$

Clearly, this $1 - \alpha$ confidence interval for σ^2 can be converted into a corresponding $1 - \alpha$ confidence interval for σ by taking square roots.

To illustrate the use of (9.5.11), suppose that in 16 test runs the gasoline consumption of an experimental engine had a standard deviation of $s = 2.2$ gallons, and that we want to construct a confidence interval for σ^2 as an indication of the "true" variability of the gasoline consumption of the engine. Suppose, furthermore, that the degree of confidence is to be 0.99. Assuming that the observed data can be looked upon as a random sample from a normal population, we substitute $n = 16$, $s = 2.2$, $\chi^2_{.005,15} = 32.801$, and $\chi^2_{.995,15} = 4.601$ into (9.5.11), getting

$$\frac{15(2.2)^2}{32.801} < \sigma^2 < \frac{15(2.2)^2}{4.601}$$

and, hence,

$$2.21 < \sigma^2 < 15.78$$

Taking square roots, we find that the corresponding 0.99 confidence limits for σ are 1.49 and 3.97 gallons.

THEORETICAL EXERCISES

1. If x is a value of a random variable having the exponential distribution (4.3.2), find k so that the interval from 0 to kx is a $1 - \alpha$ confidence interval for the parameter θ.
2. If x_1 and x_2 are the values of a random sample of size 2 from a population having the uniform density (4.3.1) with $\alpha = 0$ and $\beta = \theta$, find k so that

$$0 < \theta < k(x_1 + x_2)$$

is a $1 - \alpha$ confidence interval for θ. (*Hint:* Make use of the fact that the distribution of $\mathbf{x}_1 + \mathbf{x}_2$ is a *triangular distribution* similar to that pictured in Figure 6.5.)
3. Using the methods of Section 7.3, it can be shown that for a random sample of size $n = 2$ from the population of Exercise 2 the distribution of the *sample range* is given by

$$f(R) = \begin{cases} \dfrac{2}{\theta^2}(\theta - R) & \text{for } 0 < R < \theta \\ 0 & \text{elsewhere} \end{cases}$$

Use this result to find c so that

$$R < \theta < cR$$

is a $1 - \alpha$ confidence interval for θ.
4. Show that the confidence interval given by (9.5.2) is *shorter* than the corresponding confidence interval given by (9.5.3).
5. In Exercise 4 on page 216 the reader was asked to show that if $\bar{\mathbf{x}}_1$ and $\bar{\mathbf{x}}_2$ are the means of independent random samples of size n_1 and n_2 from normal populations having the means μ_1 and μ_2 and the variances σ_1^2 and σ_2^2, then $\bar{\mathbf{x}}_1 - \bar{\mathbf{x}}_2$ is a random variable having a normal distribution with the mean $\mu_1 - \mu_2$ and the variance $\dfrac{\sigma_1^2}{n_1} + \dfrac{\sigma_2^2}{n_2}$. Use this result to construct a $1 - \alpha$ confidence interval for $\mu_1 - \mu_2$.
6. Verify that the solution of (9.5.7) for θ leads to the confidence limits given in (9.5.8).

7. Use (9.5.9) to show that we can assert with a degree of confidence of *at least* $1 - \alpha$ that $\left|\frac{x}{n} - \theta\right|$, the absolute value of the error which we make when we use a sample proportion as an estimate of θ, is less than $z_{\alpha/2} \cdot \frac{1}{2\sqrt{n}}$.

8. If $\mathbf{x}_1$ and $\mathbf{x}_2$ are independent random variables having, respectively, binomial distributions with the parameters n_1 and θ_1 and the parameters n_2 and θ_2, use (6.2.7) and (6.2.8) to construct a $1 - \alpha$ large-sample confidence interval for $\theta_1 - \theta_2$. (*Hint:* Approximate the distribution of $\frac{\mathbf{x}_1}{n_1} - \frac{\mathbf{x}_2}{n_2}$ with a normal distribution.)

9. For large n, the sampling distribution of $\mathbf{s}$ is sometimes approximated with a normal distribution having the mean σ and the variance $\frac{\sigma^2}{2n}$ (see Exercise 10 on page 217). Show that this approximation leads to the following large-sample $1 - \alpha$ confidence interval for σ:

$$\frac{s}{1 + \frac{z_{\alpha/2}}{\sqrt{2n}}} < \sigma < \frac{s}{1 - \frac{z_{\alpha/2}}{\sqrt{2n}}}$$

APPLIED EXERCISES

10. Taking a random sample from its very extensive files, an accountant working for a department store finds that the amounts owed in 100 overdue accounts have a mean of \$27.63 and a standard deviation of \$6.74. Find a 0.95 confidence interval for the true mean of the amounts owed in all of the department store's overdue accounts. (Assume that the population is large enough to be treated as infinite.)

11. For several years, a teacher has kept records on how long it takes students to solve a rather difficult problem in mathematics. If 64 students (randomly selected from his records) took on the average 32.50 minutes with a variance of 10.89 minutes, construct a 0.99 confidence interval for the true average time it takes a student to solve this problem.

12. Given that the length of the skulls of 10 fossil skeletons of an extinct species of birds has the mean $\bar{x} = 5.73$ cm and the standard deviation $s = 0.34$ cm, find a 0.95 confidence interval for the mean length of the skulls of this species of birds.

13. In order to test the durability of a new paint, a highway department has test strips painted across heavily traveled roads in eight different locations. If, on the average, the test strips disappeared after they had been crossed by 168,479 cars and the standard deviation is 12,851 cars, construct a 0.99 confidence interval for the number of cars it will actually take on the average to wear off this paint.

14. Use the theory of Exercise 5 to construct a 0.95 confidence interval for the true difference between the average lifetimes of two kinds of light bulbs, given that a random sample of 40 light bulbs of the first kind lasted on the average 418 hours (of continuous use) with a standard deviation of 26 hours, and that a random sample of 50 light bulbs of the second kind lasted on the average 402 hours with a standard deviation of 22 hours.

15. A sample survey of registered voters in a very large city showed that among 1,000 persons interviewed 420 opposed daylight-saving time. Construct a 0.95 confidence interval for the true proportion of registered voters in this city who are opposed to daylight-saving time, using

(a) formula (9.5.8);

(b) formula (9.5.9) with the sample proportion substituted for θ inside the radicals.

16. In a random sample of 100 cans of mixed nuts (taken from a very large shipment), 24 contained no pecans. Construct a 0.99 confidence interval for the probability that there will be no pecans in a can which is randomly selected from this shipment, using

(a) formula (9.5.8);

(b) formula (9.5.9) with the sample proportion substituted for θ inside the radicals.

17. In a random sample of 600 cars stopped at a roadblock on U.S. 10 near Tucson, 114 had poorly adjusted brakes. Construct a confidence interval for the true *percentage* of cars traveling along this route which have poorly adjusted brakes, using (9.5.9) and

(a) the degree of confidence 0.95;

(b) the degree of confidence 0.99.

18. Use the theory of Exercise 7 to find the minimum sample size which will enable us to assert with the degree of confidence of *at least* 0.95 that a sample proportion (which is used to estimate the parameter θ of a binomial population) is "off" by less than 0.03.

19. Use the theory of Exercise 7 to find the minimum sample size which will enable us to assert with the degree of confidence of *at least* 0.99 that a sample proportion (which is used to estimate the parameter θ of a binomial population) is "off" by less than 0.02.

20. Referring to Exercise 12, construct a 0.95 confidence interval for the true variance of the length of the skulls of the given species of birds.

21. Referring to Exercise 13, construct a 0.99 confidence interval for the standard deviation of the population from which the sample was obtained. What does this standard deviation measure; that is, what is it indicative of?

22. Use the data of Exercise 10 and the large-sample confidence interval formula of Exercise 9 to construct a 0.95 confidence interval for the true standard deviation of the amounts owed in all of the department store's overdue accounts.

23. Use the data of Exercise 11 and the large-sample confidence interval formula of Exercise 9 to construct a 0.99 confidence interval for the true standard deviation of the time it takes students to solve the given problem.

9.6 BAYESIAN ESTIMATION

So far we have assumed in this chapter that the parameter which we want to estimate is an unknown constant; in Bayesian estimation the parameter is looked upon as a *random variable* which has a *prior distribution* reflecting either the strength of one's belief about the possible values it can assume, or collateral information. In Section 8.4.2, we already met a problem of Bayesian estimation—the parameter was that of the uniform density whose values are $\frac{1}{\theta}$ for the interval from 0 to θ, and 0 elsewhere, and its prior distribution was a gamma distribution (4.3.6) with $\alpha = 2$ and $\beta = 1$.

Now, the main problem of Bayesian estimation is that of *combining prior information about a parameter with direct sample evidence*, and in the example of Section 8.4.2 we accomplished this by determining $\varphi(\theta|x)$, namely, the conditional density of $\boldsymbol{\theta}$ given the sample value taken on by $\mathbf{x}$. In contrast to the *prior distribution* of the parameter $\boldsymbol{\theta}$, this conditional distribution which also reflects the direct sample evidence is called the *posterior distribution* of $\boldsymbol{\theta}$. More generally, if $h(\theta)$ is a value of the *prior distribution* of a parameter $\boldsymbol{\theta}$, and we want to combine the information which it conveys with direct sample evidence about $\boldsymbol{\theta}$, say, the value of a statistic $\mathbf{S}$, we determine the *posterior distribution* of $\boldsymbol{\theta}$ by means of the formula

$$\varphi(\theta|S) = \frac{f(\theta, S)}{g(S)} = \frac{h(\theta) \cdot f(S|\theta)}{g(S)} \tag{9.6.1}$$

Here $f(S|\theta)$ is a value of the sampling distribution of the statistic $\mathbf{S}$ for a given value of the parameter $\boldsymbol{\theta}$, $f(\theta, S)$ is a value of the joint distribution of $\boldsymbol{\theta}$ and $\mathbf{S}$, and $g(S)$ is a value of the marginal distribution of $\mathbf{S}$. Note that (9.6.1) is, in fact, an extension of *Bayes' rule* of Theorem 2.13 to the continuous case; hence, the term "Bayesian estimation."

Once the posterior distribution of a parameter has been obtained, it can be used to make point estimates, as in the example of Section 8.4.2, or it can be used to make probability statements about the parameter, as will be illustrated on page 285. Although the method which we have described has many other applications, we shall limit our discussion in the text to the estimation of the mean $\boldsymbol{\mu}$ of a normal population and the estimation of the parameter $\boldsymbol{\theta}$ of the binomial distribution; the Bayesian estimation of the parameter $\boldsymbol{\lambda}$ of the Poisson distribution is treated in Exercise 4 on page 286.

9.6.1 Estimation of the Binomial Parameter $\boldsymbol{\theta}$

Suppose that we want to estimate the parameter $\boldsymbol{\theta}$ of the binomial distribution on the basis of x, the observed number of "successes" in n trials, and the knowledge that the prior distribution of $\boldsymbol{\theta}$ is a *beta distribution* (4.3.7) with given values of α and β. Thus,

$$f(x|\theta) = \binom{n}{x} \theta^x (1 - \theta)^{n-x} \qquad \text{for } x = 0, 1, 2, \ldots, n$$

$$h(\theta) = \begin{cases} \dfrac{\Gamma(\alpha + \beta)}{\Gamma(\alpha)\cdot\Gamma(\beta)}\cdot\theta^{\alpha-1}(1 - \theta)^{\beta-1} & \text{for } 0 < \theta < 1 \\ 0 & \text{elsewhere} \end{cases}$$

and, hence,

$$\begin{aligned} f(\theta, x) &= \frac{\Gamma(\alpha + \beta)}{\Gamma(\alpha)\cdot\Gamma(\beta)}\cdot\theta^{\alpha-1}(1 - \theta)^{\beta-1} \times \binom{n}{x}\theta^x(1 - \theta)^{n-x} \\ &= \binom{n}{x}\cdot\frac{\Gamma(\alpha + \beta)}{\Gamma(\alpha)\cdot\Gamma(\beta)}\cdot\theta^{x+\alpha-1}(1 - \theta)^{n-x+\beta-1} \end{aligned}$$

for $0 < \theta < 1$ and $x = 0, 1, 2, \ldots, n$, and $f(\theta, x) = 0$ elsewhere. To obtain the marginal density of $\mathbf{x}$, let us make use of the fact that the integral of the beta density (4.3.7) from 0 to 1 equals 1, namely, that

$$\int_0^1 x^{\alpha-1}(1 - x)^{\beta-1}\,dx = \frac{\Gamma(\alpha)\cdot\Gamma(\beta)}{\Gamma(\alpha + \beta)} \qquad (9.6.2)$$

Thus, we get

$$g(x) = \binom{n}{x}\cdot\frac{\Gamma(\alpha+\beta)}{\Gamma(\alpha)\cdot\Gamma(\beta)}\cdot\frac{\Gamma(\alpha+x)\cdot\Gamma(n-x+\beta)}{\Gamma(n+\alpha+\beta)} \qquad \text{for } x = 0, 1, \ldots, n$$

and, hence,

$$\varphi(\theta|x) = \frac{\Gamma(n+\alpha+\beta)}{\Gamma(\alpha+x)\cdot\Gamma(n-x+\beta)}\cdot\theta^{x+\alpha-1}(1-\theta)^{n-x+\beta-1}$$

for $0 < \theta < 1$, and $\varphi(\theta|x) = 0$ elsewhere, according to (9.6.1).

Having obtained this posterior distribution, let us make use of the fact that (under very general conditions) the *mean* of the posterior distribution minimizes the Bayes risk when the loss function is *quadratic*, namely, when it is given by $L[d(x), \theta] = c[d(x) - \theta]^2$, where c is a positive constant. (Note that this is the kind of loss function which we used in the example on page 249.) Since the posterior distribution of $\boldsymbol{\theta}$ which we obtained is a beta distribution with the parameters $x + \alpha$ and $n - x + \beta$, it follows from Theorem 5.16 that

$$E(\boldsymbol{\theta}|x) = \frac{x+\alpha}{\alpha+\beta+n}$$

and we conclude that the decision function which is given by $d(x) = \frac{x+\alpha}{\alpha+\beta+n}$ minimizes the Bayes risk when the loss function is quadratic and the prior distribution of $\boldsymbol{\theta}$ is of the given form. Thus, if we "feel" that the prior distribution of $\boldsymbol{\theta}$ is a beta distribution with the parameters $\alpha = \beta = 40$ (see Exercise 2 on page 286) and we actually obtain 42 "successes" in 120 trials, our estimate of $\boldsymbol{\theta}$ is $d(42) = \frac{42+40}{40+40+120} = \frac{82}{200} = 0.41$. Without knowledge of the prior distribution of $\boldsymbol{\theta}$, our "best" estimate of θ according to the criteria of Section 9.3.1 (see Exercise 6 on page 264) would have been the sample proportion $\frac{42}{120} = 0.35$.

9.6.2 Estimation of the Mean of a Normal Population

Suppose that we want to estimate the mean $\boldsymbol{\mu}$ of a normal population with the known variance σ^2 on the basis of the mean of a random sample of size n and the knowledge that the prior distribution of $\boldsymbol{\mu}$ is a normal distribution with the mean μ_0 and the variance σ_0^2. Thus,

$$f(\bar{x}|\mu) = \frac{\sqrt{n}}{\sigma\sqrt{2\pi}}\cdot e^{-\frac{1}{2}\left(\frac{\bar{x}-\mu}{\sigma/\sqrt{n}}\right)^2} \qquad \text{for } -\infty < \bar{x} < \infty$$

according to Theorem 7.1, and

$$h(\mu) = \frac{1}{\sigma_0\sqrt{2\pi}}\cdot e^{-\frac{1}{2}\left(\frac{\mu-\mu_0}{\sigma_0}\right)^2} \qquad \text{for } -\infty < \mu < \infty$$

so that

$$\varphi(\mu|\bar{x}) = \frac{h(\mu)\cdot f(\bar{x}|\mu)}{g(\bar{x})} = \frac{\sqrt{n}}{2\pi\sigma\sigma_0 g(\bar{x})}\cdot e^{-\frac{1}{2}\left(\frac{\bar{x}-\mu}{\sigma/\sqrt{n}}\right)^2-\frac{1}{2}\left(\frac{\mu-\mu_0}{\sigma_0}\right)^2}$$

$$\text{for } -\infty < \mu < \infty$$

according to (9.6.1). Now, if we collect powers of μ in the exponent of e, we get

$$-\frac{1}{2}\left(\frac{n}{\sigma^2}+\frac{1}{\sigma_0^2}\right)\mu^2 + \left(\frac{n\bar{x}}{\sigma^2}+\frac{\mu_0}{\sigma_0^2}\right)\mu - \frac{1}{2}\left(\frac{n\bar{x}^2}{\sigma^2}+\frac{\mu_0^2}{\sigma_0^2}\right)$$

and if we let

$$\frac{1}{\sigma_1^2} = \frac{n}{\sigma^2}+\frac{1}{\sigma_0^2} \quad \text{and} \quad \mu_1 = \frac{n\bar{x}\sigma_0^2+\mu_0\sigma^2}{n\sigma_0^2+\sigma^2} \tag{9.6.3}$$

factor out $-\frac{1}{2\sigma_1^2}$, and complete the square, the exponent of e in the expression for $\varphi(\mu|\bar{x})$ becomes

$$-\frac{1}{2\sigma_1^2}(\mu-\mu_1)^2 + R$$

where R involves n, $\bar{x}$, μ_0, σ, and σ_0, *but not* μ. Thus, the posterior distribution of $\boldsymbol{\mu}$ becomes

$$\varphi(\mu|\bar{x}) = \frac{\sqrt{n}\cdot e^R}{2\pi\sigma\sigma_0 g(\bar{x})}\cdot e^{-\frac{1}{2\sigma_1^2}(\mu-\mu_1)^2} \qquad \text{for } -\infty < \mu < \infty$$

which is easily identified as a normal distribution with the mean μ_1 and the variance σ_1^2. Hence, it can be written as

$$\varphi(\mu|\bar{x}) = \frac{1}{\sigma_1\sqrt{2\pi}}\cdot e^{-\frac{1}{2}\left(\frac{\mu-\mu_1}{\sigma_1}\right)^2} \qquad \text{for } -\infty < \mu < \infty$$

where μ_1 and σ_1 are defined in (9.6.3). Note that we did not have to determine $g(\bar{x})$ in this case as it was "absorbed" in the constant in the final result.

Again making use of the fact that the mean of the posterior distribution minimizes the Bayes risk when the loss function is quadratic, we conclude that the decision function given by

$$d(\bar{x}) = \mu_1 = \frac{n\bar{x}\sigma_0^2+\mu_0\sigma^2}{n\sigma_0^2+\sigma^2}$$

minimizes the Bayes risk under these conditions when the prior distribution of $\boldsymbol{\mu}$ is a normal distribution with the mean μ_0 and the variance σ_0^2. It is of interest to note that this decision function can be written as

$$d(\bar{x}) = w \cdot \bar{x} + (1 - w) \cdot \mu_0 \quad \text{where} \quad w = \frac{n}{n + \dfrac{\sigma^2}{\sigma_0^2}} \tag{9.6.4}$$

namely, as a *weighted average* of $\bar{x}$ and μ_0, representing, respectively, the direct sample evidence and the prior information.

Since the posterior distribution of $\boldsymbol{\mu}$ given $\bar{x}$ turned out to be a normal distribution, we can assert with the probability $1 - \alpha$ that the value of $\boldsymbol{\mu}$ is between $\mu_1 - z_{\alpha/2} \cdot \sigma_1$ and $\mu_1 + z_{\alpha/2} \cdot \sigma_1$, where $z_{\alpha/2}$ is as defined on page 272. In connection with this interval it is important to note that *it is not a confidence interval,* but an interval for which we can assert with the probability $1 - \alpha$ that is contains a value of $\boldsymbol{\mu}$. Whereas in a confidence interval the limits are values of random variables, the random variable is now $\boldsymbol{\mu}$, itself. In fact, using the posterior distribution of $\boldsymbol{\mu}$ given $\bar{x}$, we can make all sorts of probability statements about $\boldsymbol{\mu}$.

To illustrate the use of all these results, suppose that a distributor of soft drink vending machines knows that in a supermarket one of his machines will sell on the average $\mu_0 = 738$ drinks per week. Of course, the mean will vary somewhat from market to market, and this variation is measured by the standard deviation $\sigma_0 = 13.4$. So far as a machine placed in a particular market is concerned, the number of drinks sold will vary from week to week, and this variation is measured by the standard deviation $\sigma = 42.5$. All this is prior and collateral information. Now suppose that the distributor has placed one of his soft drink vending machines into a new supermarket, and that during the first 10 weeks the machine has sold on the average $\bar{x} = 692$ drinks per week. *What can we infer on the basis of all this about* $\boldsymbol{\mu}$, *the "true" average weekly sales of the machine placed in the new market?*

Assuming that it is reasonable to treat the prior distribution of $\boldsymbol{\mu}$ as a normal distribution with the mean $\mu_0 = 738$ and the standard deviation $\sigma_0 = 13.4$, we find that substitution into the formula for μ_1 yields

$$\frac{10 \cdot 692(13.4)^2 + 738(42.5)^2}{10(13.4)^2 + (42.5)^2} = 715$$

This is a point estimate of the true average weekly sales of the machine placed in the new market, which accounts *in part for the prior information* and *in part for the direct evidence obtained for the particular market.*

To continue, we might ask for the probability that the value of $\boldsymbol{\mu}$ is actually between 710 and 720, and to this end we must first calculate σ_1 according to (9.6.3). Getting

$$\frac{1}{\sigma_1^2} = \frac{10}{(42.5)^2} + \frac{1}{(13.4)^2} = 0.0111$$

and, hence, $\sigma_1^2 = 90.0$ and $\sigma_1 = 9.5$, we find that the answer to our question is given by the area of the shaded region of Figure 9.1, namely, the

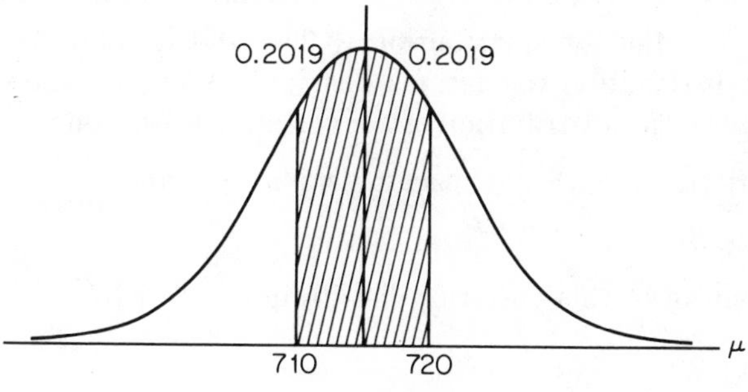

FIGURE 9.1 Posterior distribution of μ.

area under the standard normal distribution between $z = \frac{710 - 715}{9.5} = -0.53$ and $z = \frac{720 - 715}{9.5} = 0.53$. Since the entry corresponding to $z = 0.53$ in Table III is 0.2019, we find that the answer is approximately 0.40. This is the probability that the true average weekly sales of the machine placed in the particular market will be between 710 and 720 (soft drinks).

THEORETICAL EXERCISES

1. Making use of the results of Exercise 18 on page 172, show that the mean of the posterior distribution of $\boldsymbol{\theta}$ obtained on page 282 can be written as

$$E(\boldsymbol{\theta}|x) = w \cdot \frac{x}{n} + (1 - w) \cdot \theta_0$$

where θ_0 and σ_0^2 are the mean and the variance of the prior beta distribution of $\boldsymbol{\theta}$, and

$$w = \frac{n}{n + \dfrac{\theta_0(1-\theta_0)}{\sigma_0^2} - 1}$$

2. In the example on page 282, the prior distribution of the parameter $\boldsymbol{\theta}$ of the binomial distribution was a beta distribution with the parameters $\alpha = \beta = 40$. Use Theorem 5.16 to find the mean and the variance of this prior distribution and describe its shape.
3. Verify that (9.6.4) follows from the formula for μ_1 in (9.6.3).
4. If $\mathbf{x}$ has the Poisson distribution (3.3.9) and its parameter $\boldsymbol{\lambda}$ has as its prior distribution the gamma distribution (4.3.6), show that
   (a) the posterior distribution of $\boldsymbol{\lambda}$ given an observation of $\mathbf{x}$ is a gamma distribution with the parameters $\alpha + x$ and $\dfrac{\beta}{\beta + 1}$ (instead of α and β);
   (b) the mean of this posterior distribution of $\boldsymbol{\lambda}$ is $\mu_1 = \dfrac{\beta(\alpha + x)}{\beta + 1}$.

APPLIED EXERCISES

5. The output of a certain transistor production line is checked daily by inspecting a sample of 100 units. Over a long period of time, the process has maintained a yield of 80 percent, namely, a proportion defective of 20 percent, and the variation of the proportion defective from day to day is measured by a standard deviation of 0.04. If on a certain day the sample contains 38 defectives, give a point estimate of that day's true proportion defective on the basis of the direct sample evidence as well as the prior information. (Use results of Exercise 1.)
6. Records of a university (collected over many years) show that on the average 74 percent of all incoming freshmen have I.Q.'s of at least 115. Of course, the percentage varies somewhat from year to year and this variation is measured by a standard deviation of 3 percent. If a sample check of 30 freshmen entering the university in 1969 showed that only 18 of them have I.Q.'s of at least 115, estimate the true proportion of students with I.Q.'s of at least 115 in that freshman class
   (a) Using only the direct sample information;
   (b) Using the sample data as well as the prior information.
7. With reference to the illustration of Section 9.6.2 find
   (a) the probability that the true average weekly sales of the machine

placed in the given market will be between 712 and 725 (soft drinks);

(b) an interval for which we can assert with the probability 0.95 that $\boldsymbol{\mu}$ will fall inside the interval.

8. An insurance company knows that the annual losses on a certain kind of liability insurance average μ_0 = \$73.50 with a standard deviation of σ_0 = \$4.30 for different risks (that is, different policy holders). It also knows that for any given policyholder the annual losses have the standard deviation σ = \$21.72. If during the first 5 years a new policy has been in force, the losses averaged $\bar{x}$ = \$154.36, find
(a) a point estimate of this policy's true average annual losses, which is based on the direct evidence as well as the collateral information;
(b) the probability that the true average annual losses for this policy are between \$80.00 and \$100.00;
(c) an interval for which we can assert with a probability of 0.99 that the true average annual losses of the policy will fall inside the interval.

9. An office manager knows that for a certain kind of business the daily number of incoming telephone calls is a random variable having a Poisson distribution, whose parameter has a prior gamma distribution with $\alpha = 50$ and $\beta = 2$. Being told that one such business had 112 incoming calls on a given day, what would be his estimate of that particular business' average daily number of incoming calls if
(a) he considered only the direct information;
(b) he considered only the prior information;
(c) if he combined the two kinds of information.
(*Hint:* Use the theory of Exercise 4.)

REFERENCES

The sampling distribution of **s** is discussed in

Keeping, E. S., *Introduction to Statistical Inference.* Princeton, N.J.: D. Van Nostrand Co., Inc., 1962,

and the proof of Theorem 9.3 may be found in

Wilks, S. S., *Mathematical Statistics.* New York: John Wiley & Sons, Inc., 1962.

A general method for obtaining confidence intervals is given in

Mood, A. M., and Graybill, F. A., *Introduction to the Theory of Statistics.* New York: McGraw-Hill Book Company, 1963 and further criteria for judging the relative merits of confidence intervals may be found in

Lehmann, E. L., *Testing Statistical Hypotheses.* New York: John Wiley & Sons, Inc., 1959,

and in other advanced texts on mathematical statistics. Special tables for constructing 0.95 and 0.99 confidence intervals for proportions are given in the *Biometrika Tables* referred to on page 229. Various important properties of maximum likelihood estimators are discussed in the book by E. S. Keeping listed above. A derivation of the Cramér-Rao inequality and the most general conditions under which it applies may be found in

Rao, C. R., *Advanced Statistical Methods in Biometric Research,* New York: John Wiley & Sons, Inc., 1952, Chapter 4.

10

HYPOTHESIS TESTING: THEORY

10.1 INTRODUCTION

If an engineer has to decide on the basis of sample data whether the true average lifetime of a certain kind of tire is at least 22,000 miles, if an agronomist has to decide on the basis of experiments whether one kind of fertilizer produces a higher yield of soybeans than another, and if a manufacturer of pharmaceutical products has to decide on the basis of samples whether 90 percent of all patients given a new medication will recover from a certain disease, *these problems can all be translated into the language of statistical tests of hypotheses.* In the first case we might say that the engineer has to test the hypothesis that θ, the parameter of an exponential population, is at least 22,000; in the second case we might say that the agronomist has to decide whether $\mu_1 > \mu_2$, where μ_1 and μ_2 are the means of two normal populations; and in the third case we might say that the manufacturer has to decide whether θ, the parameter of a binomial population, equals 0.90. In each case it must be assumed, of course, that the chosen distribution correctly describes the experimental conditions, namely, that the distribution provides the correct *statistical model.*

As in the above examples, most tests of statistical hypotheses concern the parameters of distributions, but sometimes they also concern the type, or nature, of the distributions, themselves. For instance, in the first of our three examples the engineer may also have to decide whether he is actually dealing with a sample from an exponential population, or whether his data are values of random variables having, say, the Weibull distribution of Exercise 12 on page 117.

If a statistical hypothesis completely specifies the underlying distribution, that is, if it specifies the functional form of the distribution as well as the values of all parameters, it is referred to as a *simple hypothesis;* if not, it is referred to as a *composite hypothesis.* Thus, in the first of the above examples the hypothesis is *composite* since $\theta \geq 22{,}000$ does *not* assign a specific value to the parameter θ, but in the third example, the one dealing with the effectiveness of the new medication, the hypothesis is *simple* (provided we specify the sample size n).

To be able to construct suitable criteria for testing statistical hypotheses, it is necessary that we also formulate *alternative hypotheses.* For instance, in the example dealing with the lifetimes of the tires we might formulate the alternative hypothesis that the parameter θ of the exponential population is less than 22,000; in the example dealing with the two kinds of fertilizer we might formulate the alternative hypothesis $\mu_1 = \mu_2$; and in the example dealing with the new medication we might formulate the alternative hypothesis that the parameter θ of the given binomial population is only 0.60 (the disease's recovery rate without the new medication).

The concept of simple and composite hypotheses applies also to alternative hypotheses, and in the first example we can now say that we are testing the *composite hypothesis* $\theta \geq 22{,}000$ against the *composite alternative* $\theta < 22{,}000$, where θ is the parameter of an exponential population. Similarly, in the second example we are testing the *composite hypothesis* $\mu_1 > \mu_2$ against the *composite alternative* $\mu_1 = \mu_2$, where μ_1 and μ_2 are the means of two normal populations, and in the third example we are testing the *simple hypothesis* $\theta = 0.90$ against the *simple alternative* $\theta = 0.60$, where θ is the parameter of a binomial population for which n is given. (Note that in the second example the alternative is *composite* because we do not actually specify the values of μ_1 and μ_2; it would have been *simple* if we had used the alternative hypothesis $\mu_1 = \mu_2 = 28.6$ bushels per acre.)

Symbolically, we shall use H_0 (standing for *null hypothesis,* a term which will be explained on page 303) for whatever hypothesis we shall want to test and H_1 for the alternative hypothesis. Problems involving more than two hypotheses, that is, problems involving several alternative hypotheses, tend to be quite complicated, and they will not be studied in this book.

10.2 SIMPLE HYPOTHESES

In Section 10.2.1 we shall introduce tests of simple hypotheses against simple alternatives as 2×2 zero-sum two-person games, and discuss some of the basic ideas underlying tests of hypotheses in terms of the concepts which we studied in Chapter 8; in Section 10.2.2 we shall then give a general result, the *Neyman-Pearson lemma,* for choosing the "best" criterion for testing a given simple hypothesis against a given simple alternative.

10.2.1 Losses, Errors, and Risks

Suppose that a statistician must test the null hypothesis that the parameter of a given kind of population is θ_0 against the alternative hypothesis that the parameter is θ_1. He either accepts the null hypothesis, action a_0, or he accepts the alternative, action a_1, and depending on the true "state of Nature" and the action which he takes, his losses are as shown in the following table:

		Statistician a_0	a_1
Nature	θ_0	$L(a_0, \theta_0)$	$L(a_1, \theta_0)$
	θ_1	$L(a_0, \theta_1)$	$L(a_1, \theta_1)$

These losses can be positive or negative (reflecting penalties or rewards), and the only condition which we shall impose is that

$$L(a_0, \theta_0) < L(a_1, \theta_0) \quad \text{and} \quad L(a_1, \theta_1) < L(a_0, \theta_1)$$

namely, that in either case *the right decision is more profitable than the wrong one.*

As in the statistical games of Section 8.3, the statistician's choice will depend on the outcome of an experiment and the decision function **d**, which tells him for each possible outcome what action to take. If the null hypothesis is true and the statistician accepts the alternative hypothesis (namely, if the value of the parameter is θ_0 and the statistician takes action a_1), he is said to be committing a *Type I error*, and we shall let $\alpha(\mathbf{d})$ denote the probability that the statistician will commit a Type I error when he uses the decision function **d**. Correspondingly, if the alternative hypothesis is true and the statistician accepts the null hypothesis (namely, if the value of the parameter is θ_1 and the statistician takes action a_0), he is said to be committing a *Type II error*, and we shall let $\beta(\mathbf{d})$ denote the probability that the statistician will commit a Type II error when he uses the decision function **d**. Thus, the values of the *risk function* (8.3.1) are

$$\begin{aligned} R(\mathbf{d}, \theta_0) &= [1 - \alpha(\mathbf{d})]L(a_0, \theta_0) + \alpha(\mathbf{d})L(a_1, \theta_0) \\ &= L(a_0, \theta_0) + \alpha(\mathbf{d})[L(a_1, \theta_0) - L(a_0, \theta_0)] \end{aligned}$$

and

$$\begin{aligned} R(\mathbf{d}, \theta_1) &= \beta(\mathbf{d})L(a_0, \theta_1) + [1 - \beta(\mathbf{d})]L(a_1, \theta_1) \\ &= L(a_1, \theta_1) + \beta(\mathbf{d})[L(a_0, \theta_1) - L(a_1, \theta_1)] \end{aligned}$$

where, by assumption, the quantities in brackets are both positive. It is apparent from this (and should, perhaps, have been obvious from the beginning) that the statistician should look for a decision function which makes the probabilities of both types of errors as small as possible.

The fact that this is easier said than done is illustrated by the following example: Suppose that in the third illustration on page 289 the manufacturer of the new medication wants to test the null hypothesis $\theta_0 = 0.90$ against the alternative $\theta_1 = 0.60$ on the basis of $\mathbf{x}$, the observed number of successes in $n = 20$ trials. If he uses the decision function

$$d_1(x) = \begin{cases} a_0 & \text{for } x \geq 15 \\ a_1 & \text{for } x < 15 \end{cases}$$

the probability of a Type I error is the probability of getting fewer than 15 successes when $\theta = 0.90$, the probability of a Type II error is the probability of getting 15 or more successes when $\theta = 0.60$, and as can be seen from Table I, these probabilities are, respectively, $\alpha(\mathbf{d}_1) = 0.0114$ and $\beta(\mathbf{d}_1) = 0.1255$. To make the probability of a Type II error smaller, he could use the decision function

$$d_2(x) = \begin{cases} a_0 & \text{for } x \geq 16 \\ a_1 & \text{for } x < 16 \end{cases}$$

and it can easily be checked that this would make $\alpha(\mathbf{d}_2) = 0.0433$ and $\beta(\mathbf{d}_2) = 0.0509$. *Thus, while the probability of a Type II error has become smaller, that of a Type I error has increased.* The only way in which we could reduce the probabilities of *both* types of errors would be to increase the size of the sample, but *so long as n is held fixed, this inverse relationship between the probabilities of Type I and Type II errors is typical of statistical decision procedures.* In other words, if the probability of one type of error is reduced, that of the other type of error is increased, and this is very unsatisfactory so far as the risk function on page 291 is concerned.

If we could assign prior probabilities to θ_1 and θ_2 and if we knew the exact values of all the losses $L(a_i, \theta_j)$ in the table on page 291, we could calculate the *Bayes risk* (defined on page 247) and look for the decision function which minimizes this risk. Alternately, if we looked upon Nature as a malevolent opponent we could use the minimax criterion and choose the decision function which minimizes the maximum risk, but as must have been apparent from the applied exercises on page 240, this is not a very realistic approach to most practical situations. In the theory of hypothesis testing which is nowadays referred to as "traditional" or "classical," namely, the *Neyman-Pearson theory*, we circumvent the dependence between probabilities of Type I and Type II errors by *limiting ourselves to decision functions for which* $\alpha(\mathbf{d}) \leq \alpha$, *where* α *is a constant, and then choosing the one for which* $\beta(\mathbf{d})$ *is a minimum.* For all practical purposes this means that we hold the probability of a Type I error fixed, and then look for the decision function which minimizes the probability of a Type II error. [The inequality in $\alpha(\mathbf{d}) \leq \alpha$ is included here to take care of discrete random variables, where it may be impossible to find a decision function for which $\alpha(\mathbf{d})$ is exactly equal to α.]

10.2.2 The Neyman-Pearson Lemma

To choose a decision function for testing a null hypothesis H_0 against an alternative hypothesis H_1 is to partition the sample space (of outcomes on which the decision is based) into two sets: a *region of acceptance* for H_0 and a *region of rejection* for H_0. For instance, for the decision function $\mathbf{d}_1$ on page 292 the region of acceptance for H_0 (the null hypothesis $\theta = 0.90$) was given by $x = 15, 16, 17, 18, 19$, and 20, and the region of rejection for H_0 was given by $x = 0, 1, 2, \ldots, 13$, and 14. It is customary to refer to a region of rejection for H_0 as a *critical region*, and to the probability of obtaining an outcome which falls into a critical region *when H_0 is true* as its *size*. Thus, *the size of a critical region is simply the probability of committing a Type I error*, and for the decision function $\mathbf{d}_1$ on page 292 it equaled 0.0114.

To give another example, suppose that $\mathbf{x}_1$ and $\mathbf{x}_2$ constitute a random sample of size 2 from a normal population having the variance $\sigma^2 = 1$, and that we want to test the null hypothesis that the mean of the population is μ_0 against the alternative hypothesis that it is μ_1, where $\mu_1 > \mu_0$. Suppose, furthermore, that the null hypothesis H_0 is to be rejected if $\bar{\mathbf{x}}$ takes on a value greater than $\mu_0 + 1$, namely, if the values of the random sample are such that $x_1 + x_2 > 2\mu_0 + 2$. As is indicated in Figure 10.1, the critical region of this test criterion consists of that portion of the plane which lies above the line $x_1 + x_2 = 2\mu_0 + 2$, and in Exercise 3 on page 297 the reader will be asked to use Theorem 7.1 to verify that the size of this critical region is approximately 0.08.

Note that in this second example we specified the critical region *directly* without first mentioning the corresponding decision function, which happens to be

$$d(x_1, x_2) = \begin{cases} a_0 & \text{for } x_1 + x_2 \leq 2\mu_0 + 2 \\ a_1 & \text{for } x_1 + x_2 > 2\mu_0 + 2 \end{cases}$$

where a_0 and a_1 denote, as before, the acceptance and the rejection of H_0. *Since the specification of a critical region thus defines a corresponding decision function and vice versa, the two terms are used interchangeably in statistics.* In fact, both terms are also used interchangeably with "test criterion" or simply "test."

As we have indicated on page 292, in the Neyman-Pearson theory of

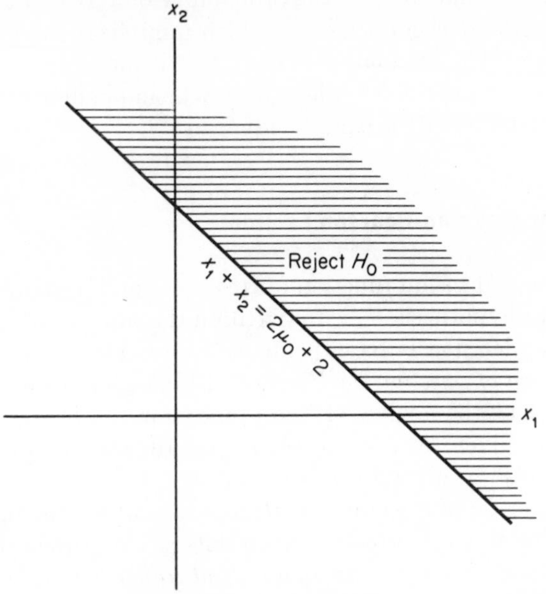

FIGURE 10.1 Critical region.

hypothesis testing we limit ourselves to decision functions for which the probability of a Type I error is less than or equal to some constant α; we can now express this by saying that *we limit ourselves to critical regions of size less than or equal to α.* With this restriction, a critical region for testing a simple null hypothesis H_0 against a simple alternative H_1 is said to be *best* or *most powerful,* if the corresponding probability of committing a Type II error is a minimum.

Theorem 10.1, to be proved below, provides sufficient conditions for the existence of a most powerful critical region of size α for testing the simple null hypothesis $\theta = \theta_0$ against the simple alternative $\theta = \theta_1$, where θ is the parameter of a population given by $f(x; \theta)$. To simplify our notation, we shall let L_0 and L_1 denote the *likelihoods* (see page 268) of a random sample of size n from the given population when its parameter θ is, respectively, θ_0 and θ_1. Symbolically,

$$L_0 = \prod_{i=1}^{n} f(x_i; \theta_0) \quad \text{and} \quad L_1 = \prod_{i=1}^{n} f(x_i; \theta_1)$$

THEOREM 10.1 (*Neyman-Pearson Lemma*)

If there exists a critical region C of size α and a constant k such that

$$\frac{L_0}{L_1} \le k \qquad \textit{inside } C$$

and

$$\frac{L_0}{L_1} \geq k \qquad \textit{outside } C$$

then C is a most powerful critical region of size α for testing $\theta = \theta_0$ against $\theta = \theta_1$.

(Intuitively speaking, it stands to reason that L_0/L_1 should be *small* for sample points inside C, which lead to Type I errors when $\theta = \theta_0$ and to correct decisions when $\theta = \theta_1$; similarly, it stands to reason that L_0/L_1 should be *large* for sample points outside C, which lead to correct decisions when $\theta = \theta_0$ and Type II errors when $\theta = \theta_1$.)

To prove Theorem 10.1, named after the two statisticians who first gave its proof, suppose that C is a critical region satisfying the conditions of the theorem and that D is some other critical region of size α. Thus,

$$\int_C \cdots \int L_0\, dx = \int_D \cdots \int L_0\, dx = \alpha$$

where dx stands for $dx_1\, dx_2 \cdots dx_n$, and the two multiple integrals are taken over the respective n-dimensional regions C and D. Now, making use of the fact that C is the union of the disjoint sets $C \cap D$ and $C \cap D'$ while D is the union of the disjoint sets $C \cap D$ and $C' \cap D$, we can write

$$\int_{C\cap D} \cdots \int L_0\, dx + \int_{C\cap D'} \cdots \int L_0\, dx = \int_{C\cap D} \cdots \int L_0\, dx + \int_{C'\cap D} \cdots \int L_0\, dx = \alpha$$

and, hence,

$$\int_{C\cap D'} \cdots \int L_0\, dx = \int_{C'\cap D} \cdots \int L_0\, dx$$

Then, since $L_1 \geq L_0/k$ inside C and $L_1 \leq L_0/k$ outside C, it follows that

$$\int_{C\cap D'} \cdots \int L_1\, dx \geq \int_{C\cap D'} \cdots \int \frac{L_0}{k}\, dx = \int_{C'\cap D} \cdots \int \frac{L_0}{k}\, dx \geq \int_{C'\cap D} \cdots \int L_1\, dx$$

and, hence, that

$$\int_{C\cap D'} \cdots \int L_1\, dx \geq \int_{C'\cap D} \cdots \int L_1\, dx$$

Finally,

$$\begin{aligned} \int_C \cdots \int L_1\, dx &= \int_{C\cap D} \cdots \int L_1\, dx + \int_{C\cap D'} \cdots \int L_1\, dx \\ &\geq \int_{C\cap D} \cdots \int L_1\, dx + \int_{C'\cap D} \cdots \int L_1\, dx = \int_D \cdots \int L_1\, dx \end{aligned}$$

so that

$$\int_C \cdots \int L_1\, dx \geq \int_D \cdots \int L_1\, dx$$

and this completes the proof of Theorem 10.1. The final inequality states that for the critical region C the probability of *not* committing a Type II error is greater than or equal to the corresponding probability for any other critical region of size α. (For the discrete case the proof is the same, with summations taking the place of integrals.)

To illustrate how Theorem 10.1 can be used to find a most powerful critical region for testing a simple hypothesis against a simple alternative, suppose that $\mathbf{x}_1, \mathbf{x}_2, \ldots, \mathbf{x}_n$ is a random sample of size n from a normal population with the variance $\sigma^2 = 1$, and that on the basis of such a sample we want to test the null hypothesis $\mu = \mu_0$ against the alternative $\mu = \mu_1$, where $\mu_1 > \mu_0$. (In Exercise 6 on page 298 the reader will be asked to solve the analogous problem where $\mu_1 < \mu_0$.) The two likelihoods are

$$L_0 = \left(\frac{1}{\sqrt{2\pi}}\right)^n \cdot e^{-\frac{1}{2}\Sigma(x_i-\mu_0)^2} \quad \text{and} \quad L_1 = \left(\frac{1}{\sqrt{2\pi}}\right)^n \cdot e^{-\frac{1}{2}\cdot\Sigma(x_i-\mu_1)^2}$$

where the summations extend from $i = 1$ to $i = n$, and after some simplifications their ratio becomes

$$\frac{L_0}{L_1} = e^{\frac{n}{2}(\mu_1^2-\mu_0^2)+(\mu_0-\mu_1)\cdot\Sigma x_i}$$

Thus, we must find a constant k and a region C of the sample space such that

$$e^{\frac{n}{2}(\mu_1^2-\mu_0^2)+(\mu_0-\mu_1)\cdot\Sigma x_i} \leq k \qquad \textit{inside } C$$

$$e^{\frac{n}{2}(\mu_1^2-\mu_0^2)+(\mu_0-\mu_1)\cdot\Sigma x_i} \geq k \qquad \textit{outside } C$$

and after taking logarithms, subtracting $\frac{n}{2}(\mu_1^2 - \mu_0^2)$, and dividing by the *negative* quantity $n(\mu_0 - \mu_1)$, these two inequalities become

$$\bar{x} \geq K \qquad \textit{inside } C$$

$$\bar{x} \leq K \qquad \textit{outside } C$$

where K is an expression in k, n, μ_0, and μ_1. *Thus, in this example* $\bar{x} \geq K$ *is a most powerful critical region.* In actual practice, constants like K are determined by making use of the size of the critical region and appropriate theory, and in our example Theorem 7.1 leads to $K = \mu_0 + z_\alpha \cdot \frac{1}{\sqrt{n}}$, where z_α is as defined on page 272. Thus, the most powerful critical region of size α for testing the null hypothesis $\mu = \mu_0$ against the alternative $\mu = \mu_1$ (with $\mu_1 > \mu_0$) for the given normal population is

$$\bar{x} \geq \mu_0 + z_\alpha \cdot \frac{1}{\sqrt{n}}$$

and it should be noted that it does not depend on μ_1; this is an important property, to which we shall refer again in Section 10.3.1.

THEORETICAL EXERCISES

1. Decide in each case whether the given hypothesis is simple or composite:
   (a) the hypothesis that a random variable has the gamma distribution (4.3.6) with $\alpha = 3$ and $\beta = 2$;
   (b) the hypothesis that a random variable has the gamma distribution (4.3.6) with $\alpha = 3$ and $\beta \neq 2$;
   (c) the hypothesis that a random variable has an exponential density;
   (d) the hypothesis that a random variable has the beta distribution with the mean $\mu = 0.50$;
   (e) the hypothesis that a random variable has a Poisson distribution with $\lambda = 1.25$;
   (f) the hypothesis that a random variable has a Poisson distribution with $\lambda > 1.25$;
   (g) the hypothesis that a random variable has a normal distribution with the mean $\mu = 100$;
   (h) the hypothesis that a random variable has the negative binomial distribution with $k = 3$ and $\theta < 0.60$.
2. A bowl contains seven marbles of which θ are red while the others are blue. In order to test the null hypothesis $\theta = 2$ against the alternative $\theta = 4$, two of the marbles are randomly drawn without replacement and the null hypothesis is rejected if and only if both are red. Find the probabilities of committing Type I and Type II errors with this criterion.
3. Verify that for the illustration on page 293 the size of the critical region is approximately 0.08.
4. Suppose that in the example on page 292 we had used the decision function

$$d_3(x) = \begin{cases} a_0 & \text{for } x \geq 14 \\ a_1 & \text{for } x < 14 \end{cases}$$

What would have been the probabilities of Type I and Type II errors?

5. A single observation of a random variable having an exponential distribution is to be used to test the null hypothesis that the mean of the distribution is $\theta = 2$ against the alternative that it is $\theta = 5$. If the null hypothesis is accepted if and only if the observed value of the random variable is less than 3, find the probabilities of Type I and Type II errors.

6. Show that if $\mu_1 < \mu_0$ in the example on page 296, the Neyman-Pearson lemma yields the critical region $\bar{x} \leq \mu_0 - z_\alpha \cdot \frac{1}{\sqrt{n}}$.

7. A random sample of size n from an exponential population is to be used to test the null hypothesis that its parameter is θ_0 against the alternative that its parameter is θ_1, where $\theta_1 > \theta_0$. Use the Neyman-Pearson lemma to find the most powerful critical region of size α, and use the result on page 192 to indicated how to evaluate the constant.

8. Use the Neyman-Pearson lemma to indicate how to construct the most powerful critical region of size α to test the null hypothesis that θ, the parameter of a binomial distribution with a given value of n, equals θ_0 against the alternative that it equals $\theta_1 < \theta_0$.

9. If $n = 80$, $\theta_0 = 0.40$, and $\theta_1 = 0.30$, and α is as large as possible without exceeding 0.05, use the normal approximation of the binomial distribution to find the probability of committing a Type II error with the criterion constructed in Exercise 8.

10. A single observation of a random variable having the geometric distribution (3.3.6) is to be used to test the null hypothesis that its parameter equals θ_0 against the alternative that it equals $\theta_1 > \theta_0$. Use the Neyman-Pearson lemma to find the best critical region of size α.

11. Given a random sample of size n from a normal population with $\mu = 0$, use the Neyman-Pearson lemma to construct the most powerful critical region of size α to test the null hypothesis $\sigma = \sigma_0$ against the alternative $\sigma = \sigma_1 > \sigma_0$.

APPLIED EXERCISES

12. An airline wants to test the null hypothesis that 60 percent of its passengers object to smoking inside the plane. Explain under what conditions they would be committing a Type I error and under what conditions they would be committing a Type II error.

13. A doctor is asked to give an executive a thorough physical checkup to test the null hypothesis that he will be able to take on additional responsibilities. Explain under what conditions the doctor would be committing a Type I error and under what conditions he would be committing a Type II error.

14. Suppose that in the example on page 292 the manufacturer of the new medication feels that the odds are 4 to 1 that with this medication the recovery rate from the disease is 0.90 rather than 0.60. With these odds, what is the probability that he will make a wrong decision using
(a) the decision function $\mathbf{d}_1$ on page 292;
(b) the decision function $\mathbf{d}_2$ on page 292;
(c) the decision function $\mathbf{d}_3$ of Exercise 4?

10.3 COMPOSITE HYPOTHESES

In actual practice, it is relatively rare that simple hypotheses are tested against simple alternatives; usually, one or the other, or both, are composite. For instance, in the third example on page 289 it may well be more realistic to test the null hypothesis that the recovery rate from the disease is $\theta \geq 0.90$ with the new medication against the alternative that $\theta < 0.90$, namely, the alternative that the new medication is not as effective as claimed.

When we deal with composite hypotheses, the problem of evaluating the merits of decision function, test criteria, or critical regions becomes much more difficult. As will be illustrated in Section 10.3.1, we will now have to consider the probabilities of making errors (wrong decisions) for all values of the parameter, or parameters, within the domains specified under the null hypothesis H_0 and the alternative hypothesis H_1.

Also, the Neyman-Pearson lemma does not apply to composite hypothesis, but in Section 10.3.2 we shall introduce a general method for constructing tests of composite hypotheses which is usually very satisfactory, specially for large samples. The resulting tests, called *likelihood ratio tests,* are based on generalizations of the concepts introduced in Section 10.2.2.

10.3.1 The Power Function of a Test

In the example on page 292 we were able to give unique values for the probabilities of committing Type I and Type II errors with the decision functions $\mathbf{d}_1$ and $\mathbf{d}_2$ because we were testing the simple null hypothesis

$\theta = 0.90$ against the simple alternative $\theta = 0.60$. Had we been interested in testing the null hypothesis

$$H_0: \quad \theta \geq 0.90$$

against the alternative

$$H_1: \quad \theta < 0.90$$

which, as we pointed out above, would be a more realistic test of the effectiveness of the medication, the problem would have been more complicated. Referring to the decision function $\mathbf{d}_1$, we would have had to calculate the probabilities of getting fewer than 15 successes for various values of $\theta \geq 0.90$, and the probabilities of getting 15 or more successes for various values of $\theta < 0.90$—the former would all be probabilities of Type I errors, and the latter would all be probabilities of Type II errors.

For instance, with the use of Table I at the end of the book we obtained the probabilities shown in the following table:

θ	*Probability of Type I Error* $\alpha(\theta)$	*Probability of Type II Error* $\beta(\theta)$
0.95	0.0003	
0.90	0.0114	
0.85		0.9326
0.80		0.8042
0.75		0.6171
0.70		0.4163
0.65		0.2455
0.60		0.1255
0.55		0.0553
0.50		0.0207

where $\alpha(\theta)$ is the probability of a Type I error when the true value of the parameter is θ, and $\beta(\theta)$ is the corresponding probability of a Type II error. *Of course, only one of these probabilities is defined for any given value of* θ.

In order to avoid the use of the two functions $\boldsymbol{\alpha}$ and $\boldsymbol{\beta}$, each defined for different values of θ, statisticians have combined them, so to speak, into one function whose values are the *probabilities of rejecting the null hypothesis* H_0 *for the various values of the parameter* θ. Symbolically, this function, called the *power function* of a given test criterion (decision function, or critical region), is given by

$$P(\theta) = \begin{cases} \alpha(\theta) & \text{for values of } \theta \text{ assumed under } H_0 \\ 1 - \beta(\theta) & \text{for values of } \theta \text{ assumed under } H_1 \end{cases}$$

and it should be observed that for values of θ assumed under H_0 it gives the probability of committing a Type I error, and for values of θ assumed under H_1 it gives the probability of *not* committing a Type II error. The above table can thus be given as

θ	*Probability of Rejecting* H_0 $P(\theta)$
0.95	0.0003
0.90	0.0114
0.85	0.0673
0.80	0.1958
0.75	0.3829
0.70	0.5837
0.65	0.7545
0.60	0.8745
0.55	0.9447
0.50	0.9793

and the graph of this power function is shown in Figure 10.2. Of course, this power function applies only to the decision function $\mathbf{d}_1$ on page 292, but it is of interest to note that the power function of an *ideal* decision criterion for the given problem would be given by the dotted lines of Figure 10.2. (In Exercise 2 on page 309 the reader will be asked to construct the power function for the decision function $\mathbf{d}_2$ given on page 292.)

Power functions play a very important role in the evaluation of statistical tests, particularly in the comparison of *several* critical regions which might all be used to test a given null hypothesis against a given alternative. Incidentally, if we had plotted in Figure 10.2 the *probabilities of accepting* H_0 (instead of the *probabilities of rejecting* H_0), we would have obtained the *operating characteristic curve*, or simply the *OC-curve*, of the given critical region. In other words, the values of the *operating characteristic function*, used mainly in industrial applications, are given by $1 - P(\theta)$.

On page 293 we indicated that in the Neyman-Pearson theory of testing hypotheses we hold α, the probability of a Type I error, fixed, and this requires that the null hypothesis H_0 be a simple hypothesis, say, $\theta = \theta_0$. As a result, the power function of any test of this null hypothesis will pass through the point (θ_0, α), the only point at which the value of a power function is the probability of making an error. This facilitates the comparison of the power functions of several critical regions, which are all designed to test the simple null hypothesis $\theta = \theta_0$ against a composite alternative, say, the alternative hypothesis $\theta \neq \theta_0$. To illustrate, con-

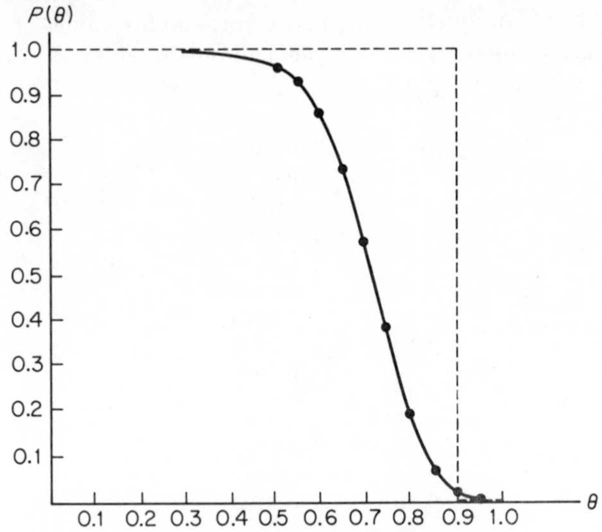

FIGURE 10.2 Power function.

sider Figure 10.3, giving the power functions of three different critical regions (test criteria, or decision functions) designed for this purpose. Since for each value of θ except θ_0 the values of power functions are probabilities of making *correct decisions*, it is desirable to have them as close to 1 as possible. Thus, it can be seen by inspection that the critical region whose power function is given by the solid curve of Figure 10.3 is preferable to the critical region whose power function is given by the curve which is dashed. The probability of not committing a Type II error with the first of these critical regions always exceeds that of the second, and we say that the first critical region is *uniformly more powerful* than the second; also, the other critical region is said to be *inadmissible*.

The same clear-cut distinction is not possible if we attempt to compare the critical regions whose power functions are given by the solid and dotted curves of Figure 10.3—in this case the first one is preferable for $\theta < \theta_0$ while the other is preferable for $\theta > \theta_0$. In situations like this we need further criteria for comparing power functions, for instance, that of Exercise 15 on page 312. (Note that if the alternative hypothesis had been $\theta > \theta_0$, the critical region whose power function is given by the dotted curve would have been *uniformly more powerful* than the critical region whose power function is given by the solid curve.)

In general, when testing a simple hypothesis against a composite alternative we specify α, the probability of committing a Type I error, and refer to one critical region of size α as *uniformly more powerful* than

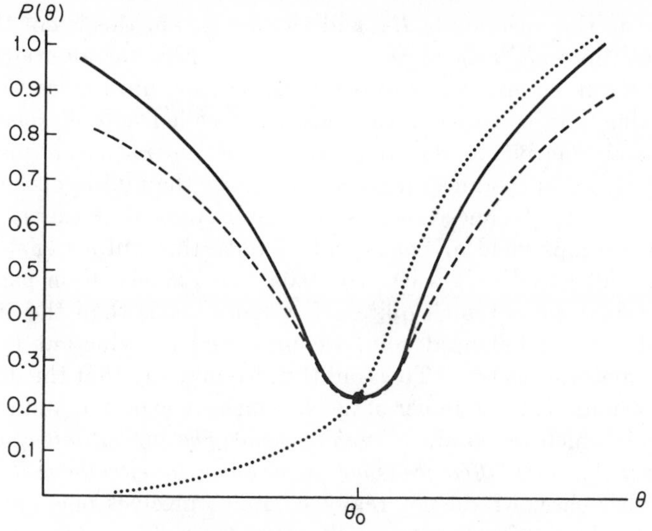

FIGURE 10.3 Power functions.

another *if the values of its power function are always greater than or equal to those of the other, with the strict inequality holding for at least one value of the parameter under consideration.* If, for a given problem, a critical region of size α is uniformly more powerful than any other critical region of size α, it is referred to as *uniformly most powerful;* unfortunately, uniformly most powerful critical regions rarely exist when we test a simple hypothesis against a composite alternative. (Of course, when we test a simple hypothesis against a simple alternative, a *most powerful* critical region of size α, as defined on page 294, is, in fact, *uniformly most powerful.*)

To be able to construct statistical tests for which the critical region is of a fixed size α, we must often hypothesize the exact opposite of what we may want to prove. For instance, if we want to show that the students in one school have a higher average I.Q. than those of another, we have to formulate the hypothesis that there is *no difference,* namely, that $\mu_1 = \mu_2$. Similarly, if we want to show that one kind of ore has a higher percentage content of uranium than another kind of ore, we have to formulate the hypothesis that the two percentages are *the same;* and if we want to show that there is a greater variability in the quality of one product than there is in the quality of another, we have to formulate the hypothesis that there is *no difference,* namely, that $\sigma_1 = \sigma_2$. In view of the assumptions of "no difference," hypotheses like these have been called *null hypotheses,* although nowadays this term is applied to any hypothesis whose false rejection is looked upon as a Type I error.

Until now we have always assumed that the acceptance of H_0 is equivalent to the rejection of H_1, and vice versa, but this is not the case, for example, in *multi-stage* or *sequential* tests, where the alternatives are to accept H_0, to accept H_1, or to defer the decision until more data have been obtained. It is also not the case in so-called *tests of significance*, where the alternative to rejecting H_0 is *reserving judgment* (instead of accepting H_1). For instance, if we want to test the null hypothesis that a coin is perfectly balanced against the alternative that this is not the case, and 100 flips yield 57 heads and 43 tails, this will *not* enable us to reject the null hypothesis when $\alpha = 0.05$ (see Exercise 6 on page 310). However, since we obtained quite a few more heads than the 50 which we expected for a balanced coin, we may well be reluctant to accept the null hypothesis as true. To avoid this, we may say that the difference between 50 and 57, the number of heads which we expected and the number of heads which we obtained, *may reasonably be attributed to chance*—or we may say that *this difference is not large enough to reject the null hypothesis.* In either case, we do not really commit ourselves one way or the other, and *so long as we do not actually accept the null hypothesis, we cannot commit a Type II error.* In a test like this (that is, in a *test of significance* where we either reject the null hypothesis or reserve judgment), the probability of a Type I error, α, is called the *level of significance.*

10.3.2 Likelihood Ratio Tests

Let us now present a general method for constructing critical regions, which in most cases have very satisfactory properties. It is called the *likelihood ratio method* and, conceptually, it is an extension of the method which we introduced in Section 10.2.2. We shall discuss this method here with reference to tests concerning one parameter θ and continuous populations, but all of our arguments can easily be extended to the multi-parameter case and to discrete populations.

To illustrate the likelihood ratio technique, let us suppose that $\mathbf{x}_1$, $\mathbf{x}_2$, . . ., and $\mathbf{x}_n$ is a random sample of size n from a population whose density is given by $f(x;\theta)$, and that Ω is the set of values which the parameter θ can take on. The null hypothesis we shall want to test is

$$H_0: \quad \theta \text{ is an element of } \omega$$

where ω is a subset of Ω, and the alternative hypothesis is

$$H_1: \quad \theta \text{ is an element of } \omega'$$

where ω' is the complement of ω with respect to Ω. Thus, the set of values which the parameter θ can take on is partitioned into the disjoint sets ω and ω'; according to the null hypothesis θ is an element of the first set, and according to the alternative hypothesis it is an element of the second. In most problems Ω is either the set of all real numbers, the set of all positive real numbers, some interval of real numbers, or a discrete set of real numbers.

When H_0 and H_1 are both simple hypotheses, ω and ω' each have only one element, and in Section 10.2.2 we constructed tests by comparing the likelihoods L_0 and L_1. In the general case, where at least one of the two hypotheses is composite, we compare instead the two quantities max L_0 and max L, where max L_0 is the maximum value of the likelihood function (see page 268) for all values of θ in ω, and max L is the maximum of the likelihood function for all values of θ in Ω. In other words, if we have a random sample of size n from a population which is given by $f(x; \theta)$, $\hat{\theta}$ is the maximum likelihood estimate of θ subject to the restriction that θ must be an element of ω, and $\hat{\hat{\theta}}$ is the maximum likelihood estimate of θ for all values of θ in Ω, then

$$\max L_0 = \prod_{i=1}^{n} f(x_i; \hat{\theta}) \tag{10.3.1}$$

and

$$\max L = \prod_{i=1}^{n} f(x_i; \hat{\hat{\theta}}) \tag{10.3.2}$$

These quantities are both values of random variables, they depend on the observed sample values $x_1, x_2, \ldots, x_n$, and their ratio

$$\lambda = \frac{\max L_0}{\max L} \tag{10.3.3}$$

is referred to as a value of the *likelihood ratio statistic* λ.

Since max L_0 is apt to be small compared to max L when the null hypothesis is *false*, it stands to reason that the null hypothesis should be rejected when λ is small. Indeed, the critical region

$$\lambda \leq k \tag{10.3.4}$$

defines the *likelihood ratio test* of the null hypothesis that θ is an element of ω against the alternative hypothesis that θ is an element of ω'. If H_0 is a simple hypothesis, k is chosen so that the size of the critical regions equals α; if H_0 is composite, k is chosen so that the probability of a Type I error is less than or equal to α for all θ in ω, and equal to α (if possible) for at least one value of θ in ω. Thus, if H_0 is a simple hypothesis and $g(\lambda)$ is the density of λ when H_0 is true, then k must be such that

$$\int_0^k g(\lambda)\, d\lambda = \alpha \tag{10.3.5}$$

In the discrete case, the integral (10.3.5) is replaced by a sum and k is taken to be the largest value for which the sum is less than or equal to α.

To illustrate the likelihood ratio technique, suppose we have a random sample of size n from a normal population with the *known* variance σ^2, and that we want to test the simple null hypothesis

$$H_0: \quad \mu = \mu_0$$

against the composite alternative

$$H_1: \quad \mu \neq \mu_0$$

where μ is, of course, the mean of the population.

Since ω contains only μ_0, it follows that $\hat{\mu} = \mu_0$, and since Ω is the set of all real numbers, it follows by the method of Section 9.4.2 that $\hat{\hat{\mu}} = \bar{x}$. Thus

$$\max L_0 = \left(\frac{1}{\sigma\sqrt{2\pi}}\right)^n \cdot e^{-\frac{1}{2\sigma^2}\cdot\sum(x_i-\mu_0)^2}$$

and

$$\max L = \left(\frac{1}{\sigma\sqrt{2\pi}}\right)^n \cdot e^{-\frac{1}{2\sigma^2}\cdot\sum(x_i-\bar{x})^2}$$

where the summations extend from $i = 1$ to $i = n$, and the value of the likelihood ratio statistic becomes

$$\lambda = \frac{e^{-\frac{1}{2\sigma^2}\cdot\sum(x_i-\mu_0)^2}}{e^{-\frac{1}{2\sigma^2}\cdot\sum(x_i-\bar{x})^2}}$$

$$= e^{-\frac{n}{2\sigma^2}(\bar{x}-\mu_0)^2} \tag{10.3.6}$$

after suitable simplifications which the reader will be asked to verify in Exercise 7 on page 310. Hence, the critical region of the likelihood ratio test is

$$e^{-\frac{n}{2\sigma^2}(\bar{x}-\mu_0)^2} \leq k$$

and, after taking logarithms and dividing by $-\frac{n}{2\sigma^2}$, it becomes

$$(\bar{x}-\mu_0)^2 \geq -\frac{2\sigma^2}{n}\cdot\ln k$$

or

$$|\bar{x} - \mu_0| \geq K$$

where K will have to be determined so that the size of the critical region is α. [Note that $\ln k$ is negative in view of (10.3.3) and (10.3.5).] Making use of Theorem 7.1, we find that the critical region of this likelihood ratio test is

$$|\bar{x} - \mu_0| \geq z_{\alpha/2} \cdot \frac{\sigma}{\sqrt{n}}$$

where $z_{\alpha/2}$ is as defined on page 272. In other words, the null hypothesis is to be rejected if $\bar{\mathbf{x}}$ takes on a value greater than or equal to $\mu_0 + z_{\alpha/2} \cdot \frac{\sigma}{\sqrt{n}}$ or a value less than or equal to $\mu_0 - z_{\alpha/2} \cdot \frac{\sigma}{\sqrt{n}}$.

In the preceding example it was easy to find the constant which made the size of the critical region equal to α, because we were able to refer to the known distribution of $\bar{\mathbf{x}}$ instead of the distribution of the likelihood ratio statistic $\boldsymbol{\lambda}$, itself. Since the distribution of $\boldsymbol{\lambda}$ is generally rather complicated, which makes it difficult to evaluate k with the use of (10.3.5), it is often convenient to use an approximation based on the following theorem, whose proof is referred to on page 313:

Theorem 10.2

For large n, the distribution of $-2 \cdot \ln \boldsymbol{\lambda}$ *approaches, under very general conditions, the chi-square distribution with* 1 *degree of freedom.*

We should add that this theorem applies only to the *one-parameter case;* if the population involves more than one unknown parameter, upon which the null hypothesis imposes r restrictions, the number of degrees of freedom in the chi-square approximation of the distribution of $-2 \cdot \ln \boldsymbol{\lambda}$ is equal to r. Thus, if we wanted to test the null hypothesis that the unknown mean and variance of a normal population are, respectively, $\mu = \mu_0$ and $\sigma^2 = \sigma_0^2$ against the alternative hypothesis that $\mu \neq \mu_0$ and $\sigma^2 \neq \sigma_0^2$, the number of degrees of freedom in the chi-square approximation of the distribution of $-2 \cdot \ln \boldsymbol{\lambda}$ would be 2; the two restrictions are $\mu = \mu_0$ and $\sigma^2 = \sigma_0^2$.

Using Theorem 10.2, we can write the critical region of this approximate likelihood ratio test as

$$-2 \cdot \ln \lambda \geq \chi^2_{\alpha,1} \tag{10.3.7}$$

where $\chi^2_{\alpha,1}$ is as defined on page 215. In connection with the example on page 306, we find that according to Theorems 7.1 and 7.2 and (10.3.6)

$$-2\cdot\ln\lambda = \frac{n}{\sigma^2}(\bar{x} - \mu_0)^2 = \left(\frac{\bar{x} - \mu_0}{\sigma/\sqrt{n}}\right)^2$$

actually *is* a value of a random variable having the chi-square distribution with 1 degree of freedom.

As we indicated on page 304, the likelihood ratio technique will generally produce satisfactory results. That this is not always the case is illustrated by the following example, which is somewhat out of the ordinary: Suppose that a random variable **x** can take on the values 1, 2, 3, 4, 5, 6, and 7, and that we want to test the *simple* null hypothesis that its probability function is

$$f(1) = f(2) = f(3) = \tfrac{1}{12}, \qquad f(4) = \tfrac{1}{4}, \qquad f(5) = f(6) = f(7) = \tfrac{1}{6}$$

against the *composite* alternative that its probability function is

$$g(1) = \frac{a}{3}, \quad g(2) = \frac{b}{3}, \quad g(3) = \frac{c}{3}, \quad g(4) = \frac{2}{3}, \quad g(5) = g(6) = g(7) = 0$$

where $a + b + c = 1$. (This alternative hypothesis *is* composite because it includes all the probability functions which we get by assigning different values from 0 to 1 to a, b, and c, subject only to the restriction that $a + b + c = 1$.)

Supposing, furthermore, that the test is to be based on a single observation, we find that for $x = 1$ we get max $L_0 = \frac{1}{12}$, max $L = \frac{1}{3}$ (corresponding to $a = 1$), and, hence, $\lambda = \frac{1}{4}$. Determining λ for the other values of **x** in the same way, we get the results shown in the following table:

x	1	2	3	4	5	6	7
λ	$\frac{1}{4}$	$\frac{1}{4}$	$\frac{1}{4}$	$\frac{3}{8}$	1	1	1

and if the size of the critical region is to be $\alpha = 0.25$, we find that the likelihood ratio technique yields the critical region for which the null hypothesis is rejected when $\lambda = \frac{1}{4}$, namely, when $x = 1$, $x = 2$, or $x = 3$; clearly, $f(1) + f(2) + f(3) = \frac{1}{12} + \frac{1}{12} + \frac{1}{12} = 0.25$. The corresponding probability of a Type II error is given by $g(4) + g(5) + g(6) + g(7)$, and, hence, it equals $\frac{2}{3}$.

Now let us consider the critical region for which the null hypothesis is rejected only when $x = 4$. Its size is also $\alpha = 0.25$ since $f(4) = \frac{1}{4}$, but the corresponding probability of a Type II error is

$$g(1) + g(2) + g(3) + g(5) + g(6) + g(7) = \frac{a}{3} + \frac{b}{3} + \frac{c}{3} + 0 + 0 + 0 = \frac{1}{3}$$

and since this is *less than* $\frac{2}{3}$, the critical region obtained by means of the likelihood ratio technique is *inadmissible*. Of course, as we pointed out at the beginning, this example *is* somewhat out of the ordinary.

THEORETICAL EXERCISES

1. A bowl contains 7 marbles of which θ are red while the others are blue. In order to test the null hypothesis $\theta \leq 2$ against the alternative $\theta > 2$, two of the marbles are randomly drawn without replacement and the null hypothesis is rejected if and only if both are red.
   (a) Find the probabilities of committing Type I errors when $\theta = 0$, 1, and 2.
   (b) Find the probabilities of committing Type II errors when $\theta = 3$, 4, 5, 6, and 7.
   (c) Plot the graph of the power function.
2. Suppose that on page 301 we had wanted to test the null hypothesis $\theta \geq 0.90$ against the alternative hypothesis $\theta < 0.90$ with the use of the decision function $\mathbf{d}_2$ on page 301. Construct the power function by calculating its value for the same values of θ as in the table on page 301.
3. A single observation is to be used to test the null hypothesis that the parameter of the exponential distribution (4.3.2) equals 10 against the alternative hypothesis that it does not equal 10. If the null hypothesis is to be rejected if and only if the observed value is less than 8 or greater than 12, find
   (a) the probability of a Type I error;
   (b) the probabilities of Type II errors when $\theta = 2, 4, 8, 16$, and 20.
   Also plot the graph of the power function.
4. A random sample of size 64 is to be used to test the null hypothesis that the mean of a normal population with the variance $\sigma^2 = 256$ is less than or equal to 40 against the alternative hypothesis that it is greater than 40. If the null hypothesis is to be rejected if and only if the mean of the random sample exceeds 43, find
   (a) the probabilities of Type I errors when $\mu = 37, 38, 39$, and 40;
   (b) the probabilities of Type II errors when $\mu = 41, 42, 43, 44, 45$, 46, 47, and 48.
   Also plot the graph of the power function.

5. The sum of the values obtained in a random sample of size 5 from a Poisson population is to be used to test the null hypothesis that the mean of the population is greater than 2 against the alternative hypothesis that it is less than or equal to 2. If the null hypothesis is to be rejected if and only if the sum of the observations is 5 or less, find
   (a) the probabilities of Type I errors when the mean of the population is 2.2, 2.4, 2.6, 2.8, and 3.0;
   (b) the probabilities of Type II errors when the mean of the population is 2.0, 1.5, 1.0, and 0.5.

   Also plot the graph of the power function. (*Hint:* Use the result obtained in the illustration of Theorem 6.2 on page 193.)
6. Verify the statement on page 304 that 57 heads and 43 tails in 100 flips of a coin does not enable us to reject the null hypothesis that the coin is perfectly balanced (against the alternative that it is not perfectly balanced) at the level of significance $\alpha = 0.05$. (*Hint:* Use the normal approximation of the binomial distribution.)
7. Verify the final step which led to (10.3.6) on page 306.
8. The number of successes in n trials is to be used to test the null hypothesis that the parameter θ of a binomial population equals $\frac{1}{2}$ against the alternative that it does not equal $\frac{1}{2}$.
   (a) Find an expression for the likelihood ratio statistic.
   (b) Use the result of part (a) to show that the critical region of the likelihood ratio test can be written as

$$x \cdot \ln x + (n - x) \cdot \ln (n - x) > k$$

   where x is the observed number of successes.
   (c) Studying the graph of $f(x) = x \cdot \ln x + (n - x) \cdot \ln (n - x)$, its minimum, and its symmetry, show that the critical region of this likelihood ratio test can also be written as $\left| x - \frac{n}{2} \right| > c$, where c is a constant which depends on the size of the critical region.
9. A random sample of size n is to be used to test the null hypothesis that the parameter θ of an exponential population equals θ_0 against the alternative that it does not equal θ_0.
   (a) Find an expression for the likelihood ratio statistic.
   (b) Use the result of part (a) to show that the critical region of the likelihood ratio test can be written as

$$\bar{x} \cdot e^{-\bar{x}/\theta_0} < K$$

(c) Studying the graph of $g(\bar{x}) = \bar{x} \cdot e^{-\bar{x}/\theta_0}$, its maximum, and whether there is any symmetry, show that the critical region of the likelihood ratio test can *not* be written as

$$|\bar{x} - \theta_0| > c$$

but that it can be written as

$$\bar{x} < c_1 \quad \text{or} \quad \bar{x} > c_2$$

where c_1 and c_2 are constants which depend on θ_0 and α.

10. Given a random sample of size n from a normal population with the mean $\mu = 0$, find an expression for the likelihood ratio statistic for testing the null hypothesis $\sigma = \sigma_0$ against the alternative hypothesis $\sigma \neq \sigma_0$. (*Hint:* See Exercise 5 on page 271.)

11. A random sample of size n from a normal population with unknown mean and variance is to be used to test the null hypothesis $\mu = \mu_0$ against the alternative $\mu \neq \mu_0$. Using the simultaneous maximum likelihood estimates of μ and σ^2 obtained in Section 9.4.2, show that the values of the likelihood ratio statistic can be written in the form

$$\lambda = \left(1 + \frac{t^2}{n-1}\right)^{-\frac{n}{2}}$$

where $t = \dfrac{\bar{x} - \mu_0}{s/\sqrt{n}}$. Note that the likelihood ratio test can, thus, be based on the t distribution of Section 7.2.4.

12. Independent random samples of size $n_1, n_2, \ldots,$ and n_k from k normal populations with unknown means and variances are to be used to test the null hypothesis $\sigma_1^2 = \sigma_2^2 = \cdots = \sigma_k^2$ against the alternative that these variances are not all equal.

(a) Show that under the null hypothesis the maximum likelihood estimates of the means μ_i and the variances σ_i^2 are

$$\hat{\mu}_i = \bar{x}_i \quad \text{and} \quad \hat{\sigma}_i^2 = \sum_{i=1}^{k} \frac{(n_i - 1)s_i^2}{n}$$

where $n = \sum_{i=1}^{k} n_i$, while without restrictions the maximum likelihood estimates of the means μ_i and the variances σ_i^2 are

$$\hat{\hat{\mu}}_i = \bar{x}_i \quad \text{and} \quad \hat{\hat{\sigma}}_i^2 = \frac{(n_i - 1)s_i^2}{n_i}$$

This follows directly from the results obtained in Section 9.4.2.

(b) Using the results of part (a), show that the likelihood ratio statistic can be written as

$$\lambda = \frac{\prod_{i=1}^{k}\left[\frac{(n_i - 1)s_i^2}{n_i}\right]^{\frac{n_i}{2}}}{\left[\sum_{i=1}^{k}\frac{(n_i - 1)s_i^2}{n}\right]^{\frac{n}{2}}}$$

(c) If $n_1 = 8$, $s_1^2 = 16$, $n_2 = 10$, $s_2^2 = 25$, $n_3 = 6$, $s_3^2 = 12$, $n_4 = 8$, and $s_4^2 = 24$ are the sample sizes and the variances of four independent random samples from four normal populations, use the result of part (b) to calculate $-2 \cdot \ln \lambda$ and then test the null hypothesis stated at the beginning of this exercise. (Note that the number of degrees of freedom for this approximate chi-square test is 3, since $\sigma_1^2 = \sigma_2^2 = \sigma_3^2 = \sigma_4^2$ imposes 3 restrictions on the parameters.)

13. Show that for $k = 2$ the likelihood ratio statistic of Exercise 12 can be expressed in terms of the ratio of the two sample variances and that the likelihood ratio test can, therefore, be based on the F distribution.

14. If 15, 28, 3, 12, 42, 19, 20, 2, 25, 30, 62, 12, 18, 16, 44, 65, 33, 51, 4, and 28 are the values of a random sample from an exponential population, use part (a) of Exercise 9 and Theorem 10.2 to test the null hypothesis that the mean of the population is 15 against the composite alternative that it is not equal to 15. Let α, the size of the critical region, be 0.05.

15. UNBIASED CRITICAL REGIONS When we test a simple null hypothesis against a composite alternative, a critical region is said to be *unbiased* if the corresponding power function takes on its *minimum value* at the value of the parameter assumed under the null hypothesis. In other words, *a critical region is unbiased if the probability of rejecting the null hypothesis is least when the null hypothesis is true.* Given a single observation of the random variable $\mathbf{x}$ having the density

$$f(x) = \begin{cases} 1 + \theta^2(\frac{1}{2} - x) & \text{for } 0 < x < 1 \\ 0 & \text{elsewhere} \end{cases}$$

where $-1 \leq \theta \leq 1$, show that the critical region $x \leq \alpha$ provides an *unbiased* and *uniformly most powerful* critical region of size α for testing the null hypothesis $\theta = 0$ against the alternative hypothesis $\theta \neq 0$.

REFERENCES

Discussions of various properties of likelihood ratio tests, particularly their large-sample properties, and a proof of Theorem 10.2, may be found in most advanced textbooks on the theory of statistics, for example, in the book by S. S. Wilks referred to on page 287. Much of the original research done in this area is reproduced in

Selected Papers in Statistics and Probability by Abraham Wald. Stanford, Calif.: Stanford University Press, 1957.

11

HYPOTHESIS TESTING: APPLICATIONS

11.1 INTRODUCTION

In Chapter 10 we discussed some of the theory which underlies statistical tests, and in this chapter we shall present some of the "standard" tests that are most widely used in applications. Most of these tests, at least those based on known population distributions, can be obtained by the likelihood ratio technique. To explain some of the terminology which we shall use, let us point out that the critical region $|\bar{x} - \mu_0| \geq z_{\alpha/2} \cdot \frac{\sigma}{\sqrt{n}}$ which we obtained for the example on page 307 is referred to as *two-sided*, or as a *two-tail test*. As is pictured in Figure 11.1, the null hypothesis

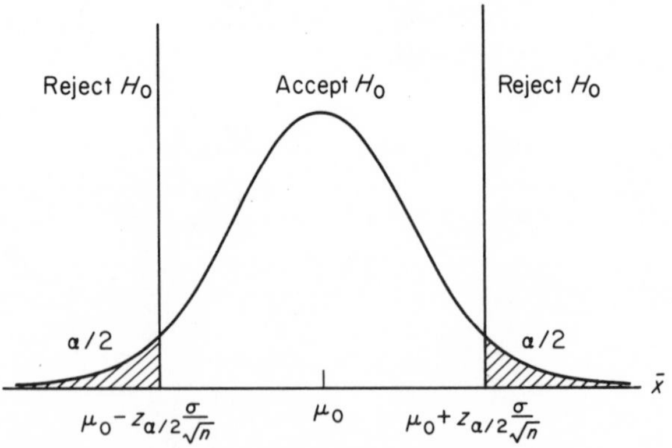

FIGURE 11.1 Critical region for alternative hypothesis $\mu \neq \mu_0$.

$\mu = \mu_0$ is rejected for values of $\bar{\mathbf{x}}$ falling into both "tails" of its sampling distribution, which stands to reason since we are testing this null hypothesis against the *two-sided alternative* $\mu \neq \mu_0$.

Had we used the *one-sided alternative* $\mu > \mu_0$, the likelihood ratio technique would have led to the *one-tail test* pictured in Figure 11.2, and if

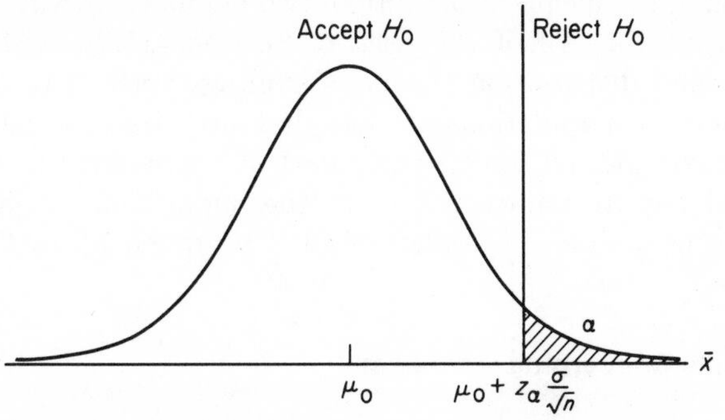

FIGURE 11.2 Critical region for alternative hypothesis $\mu > \mu_0$.

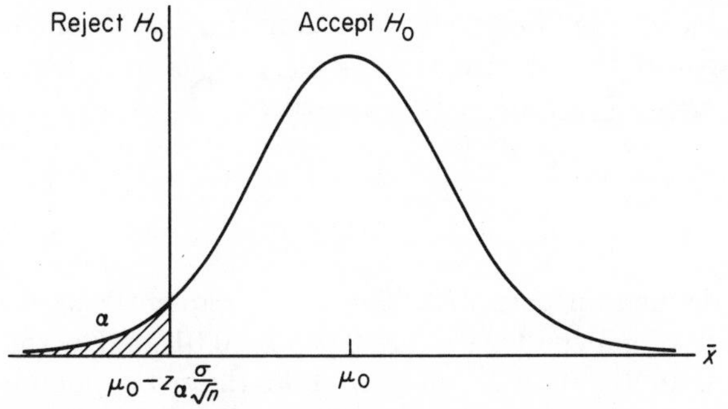

FIGURE 11.3 Critical region for alternative hypothesis $\mu < \mu_0$.

we had used the *one-sided alternative* $\mu < \mu_0$, the likelihood ratio technique would have led to the one-tail test pictured in Figure 11.3. It certainly stands to reason that in the first case we would reject the null hypothesis and accept the alternative only when $\bar{x}$ is large, namely, when it falls into the *right-hand tail* of its sampling distribution, and that the opposite is true when the alternative hypothesis is $\mu < \mu_0$. Although there are exceptions to this rule (see Exercise 1 on page 321), *two-sided alternatives usually lead to two-tail tests and one-sided alternatives usually lead to one-tail tests.*

11.2 TESTS CONCERNING MEANS

In Section 11.2.1 we shall discuss the most commonly used tests concerning the mean of a population, and in Section 11.2.2 we shall discuss corresponding tests concerning the means of two populations;

tests concerning the means of more than two populations will be taken up later in Chapter 14. All of the tests of Sections 11.2.1 and 11.2.2 are based on normal distribution theory, assuming either that the samples come from normal populations or that they are large enough to justify normal approximations; some *nonparametric* alternatives to these tests, which do not require knowledge about the population (or populations) from which the samples are obtained, will be taken up in Chapter 12.

11.2.1 Tests Concerning One Mean

The first test which we shall consider in this section is the one which we used to illustrate the likelihood ratio technique is Section 10.3.2. Given a random sample of size n from a normal population with the known variance σ^2, the "best" critical regions for testing the null hypothesis $\mu = \mu_0$ against the alternatives $\mu \neq \mu_0$, $\mu > \mu_0$, or $\mu < \mu_0$ are, respectively, $|z| \geq z_{\alpha/2}$, $z \geq z_\alpha$, and $z \leq -z_\alpha$, where

$$z = \frac{\bar{x} - \mu_0}{\sigma/\sqrt{n}} \tag{11.2.1}$$

and z_α is as defined on page 272. The most frequently used values of α, the probability of a Type I error, are 0.05 and 0.01, and as the reader was asked to show in Exercise 28 on page 173, the corresponding values of z_α and $z_{\alpha/2}$ are $z_{.05} = 1.64$, $z_{.01} = 2.33$, $z_{.025} = 1.96$, and $z_{.005} = 2.58$. Incidentally, it is easy to check that the above criteria are identical with the ones pictured in Figures 11.1, 11.2, and 11.3.

It should be noted that the critical region $z \geq z_\alpha$ can also be used to test the null hypothesis $\mu = \mu_0$ against the simple alternative $\mu = \mu_1 > \mu_0$, or to test the composite null hypothesis $\mu \leq \mu_0$ against the composite alternative $\mu > \mu_0$. In the first case we would be testing a simple hypothesis against a simple alternative as in Section 10.2.2 (see page 296, where we studied this test for $\sigma = 1$), and in the second case α would be the *maximum* probability of committing a Type I error for any value of μ assumed under the null hypothesis. Of course, similar arguments apply to the critical region $z \leq z_\alpha$.

To give an example where we use this kind of test, suppose that it is known from experience that the standard deviation of the weight of 8-ounce packages of cookies made by a certain bakery is 0.16 ounces. To check whether its production is under control on a given day, namely, to check whether the true average weight of the packages is 8 ounces, they select a random sample of 25 packages and find that their average weight is $\bar{x} = 8.112$ ounces. The null hypothesis which they shall want to test is

$\mu = 8$, and since they stand to lose money when $\mu > 8$ and the customers lose out when $\mu < 8$, the appropriate alternative is $\mu \neq 8$. Choosing $\alpha = 0.01$, they get

$$z = \frac{\bar{x} - \mu_0}{\sigma/\sqrt{n}} = \frac{8.112 - 8}{0.16/\sqrt{25}} = \frac{0.112}{0.032} = 3.5$$

which exceeds $z_{.005} = 2.58$. Hence, the null hypothesis will have to be rejected and the production process will have to be adjusted. (In order to perform this test it was necessary, of course, to look upon the 25 weights as a random sample from a normal population.)

When we deal with a *large sample* from a population which may not be normal but has the finite variance σ^2, we can use the central limit theorem to justify use of the tests which we have discussed, and when σ is unknown, we can even approximate σ with s and base the test on the value of

$$z = \frac{\bar{x} - \mu_0}{s/\sqrt{n}} \tag{11.2.2}$$

To illustrate the use of such an approximate *large-sample test* (for which n must be at least 30), suppose we want to test the null hypothesis that a certain kind of tire will last on the average 22,000 miles against the alternative that this figure is *too high;* that is, we want to test the null hypothesis $\mu = 22{,}000$ against the alternative $\mu < 22{,}000$. Suppose, furthermore, that 100 tires of the given kind lasted on the average 21,431 miles with a standard deviation of 1,295 miles, and that α is to be 0.05. Substituting the given values of $\bar{x}$, μ_0, s, and n into (11.2.2), we get

$$z = \frac{21{,}431 - 22{,}000}{1{,}295/\sqrt{100}} = -4.39$$

and since this is less than $-z_{.05} = -1.64$, the null hypothesis will have to be rejected. In other words, *the tires are not as good as claimed.* To perform this test we did not have to assume that the data constitute a random sample from a normal population—only that they constitute a random sample from a population to which the central limit theorem applies.

When n is small (less than 30) and σ is unknown, the tests which we have been discussing in this section cannot be used. However, in Exercise 11 on page 311 we saw that for random samples from normal populations the likelihood ratio technique yields a corresponding test based on

$$t = \frac{\bar{x} - \mu_0}{s/\sqrt{n}} \tag{11.2.3}$$

which, according to Theorem 7.8, is a value of a random variable having the t distribution with $n - 1$ degrees of freedom. Thus, critical regions of size α for testing the null hypothesis $\mu = \mu_0$ against the alternatives $\mu \neq \mu_0$, $\mu > \mu_0$, or $\mu < \mu_0$, are, respectively, $|t| \geq t_{\alpha/2,n-1}$, $t \geq t_{\alpha,n-1}$, and $t \leq -t_{\alpha,n-1}$, where t is given by (11.2.3). Note that the comments made on page 316 in connection with the alternative hypothesis $\mu_1 > \mu_0$ and the test of the null hypothesis $\mu \leq \mu_0$ against the alternative $\mu > \mu_0$ apply also in this case.

To illustrate these *small-sample tests*, as they are usually called, suppose that the specifications for a certain kind of ribbon call for a mean breaking strength of 185 pounds, and that five pieces (randomly selected from different rolls) have a mean breaking strength of 183.1 pounds with a standard deviation of 8.2 pounds. To test the null hypothesis $\mu = 185$ against the alternative hypothesis $\mu < 185$, the appropriate critical region of size 0.05 is $t < -t_{.05,4}$ (provided, of course, that we can look upon the data as a random sample from a normal population). Substituting $\bar{x} = 183.1$, $\mu_0 = 185$, $s = 8.2$, and $n = 5$ into (11.2.3), we get

$$t = \frac{183.1 - 185}{8.2/\sqrt{5}} = -0.49$$

which is *greater than* $-t_{.05,4} = -2.132$. Hence, the null hypothesis *cannot be rejected*, and if we have to go beyond this and say that the rolls of ribbon from which the sample was selected *meet specifications*, we are, of course, exposed to the unknown risk of committing a Type II error.

11.2.2 Differences Between Means

In applied research, there are many problems in which we are interested in hypotheses concerning differences between the means of two populations. For instance, we may want to decide upon the basis of suitable samples whether men can perform a certain task as fast as women, or we may want to decide on the basis of an appropriate sample survey whether the average weekly food expenditures of families in one city exceed those of families in another city by at least $5.00.

To be able to use the results of Theorem 6.7 and Exercise 3 on page 216 (or simply the results of Exercise 4 on page 216), let us suppose that we are dealing with *independent* random samples of size n_1 and n_2 from two normal populations having the means μ_1 and μ_2 and the *known* variances σ_1^2 and σ_2^2, and that we want to test the null hypothesis $\mu_1 - \mu_2 = \delta$, where δ is a given constant, against one of the alternatives $\mu_1 - \mu_2 \neq \delta$,

$\mu_1 - \mu_2 > \delta$, or $\mu_1 - \mu_2 < \delta$. Applying the likelihood ratio technique, we will arrive at a test based on $\bar{x}_1 - \bar{x}_2$, and, referring to the results cited at the beginning of this paragraph, we find that the respective critical regions can be written as $|z| \geq z_{\alpha/2}$, $z \geq z_\alpha$, and $z \leq -z_\alpha$, where

$$z = \frac{\bar{x}_1 - \bar{x}_2 - \delta}{\sqrt{\dfrac{\sigma_1^2}{n_1} + \dfrac{\sigma_2^2}{n_2}}} \tag{11.2.4}$$

To illustrate this kind of test, suppose that the nicotine contents of two kinds of cigarettes have variabilities of $\sigma_1 = 1.2$ and $\sigma_2 = 1.4$ milligrams, and that in an experiment designed to test the claim that $\mu_1 - \mu_2 = 2$, fifty cigarettes of the first kind had an average nicotine content of $\bar{x} = 26.1$ milligrams, while forty cigarettes of the second kind had an average nicotine content of $\bar{x}_2 = 23.8$ milligrams. To test the null hypothesis $\mu_1 - \mu_2 = 2$ against the alternative $\mu_1 - \mu_2 \neq 2$ with $\alpha = 0.05$, the critical region is $|z| \geq 1.96$, and since

$$z = \frac{26.1 - 23.8 - 2}{\sqrt{\dfrac{(1.2)^2}{50} + \dfrac{(1.4)^2}{40}}} = 1.08$$

we find that *the null hypothesis cannot be rejected.* Hence, we either *accept the null hypothesis,* or we merely say that the difference between $26.1 - 23.8 = 2.3$ and 2 is *not large enough to reject the null hypothesis,* and let it go at that.

When we deal with independent samples from populations with unknown variances which may not even be normal, we can still use the test which we have just described with s_1 substituted for σ_1 and s_2 substituted for σ_2 *so long as both samples are large enough for the central limit theorem to be invoked.*

When n_1 and n_2 are small and σ_1 and σ_2 are unknown, the tests which we have been discussing cannot be used. However, *for independent random samples from two normal populations having the same unknown variance* σ^2, the likelihood ratio technique yields a test based on

$$t = \frac{\bar{x}_1 - \bar{x}_2 - \delta}{\sqrt{\dfrac{(n_1 - 1)s_1^2 + (n_2 - 1)s_2^2}{n_1 + n_2 - 2}}\sqrt{\dfrac{1}{n_1} + \dfrac{1}{n_2}}} \tag{11.2.5}$$

and the t distribution. To show that under the given assumptions and the null hypothesis $\mu_1 - \mu_2 = \delta$, the expression given by (11.2.5) is a value of a random variable having the t distribution with $n_1 + n_2 - 2$ degrees of freedom, let us point out first that according to Theorem 6.7 and Exercise 3 on page 216

$$\frac{\bar{\mathbf{x}}_1 - \bar{\mathbf{x}}_2 - \delta}{\sigma\sqrt{\frac{1}{n_1} + \frac{1}{n_2}}} \tag{11.2.6}$$

has the *standard normal distribution*, and that according to Theorems 7.4 and 7.6

$$\frac{(n_1 - 1)\mathbf{s}_1^2 + (n_2 - 1)\mathbf{s}_2^2}{\sigma^2} \tag{11.2.7}$$

has the *chi-square distribution* with $n_1 + n_2 - 2$ degrees of freedom. Assuming that the two random variables given by (11.2.6) and (11.2.7) are independent, a fact which we shall not be able to prove (see references on page 342), it follows from (7.2.10) that (11.2.5) is, indeed, a value of a random variable having the t distribution with $n_1 + n_2 - 2$ degrees of freedom. Thus, the appropriate critical region of size α for testing the null hypothesis $\mu_1 - \mu_2 = \delta$ against the alternative $\mu_1 - \mu_2 \neq \delta$, $\mu_1 - \mu_2 > \delta$, or $\mu_1 - \mu_2 < \delta$ *under the given assumptions* are, respectively,

$$|t| \geq t_{\alpha/2, n_1+n_2-2},\ t \geq t_{\alpha, n_1+n_2-2}, \text{ and } t \leq -t_{\alpha, n_1+n_2-2}$$

where t is given by (11.2.5).

To illustrate this *two-sample t test*, suppose that in the comparison of two kinds of paint, a consumer testing service finds that four one-gallon cans of one brand cover on the average 512 square feet with a standard deviation of 31 square feet, while four one-gallon cans of another brand cover on the average 492 square feet with a standard deviation of 26 square feet. To test the null hypothesis $\mu_1 - \mu_2 = 0$ against the alternative hypothesis $\mu_1 - \mu_2 \neq 0$, we first calculate t, getting

$$t = \frac{512 - 492 - 0}{\sqrt{\frac{3(31)^2 + 3(26)^2}{6}}\sqrt{\frac{1}{4} + \frac{1}{4}}} = 0.99$$

Using the level of significance $\alpha = 0.05$, we find that $t_{.025,6} = 2.447$, which exceeds $t = 0.99$, the value obtained in the experiment. Thus, *the null hypothesis cannot be rejected*—even though the difference between the two sample means seems to be fairly large, the samples are so small that the results are not conclusive; that is, the difference may well be due to chance.

If the assumption $\sigma_1 = \sigma_2$ is untenable, there are several alternative methods that can be used. A relatively simple one consists of randomly pairing the values obtained in the two samples and then looking upon their differences as a random sample of size n_1 or n_2, whichever is smaller, from a normal population which (under the null hypothesis) has the mean $\mu = \delta$. Then we test this null hypothesis against the appropriate alternative by means of the methods of Section 11.2.1. This is a good reason for having $n_1 = n_2$, but there exist alternate techniques for handling the case where $n_1 \neq n_2$—one of these, the *Smith-Satterthwaite* test, is referred to on page 342.

So far we have limited our discussion to random samples that are *independent*, and the methods which we have introduced in this section cannot be used, for example, to decide on the basis of weights "before and after" whether a certain diet is really effective, or whether an observed difference between the average I.Q.'s of husbands and their wives is really significant. In both of these examples the samples are not independent because the data are actually *paired.* A common way of handling this kind of problem is to proceed as in the preceding paragraph, namely, to work with the differences between the paired measurements or observations. If n is large, we can then use the test described on page 316 to test the null hypothesis $\mu_1 - \mu_2 = \delta$ against the appropriate alternative, and if n is small, we can use the t test described on page 318 (provided the differences can be looked upon as a random sample from a normal population).

THEORETICAL EXERCISES

1. Given a random sample of size n from a normal population with the known variance σ^2, show that the null hypothesis $\mu = \mu_0$ can be tested against the alternative $\mu \neq \mu_0$ with the use of a *one-tail criterion* based on the chi-square distribution.
2. Suppose that a random sample from a normal population with the known variance σ^2 is to be used to test the null hypothesis $\mu = \mu_0$ against the alternative hypothesis $\mu = \mu_1$ (with $\mu_1 > \mu_0$), and that the probabilities of Type I and Type II errors are to have the preassigned values α and β. Show that the required size of the sample is given by

$$n = \frac{\sigma^2(z_\alpha + z_\beta)^2}{(\mu_1 - \mu_0)^2}$$

   Also use this formula to find n when $\sigma = 12$, $\mu_0 = 28$, $\mu_1 = 32$, $\alpha = 0.05$ and $\beta = 0.01$.
3. Suppose that independent random samples of size n from two normal populations with the known variances σ_1^2 and σ_2^2 are to be used to test the null hypothesis $\mu_1 - \mu_2 = \delta$ against the alternative hypothesis $\mu_1 - \mu_2 = \delta'$, and that the probabilities of Type I and Type II errors

are to have the preassigned values α and β. Show that the required size of the samples is given by

$$n = \frac{(\sigma_1^2 + \sigma_2^2)(z_\alpha + z_\beta)^2}{(\delta - \delta')^2}$$

Also use this formula to find n when $\sigma_1 = 12$, $\sigma_2 = 15$, $\delta = 110$, $\delta' = 113$, $\alpha = 0.01$ and $\beta = 0.01$.

APPLIED EXERCISES

4. According to the norms established for an aptitude test, college seniors should average a score of 83.9 with a standard deviation of 8.4. What can we conclude about the seniors attending a large university if 49 of them, randomly selected, averaged 86.4? Use $\alpha = 0.01$.
5. A real estate broker who is anxious to sell a piece of property to a motel chain assures them that during the summer months *on the average* 4,200 cars pass by the property each day. Being suspicious that this figure might be a bit high, the management of the motel chain conducts its own study and obtains a mean of 4,038 cars a day and a standard deviation of 512 cars a day for observations made over 36 days. What can they conclude at the level of significance $\alpha = 0.05$?
6. In 12 test runs over a marked course, a newly-designed motorboat averaged 33.6 seconds with a standard deviation of 2.3 seconds. Assuming that it is reasonable to treat the data as a random sample from a normal population, test the null hypothesis $\mu = 35$ against the alternative $\mu < 35$ at the level of significance $\alpha = 0.05$.
7. The alfalfa yields of six test plots are, respectively, 1.5, 1.9, 1.2, 1.4, 2.3, and 1.3 tons per acre. Use a critical region of size 0.05 to test the null hypothesis $\mu = 1.8$ tons per acre against the alternative hypothesis $\mu \neq 1.8$. (Assume that the yields have normal distribution.)
8. Referring to the numerical example on page 319, the one dealing with the nicotine content of the two kinds of cigarettes, for what range of values of $\bar{x}_1 - \bar{x}_2$ would the null hypothesis have been rejected? Also find the probabilities of committing Type II errors with the given criterion if (a) $\mu_1 - \mu_2 = 1.2$, (b) $\mu_1 - \mu_2 = 1.6$, (c) $\mu_1 - \mu_2 = 2.4$, and (d) $\mu_1 - \mu_2 = 2.8$.
9. A company claims that its light bulbs are superior to those of a competitor on the basis of a study which showed that a sample of 40 of its bulbs had an average "lifetime" of 522 hours (of continuous use) with a standard deviation of 28 hours, while a sample of 30 bulbs made by the competitor had an average "lifetime" of 513 hours (of con-

ded the following results:

$$n_1 = 13, \qquad s_1^2 = 19.2, \qquad n_2 = 16, \qquad s_2^2 = 3.5$$

e units of measurement are 1,000 pounds per square inch). greater than s_2^2, the statistic on which we base the test of the null s $\sigma_1^2 = \sigma_2^2$ against the alternative $\sigma_1^2 \neq \sigma_2^2$ at the level of signif- = 0.02 has the value

$$\frac{s_1^2}{s_2^2} = \frac{19.2}{3.5} = 5.49$$

this *exceeds* $F_{.01,12,15} = 3.67$, the null hypothesis will have to be In other words, we conclude that the variability of the tensile of the two kinds of structural steel is *not* the same.

THEORETICAL EXERCISES

g use of the fact that the chi-square distribution can be approx- with a normal distribution when ν (the number of degrees of m) is large, show that for *large samples* from normal populations

$$s^2 \geq \sigma_0^2 \left[1 + z_\alpha \sqrt{\frac{2}{n-1}}\right]$$

pproximate critical region of size α for testing the null hypothesis $^2_0$ against the alternative hypothesis $\sigma^2 > \sigma_0^2$. Also construct ponding critical regions for testing this null hypothesis against ernatives $\sigma^2 < \sigma_0^2$ and $\sigma^2 \neq \sigma_0^2$. (See Exercise 8 on page 216.)

g use of the result of Exercise 10 on page 217, show that for random samples from normal populations, tests of the null hesis $\sigma^2 = \sigma_0^2$ can be based on the statistic

$$\left(\frac{s}{\sigma_0} - 1\right)\sqrt{2(n-1)}$$

has approximately the standard normal distribution.

APPLIED EXERCISES

fetimes of certain batteries are supposed to have a variance of at 225 hours. Test the null hypothesis $\sigma^2 = 225$ against the alter- $\sigma^2 > 225$ at $\alpha = 0.05$ if the lifetimes of 20 of these batteries

tinuous use) with a standard deviation of 24 hours. Test the null hypothesis $\mu_1 - \mu_2 = 0$ against a suitable one-sided alternative to check whether this claim is justified. Use $\alpha = 0.05$.

10. Sample surveys conducted in a certain large county in 1950 and again in 1965 showed that in 1950 the average height of 400 ten-year-old boys was 53.2 inches with a standard deviation of 2.4 inches, while in 1965 the average height of 500 ten-year-old boys was 53.9 inches with a standard deviation of 2.5 inches. Test the null hypothesis $\mu_1 - \mu_2 = -0.5$ inches against the one-sided alternative $\mu_1 - \mu_2 < -0.5$ inches at the level of significance $\alpha = 0.05$.

11. To find out whether the inhabitants of two South Pacific islands may be regarded as having the same racial ancestry, an anthropologist determines the cephalic indices of six adult males from each island, getting $\bar{x}_1 = 77.4$, $\bar{x}_2 = 72.2$, and the corresponding standard deviations $s_1 = 3.3$ and $s_2 = 2.1$. Use $\alpha = 0.01$ to check whether the difference between the two sample means can reasonably be attributed to chance. (Assume that populations are normal and have the same variance.)

12. If 8 short-range rockets of one kind have a mean target error of $\bar{x}_1 = 98$ feet with a standard deviation of $s_1 = 18$ feet while 10 short-range rockets of another kind have a mean target error of $\bar{x}_2 = 76$ feet with a standard deviation of $s_2 = 15$ feet, test the null hypothesis $\mu_1 - \mu_2 = 15$ against the alternative hypothesis $\mu_1 - \mu_2 > 15$. Let the size of the critical region be $\alpha = 0.05$ and assume that populations are normal and have the same variance.

13. The following are the "before and after" weights of 10 women who have been on a certain reducing diet for five weeks: 148 and 144, 179 and 162, 125 and 126, 149 and 131, 147 and 132, 151 and 146, 145 and 145, 169 and 152, 138 and 127, 120 and 118. Use these figures to test the null hypothesis that with this diet a woman will *on the average* lose 10 pounds in five weeks against the alternative that the diet is *not that good*. Use $\alpha = 0.05$. (Assume that differences have normal distribution.)

14. To determine the effectiveness of an industrial safety program, the following data were collected over a period of a year on the average weekly loss of man hours due to accidents in 12 plants "before and after" the program was put into operation: 50 and 41, 87 and 75, 37 and 35, 141 and 129, 59 and 60, 65 and 53, 24 and 26, 88 and 85, 25 and 29, 36 and 31, 50 and 48, 35 and 37. Use $\alpha = 0.01$ to test the null hypothesis that the safety program is *not* effective against a suitable one-sided alternative. (Assume that differences have normal distribution.)

11.3 TESTS CONCERNING VARIANCES

There are several reasons why it is important to test hypotheses concerning the variances of populations. So far as *direct* applications are concerned, a manufacturer who has to meet rigid specifications will have to perform tests about the variability of his product, a teacher may want to know whether certain statements are true about the variability which he can expect in the performance of a student, and a pharmacist may have to check whether the variation in the potency of a medicine is within permissible limits. So far as *indirect* applications are concerned, tests about variances are often prerequisites for tests concerning other parameters. For instance, the *two-sample t test* described on page 320 requires that the two population variances are equal, and in practice this means that we may have to check on the reasonableness of this assumption before we perform the test concerning the means.

The tests which we shall study in this section include a test of the null hypothesis that the variance of a normal population equals a given constant, and the likelihood ratio test of the equality of the variances of two normal populations (which was referred to in Exercise 13 on page 312).

The first of these tests is essentially that of Exercise 10 on page 311, although we shall not make the simplifying assumption that $\mu = 0$. Given a random sample of size n from a normal population, we shall want to test the null hypothesis $\sigma^2 = \sigma_0^2$ against one of the alternatives $\sigma^2 \neq \sigma_0^2$, $\sigma^2 > \sigma_0^2$, or $\sigma^2 < \sigma_0^2$, and, as in Exercise 10 on page 311, the likelihood ratio technique leads to a test based on s^2, the value of the sample variance. Using Theorem 7.6, we can thus write the critical regions for testing the null hypothesis against the two *one-sided alternatives* as $\chi^2 \geq \chi^2_{\alpha,n-1}$ and $\chi^2 \leq \chi^2_{1-\alpha,n-1}$, where

$$\chi^2 = \frac{(n-1)s^2}{\sigma_0^2} \tag{11.3.1}$$

and $\chi^2_{\alpha,n-1}$ and $\chi^2_{1-\alpha,n-1}$ are as defined on page 215. So far as the *two-sided alternative* is concerned, we reject the null hypothesis if

$$\chi^2 \geq \chi^2_{\alpha/2,n-1} \quad \text{or} \quad \chi^2 \leq \chi^2_{1-\alpha/2,n-1}$$

and the size of all these critical regions is, of course, equal to α.

To illustrate this kind of test, suppose that the thickness of a part used in a semiconductor is its critical dimension and that measurements of the thickness of a random sample of 18 such parts have the variance $s^2 = 0.68$, where the measurements are in thousandths of an inch. The process is considered to be under control if the variation of these thicknesses is given by a variance not greater than 0.36, so that we shall want to test the

null hypothesis $\sigma^2 = 0.36$ against the alte
say, at $\alpha = 0.05$. Substituting into (11.3.1)

$$\chi^2 = \frac{17(0.68)}{0.36} = 3$$

and since this exceeds $\chi^2_{.05,17} = 27.587$, the n
rejected at this level of significance (and th
facture of these parts will have to be adjusted
null hypothesis could *not* have been rejected
to indicate that the choice of α is somethi
decided upon beforehand. (This will save us f
a level of significance which happens to suit

In Exercise 13 on page 312 the reader w
likelihood ratio statistic for testing the equal
normal populations can be expressed in tern
sample variances. Given independent random
from two normal populations with the varian
that in accordance with the results obtained
critical regions of size α for testing the null hyp
one-sided alternatives $\sigma_1^2 > \sigma_2^2$ or $\sigma_1^2 < \sigma_2^2$ are, res

$$\frac{s_1^2}{s_2^2} \geq F_{\alpha,n_1-1,n_2-1} \quad \text{and} \quad \frac{s_2^2}{s_1^2} \geq F$$

where F_{α,n_1-1,n_2-1} and F_{α,n_2-1,n_1-1} are as defined
priate critical region for testing the null hypoth
alternative $\sigma_1^2 \neq \sigma_2^2$ is

$$\frac{s_1^2}{s_2^2} \geq F_{\alpha/2,n_1-1,n_2-1} \qquad \text{if } s_1^2 \geq$$

and

$$\frac{s_2^2}{s_1^2} \geq F_{\alpha/2,n_2-1,n_1-1} \qquad \text{if } s_1^2 <$$

Note that this test is based entirely on the ri
distribution, which is made possible by the result o
namely, the fact that if the random variable **x** has
ν_1 and ν_2 degrees of freedom, then $1/\mathbf{x}$ has an F d
ν_1 degrees of freedom.

To illustrate this F test for the equality of the v
populations, suppose that we are interested in con
of the tensile strength of two kinds of structural ste

ment yi

(where
Since s_1^2
hypoth
icance

and sin
rejecte
streng

1. M
im
fre

is
σ^2
c
t

2. M
l
h

3.

(which constitute a random sample from a normal population) have the sample variance $s^2 = 397$.

4. In a random sample, the weights of 24 Black Angus steers of a certain age have a standard deviation of 238 pounds. Test the null hypothesis $\sigma = 250$ pounds against the two-sided alternative $\sigma \neq 250$ pounds at the level of significance $\alpha = 0.01$.

5. In a random sample, the time which 30 women took to complete the written test for their driver's license had a variance of 6.4 minutes. Test the null hypothesis $\sigma^2 = 8$ against the alternative hypothesis $\sigma^2 < 8$ at the level of significance $\alpha = 0.05$ using
   (a) the method described in the text;
   (b) the method of Exercise 2.

6. Test at the level of significance $\alpha = 0.02$ whether it was reasonable to assume in the example on page 320 dealing with the two kinds of paint that the two (normal) populations have the same variance.

7. Test at the level of significance $\alpha = 0.10$ whether it was reasonable to assume in Exercise 11 on page 323 that $\sigma_1^2 = \sigma_2^2$.

8. Test at the level of significance $\alpha = 0.02$ whether it was reasonable to assume in Exercise 12 on page 323 dealing with the two kinds of rockets that $\sigma_1^2 = \sigma_2^2$.

9. The following are the scores obtained in a personality test by samples of nine married women and nine unmarried women:

Unmarried	88	68	77	82	63	80	78	71	72
Married	73	77	67	74	74	64	71	71	72

   Assuming that these data can be looked upon as independent random samples from two normal populations, test the null hypothesis $\sigma_1^2 = \sigma_2^2$ against the one-sided alternative $\sigma_1^2 > \sigma_2^2$ at the level of significance $\alpha = 0.05$. (σ_1^2 and σ_2^2 are, respectively, the variance of the scores of unmarried women and the variance of the scores of married women.)

11.4 TESTS BASED ON COUNT DATA

If the outcome of an experiment is the number of votes which a candidate receives in a poll, the number of imperfections found in a piece of cloth, the number of children who are absent from school, ..., we refer to these data as *count data,* in contrast to measurements that are given on a continuous scale. Appropriate models for the analysis of count data are the binomial distribution, the Poisson distribution, the multinomial distribution, and the many other discrete distributions which we studied in Chapter 3. In the remainder of this chapter we shall present some of the most common tests based on count data, primarily those concerning the parameters of binomial and multinomial distributions.

11.4.1 Tests Concerning Proportions

The parameter θ of the binomial distribution (3.3.3) is the probability of a success on an individual trial and, hence, the *proportion* of successes one can expect in the long run. To test on the basis of a sample whether the "true" proportion of cures from a certain disease is 0.90 or whether the "true" proportion of defectives coming off an assembly line is 0.02 is, thus, equivalent to testing hypotheses about the parameter θ of binomial distributions.

In Exercise 8 on page 298 the reader was asked to show that the most powerful critical region for testing the null hypothesis $\theta = \theta_0$ against the alternative $\theta = \theta_1 < \theta_0$, where θ is the parameter of a binomial population, is based on the value of $\mathbf{x}$, the number of "successes" obtained in n trials. When it comes to composite alternatives, the likelihood ratio technique also yields tests based on the observed number of successes (as we saw in Exercise 8 on page 310 for the special case where $\theta_0 = \frac{1}{2}$). In fact, if we want to test the null hypothesis $\theta = \theta_0$ against the *one-sided alternative* $\theta > \theta_0$, the critical region of size α of the likelihood criterion is

$$x \geq k_\alpha$$

where k_α is the *smallest* integer for which

$$\sum_{y=k_\alpha}^{n} b(y; n, \theta_0) \leq \alpha \tag{11.4.1}$$

and $b(y; n, \theta_0)$ is the probability of getting y successes in n binomial trials when $\theta = \theta_0$. The size of this critical region (as well as the ones which follow) is, thus, as close as possible to α witho..t exceeding α.

The corresponding critical region for testing the null hypothesis $\theta = \theta_0$ against the *one-sided alternative* $\theta < \theta_0$ is

$$x \leq k'_\alpha$$

where k'_α is the *largest* integer for which

$$\sum_{y=0}^{k_\alpha'} b(y; n, \theta_0) \leq \alpha \tag{11.4.2}$$

and, finally, the critical region for testing the null hypothesis $\theta = \theta_0$ against the *two-sided alternative* $\theta \neq \theta_0$ is

$$x \geq k_{\alpha/2} \quad \text{or} \quad x \leq k'_{\alpha/2}$$

To illustrate this kind of test concerning the parameter θ of a binomial population, suppose we want to test the null hypothesis $\theta = 0.50$ against the two-sided alternative $\theta \neq 0.50$ at the level of significance $\alpha = 0.05$. If n is to be 20, we find from Table I that $k_{.025} = 15$, $k'_{.025} = 5$, and, hence, that the null hypothesis will have to be rejected if the number of "successes" in 20 trials exceeds 14 or is less than 6. It will be left to the reader to verify that the corresponding probability of committing a Type I error is actually 0.0414.

Unless n is *very small,* the tests which we have described require the use of a table of binomial probabilities. For $n \leq 20$ we can use Table I at the end of this book, and for values of n up to 100 we can use the tables referred to on page 100. For larger values of n we make use of the normal approximation of the binomial distribution and treat

$$z = \frac{x - n\theta}{\sqrt{n\theta(1 - \theta)}} \tag{11.4.3}$$

as a value of a random variable having the standard normal distribution. For large n, we can thus test the null hypothesis $\theta = \theta_0$ against the alternatives $\theta > \theta_0$, $\theta < \theta_0$, or $\theta \neq \theta_0$ using, respectively, the critical regions $z \geq z_\alpha$, $z \leq -z_\alpha$, and $|z| \geq z_{\alpha/2}$, where z is given by (11.4.3) with θ_0 substituted for θ.

11.4.2 Differences Among k Proportions

In applied research there are many problems in which we must decide whether observed differences among sample proportions, or percentages, are significant or whether they can be attributed to chance. For instance, if 6 percent of the frozen chicken in a sample from one supplier fails to meet certain standards and only 4 percent in a sample from another supplier, we may want to decide whether the difference between these two percentages is significant. Similarly, we may want to judge on the basis of sample proportions whether the "true" proportion of voters who favor a certain candidate is the same in four different cities.

To indicate a general method for handling problems of this kind, suppose that $x_1, x_2, \ldots,$ and x_k are observed values of a set of independent random variables $\mathbf{x}_1, \mathbf{x}_2, \ldots,$ and $\mathbf{x}_k$ having binomial distributions with the respective parameters n_1 and θ_1, n_2 and θ_2, $\ldots$, and n_k and θ_k. If the n's are sufficiently large, we can approximate the distributions of the random variables

$$\frac{\mathbf{x}_i - n_i\theta_i}{\sqrt{n_i\theta_i(1 - \theta_i)}} \qquad \text{for } i = 1, 2, \ldots, k$$

with independent standard normal distributions, and, according to Theorem 7.3, we can then look upon

$$\sum_{i=1}^{k} \frac{(x_i - n_i\theta_i)^2}{n_i\theta_i(1 - \theta_i)} \tag{11.4.4}$$

as a value of a random variable having the chi-square distribution with k degrees of freedom. To test the null hypothesis $\theta_1 = \theta_2 = \cdots = \theta_k = \theta_0$ (against the alternative that at least one of the θ's does not equal θ_0) we can thus use the critical region $\chi^2 \geq \chi^2_{\alpha,k}$, where χ^2 is given by (11.4.4) with θ_0 substituted for the θ_i.

When θ_0 cannot be specified, that is, when we are interested only in the null hypothesis $\theta_1 = \theta_2 = \cdots = \theta_k$, it is customary to substitute in (11.4.4) the *pooled estimate*

$$\hat{\theta} = \frac{x_1 + x_2 + \cdots + x_k}{n_1 + n_2 + \cdots + n_k} \tag{11.4.5}$$

and as a result of this the critical region becomes $\chi^2 \geq \chi^2_{\alpha,k-1}$, where

$$\chi^2 = \sum_{i=1}^{k} \frac{(x_i - n_i\hat{\theta})^2}{n_i\hat{\theta}(1 - \hat{\theta})} \tag{11.4.6}$$

The loss of one degree of freedom, namely, the change in the critical region from $\chi^2_{\alpha,k}$ to $\chi^2_{\alpha,k-1}$, is due to the fact that an estimate is substituted for the unknown parameter θ_0; a formal discussion of this is referred to on page 342.

Before we give an example to illustrate this kind of test, let us first present an alternate form of (11.4.6) which, as we shall see in Section 11.4.3, lends itself more readily to other applications. Arranging the data as in the following table,

	Successes	*Failures*
Sample 1	x_1	$n_1 - x_1$
Sample 2	x_2	$n_2 - x_2$
	$\cdots$	$\cdots$
Sample k	x_k	$n_k - x_k$

let us refer to its entries as the *observed cell frequencies* f_{ij}, where the first subscript indicates the row and the second subscript indicates the column of this $k \times 2$ table.

Under the null hypothesis $\theta_1 = \theta_2 = \cdots = \theta_k = \theta_0$ the *expected cell frequencies* for the first column are $n_i\theta_0$ for $i = 1, 2, \ldots, k$, and those for the second column are $n_i(1 - \theta_0)$. When θ_0 is unknown, we substitute as before $\hat{\theta}$ for θ_0, where $\hat{\theta}$ is given by (11.4.5), and *estimate* the expected cell frequencies as

$$e_{i1} = n_i\hat{\theta} \quad \text{and} \quad e_{i2} = n_i(1 - \hat{\theta}) \tag{11.4.7}$$

for $i = 1, 2, \ldots, k$. It will be left to the reader to show in Exercise 1 on page 332 that (11.4.6) can thus be written as

$$\chi^2 = \sum_{j=1}^{2} \sum_{i=1}^{k} \frac{(f_{ij} - e_{ij})^2}{e_{ij}} \tag{11.4.8}$$

To illustrate this kind of test, suppose that we want to determine, on the basis of the *sample data* shown in the following table, whether the true proportion of housewives favoring Detergent A over Detergent B is the same in all three cities:

	Number Favoring Detergent A	*Number Favoring Detergent B*	
Los Angeles	232	168	$n_1 = 400$
San Diego	260	240	$n_2 = 500$
Fresno	197	203	$n_3 = 400$

Thus, the pooled estimate of θ_0, the true proportion of housewives favoring Detergent A over Detergent B in all three cities, is given by

$$\hat{\theta} = \frac{232 + 260 + 197}{400 + 500 + 400} = \frac{689}{1{,}300} = 0.53$$

(which assumes, of course, that the null hypothesis $\theta_1 = \theta_2 = \theta_3$ is true). Then, using (11.4.7), we estimate the expected cell frequencies as

$$e_{11} = 400(0.53) = 212 \quad \text{and} \quad e_{12} = 400(0.47) = 188$$

$$e_{21} = 500(0.53) = 265 \quad \text{and} \quad e_{22} = 500(0.47) = 235$$

$$e_{31} = 400(0.53) = 212 \quad \text{and} \quad e_{32} = 400(0.47) = 188$$

and substitution into (11.4.8) yields

$$\chi^2 = \frac{(232 - 212)^2}{212} + \frac{(260 - 265)^2}{265} + \frac{(197 - 212)^2}{212} + \frac{(168 - 188)^2}{188} + \frac{(240 - 235)^2}{235} + \frac{(203 - 188)^2}{188}$$

$$= 6.48$$

Since this exceeds $\chi^2_{.05,2} = 5.991$, the null hypothesis $\theta_1 = \theta_2 = \theta_3$ will have to be rejected; in other words, the true proportions of women favoring Detergent A over Detergent B in the three cities are not the same.

THEORETICAL EXERCISES

1. Show that the two formulas for χ^2 given by (11.4.6) and (11.4.8) are equivalent.

2. Modify the criteria on pages 328 and 329 so that they can be used to test the null hypothesis $\lambda = \lambda_0$, where λ is the parameter of the Poisson distribution (3.3.9), on the basis of n observations. (*Hint:* Refer to the result obtained on page 193.) Also use Table II to find values corresponding to $k_{.025}$ and $k'_{.025}$ to test the null hypothesis $\lambda = 3.6$ against the alternative $\lambda \neq 3.6$ at $\alpha = 0.05$ on the basis of *five* observations.

3. Show that for $k = 2$ the expression (11.4.6) can be written as

$$\chi^2 = \frac{(n_1 + n_2)(n_2x_1 - n_1x_2)^2}{n_1n_2(x_1 + x_2)[(n_1 + n_2) - (x_1 + x_2)]}$$

4. Show that for $k = 2$ and *large samples* the null hypothesis $\theta_1 = \theta_2$, where θ_1 and θ_2 are the parameters of two binomial populations, can also be tested by looking upon

$$z = \frac{\frac{x_1}{n_1} - \frac{x_2}{n_2}}{\sqrt{\hat{\theta}(1 - \hat{\theta})\left(\frac{1}{n_1} + \frac{1}{n_2}\right)}}$$

where $\hat{\theta} = \frac{x_1 + x_2}{n_1 + n_2}$, as a value of a random variable having the standard normal distribution. [*Hint:* Refer to (6.2.7) and (6.2.8).]

5. Show that the square of the expression given for z in Exercise 4 equals the one given for χ^2 in (11.4.6) with $k = 2$, so that the corresponding tests are actually *equivalent*.

APPLIED EXERCISES

6. The null hypothesis $\theta = 0.45$ is to be tested against the alternative $\theta < 0.45$ at $\alpha = 0.05$, where θ is the parameter of a binomial population with $n = 19$. Use Table I to find $k'_{.05}$ and the probabilities of committing Type II errors with this criterion when $\theta = 0.35$, $\theta = 0.30$, and $\theta = 0.25$.
7. The null hypothesis $\theta = 0.25$ is to be tested against the alternative $\theta > 0.25$ at $\alpha = 0.01$, where θ is the parameter of a binomial population with $n = 20$. Use Table I to find $k_{.01}$ and the probabilities of committing Type II errors with this criterion when $\theta = 0.35$, $\theta = 0.40$, and $\theta = 0.45$.
8. The null hypothesis $\theta = 0.70$ is to be tested against the alternative $\theta \neq 0.70$ at $\alpha = 0.05$, where θ is the parameter of a binomial population with $n = 18$. Use Table I to find $k_{.025}$ and $k'_{.025}$ and the probabilities of committing Type II errors with this criterion when $\theta = 0.60$, $\theta = 0.65$, $\theta = 0.75$, and $\theta = 0.80$.
9. The null hypothesis $\theta = 0.40$ is to be tested against the alternative $\theta \neq 0.40$ at $\alpha = 0.01$, where θ is the parameter of a binomial population with $n = 16$. Use Table I to find $k_{.005}$ and $k'_{.005}$ and the probabilities of committing Type II errors with this criterion when $\theta = 0.30$, $\theta = 0.35$, $\theta = 0.45$, and $\theta = 0.50$.
10. A large TV retailer in San Francisco claims that 80 percent of all service calls on color television sets are concerned with the small receiving tube. Test this claim against the alternative $\theta \neq 0.80$ at $\alpha = 0.05$ if a random sample of 225 service calls on color television sets included 167 which were concerned with the small receiving tube.
11. A college magazine maintains that 70 percent of all college men prefer living off campus to living in a dormitory. Test this claim against the alternative $\theta < 0.70$ at $\alpha = 0.01$ if in a random sample of 400 college men there were 266 who expressed this preference.
12. A distributor of processed foods claims that 60 percent of all housewives are willing to pay 15 cents more for an 8-ounce package of pitted prunes than for an 8-ounce package of unpitted prunes. Test this claim against the alternative $\theta > 0.60$ if in a random sample of 200 housewives interviewed in a supermarket 132 said that they are willing to pay the extra 15 cents while the other 68 said that they are not. Use $\alpha = 0.01$.
13. To test the effectiveness of a new pain-relieving drug, 120 patients at a clinic are given a pill containing the drug while 120 others are given a placebo. What can we conclude about the effectiveness of the

drug (at the level of significance $\alpha = 0.05$) if in the first group 79 of the patients felt a beneficial effect while 56 of those who received the placebo felt a beneficial effect. Use the statistic of Exercise 4 and a suitable one-sided alternative to test the null hypothesis $\theta_1 = \theta_2$.

14. If 74 of 250 persons who watched a certain television program in black and white and 92 of 250 persons who watched the same program in color remembered two hours later what products were advertized, test the null hypothesis that there is no difference between the corresponding "true" proportions at the level of significance $\alpha = 0.01$.

15. To landscape its grounds, a bank purchased 400 tulip bulbs from one nursery and 200 from another. If 46 of the 400 bulbs from the first nursery failed to bloom while 18 of the 200 bulbs from the other nursery failed to bloom, test the null hypothesis that there is no difference between the corresponding "true" proportions at the level of significance $\alpha = 0.05$.

16. If 26 of 200 tires of Brand A failed to last 20,000 miles, while the corresponding figures for 200 tires of Brands B, C, and D were 23, 15, and 32, test the null hypothesis that there is no difference in the quality of the four kinds of tires. Use $\alpha = 0.05$.

17. In a random sample of 250 persons with low-incomes 155 are for a certain piece of legislation, while in random samples of 200 persons with average incomes and 150 persons with high incomes there are, respectively, 118 and 87 who favor the legislation. Use $\alpha = 0.05$ to test the null hypothesis that the proportion of persons favoring the legislation is the same for all three groups.

18. In January, 1970, Salesman A contacted 110 customers and sold insurance policies to 13, Salesman B contacted 80 customers and sold insurance policies to 17, while Salesman C contacted 160 customers and sold insurance policies to 20. Looking at these data as random samples of the salesmen's performance, use a level of significance of $\alpha = 0.01$ to check whether the differences among the proportions of sales are significant.

11.4.3 Contingency Tables

Chi-square criteria like that based on (11.4.8) play an important role in many problems dealing with the analysis of count data. In this section we shall use it to analyze *contingency tables* like the following 3×3 table obtained in a study of the relationship between a person's ability in mathematics and his interest in statistics:

		Ability in Mathematics		
		Low	*Average*	*High*
Interest in Statistics	*Low*	63	42	15
	Average	58	61	31
	High	14	47	29

The notation which is generally used in the analysis of an $r \times k$ table is as shown below, where f_{ij} is the *observed cell frequency* for the cell belonging to the ith row and the jth column, the row and column totals are, respectively,

$$f_{i.} = \sum_{j=1}^{k} f_{ij} \quad \text{and} \quad f_{.j} = \sum_{i=1}^{r} f_{ij}$$

and f is the *grand total*, namely, the sum of all the cell frequencies f_{ij}. Thus, if one variable has the k categories $A_1, A_2, \ldots, A_k$, and the other variable has the r categories $B_1, B_2, \ldots, B_r$, we are faced with the following scheme:

	A_1	A_2		A_k	
B_1	f_{11}	f_{12}	$\cdots$	f_{1k}	$f_{1.}$
B_2	f_{21}	f_{22}	$\cdots$	f_{2k}	$f_{2.}$
$\cdots$	$\cdots$	$\cdots$	$\cdots$	$\cdots$	$\cdots$
B_r	f_{r1}	f_{r2}	$\cdots$	f_{rk}	$f_{r.}$
	$f_{.1}$	$f_{.2}$		$f_{.k}$	f

The null hypothesis we shall want to test is that the two variables are *independent*. More specifically, if θ_{ij} is the probability that an item will fall into the cell belonging to the ith row and the jth column, $\theta_{i.}$ is the probability that an item will fall into the ith row, and $\theta_{.j}$ is the probability that an item will fall into the jth column, the null hypothesis which we shall want to test is that

$$\theta_{ij} = (\theta_{i.})(\theta_{.j}) \qquad \text{for } i = 1, 2, \ldots, r, \text{ and } j = 1, 2, \ldots, k$$

To test this null hypothesis against the alternative that θ_{ij} does *not* equal the product of $\theta_{i.}$ and $\theta_{.j}$ for at least one pair of values of i and j, we *estimate* the probabilities $\theta_{i.}$ and $\theta_{.j}$ as

$$\hat{\theta}_{i.} = \frac{f_{i.}}{f} \quad \text{and} \quad \hat{\theta}_{.j} = \frac{f_{.j}}{f}$$

and, hence, the *expected cell frequencies* as

$$e_{ij} = (\hat{\theta}_{i.})(\hat{\theta}_{.j}) \cdot f = \frac{f_{i.}}{f} \cdot \frac{f_{.j}}{f} \cdot f = \frac{(f_{i.})(f_{.j})}{f} \tag{11.4.9}$$

Then, we base our decision on the value of

$$\chi^2 = \sum_{i=1}^{r} \sum_{j=1}^{k} \frac{(f_{ij} - e_{ij})^2}{e_{ij}} \tag{11.4.10}$$

and reject the null hypothesis if the value which we obtain exceeds $\chi^2_{\alpha,(r-1)(k-1)}$. Since the statistic whose value is given by (11.4.10) has only *approximately* a chi-square distribution with $(r - 1)(k - 1)$ degrees of freedom, it is customary to use this test only when none of the e_{ij} is less than 5; this requires at times that we combine some of the cells, as in the illustration on page 337, with a corresponding loss in the number of degrees of freedom.

To justify the fact that in this case the number of degrees of freedom is $(r - 1)(k - 1)$, let us make the following observation: Whenever the expected cell frequencies in formulas like (11.4.8) and (11.4.10) are estimated on the basis of sample data, the number of degrees of freedom is $s - t - 1$, where s is the number of terms in the summation and t is the number of *independent* parameters replaced by estimates. In the use of (11.4.8) we had $s = 2k$, $t = k$ since we had to estimate the k parameters $\theta_1, \theta_2, \ldots, \theta_k$, and the number of degrees of freedom was $2k - k - 1 = k - 1$. In the analysis of an $r \times k$ contingency table $s = rk$, and only $t = r + k - 2$ of the parameters are independent since the r parameters $\theta_{i.}$ and the k parameters $\theta_{.j}$ are subject to the *two* restrictions that their respective sums must equal 1; hence, $s - t - 1 = rk - (r + k - 2) - 1 = (r - 1)(k - 1)$.

Returning now to our numerical example, we find that the expected cell frequencies for the first row are 45, 50, 25, those of the second row are 56.25, 62.5, 31.25, and those of the third row are 33.75, 37.5, 18.75.

Rounding these figures to the nearest integers, we find that

$$\chi^2 = \frac{(63-45)^2}{45} + \frac{(42-50)^2}{50} + \frac{(14-25)^2}{25}$$
$$+ \frac{(58-56)^2}{56} + \frac{(61-62)^2}{62} + \frac{(31-31)^2}{31}$$
$$+ \frac{(14-34)^2}{34} + \frac{(47-38)^2}{38} + \frac{(29-19)^2}{19}$$
$$= 32.51$$

and since this exceeds $\chi^2_{.01,4} = 13.277$, the null hypothesis will have to be rejected and we conclude that there is a relationship between a person's ability in mathematics and his interest in statistics. (It did not matter in this case that we rounded the expected cell frequencies to the nearest integer, and it is common practice to round them to the nearest integer or to one decimal.)

11.4.4 Goodness of Fit

The term "goodness of fit" has two different connotations in statistics, and as it is to be used here it applies to tests in which we want to determine *whether a set of data may be looked upon as values of a random variable having a given distribution;* the other kind of "goodness of fit" will be discussed in Chapter 13. To illustrate, suppose that we want to decide on the basis of the data (observed frequencies) shown in the following table whether the number of errors a compositor makes in setting a galley of type is a random variable having a Poisson distribution:

Number of Errors	*Observed Frequencies* f_i	*Poisson Probabilities with* $\lambda = 3$	*Expected Frequencies* e_i
0	18	0.0498	21.9
1	53	0.1494	65.7
2	103	0.2240	98.6
3	107	0.2240	98.6
4	82	0.1680	73.9
5	46	0.1008	44.0
6	18	0.0504	22.2
7	10 }	0.0216	9.5 }
8	2 } 13	0.0081	3.6 } 14.3
9	1 }	0.0027	1.2 }
	440		

Using the mean of the observed distribution to estimate λ, the parameter of the Poisson distribution, we get $\hat{\lambda} = \frac{1{,}341}{440} = 3.05$ or, approximately, $\hat{\lambda} = 3$. Then copying the Poisson probabilities for $\lambda = 3$ from Table II and multiplying by 440, the total frequency, we get the *expected frequencies* shown in the righthand column.

An appropriate test of the null hypothesis that the observed data come from a population having a Poisson distribution (against the alternative that it has some other distribution) is to reject the null hypothesis if $\chi^2 \geq \chi^2_{\alpha,m-2}$, where

$$\chi^2 = \sum_{i=1}^{m} \frac{(f_i - e_i)^2}{e_i} \tag{11.4.11}$$

and s and t in the formula for the number of degrees of freedom are, respectively, m (the number of terms added) and 1 (only one parameter is estimated on the basis of the data). Thus, the number of degrees of freedom is $m - 1 - 1 = m - 2$.

In order to apply this test to the given data, we shall have to combine (pool) the last three classes; thus, making sure that each expected frequency is at least 5, we get

$$\begin{aligned}\chi^2 &= \frac{(18 - 21.9)^2}{21.9} + \frac{(53 - 65.7)^2}{65.7} + \frac{(103 - 98.6)^2}{98.6} + \frac{(107 - 98.6)^2}{98.6}\\ &\quad + \frac{(82 - 73.9)^2}{73.9} + \frac{(46 - 44.0)^2}{44.0} + \frac{(18 - 22.2)^2}{22.2} + \frac{(13 - 14.3)^2}{14.3}\\ &= 5.95\end{aligned}$$

which is less than $\chi^2_{.05,6} = 12.592$. Hence, the null hypothesis cannot be rejected, and the fairly close agreement between the observed and expected frequencies makes it reasonable to accept the null hypothesis that the data come from a Poisson population. In other words, the Poisson distribution provides a "good fit."

The method illustrated by means of this example applies equally well to testing the fit of other kinds of distributions, and in each case the number of degrees of freedom for the chi-square criterion must be determined by means of the formula $s - t - 1$, as explained on page 336.

THEORETICAL EXERCISES

1. Show that the expected cell frequencies given by (11.4.9) satisfy the equations $\sum_{i=1}^{r} e_{ij} = f_{.j}$ and $\sum_{j=1}^{k} e_{ij} = f_{i.}$, so that only $(r - 1)(k - 1)$

of the e_{ij} have to be calculated with the use of (11.4.9) and the others may then be obtained by subtraction from the totals of the appropriate rows or columns.

2. If the analysis of a contingency table shows that there *is* a relationship between the two variables under consideration, the *strength* of this relationship may be measured by means of the *contingency coefficient*

$$C = \sqrt{\frac{\chi^2}{\chi^2 + f}}$$

where χ^2 is the value obtained with (11.4.10) and f is the grand total as defined on page 335. Show that

(a) for a 2×2 contingency table the maximum value of C is $\frac{1}{2}\sqrt{2}$;

(b) for a 3×3 contingency table the maximum value of C is $\frac{1}{3}\sqrt{6}$.

APPLIED EXERCISES

3. A sample survey, designed to show where persons living in different parts of the country buy non-prescribed medicines, yielded the following results:

	Northeast	*North Central*	*South*	*West*
Drugstores	219	200	181	180
Grocery Stores	39	52	89	60
Others	42	48	30	60

Test the null hypothesis that, so far as non-prescribed medicines are concerned, the buying habits of persons living in the given parts of the country are the same. Use the level of significance $\alpha = 0.05$.

4. The following table is based on a study of the relationship between race and blood type in a Near Eastern country:

	Blood Type			
	O	A	B	AB
Race 1	176	148	96	72
Race 2	78	50	45	12
Race 3	15	19	8	7

Use $\alpha = 0.01$ to test the null hypothesis that there is no relationship between race and blood type in the country under consideration.

5. The following sample data pertain to the shipments received by a large firm from three different vendors:

	Number Rejected	*Number Imperfect but Acceptable*	*Number Perfect*
Vendor A	12	23	89
Vendor B	8	12	62
Vendor C	21	30	119

Test at the level of significance $\alpha = 0.01$ whether the three vendors ship products of equal quality.

6. In order to check whether a die is balanced, it was rolled 240 times and the following results were obtained: 1 occurred 33 times, 2 occurred 51 times, 3 occurred 49 times, 4 occurred 36 times, 5 occurred 32 times, and 6 occurred 39 times. Test the null hypothesis that the die is balanced at the level of significance $\alpha = 0.05$.

7. It is desired to test whether the number of gamma rays emitted per second by a certain radioactive substance is a random variable having the Poisson distribution with $\lambda = 2.4$. Test this null hypothesis at $\alpha = 0.05$ on the basis of the following results obtained in 300 1-second intervals:

Number of Gamma Rays	*Frequency*
0	19
1	48
2	66
3	74
4	44
5	35
6	10
7	4

8. With reference to the data of Exercise 7, use the level of significance $\alpha = 0.05$ to test the null hypothesis that the data constitute a random sample from a population having a Poisson distribution.

9. The following is the distribution of the number of hours which a certain kind of machine part lasted while in continuous use:

Number of Hours	*Frequency*
0–5	37
5–10	20
10–15	17
15–20	13
20–25	8
25–30	5

Use $\alpha = 0.05$ to test the null hypothesis that these data constitute a random sample from an exponential population (4.3.2). (*Hint:* First calculate the mean of the distribution, and then determine the exponential probabilities associated with the various intervals, changing the last one to read "25 or more.")

10. The following is the distribution of the readings obtained with a Geiger counter of the number of particles emitted by a radioactive substance in 100 successive 40-second intervals:

Number of Particles	*Frequency*
5–9	1
10–14	10
15–19	37
20–24	36
25–29	13
30–34	2
35–39	1

(a) Verify that the mean of this distribution is $\bar{x} = 20$ and that its standard deviation is $s = 5$.

(b) Find the probabilities that a random variable having a normal distribution with $\mu = 20$ and $\sigma = 5$ takes on a value between 4.5 and 9.5, 9.5 and 14.5, 14.5 and 19.5, 19.5 and 24.5, 24.5 and 29.5, 29.5 and 34.5, and 34.5 and 39.5.

(c) Find the expected *normal curve frequencies* for the various classes by multiplying the corresponding probabilities obtained in part (b) by the total frequency, and then test the null hypothesis that the data constitute a random sample from a normal population. Use the level of significance $\alpha = 0.05$.

REFERENCES

For a proof of the independence of the random variables given by (11.2.6) and (11.2.7) see

Brunk, H. D., *An Introduction to Mathematical Statistics*, 2nd ed. New York: Blaisdell Publishing Co., 1965,

and the problem of determining the appropriate number of degrees of freedom for various uses of the chi-square statistic is discussed in

Cramér, H., *Mathematical Methods of Statistics*. Princeton, N.J.: Princeton University Press, 1946.

The *Smith-Satterthwaite* test of the null hypothesis that two normal populations with unequal variances have the same mean is given in

Miller, I., and Freund, J. E., *Probability and Statistics for Engineers*. Englewood Cliffs, N.J.: Prentice-Hall, Inc., 1965.

12

HYPOTHESIS TESTING: NONPARAMETRIC METHODS

12.1 INTRODUCTION

To handle problems in which the various assumptions underlying standard tests cannot be met, statisticians have developed many alternate techniques which have become known as *nonparametric methods.* Thus, the term is used somewhat loosely to include *distribution-free methods* (like the tolerance limits of Exercise 10 on page 228) where we make no assumptions about the populations from which we are sampling, except perhaps that they must be continuous, as well as methods which are nonparametric only in the sense that we are not concerned with the parameters of populations *of a given kind.*

Many of these nonparametric methods have great intuitive appeal, and they can often be described as "quick and easy" or "short-cut" techniques, for they are usually easy to perform and require fewer computations than the corresponding "standard" tests which they replace. As should be expected, though, by making fewer assumptions we expose ourselves to greater risks, and *the advantage of nonparametric methods that they apply under much more general conditions is offset by the fact that they are generally less efficient.* Since the development of nonparametric methods has been enormous in recent years, enough to fill volumes many times the size of this book, we shall present here only a few of the nonparametric tests which have become fairly popular in recent years; mainly, they will serve as an introduction to what this branch of statistics is all about.

12.2 THE SIGN TEST

Except for the large-sample tests, all of the tests concerning the means of populations which we studied in Chapter 11 were based on the assumption that we are sampling from normal populations. When this assumption cannot be met, these tests can be replaced by various nonparametric alternatives, among them the *sign test,* which we shall describe briefly in this section. To illustrate, let us assume that we are dealing with a *symmetrical* distribution so that the probability of getting a value less than the mean equals the probability of getting a value greater than the mean; we can then test the null hypothesis $\mu = \mu_0$ against an appropriate alternative by replacing each sample value exceeding μ_0 by a *plus sign,* each sample value less than μ_0 by a *minus sign,* and then test the null hypothesis that these plus and minus signs constitute a random sample from a binomial population (3.3.3) with $\theta = \frac{1}{2}$. (If a sample value happens to equal μ_0, it is discarded, and if the population distribution is *not symmetrical,* the technique can still be used to test the null hypothesis $\tilde{\mu} = \tilde{\mu}_0$, where $\tilde{\mu}$ is the population median.)

To illustrate, suppose that we want to test the hypothesis that the average breaking strength of a certain kind of 2-inch cotton ribbon is $\mu = 160$ pounds against the alternative $\mu > 160$ pounds at $\alpha = 0.05$, and that 20 measurements yielded the following results:

163	165	160	189	161	171	158	151	169	162
+	+		+	+	+	−	−	+	+
163	139	172	165	148	166	172	163	187	173
+	−	+	+	−	+	+	+	+	+

Since 15 of these measurements exceed 160, one equals 160, and four are less than 160, we shall have to see whether "15 successes in 19 trials" supports the null hypothesis $\theta = \frac{1}{2}$ or the alternative hypothesis $\theta > \frac{1}{2}$. Using Table I we find that $k_{.05} = 14$ for $n = 19$, where $k_{.05}$ is as defined on page 328, and, hence, that the null hypothesis will have to be rejected. (Of course, if the sample had been large, we would have used the normal approximation to the binomial distribution instead of Table I.)

The sign test has many other applications; for instance, it can be used as a nonparametric alternative for the *paired-sample test* described on page 321. In fact, in Exercise 6 on page 350 the reader will be asked to test the null hypothesis $\mu_1 = \mu_2$ against an appropriate alternative by replacing the differences between paired observations by their signs and then proceeding as above.

12.3 THE MEDIAN TEST

There are several nonparametric methods for testing the null hypothesis that two samples come from identical populations. One that is particularly sensitive to the alternative that the populations have unequal medians or means is the *median test*, in which the samples from two populations are looked upon as *one sample*, and the ultimate decision is based on the number of values from each population which fall above and below the median of this combined sample. To avoid the theoretical complications which result from *ties*, it will be assumed that the populations with which we are dealing are continuous. Also, to simplify matters it will be assumed that $n_1 + n_2 = 2k$, where n_1 and n_2 are the sizes of the two samples and k is a positive integer. If x is the number of values of the first sample which fall below the median of the combined data, we are thus faced with the following situation:

	Population 1	*Population 2*	
Below Median	x	$k - x$	k
Above Median	$n_1 - x$	$n_2 - k + x$	k
	n_1	n_2	$2k$

This table differs from those of Section 11.4.3 in that the row totals as well as the column totals are fixed, but if n_1 and n_2 are large, it can nevertheless be analyzed in the same way. In other words, we calculate χ^2 as in the analysis of an $r \times k$ table and reject the null hypothesis that the two samples come from identical populations if the value which we obtain exceeds $\chi^2_{\alpha,1}$.

To illustrate this technique, suppose that the following are the grades which random samples of eighth graders from two different schools obtained in an achievement test in mathematics:

School 1: 43, 80, 99, 86, 68, 70, 85, 93, 98, 96,
75, 81, 32, 92, 96, 64, 79, 97, 76, 80

School 2: 76, 65, 73, 95, 77, 99, 55, 35, 72, 83,
70, 65, 86, 60, 62, 90, 71, 65, 89, 71,
80, 76, 93, 94

Since the median of an even number of measurements is *by definition* the *mean of the two middle values* (after the data have been arranged according to size), it can easily be verified that the median of the forty-four grades is 78 and that

	School 1	*School 2*	
Below Median	7	15	22
Above Median	13	9	22
	20	24	44

Calculating the expected cell frequency e_{11} according to (11.4.9), we obtain $e_{11} = \frac{22 \cdot 20}{44} = 10$; then we find by subtraction that $e_{12} = 22 - 10 = 12$, $e_{21} = 20 - 10 = 10$, and $e_{22} = 24 - 12 = 12$. Thus,

$$\chi^2 = \frac{(7-10)^2}{10} + \frac{(15-12)^2}{12} + \frac{(13-10)^2}{10} + \frac{(9-12)^2}{12}$$
$$= 3.30$$

and the null hypothesis that the two samples come from identical populations *cannot be rejected* at the level of significance $\alpha = 0.05$ since $\chi^2_{.05,1} = 3.841$. (It may well be that the t test of Section 11.2.1 would have led to the rejection of this null hypothesis, but it is apparent from the data, particularly the clustering of high grades in both schools, that we can hardly look upon the data as random samples from normal populations.)

When the samples are small, we cannot use this approximate chi-square criterion; instead, we regard as *equally likely* the $\binom{2k}{n_1}$ ways in which n_1 of the $2k$ ordered observations can be ascribed to Population 1, and base our decision on the distribution of $\mathbf{x}$, the number of values from Population 1 which fall below the median of the combined data. Since x of the k values *below* the median can be chosen in $\binom{k}{x}$ ways and $n_1 - x$ of the k values *above* the median can be chosen in $\binom{k}{n_1 - x}$ ways, x of the k values below the median and $n_1 - x$ of the k values above the median can be ascribed to Population 1 in $\binom{k}{x}\binom{k}{n_1 - x}$ ways, and it follows that the distribution of $\mathbf{x}$ is given by

$$f(x) = \frac{\binom{k}{x}\binom{k}{n_1 - x}}{\binom{2k}{n_1}} \tag{12.3.1}$$

for $x = 0, 1, 2, \ldots, n_1$, subject to the restriction that neither x nor $n_1 - x$ can exceed k. This distribution is readily identified as the *hypergeometric distribution* (3.3.5), and in Exercise 2 on page 349 the reader will be asked to show that (12.3.1) is equivalent to

$$f(x) = \frac{\binom{n_1}{x}\binom{n_2}{k - x}}{\binom{n_1 + n_2}{k}} \tag{12.3.2}$$

in which form it is given in most other texts. To illustrate the use of (12.3.1), suppose that in a given example we have $n_1 = 10$ and $n_2 = 8$, and that we want to know for what values of $\mathbf{x}$ we would reject the null hypothesis that the two samples come from identical populations at the level of significance $\alpha = 0.05$. Since $f(1) = f(9) = 0.0002$, $f(2) = f(8) = 0.0074$, and $f(3) = f(7) = 0.0691$, as the reader will be asked to verify in Exercise 8 on page 350, it follows that the critical region consists of $x = 1$, $x = 2$, $x = 8$, and $x = 9$. Note that it is impossible in this example for x to equal 0 or 10.

12.4 TESTS BASED ON RANK SUMS

In this section we shall describe another nonparametric test which can be used to test the null hypothesis that two samples come from identical populations. It is called the *Mann-Whitney test* (the *Wilcoxon test*, or the *U test*), and to illustrate it suppose that we want to compare two kinds of emergency flares on the basis of the following burning times (rounded to the nearest tenth of a minute):

Brand A: 14.9, 11.3, 13.2, 16.6, 17.0, 14.1, 15.4, 13.0, 16.9

Brand B: 15.2, 19.8, 14.7, 18.3, 16.2, 21.2, 18.9, 12.2, 15.3, 19.4

Arranging these values *jointly* (as if they were one sample) in an increasing order of magnitude and assigning them in this order the ranks 1, 2, 3, ..., and 19, we find that the values of the first sample (Brand *A*) occupy ranks 1, 3, 4, 5, 7, 10, 12, 13, and 14, while those of the second sample (Brand *B*) occupy ranks 2, 6, 8, 9, 11, 15, 16, 17, 18, and 19. (Had there been *ties*, we would have assigned to each of the tied observations the mean of the ranks which they jointly occupy; thus, if the third and fourth values had been the same we would have assigned each the rank 3.5, and if the fifth, sixth, and seventh values had been the same we would have assigned each the rank 6.)

If there is an appreciable difference between the means of the two populations, most of the lower ranks are apt to be occupied by the values of one sample while most of the higher ranks are apt to be occupied by the values of the other sample. Thus, in the *Mann-Whitney test* we base our decision (about the null hypothesis that the two samples come from identical populations) on the ranks occupied by the values of the two samples; more specifically, we base our decision on the value of

$$\mathbf{U} = n_1 n_2 + \frac{n_1(n_1 + 1)}{2} - \mathbf{R}_1 \tag{12.4.1}$$

where n_1 and n_2 are the sizes of the two samples and $\mathbf{R}_1$ is the sum of the ranks assigned to the values of the first sample. (In practice, we find whichever rank sum is most easily obtained and call it R_1, as it is immaterial which sample is referred to as the "first.") In our example we have $n_1 = 9, n_2 = 10, R_1 = 1 + 3 + 4 + 5 + 7 + 10 + 12 + 13 + 14 = 69$, and, hence,

$$U = 9 \cdot 10 + \frac{9 \cdot 10}{2} - 69 = 66$$

Under the null hypothesis that the two samples come from identical populations (which we shall assume to be continuous so that the probability is zero that there will be any ties), the random variable $\mathbf{R}_1$ is the sum of n_1 positive integers selected at random from among the first $n_1 + n_2$. Making use of the result of Exercise 4 on page 205 with $n = n_1$ and $N = n_1 + n_2$, we thus find that

$$E(\mathbf{R}_1) = \frac{n_1(n_1 + n_2 + 1)}{2}$$

and

$$\text{var}(\mathbf{R}_1) = \frac{n_1 n_2(n_1 + n_2 + 1)}{12}$$

and, hence, that

$$E(\mathbf{U}) = \frac{n_1 n_2}{2} \quad \text{and} \quad \text{var}(\mathbf{U}) = \frac{n_1 n_2(n_1 + n_2 + 1)}{12} \tag{12.4.2}$$

according to the various properties of mathematical expectations which we studied in Chapter 5.

When n_1 and n_2 are small, we base the Mann-Whitney test on the exact distribution of $\mathbf{U}$ and specially constructed tables like those referred to on page 357. However, it can be shown that when n_1 and n_2 are both greater than 8 the distribution of $\mathbf{U}$ can be approximated closely with a normal distribution, namely, that

$$z = \frac{U - E(\mathbf{U})}{\sqrt{\text{var}(\mathbf{U})}} \tag{12.4.3}$$

can be looked upon as a value of a random variable having the standard normal distribution. Since high ranks lead to small values of $\mathbf{U}$, the critical region for testing the null hypothesis against the alternative $\mu_1 > \mu_2$ is $z \leq -z_\alpha$, the critical region for testing the null hypothesis against the alternative $\mu_1 < \mu_2$ is $z \geq z_\alpha$, and the critical region for testing the null hypothesis against the alternative $\mu_1 \neq \mu_2$ is $|z| \geq z_{\alpha/2}$. Note that the subscripts of the μ's will have to match those in (12.4.1).

Returning now to our numerical example where we had $n_1 = 9$, $n_2 = 10$, and $U = 66$, we find that

$$E(\mathbf{U}) = \frac{9 \cdot 10}{2} = 45, \quad \text{var}(\mathbf{U}) = \frac{9 \cdot 10 \cdot 20}{12} = 150$$

and, hence, that

$$z = \frac{66 - 45}{\sqrt{150}} = 1.71$$

Since $z_{.025} = 1.96$, it follows that the null hypothesis (that the two populations are identical) *cannot be rejected* against the alternative $\mu_1 \neq \mu_2$ at the level of significance $\alpha = 0.05$. (Note that it could have been rejected at this level of significance against the alternative $\mu_1 < \mu_2$.)

An interesting feature of the Mann-Whitney test is that, with a slight modification, it can also be used to test the null hypothesis that the two samples come from identical populations against the alternative that the two populations have *unequal dispersions*, namely, that they differ in variability or spread. As before, the values of the two samples are arranged jointly in an increasing order of magnitude, but now they are ranked *from both ends toward the middle*. We assign rank 1 to the smallest value, ranks 2 and 3 to the largest and second largest values, ranks 4 and 5 to the second and third smallest values, ranks 6 and 7 to the third and fourth largest values, and so on. Subsequently, the calculation of U and z is the same as before, but a small value of $\mathbf{R}_1$ and, hence, a large value of $\mathbf{U}$ tends to indicate that the first population has a greater variation then the second.

THEORETICAL EXERCISES

1. Show that for the 2 × 2 table of the large-sample median test the value of the chi-square statistic (11.4.10) can be calculated by means of the formula

$$\chi^2 = \frac{2k}{n_1 n_2}(2x - n_1)^2$$

Verify that this formula also yields the value 3.3 for the illustration given in the text.

2. Verify that the probability functions given by (12.3.1) and (12.3.2) are equivalent.

APPLIED EXERCISES

3. The following are the number of minutes which it took a random sample of 20 technicians to perform a certain task: 18.1, 20.3, 18.3, 15.6, 22.5, 16.8, 17.6, 16.9, 18.1, 16.9, 19.1, 16.7, 19.5, 18.5, 20.1, 18.8, 19.0, 17.5, 18.4, and 18.1. Use the sign test at the level of significance $\alpha = 0.05$ to test the null hypothesis that these measurements constitute a random sample from a population having the mean $\mu = 19.4$ against the two-sided alternative $\mu \neq 19.4$.

4. The following are sample data on the amount of money (in dollars) which a person spends visiting a certain amusement park: 7.25, 8.40, 9.75, 10.65, 13.50, 11.75, 9.25, 8.50, 10.10, 10.45, 11.20, 10.45, 8.60, 9.85, 9.00, 10.25, 6.40, 10.80, 15.35, 11.60, 10.85, 8.45, 7.25, 10.75, 13.60, 11.75, 9.60, 10.80, 9.90, 7.70, 9.75, 10.80, 11.15, 8.95, 10.20, 9.60, 10.25, 9.30, 11.30, and 13.65. Use the sign test and a level of significance of $\alpha = 0.05$ to test the null hypothesis that on the average a person spends \$9.50 at the park (against the alternative that this figure is too low).

5. Referring to the data of Exercise 13 on page 323, use the sign test and $\alpha = 0.05$ to test the null hypothesis that with the given diet a woman will on the average lose 10 pounds in five weeks against the alternative that the diet is not that good.

6. Apply the sign test to the data of Exercise 14 on page 323.

7. The following are the number of defective pieces produced on 30 days by two machines: 2 and 5, 6 and 8, 5 and 6, 9 and 9, 6 and 10, 7 and 4, 1 and 5, 4 and 8, 3 and 6, 5 and 6, 6 and 10, 8 and 6, 2 and 5, 1 and 7, 5 and 9, 3 and 3, 6 and 7, 4 and 9, 4 and 2, 5 and 7, 1 and 2, 7 and 9, 4 and 6, 7 and 5, 6 and 8, 7 and 10, 2 and 4, 5 and 4, 6 and 8, and 3 and 5. Use the sign test at the level of significance $\alpha = 0.01$ to test the null hypothesis that the two machines perform equally well against the alternative that the first machine is better.

8. Verify the probabilities on page 347 which led to the critical region for the median test when $n_1 = 10$ and $n_2 = 8$.

9. For comparison, two kinds of feed are fed to samples of pigs, and the following are their gains in weight (in ounces) after a fixed period of time:

Feed C: 10.4, 8.8, 9.2, 10.0, 10.8, 12.6, 11.2

Feed D: 12.0, 18.2, 8.0, 9.6, 11.8, 7.6, 14.0, 13.0, 11.4

Use the small-sample median test at $\alpha = 0.05$ to test the null hypothesis that the two samples come from identical populations against the two-sided alternative that the medians of the two populations are not the same.

10. The following are samples of the *down-times* (periods in which they were inoperative due to failures) of two different computers, in minutes:

Computer 1: 73, 82, 49, 72, 68, 72, 69, 61, 63, 91, 67, 58, 66, 43, 60, 81, 52, 97, 63, 59, 70, 65

Computer 2: 64, 59, 57, 60, 93, 52, 61, 47, 66, 60, 61, 33, 55, 73, 85, 43, 60, 70, 58, 60, 70, 57, 82, 58

Use the large-sample median test at $\alpha = 0.05$ to test the null hypothesis that the two samples come from identical populations. Check the value of χ^2 obtained for the 2×2 table by means of the formula of Exercise 1.

11. The following are the Rockwell hardness numbers obtained for aluminum die castings randomly selected from two different lots:

Lot A: 54, 97, 89, 76, 59, 78, 87, 56, 74, 70, 81, 68, 79, 53, 71

Lot B: 75, 82, 67, 73, 72, 86, 55, 77, 94, 88, 52, 65

(a) Use the Mann-Whitney test at $\alpha = 0.05$ to test the null hypothesis that the two samples come from identical populations against the alternative that the two populations have *unequal means.*

(b) Use the Mann-Whitney test (modified as suggested on page 349) at $\alpha = 0.05$ to test the null hypothesis that the two samples come from identical populations against the alternative that the two populations have unequal dispersions.

12. Use the Mann-Whitney test to decide whether the two sets of grades given on page 345 can be looked upon as random samples from identical populations. Treat ties between observations belonging to different samples as indicated on page 347.

13. Use the Mann-Whitney test instead of the median test to rework Exercise 10. Treat ties between observations belonging to different samples as indicated on page 347.

14. An agricultural experiment was performed to compare the average yield of two varieties of corn. Use the Mann-Whitney test with $\alpha = 0.05$ to decide whether it is reasonable to treat the following two samples as coming from identical populations:

Variety A: 91.1, 89.4, 93.5, 85.6, 93.6, 94.2, 92.3, 87.2, 87.7

Variety B: 104.0, 89.8, 102.2, 88.0, 90.5, 91.8, 84.9, 98.1, 96.0, 89.7

15. It is desired to compare the variability of the mileage yield of two kinds of gasoline on the basis of the following figures (in miles per gallon):

Gasoline X: 18.6, 19.1, 17.7, 19,3. 18.4, 16.9, 18.0, 18.3, 17.5, 16.3, 19.0, 15.5, 17.3

Gasoline Y: 19.2, 18.9, 17.8, 18.1, 18.2, 19.4, 16.8, 18.5, 17.2, 17.6, 18.7

Use the Mann-Whitney test (modified as suggested on page 349) at $\alpha = 0.05$ to test the null hypothesis that the samples come from identical populations against the alternative that the two populations have unequal dispersions.

12.5 TESTS BASED ON RUNS

Several methods have been developed in the last few decades which make it possible to judge the randomness of observed data on the basis of the order in which the observations were obtained. We can thus test whether patterns that look suspiciously non-random can be attributed to chance and, what is most important, this is done *after the data have been collected.* The technique which we shall describe in this section is based on the *theory of runs,* where *a run is a succession of identical letters (or other kinds of symbols) which is followed and preceded by different letters or no letters at all.* To illustrate, let us consider the following arrangement of *defective, d,* and *nondefective, n,* pieces produced in the given order by a certain machine:

$$\underbrace{n\ n\ n\ n\ n}\ \underbrace{d\ d\ d\ d}\ \underbrace{n\ n\ n\ n\ n\ n\ n\ n\ n\ n}\ \underbrace{d\ d}\ \underbrace{n\ n}\ \underbrace{d\ d\ d\ d}\ \underbrace{n}\ \underbrace{d\ d}\ \underbrace{n\ n}$$

Using braces to combine the letters which constitute a run, we find that there is first a run of five n's, then a run of four d's, then a run of ten n's, . . ., and finally a run of two n's; in all, there are *nine* runs of varying lengths.

The total number of runs appearing in an arrangement of this kind is often a good indication of a possible lack of randomness. If there are *too few runs,* we might suspect a definite grouping or clustering, or perhaps a trend; if there are *too many runs,* we might suspect some sort of repeated alternating pattern. In our example there seems to be a definite clustering, the defective pieces seem to come in groups, but it will have to be seen whether this is significant or whether it can be attributed to chance.

To find the probability that n_1 letters of one kind and n_2 letters of another kind will form u runs when each of the $\binom{n_1 + n_2}{n_1}$ possible arrangements of these letters is regarded as equally likely, let us investigate first the case where u is *even*, namely, where $u = 2k$ and k is a positive integer. In that case there will have to be k runs of each kind alternating with one another. To find the number of ways in which n_1 letters can form k runs, let us first consider the very simple case where we have five letters c which are to be divided up into three runs. Using vertical bars to separate the five letters into three runs, we find that there are the *six* possibilities

$$c\,|c\,|c\;c\;c \qquad c\,|c\;c\,|c\;c \qquad c\,|c\;c\;c\,|c$$

$$c\;c\,|c\,|c\;c \qquad c\;c\,|c\;c\,|c \qquad c\;c\;c\,|c\,|c$$

corresponding to the $\binom{4}{2}$ ways in which we can put two vertical bars into two of the four spaces between the five c's. By the same token there are $\binom{n_1 - 1}{k - 1}$ ways in which the n_1 letters of the first kind can form k runs, $\binom{n_2 - 1}{k - 1}$ ways in which the n_2 letters of the second kind can form k runs, and it follows that there are altogether $2\binom{n_1 - 1}{k - 1}\binom{n_2 - 1}{k - 1}$ ways in which these $n_1 + n_2$ letters can form $2k$ runs. The factor 2 is accounted for by the fact that when we combine the two kinds of runs so that they alternate, we can begin either with a run of the first kind of letter or with a run of the second kind. Thus, when $u = 2k$ (where k is a positive integer), the probability of getting that many runs is

$$f(u) = \frac{2\binom{n_1 - 1}{k - 1}\binom{n_2 - 1}{k - 1}}{\binom{n_1 + n_2}{n_1}} \tag{12.5.1}$$

and it will be left to the reader to show in Exercise 1 on page 355 that similar arguments lead to

$$f(u) = \frac{\binom{n_1 - 1}{k}\binom{n_2 - 1}{k - 1} + \binom{n_1 - 1}{k - 1}\binom{n_2 - 1}{k}}{\binom{n_1 + n_2}{n_1}} \tag{12.5.2}$$

when $u = 2k + 1$ (where k is a positive integer).

When n_1 and n_2 are small, tests of the null hypothesis of randomness are based on specially constructed tables like those referred to on page 357. However, when n_1 and n_2 are both 10 or more, the sampling distribution of $\mathbf{u}$ can be approximated with a normal distribution, and to make use of this approximation let us state, without proof, that

$$E(\mathbf{u}) = \frac{2n_1n_2}{n_1 + n_2} + 1 \text{ and } \operatorname{var}(\mathbf{u}) = \frac{2n_1n_2(2n_1n_2 - n_1 - n_2)}{(n_1 + n_2)^2(n_1 + n_2 - 1)} \tag{12.5.3}$$

[These results can be obtained directly from (12.5.1) and 12.5.2), but a much easier way is given in the book by H. D. Brunk listed on page 357.]

Thus, for sufficiently large values of n_1 and n_2 we base our decision on the value which we obtain for

$$z = \frac{u - E(\mathbf{u})}{\sqrt{\operatorname{var}(\mathbf{u})}} \tag{12.5.4}$$

and reject the null hypothesis of randomness when $|z| \geq z_{\alpha/2}$. (Of course, if we wanted to test the null hypothesis of randomness against a *specific kind of non-randomness*, say, a trend or a repeated cyclical pattern, we would reject the null hypothesis when there are, respectively, too few runs or too many runs, namely, for $z \leq -z_\alpha$ or $z \geq z_\alpha$.)

Returning now to the numerical example on page 352, the one dealing with the defective and nondefective pieces turned out by the given machine, we find that $n_1 = 20$, $n_2 = 12$, $u = 9$, and hence that

$$E(\mathbf{u}) = \frac{2\cdot 20\cdot 12}{20 + 12} + 1 = 16$$

$$\operatorname{var}(\mathbf{u}) = \frac{2\cdot 20\cdot 12(2\cdot 20\cdot 12 - 20 - 12)}{(20 + 12)^2(20 + 12 - 1)} = 6.77$$

and

$$z = \frac{9 - 16}{\sqrt{6.77}} = -2.63$$

Since this value is less than $-z_{.005} = -2.58$, the null hypothesis of randomness can be rejected at the level of significance $\alpha = 0.01$. In fact, the total number of runs is much smaller than expected, and there is a strong indication that the defective pieces appear in clusters (or groups).

12.5.1 Runs Above and Below the Median

The method of the preceding section is not limited to tests of the randomness of series of attributes (such as the d's and n's of our numerical example). Any sample which consists of numerical measurements or observations can be treated similarly by using the letters a and b to denote, respectively, values falling above and below the median of the sample. (Numbers equaling the median are omitted.) The resulting series of a's and b's can then be tested for randomness on the basis of the total number of runs of a's and b's, namely the total number of *runs above and below the median.*

To illustrate this technique, let us consider the following data constituting a sample of the speeds (in miles per hour) at which every fifth passenger car was timed at a certain check-point in the given order: 46, 58, 60, 56, 70, 66, 48, 54, 62, 41, 39, 52, 45, 62, 53, 69, 65, 65, 67, 76, 52, 52, 59, 59, 67, 51, 46, 61, 40, 43, 42, 77, 67, 63, 59, 63, 63, 72, 57, 59, 42, 56, 47, 62, 67, 70, 63, 66, 69, and 73. The median of these speeds is 59.5, and the given data can, therefore, be represented by the following arrangement of a's and b's:

$$b\ b\ a\ b\ a\ a\ b\ b\ a\ b\ b\ b\ b\ a\ b\ a\ a\ a\ a\ a\ b\ b\ b\ b\ a$$
$$b\ b\ a\ b\ b\ b\ a\ a\ a\ b\ a\ a\ a\ b\ b\ b\ b\ b\ a\ a\ a\ a\ a\ a\ a$$

Since $n_1 = 25$, $n_2 = 25$, and $u = 20$, we get

$$E(\mathbf{u}) = \frac{2\cdot 25\cdot 25}{25 + 25} + 1 = 26$$

$$\text{var}(\mathbf{u}) = \frac{2\cdot 25\cdot 25(2\cdot 25\cdot 25 - 25 - 25)}{(25 + 25)^2(25 + 25 - 1)} = 12.2$$

and

$$z = \frac{20 - 26}{\sqrt{12.2}} = -1.72$$

which falls between $-z_{.025} = -1.96$ and $z_{.025} = 1.96$. Hence, the null hypothesis of randomness cannot be rejected at the level of significance $\alpha = 0.05$.

The method of runs above and below the median is especially useful in detecting trends and cyclical patterns in economic data. If there is a trend, there will be first mostly a's and later mostly b's (or vice versa), and if there is a repeated cyclical pattern there will be a systematic alternation of a's and b's and, probably, too many runs.

THEORETICAL EXERCISES

1. Verify (12.5.2) making use of the fact that when there are $2k + 1$ runs, k runs must be of one kind and $k + 1$ runs must be of the other kind.
2. If a person gets 7 heads and 3 tails in 10 flips of a coin, find the probabilities for 2, 3, 4, 5, 6, and 7 runs.

3. If $n_1 = 6$ and $n_2 = 5$ find the probability of getting at least 8 runs.
4. If $n_1 = n_2 = 8$ find the critical region (the values of $\mathbf{u}$) for rejecting the null hypothesis of randomness at the level of significance $\alpha = 0.01$.

APPLIED EXERCISES

5. Simulate *mentally* 100 flips of a coin by writing down a series of one hundred H's and T's (representing *heads* and *tails*). Test for randomness at the level of significance $\alpha = 0.01$ on the basis of the total number of runs.
6. Choose any four complete columns of random digits from Table IX (200 digits in all), represent each 0, 2, 4, 6, or 8 by the letter E, each 1, 3, 5, 7, or 9 by the letter O, and test for randomness at $\alpha = 0.01$ on the basis of the total number of runs.
7. The following is the order in which healthy, H, and diseased, D, pine trees were observed in a survey conducted by the Forestry Service:

   $H\ H\ H\ D\ H\ H\ H\ H\ H\ D\ D\ D\ H\ H\ H\ H\ H\ D\ D\ D\ D$

   (Cont.) $H\ H\ H\ H\ D\ D\ H\ H\ H\ H\ H\ H\ H\ H\ D\ D\ D\ D\ H\ H$

   Test for randomness at the level of significance $\alpha = 0.05$.
8. The theory of runs can also be used as an alternative to the median test or the Mann-Whitney test, namely, to test the null hypothesis that two samples come from identical populations. We simply rank the data jointly, write a 1 below each value belonging to the first sample, a 2 below each value belonging to the second sample, and then test the randomness of this arrangement of 1's and 2's. If there are too few runs, this may well be accounted for by the fact that the two samples come from populations with unequal means.
   (a) Apply this technique to the illustration (dealing with the burning times of the emergency flares) on page 347.
   (b) Apply this technique to the illustration (dealing with the grades obtained by the eighth graders from the two schools) on page 345. Where necessary, resolve ties by flipping a coin (or using random numbers.)
9. Use the method of runs above and below the median and $\alpha = 0.05$ to test whether the sample of Exercise 4 on page 350 may be regarded as a random sample.
10. The following are the number of defective pieces produced by a machine on fifty consecutive days: 7, 14, 17, 10, 18, 19, 23, 19, 14, 10,

12, 18, 19, 13, 24, 26, 9, 16, 19, 14, 19, 10, 15, 22, 25, 24, 20, 9, 17, 28, 29, 19, 25, 23, 24, 28, 31, 19, 24, 30, 27, 24, 39, 35, 23, 26, 28, 31, 37, and 40. Use the method of runs above and below the median and $\alpha = 0.05$ to test the null hypothesis of randomness against the alternative that there is a trend.

REFERENCES

The tables that are needed to perform the Mann-Whitney test and the run test of Section 12.5 for *small samples* may be found in

Owen, D. B., *Handbook of Statistical Tables.* Reading, Mass.: Addison-Wesley Publishing Co., Inc., 1962.

and in most books which deal specifically with nonparametric methods. The proof of (12.5.3) referred to on page 354 is given in

Brunk, H. D., *An Introduction to Mathematical Statistics,* 2nd ed. New York: Blaisdell Publishing Co., 1965.

13

REGRESSION AND CORRELATION

13.1 REGRESSION

A major objective of many statistical investigations is to establish relationships which make it possible to predict one or more variables in terms of others. Thus, studies are made to predict the potential sales of a new product in terms of its price, a patient's weight in terms of the number of weeks he or she has been on a diet, family expenditures on entertainment in terms of family income, the per capita consumption of certain foods in terms of their nutritional values and the amount of money spent advertising them on television, and so forth.

Although it is, of course, desirable to be able to predict one quantity *exactly* in terms of others, this is seldom possible, and in most instances we have to be satisfied with predicting averages or expected values. Thus, we may not be able to predict *exactly* how much money Mr. Brown will make ten years after graduating from college, but (given suitable data) we can predict the *average* income of a college graduate in terms of the number of years he has been out of college. Similarly, we can at best predict the *average* yield of a given variety of wheat in terms of data on the rainfall in July, and we can at best predict the *average* performance of students starting college in terms of their I. Q.'s.

Formally, if we are given the joint distribution of two random variables $\mathbf{x}$ and $\mathbf{y}$ and $\mathbf{x}$ is known to take on the value x, the basic problem of *bivariate regression* is that of determining the conditional expectation $E(\mathbf{y}|x)$, namely, the "average" value of $\mathbf{y}$ for the given value of $\mathbf{x}$. [The term "regression," as it is used here, dates back to Francis Galton, who employed it first in connection with a study of the heights of fathers and sons, in which he observed a regression (a "turning back") from the heights of sons to the heights of their fathers.] In problems involving more than two random variables, that is, in *multivariate regression*, we are correspondingly concerned with quantities such as $E(\mathbf{z}|x, y)$, the expected value of $\mathbf{z}$ for given values of $\mathbf{x}$ and $\mathbf{y}$, $E(\mathbf{x}_4|x_1, x_2, x_3)$, the expected value of $\mathbf{x}_4$ for given values of $\mathbf{x}_1$, $\mathbf{x}_2$, and $\mathbf{x}_3$, and so on.

If the joint density of two random variables **x** and **y** is given by $f(x, y)$, the problem of bivariate regression is simply that of determining the *conditional density* $\varphi(y|x)$ by means of (4.5.3) and then evaluating the integral

$$E(\mathbf{y}|x) = \int y \cdot \varphi(y|x)\, dy \tag{13.1.1}$$

Alternately, we might be interested in the conditional expectation

$$E(\mathbf{x}|y) = \int x \cdot \pi(x|y)\, dx \tag{13.1.2}$$

where the conditional density $\pi(x|y)$ is also determined in accordance with (4.5.3). [In the discrete case, where we are dealing with probability functions instead of probability densities, the integrals in (13.1.1) and (13.1.2) are simply replaced by sums.] If we are not given the joint density of the two random variables, or at least not all of its parameters, the determination of $E(\mathbf{y}|x)$ or $E(\mathbf{x}|y)$ becomes a problem of *estimation* based on sample data; this is an entirely different problem, which we shall discuss in Section 13.1.2 and in Section 13.2.2.

To illustrate the determination of $E(\mathbf{y}|x)$, the conditional expectation of **y** for a given value of **x**, let us consider the case where these random variables have the joint density

$$f(x, y) = \begin{cases} x \cdot e^{-x(1+y)} & \text{for } x > 0 \text{ and } y > 0 \\ 0 & \text{elsewhere} \end{cases}$$

Integrating out y we find that the marginal density of **x** is given by

$$g(x) = \begin{cases} e^{-x} & \text{for } x > 0 \\ 0 & \text{for } x \leq 0 \end{cases}$$

and, hence, the conditional density of **y** given x is given by

$$\varphi(y|x) = \frac{f(x, y)}{g(x)} = \frac{x \cdot e^{-x(1+y)}}{e^{-x}} = x \cdot e^{-xy} \tag{13.1.3}$$

for $y > 0$ and $\varphi(y|x) = 0$ for $y \leq 0$. Hence,

$$E(\mathbf{y}|x) = \int_0^\infty y \cdot x \cdot e^{-xy}\, dy = \frac{1}{x}$$

and we could have saved ourselves the trouble of evaluating this integral by observing that (13.1.3) is an exponential density (4.3.2) with $\theta = \frac{1}{x}$. Thus, according to Theorem 5.14

$$E(\mathbf{y}|x) = \theta = \frac{1}{x} \quad \text{and} \quad \text{var}(\mathbf{y}|x) = \theta^2 = \frac{1}{x^2}$$

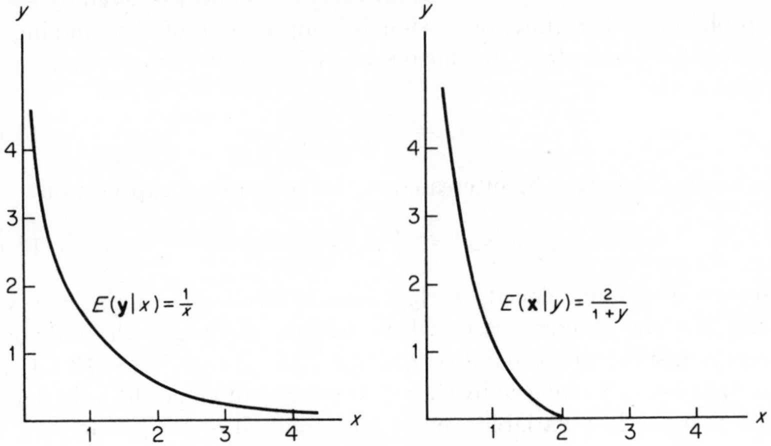

FIGURE 13.1 Regression curves.

where var($\mathbf{y}|x$) is the *conditional variance* of $\mathbf{y}$ for a given value of $\mathbf{x}$. The graph of the relationship between x and $E(\mathbf{y}|x)$, the *regression curve of* $\mathbf{y}$ *on* $\mathbf{x}$, is shown in the first diagram of Figure 13.1; the second diagram shows the *regression curve of* $\mathbf{x}$ *on* $\mathbf{y}$, which is given by

$$E(\mathbf{x}|y) = \frac{2}{1+y}$$

as the reader will be asked to verify in Exercise 1 on page 367.

To give an illustration dealing with the discrete case, let us find $E(\mathbf{y}|x)$ for a pair of random variables $\mathbf{x}$ and $\mathbf{y}$ having the *multinomial distribution*

$$f(x, y) = \frac{n!}{x!y!(n-x-y)!} \cdot \theta_1^x \theta_2^y (1-\theta_1-\theta_2)^{n-x-y}$$

for $x = 0, 1, 2, \ldots, n$, $y = 0, 1, 2, \ldots, n$, subject to the restriction that $x + y \leq n$. As we saw in Exercise 9 on page 98, the marginal distribution of $\mathbf{x}$ is a binomial distribution with the parameters n and θ_1, so that (3.4.4) yields

$$\varphi(y|x) = \frac{\binom{n-x}{y} \theta_2^y (1-\theta_1-\theta_2)^{n-x-y}}{(1-\theta_1)^{n-x}}$$

for $y = 0, 1, 2, \ldots, n-x$. Rewriting this formula as

$$\varphi(y|x) = \binom{n-x}{y} \left(\frac{\theta_2}{1-\theta_1}\right)^y \left(\frac{1-\theta_1-\theta_2}{1-\theta_1}\right)^{n-x-y}$$

we find by inspection that the conditional distribution of $\mathbf{y}$ for a given value of $\mathbf{x}$ is a binomial distribution with the parameters $n - x$ and $\frac{\theta_2}{1-\theta_1}$, so that

$$E(\mathbf{y}|x) = \frac{(n-x)\theta_2}{1-\theta_1}$$

according to Theorem 5.8. Thus, if $\mathbf{x}$ is the number of times an *even* number comes up in 30 rolls of a balanced die, while $\mathbf{y}$ is the number of times the result is a *five*, we have

$$E(\mathbf{y}|x) = \frac{(30-x)\frac{1}{6}}{1-\frac{1}{2}} = \frac{1}{3}(30-x)$$

This stands to reason, because for each of the $30 - x$ outcomes that are *not even*, there are the three equally likely possibilities 1, 3, or 5.

To illustrate the *multivariate case*, let us refer to the example on page 130, where three random variables $\mathbf{x}_1$, $\mathbf{x}_2$, and $\mathbf{x}_3$ had the joint density

$$f(x_1, x_2, x_3) = \begin{cases} (x_1 + x_2)e^{-x_3} & \text{for } 0 < x_1 < 1,\ 0 < x_2 < 1,\ x_3 > 0 \\ 0 & \text{elsewhere} \end{cases}$$

As was shown on page 132, the joint marginal density of $\mathbf{x}_1$ and $\mathbf{x}_3$ is given by

$$g(x_1, x_3) = \begin{cases} (x_1 + \frac{1}{2})e^{-x_3} & \text{for } 0 < x_1 < 1 \text{ and } x_3 > 0 \\ 0 & \text{elsewhere} \end{cases}$$

so that

$$E(\mathbf{x}_2|x_1, x_3) = \int x_2 \frac{f(x_1, x_2, x_3)}{g(x_1, x_3)}\,dx_2 = \int_0^1 \frac{x_2(x_1 + x_2)}{(x_1 + \frac{1}{2})}\,dx_2 = \frac{x_1 + \frac{2}{3}}{2x_1 + 1}$$

Note that this conditional expectation depends on x_1 but not on x_3, which should really have been expected since we indicated on page 135 that there is pairwise independence between $\mathbf{x}_2$ and $\mathbf{x}_3$.

13.1.1 Linear Regression

An important feature of the discrete example of Section 13.1 (which dealt with the multinomial distribution) is that the regression equation is *linear*, namely, that it is of the form

$$E(\mathbf{y}|x) = \alpha + \beta x \tag{13.1.4}$$

where α and β are constants, called the *regression coefficients*. There are several reasons why linear regression equations are of special interest:

First, they lend themselves readily to further mathematical treatment; then, they often provide good approximations to otherwise complicated regression equations; and, finally, in the case of the *bivariate normal distribution*, which we shall study in Section 13.2, the regression equations are, in fact, linear.

To simplify the study of linear regression equations, let us express the regression coefficients α and β of (13.1.4) in terms of some of the lower moments of the joint distribution of $\mathbf{x}$ and $\mathbf{y}$, namely, in terms of $E(\mathbf{x}) = \mu_1$, $E(\mathbf{y}) = \mu_2$, $\text{var}(\mathbf{x}) = \sigma_1^2$, $\text{var}(\mathbf{y}) = \sigma_2^2$, and $\text{cov}(\mathbf{x}, \mathbf{y}) = \sigma_{12}$. Substituting $\alpha + \beta x$ for $E(\mathbf{y}|x)$ in (13.1.1), we get

$$\int y \cdot \varphi(y|x)\, dy = \alpha + \beta x \tag{13.1.5}$$

and if we then multiply the expression on both sides of this equation by $g(x)$, the corresponding value of the marginal density of $\mathbf{x}$, and integrate on x, we obtain

$$\iint y \cdot \varphi(y|x) g(x)\, dy\, dx = \alpha \int g(x)\, dx + \beta \int x \cdot g(x)\, dx$$

or

$$\mu_2 = \alpha + \beta\mu_1 \tag{13.1.6}$$

since $\varphi(y|x)g(x) = f(x, y)$. Had we multiplied the expressions on both sides of (13.1.5) also by x before integrating on x, we would have obtained

$$\iint xy \cdot f(x, y)\, dy\, dx = \alpha \int x \cdot g(x)\, dx + \beta \int x^2 \cdot g(x)\, dx$$

or

$$E(\mathbf{xy}) = \alpha\mu_1 + \beta E(\mathbf{x}^2) \tag{13.1.7}$$

Solving (13.1.6) and (13.1.7) for α and β and making use of the fact that $E(\mathbf{xy}) = \sigma_{12} + \mu_1\mu_2$ and $E(\mathbf{x}^2) = \sigma_1^2 + \mu_1^2$, we find that

$$\alpha = \mu_2 - \frac{\sigma_{12}}{\sigma_1^2} \cdot \mu_1 \quad \text{and} \quad \beta = \frac{\sigma_{12}}{\sigma_1^2} \tag{13.1.8}$$

and, hence, that (13.1.4) can be written as

$$E(\mathbf{y}|x) = \mu_2 + \frac{\sigma_{12}}{\sigma_1^2}(x - \mu_1) \tag{13.1.9}$$

Similarly, when the regression equation of $\mathbf{x}$ on $\mathbf{y}$ is linear, it can be written as

$$E(\mathbf{x}|y) = \mu_1 + \frac{\sigma_{12}}{\sigma_2^2}(y - \mu_2) \tag{13.1.10}$$

A quantity which is closely related to the covariance is the *correlation coefficient*, ρ(*rho*); it is defined as

$$\rho = \frac{\sigma_{12}}{\sigma_1 \cdot \sigma_2} \tag{13.1.11}$$

and it enables us to write the linear regression equations (13.1.9) and (13.1.10) as

$$E(\mathbf{y}|x) = \mu_2 + \rho \frac{\sigma_2}{\sigma_1}(x - \mu_1)$$

and (13.1.12)

$$E(\mathbf{x}|y) = \mu_1 + \rho \frac{\sigma_1}{\sigma_2}(y - \mu_2)$$

It follows that if a regression equation is linear and $\rho = 0$, then $E(\mathbf{y}|x)$ does not depend on x [or $E(\mathbf{x}|y)$ does not depend on y.]

When $\rho = 0$ and, hence, $\sigma_{12} = 0$, the two random variables $\mathbf{x}$ and $\mathbf{y}$ are said to be *uncorrelated*, and we can now paraphrase the assertions which we made on page 179 by saying that *if two random variables are independent they are also uncorrelated, but if two random variables are uncorrelated they are not necessarily independent;* the latter is again illustrated in Exercise 9 on page 368.

The correlation coefficient and its estimates are of importance in many statistical investigations, and they will be discussed in some detail in Section 13.2.1. At this time, let us merely point out that $-1 \leq \rho \leq +1$, as the reader will be asked to prove in Exercise 11 on page 368, and that the sign of ρ tells us directly whether the slope of a regression line is upward or downward.

13.1.2 The Method of Least Squares

In the preceding sections we have discussed the problem of regression only in connection with random variables having known joint distributions. In actual practice, there are many problems where a set of *paired data* (that is, the paired values of two random variables $\mathbf{x}$ and $\mathbf{y}$) gives the indication that the regression is linear, where we do *not* know the joint distribution of the two random variables, but nevertheless want to *estimate* the regression coefficients α and β. Problems of this kind are usually handled with the use of the *method of least squares*, a method of curve fitting suggested early in the nineteenth century by the French mathematician Adrien Legendre.

To illustrate this technique, let us consider the following data on the weight losses of 10 persons and the number of months they have been on a special reducing diet:

Months on Diet x	*Weight Loss (pounds)* y
4	17
17	64
14	53
1	1
10	45
22	71
9	38
12	40
4	11
7	24

Plotting these data as in Figure 13.2, we get the impression that a straight line provides a reasonably good fit—the points do not all fall on a straight line, but the over-all pattern seems to indicate that there may very well be a linear regression.

Once we have decided that a straight line will give a good approximation to the regression of weight loss on the number of months a person has

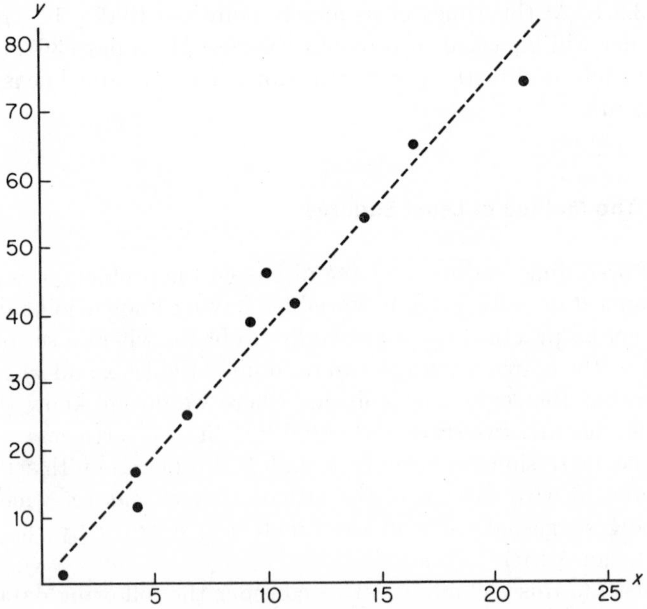

FIGURE 13.2 Data on length of diet and weight loss.

been on this diet, we face the problem of determining the equation of the line which in some sense provides the *best possible fit* to the given data; in other words, we face the problem of *estimating* the coefficients α and β of the regression line $y = \alpha + \beta x$. The *least squares criterion* on which we shall base this "goodness of fit" requires that we *minimize the sum of the squares of the vertical deviations from the points to the line,* as indicated in Figure 13.3. Thus, if we are given a set of paired data (x_1, y_1), (x_2, y_2),

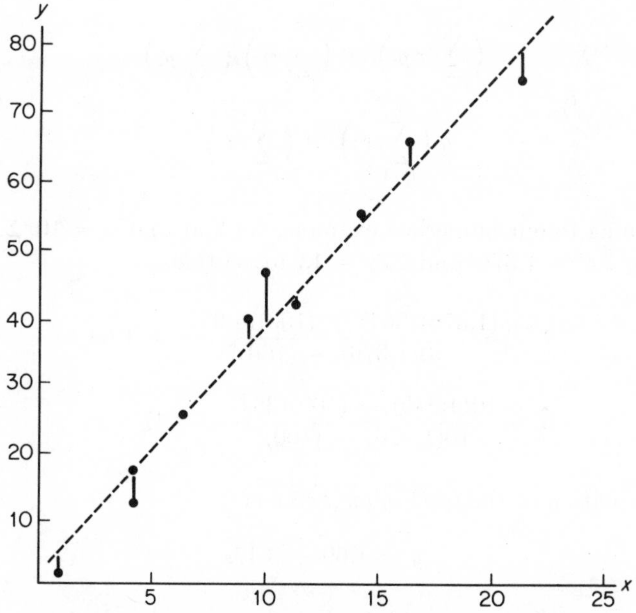

FIGURE 13.3 Data on length of diet and weight loss.

. . ., and (x_n, y_n), the *least squares estimates* of the regression coefficients are those values of α and β for which the quantity

$$\sum_{i=1}^{n} [y_i - (\alpha + \beta x_i)]^2 \tag{13.1.13}$$

is a minimum. Differentiating partially with respect to α and β, and equating these partial derivatives to zero, we obtain

$$\sum_{i=1}^{n} (-2)[y_i - (\alpha + \beta x_i)] = 0 \tag{13.1.14}$$

and

$$\sum_{i=1}^{n} (-2)x_i[y_i - (\alpha + \beta x_i)] = 0 \tag{13.1.15}$$

Then, solving this system of equations for α and β, we find that the *least squares estimates* of the regression coefficients are

$$\hat{\alpha} = \frac{\left(\sum_{i=1}^{n} x_i^2\right)\left(\sum_{i=1}^{n} y_i\right) - \left(\sum_{i=1}^{n} x_i\right)\left(\sum_{i=1}^{n} x_i y_i\right)}{n\left(\sum_{i=1}^{n} x_i^2\right) - \left(\sum_{i=1}^{n} x_i\right)^2} \tag{13.1.16}$$

and

$$\hat{\beta} = \frac{n\left(\sum_{i=1}^{n} x_i y_i\right) - \left(\sum_{i=1}^{n} x_i\right)\left(\sum_{i=1}^{n} y_i\right)}{n\left(\sum_{i=1}^{n} x_i^2\right) - \left(\sum_{i=1}^{n} x_i\right)^2} \tag{13.1.17}$$

Returning to our numerical example, we find that $n = 10$, $\Sigma x = 100$, $\Sigma y = 364$, $\Sigma x^2 = 1{,}376$, and $\Sigma xy = 4{,}945$, so that

$$\hat{\alpha} = \frac{(1{,}376)(364) - (100)(4{,}945)}{10(1{,}376) - (100)^2} = 1.69$$

$$\hat{\beta} = \frac{10(4{,}945) - (100)(364)}{10(1{,}376) - (100)^2} = 3.47$$

and the equation of the *least squares line* is

$$y = 1.69 + 3.47x$$

Using this equation, we could predict, for example, that a person who stays on this diet for 14 months will lose $1.69 + 3.47(14) = 50.27$ pounds, but *not having made any assumptions about the joint distribution of the random variables with which we are concerned, we cannot judge the "goodness" of this prediction, and we cannot judge the "goodness" of the estimates* $\hat{\alpha} = 1.69$ *and* $\hat{\beta} = 3.47$.

In actual practice we seldom use the formulas (13.1.16) and (13.1.17); instead we obtain the least squares estimates of the regression coefficients by solving the so-called system of *normal equations*

$$\begin{aligned} \sum_{i=1}^{n} y_i &= \alpha n + \beta \cdot \sum_{i=1}^{n} x_i \\ \sum_{i=1}^{n} x_i y_i &= \alpha \cdot \sum_{i=1}^{n} x_i + \beta \cdot \sum_{i=1}^{n} x_i^2 \end{aligned} \tag{13.1.18}$$

which, as the reader will be asked to verify in Exercise 14 on page 369, follow immediately from (13.1.14) and (13.1.15).

The least squares criterion, or in other words the *method of least squares*, is used in many problems of curve fitting which are more general than the one treated in this section. In Exercise 13 on page 369 the reader will be asked to use the method of least squares to estimate the coefficients β_0, β_1, and β_2 in the parabolic regression equation $y = \beta_0 + \beta_1 x + \beta_2 x^2$, and in Exercises 19 and 20 he will be asked to estimate the constants of regression equations representing exponential functions and power functions. Also, in Exercise 22 on page 372 the reader will be asked to use the method of least squares to estimate the coefficients of a *multiple regression equation* of the form $z = \alpha + \beta_1 x + \beta_2 y$.

THEORETICAL EXERCISES

1. With reference to the first example of Section 13.1 verify the value of $E(\mathbf{x}|y)$ given on page 360. (*Hint:* Identify the conditional density of $\mathbf{x}$ given y.)
2. Given the joint density

$$f(x, y) = \begin{cases} \frac{2}{5}(2x + 3y) & \text{for } 0 < x < 1 \text{ and } 0 < y < 1 \\ 0 & \text{elsewhere} \end{cases}$$

   find $E(\mathbf{y}|x)$ and $E(\mathbf{x}|y)$.
3. Given the joint density

$$f(x, y) = \begin{cases} 6x & \text{for } 0 < x < y < 1 \\ 0 & \text{elsewhere} \end{cases}$$

   find $E(\mathbf{y}|x)$, $\text{var}(\mathbf{y}|x)$, and $E(\mathbf{x}|y)$.
4. Given the joint density

$$f(x, y) = \begin{cases} \dfrac{2x}{(1 + x + xy)^3} & \text{for } x > 0 \text{ and } y > 0 \\ 0 & \text{elsewhere} \end{cases}$$

   show that $E(\mathbf{y}|x) = 1 + \dfrac{1}{x}$ and that $\text{var}(\mathbf{y}|x)$ does not exist.
5. Referring to Exercise 1 on page 96, use the results of parts (c) and (d) to find $E(\mathbf{x}|2)$ and $E(\mathbf{y}|1)$.

6. Referring to Exercise 2 on page 97, use the result of part (d) to find an expression for $E(\mathbf{y}|x)$.
7. Given the joint density

$$f(x, y) = \begin{cases} 2 & \text{for } 0 < y < x < 1 \\ 0 & \text{elsewhere} \end{cases}$$

show that

(a) $E(\mathbf{y}|x) = \frac{1}{2}x$ and $E(\mathbf{x}|y) = \frac{1}{2}(1 + y)$;

(b) $E(\mathbf{x}^m\mathbf{y}^n) = \dfrac{2}{(n + 1)(m + n + 2)}$;

and verify the results of part (a) by substituting the appropriate moments, obtained with the formula of part (b), into (13.1.9) and (13.1.10).
8. Given the joint density

$$f(x, y) = \begin{cases} 24xy & \text{for } x > 0,\ y > 0, \text{ and } x + y < 1 \\ 0 & \text{elsewhere} \end{cases}$$

show that $E(\mathbf{y}|x) = \frac{2}{3}(1 - x)$ and verify this result by determining the necessary moments and substituting them into (13.1.9).
9. Given the joint density

$$f(x, y) = \begin{cases} 1 & \text{for } -y < x < y \text{ and } 0 < y < 1 \\ 0 & \text{elsewhere} \end{cases}$$

show that the random variables $\mathbf{x}$ and $\mathbf{y}$ are *uncorrelated* but *not independent.*
10. Show that if $E(\mathbf{y}|x)$ is linear in x and $\text{var}(\mathbf{y}|x)$ is constant, then $\text{var}(\mathbf{y}|x) = \sigma_2^2(1 - \rho^2)$.
11. Given a pair of random variables $\mathbf{x}$ and $\mathbf{y}$ having the variances σ_1^2 and σ_2^2 and the correlation coefficient ρ, use Theorem 6.3 to express $\text{var}\left(\dfrac{\mathbf{x}}{\sigma_1} + \dfrac{\mathbf{y}}{\sigma_2}\right)$ and $\text{var}\left(\dfrac{\mathbf{x}}{\sigma_1} - \dfrac{\mathbf{y}}{\sigma_2}\right)$ in terms of σ_1, σ_2, and ρ. Then, making use of the fact that variances cannot be negative, show that $-1 \leq \rho \leq +1$.
12. Given the random variables $\mathbf{x}_1$, $\mathbf{x}_2$, and $\mathbf{x}_3$ having the joint density $f(x_1, x_2, x_3)$, show that if the regression of $\mathbf{x}_3$ on $\mathbf{x}_1$ and $\mathbf{x}_2$ is *linear* and written as

$$E(\mathbf{x}_3|x_1, x_2) = \alpha + \beta_1(x_1 - \mu_1) + \beta_2(x_2 - \mu_2)$$

then

$$\alpha = \mu_3$$

$$\beta_1 = \frac{\sigma_{13}\sigma_2^2 - \sigma_{12}\sigma_{23}}{\sigma_1^2\sigma_2^2 - \sigma_{12}^2}$$

$$\beta_2 = \frac{\sigma_{23}\sigma_1^2 - \sigma_{12}\sigma_{13}}{\sigma_1^2\sigma_2^2 - \sigma_{12}^2}$$

where $\mu_i = E(\mathbf{x}_i)$, $\sigma_i^2 = \text{var}(\mathbf{x}_i)$, and $\sigma_{ij} = \text{cov}(\mathbf{x}_i, \mathbf{x}_j)$. [*Hint:* Proceed as on page 362, multiplying by $(x_1 - \mu_1)$ and $(x_2 - \mu_2)$, respectively, to obtain the second and third equations.]

13. In some problems it is desired to fit a parabola of the form $y = \beta_0 + \beta_1 x + \beta_2 x^2$ to a set of paired data (x_1, y_1), (x_2, y_2), $\ldots$, (x_n, y_n). Minimizing

$$\sum_{i=1}^{n} [y_i - (\beta_0 + \beta_1 x_i + \beta_2 x_i^2)]^2$$

with respect to β_0, β_1, and β_2, derive a set of normal equations, whose solution will yield the least squares estimates of β_0, β_1, and β_2.

14. Verify that the system of equations (13.1.18) follows directly from (13.1.14) and (13.1.15)

15. In some problems it is desired to fit a *plane* having the equation $z = \alpha + \beta_1 x + \beta_2 y$ to a set of data consisting of the triplets (x_i, y_i, z_i) for $i = 1, 2, \ldots,$ and n. Minimizing

$$\sum_{i=1}^{n} [z_i - (\alpha + \beta_1 x_i + \beta_2 y_i)]^2$$

with respect to α, β_1, and β_2, derive a set of normal equations, whose solution will yield the least squares estimates of α, β_1, and β_2.

APPLIED EXERCISES

16. Rework the example on page 366 by calculating the regression coefficients with the use of the normal equations (13.1.18) instead of the formulas (13.1.16) and (13.1.17).

17. Various doses of a poison were given to groups of 25 mice and the following results were observed:

Dose in mg x	*Number of Deaths* y
4	1
6	3
8	6
10	8
12	14
14	16
16	20
18	21

Find the equation of the least squares line fit to these data
(a) by using formulas (13.1.16) and (13.1.17);
(b) by solving the system of normal equations (13.1.18).

18. The following are the grades which 12 students obtained in the mid-term and final examinations in a course in statistics:

Mid-Term Examination x	*Final Examination* y
71	83
49	62
80	76
73	77
93	89
85	74
58	48
82	78
64	76
32	51
87	73
80	89

Find the equation of the least squares line which will enable us to predict a student's final examination grade in this course on the basis of his mid-term grade. Also predict the final examination grade of a student who received an 81 on the mid-term examination.

19. If a set of paired data gives the indication that the regression equation is of the form $y = \alpha \cdot \beta^x$, it is customary to estimate α and β by fitting a line of the form $\log y = \log \alpha + x \cdot \log \beta$ to the points $(x_i, \log y_i)$ by the method of least squares. Use this technique to fit an *exponential curve* like this to the following data on the growth of cactus grafts under controlled environmental conditions:

Weeks after Grafting x	*Height (Inches)* y
1	2.0
2	2.4
4	5.1
5	7.3
6	9.4
8	18.3

20. If a set of paired data gives the indication that the regression equation is of the form $y = \alpha \cdot x^{\beta}$, it is customary to estimate α and β by fitting a line of the form $\log y = \log \alpha + \beta \cdot \log x$ to the points $(\log x_i, \log y_i)$ by the method of least squares. Use this technique to fit a *power function* like this to the following data on the unit cost of producing certain electronic components and the number of units produced:

Lot Size x	*Unit Cost* y
50	\$108
100	\$ 53
250	\$ 24
500	\$ 9
1,000	\$ 5

Also use the result to estimate the unit cost for a lot of 400 components.

21. Use the theory of Exercise 13 to fit a *parabola* of the form $y = \beta_0 + \beta_1 x + \beta_2 x^2$ to the following data pertaining to the drying time of a varnish and the amount of a certain chemical that is added:

Amount of Additive (Grams) x	*Drying Time (Hours)* y
1	8.5
2	8.0
3	6.0
4	5.0
5	6.0
6	5.5
7	6.5
8	7.0

Also estimate the drying time of the varnish when 3.5 grams of the chemical are added.

22. Use the theory of Exercise 15 to fit a *plane* of the form $z = \alpha + \beta_1 x + \beta_2 y$ to the following data pertaining to the grades which 10 students obtained in an examination, their I.Q.'s, and the number of hours they spent studying for the examination:

Grade z	*I.Q.* x	*Hours Studied* y
79	112	5
97	126	13
51	100	3
65	114	7
82	112	11
93	121	9
81	110	8
38	103	4
60	111	6
86	124	2

Also predict the grade of a student with an I.Q. of 108 who studies 5 hours for this examination.

13.2 THE BIVARIATE NORMAL DISTRIBUTION

If a random variable has a normal distribution, its density is of the form

$$f(x) = k \cdot e^{-(ax^2+bx+c)} \qquad \text{for } -\infty < x < \infty$$

where a, b, c, and k are constants, $k > 0$, and the second degree polynomial $ax^2 + bx + c$ is positive or zero for all values of the random variable $\mathbf{x}$. Ordinarily, the constants a, b, c, and k are expressed in terms of the moments of the distribution; that is, $a = \frac{1}{2\sigma^2}$, $b = -\frac{\mu}{\sigma^2}$, $c = \frac{\mu^2}{2\sigma^2}$, and $k = \frac{1}{\sigma\sqrt{2\pi}}$, as can easily be verified by comparison with (5.3.5).

A logical way to extent the normal distribution to the *bivariate case*, namely, to the joint distribution of two random variables, would be to consider the joint density

$$f(x, y) = k \cdot e^{-(ax^2+by^2+cxy+dx+ey+f)}$$

for $-\infty < x < \infty$ and $-\infty < y < \infty$, where a, b, c, d, e, f, and k are constants, $k > 0$, and the second degree polynomial $ax^2 + by^2 + cxy + dx + ey + f$ is positive or zero for all values of the random variables $\mathbf{x}$ and $\mathbf{y}$.

Although we could proceed as in Section 5.3.6 and express the various constants in terms of moments, we shall instead state the result and then verify later that the symbols μ_1, μ_2, σ_1^2, σ_2^2, and ρ appearing in (13.2.1) are, indeed, the two means, the two variances, and the correlation coefficient of $\mathbf{x}$ and $\mathbf{y}$. Thus, let us define the *bivariate normal density* as

$$f(x, y) = \frac{e^{-\frac{1}{2(1-\rho^2)}\left[\left(\frac{x-\mu_1}{\sigma_1}\right)^2 - 2\rho\left(\frac{x-\mu_1}{\sigma_1}\right)\left(\frac{y-\mu_2}{\sigma_2}\right) + \left(\frac{y-\mu_2}{\sigma_2}\right)^2\right]}}{2\pi\sigma_1\sigma_2\sqrt{1-\rho^2}} \tag{13.2.1}$$

for $-\infty < x < \infty$, $-\infty < y < \infty$, $\sigma_1 > 0$, $\sigma_2 > 0$, and $-1 < \rho < +1$.

To show that the symbols μ_1 and σ_1^2 in (13.2.1) actually represent $E(\mathbf{x})$ and $\text{var}(\mathbf{x})$, let us first find the marginal density of $\mathbf{x}$ by integrating out y. We thus get

$$g(x) = \frac{e^{-\frac{1}{2(1-\rho^2)}\left(\frac{x-\mu_1}{\sigma_1}\right)^2}}{2\pi\sigma_1\sigma_2\sqrt{1-\rho^2}} \int_{-\infty}^{\infty} e^{-\frac{1}{2(1-\rho^2)}\left[\left(\frac{y-\mu_2}{\sigma_2}\right)^2 - 2\rho\left(\frac{x-\mu_1}{\sigma_1}\right)\left(\frac{y-\mu_2}{\sigma_2}\right)\right]}\, dy$$

and to simplify the notation let us make the substitution $u = \dfrac{x-\mu_1}{\sigma_1}$. Also changing the variable of integration by letting $v = \dfrac{y-\mu_2}{\sigma_2}$, we obtain

$$g(x) = \frac{e^{-\frac{1}{2(1-\rho^2)}u^2}}{2\pi\sigma_1\sqrt{1-\rho^2}} \int_{-\infty}^{\infty} e^{-\frac{1}{2(1-\rho^2)}(v^2 - 2\rho uv)}\, dv$$

and after completing the square by letting $v^2 - 2\rho uv = (v - \rho u)^2 - \rho^2 u^2$ and collecting terms, we get

$$g(x) = \frac{e^{-\frac{1}{2}u^2}}{\sigma_1\sqrt{2\pi}} \left\{\frac{1}{\sqrt{2\pi}\sqrt{1-\rho^2}} \int_{-\infty}^{\infty} e^{-\frac{1}{2}\left(\frac{v-\rho u}{\sqrt{1-\rho^2}}\right)^2}\, dv\right\}$$

Finally, identifying the quantity in parentheses as the integral of a normal density from $-\infty$ to ∞ and, hence, equaling 1, we obtain

$$g(x) = \frac{e^{-\frac{1}{2}u^2}}{\sigma_1\sqrt{2\pi}} = \frac{1}{\sigma_1\sqrt{2\pi}}\, e^{-\frac{1}{2}\left(\frac{x-\mu_1}{\sigma_1}\right)^2} \tag{13.2.2}$$

for $-\infty < x < \infty$. It follows by inspection that the marginal density of $\mathbf{x}$ is normal with the mean μ_1 and the standard deviation σ_1, which justifies the use of these two symbols in (13.2.1). By symmetry, it also follows that the marginal density of **y** is normal with the mean μ_2 and the standard deviation σ_2, so that the only thing that remains to be shown is that the symbol ρ in (13.2.1) is, indeed, the correlation coefficient as defined in (13.1.11). To this end, let us first find the conditional density $\varphi(y|x)$ by substituting into $\varphi(y|x) = \dfrac{f(x, y)}{g(x)}$ the expressions for $f(x, y)$ and $g(x)$ given in (13.2.1) and 13.2.2). Letting $u_1 = \dfrac{x - \mu_1}{\sigma_1}$ and $u_2 = \dfrac{y - \mu_2}{\sigma_2}$, we thus get

$$\varphi(y|x) = \frac{\dfrac{1}{2\pi\sigma_1\sigma_2\sqrt{1-\rho^2}}\, e^{-\frac{1}{2(1-\rho^2)}[u_1^2 - 2\rho u_1 u_2 + u_2^2]}}{\dfrac{1}{\sqrt{2\pi}\sigma_1}\, e^{-\frac{1}{2}u_1^2}}$$

$$= \frac{1}{\sqrt{2\pi}\sigma_2\sqrt{1-\rho^2}}\, e^{-\frac{1}{2(1-\rho^2)}[u_2^2 - 2\rho u_1 u_2 + \rho^2 u_1^2]}$$

$$= \frac{1}{\sqrt{2\pi}\sigma_2\sqrt{1-\rho^2}}\, e^{-\frac{1}{2}\left[\frac{u_2 - \rho u_1}{\sqrt{1-\rho^2}}\right]^2}$$

Finally, expressing this result in terms of the original variables, we obtain

$$\varphi(y|x) = \frac{1}{\sigma_2\sqrt{2\pi}\sqrt{1-\rho^2}}\, e^{-\frac{1}{2}\left[\frac{y - \left\{\mu_2 + \rho\frac{\sigma_2}{\sigma_1}(x - \mu_1)\right\}}{\sigma_2\sqrt{1-\rho^2}}\right]^2} \tag{13.2.3}$$

for $-\infty < y < \infty$, and it can be seen by inspection that this is a normal distribution with the mean

$$E(\mathbf{y}|x) = \mu_2 + \rho\frac{\sigma_2}{\sigma_1}(x - \mu_1) \tag{13.2.4}$$

and the variance

$$\text{var}(\mathbf{y}|x) = \sigma_2^2(1 - \rho^2) \tag{13.2.5}$$

We have thus shown that the regression of **y** on **x** is *linear*, and by comparing (13.2.4) with (13.1.12) we find that the constant ρ in (13.2.1) is, indeed, the correlation coefficient. By symmetry, the regression of **x** on **y** is also linear, with

$$E(\mathbf{x}|y) = \mu_1 + \rho\frac{\sigma_1}{\sigma_2}(y - \mu_2) \tag{13.2.6}$$

and $\text{var}(\mathbf{x}|y) = \sigma_1^2(1 - \rho^2)$.

The bivariate normal distribution has many interesting properties, some of which are statistical while others are of a purely mathematical nature. Among the statistical properties, we find, for example, that *two random variables having the bivariate normal distribution are independent if and only if they are uncorrelated* (see Exercise 1 on page 376). Also, we have shown that for two random variables having the bivariate normal distribution the marginal distributions are normal, but the converse is not necessarily true. To illustrate, let us consider the bivariate density

$$f^*(x, y) = \begin{cases} 2f(x, y) & \text{inside square 2 and 4 of Figure 13.4} \\ 0 & \text{inside squares 1 and 3 of Figure 13.4} \\ f(x, y) & \text{elsewhere} \end{cases}$$

where $f(x, y)$ is the bivariate normal density with zero means and $\rho = 0$. It is easy to see that even though $f^*(x, y)$ is *not* a bivariate normal density, the corresponding marginal densities of **x** and **y** are normal.

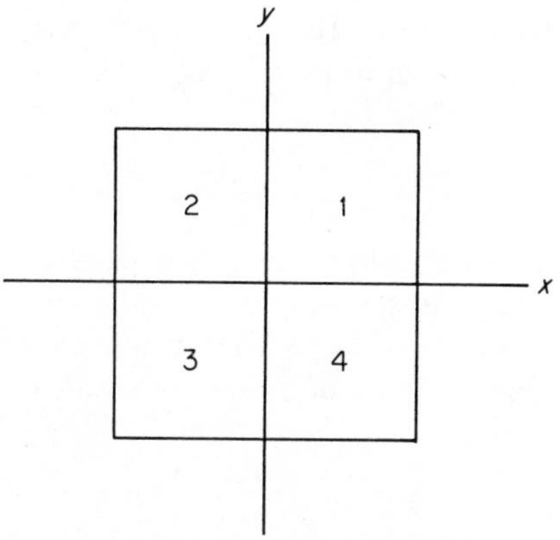

FIGURE 13.4 Sample space for the bivariate density $f^*(x, y)$.

Many interesting mathematical properties of (13.2.1) are obtained by studying the *bivariate normal surface* whose equation is $z = f(x, y)$, where $f(x, y)$ is given by (13.2.1). This surface, illustrated in Figure 13.5, is also referred to as a *normal regression surface*, and it will be left to the reader to verify that this surface has a maximum at $x = \mu_1$ and $y = \mu_2$, that any plane parallel to the z-axis intersects the surface in a curve having the shape of a normal distribution, and that planes parallel to the xy-plane intersect the surface in ellipses, which are called *contours of constant probability density*. The axes of these ellipses do not coincide with the regression lines except in the special case where $\rho = 0$. When $\rho = 0$ and $\sigma_1 = \sigma_2$, the contours of constant probability density are *circles*, and it is customary to refer to the corresponding joint density as the *circular normal distribution*.

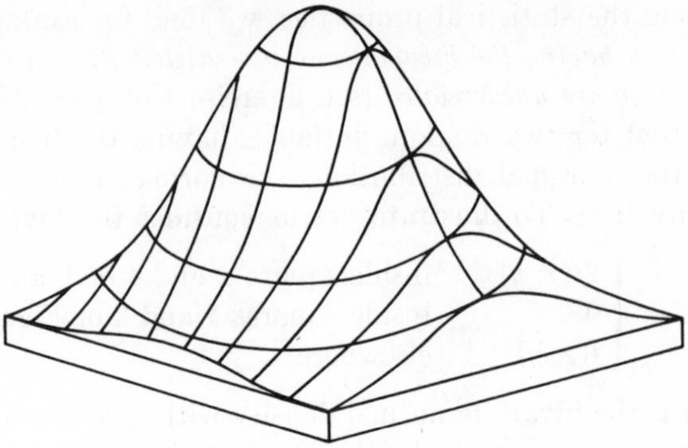

FIGURE 13.5 Normal regression surface.

The study of the elliptical contours of constant probability density is of special interest when $\rho \to 1$. In that case, the major axes of the ellipses as well as the two regression lines *approach* the line

$$y = \mu_2 + \frac{\sigma_2}{\sigma_1}(x - \mu_1)$$

the major axes get longer while the minor axes get shorter, and in the limiting case $\rho = 1$ the ellipses *degenerate* to the above line. From a probabilistic point of view this means that when $\rho \to 1$ the probability of obtaining points (that is, pairs of values of $\mathbf{x}$ and $\mathbf{y}$) outside any narrow strip containing the line

$$y = \mu_2 + \frac{\sigma_2}{\sigma_1}(x - \mu_1)$$

approaches 0. In fact, in the limiting case where $\rho = 1$, the entire probability is concentrated along this line and the bivariate distribution degenerates to the *univariate* (one variable) case. This argument applies also when $\rho \to -1$, and this is why $\rho = -1$ and $\rho = +1$ are excluded in the definition of the bivariate normal distribution.

THEORETICAL EXERCISES

1. Show that if two random variables have the bivariate normal distribution they are independent if and only if they are uncorrelated.
2. Show that any plane perpendicular to the xy-plane intersects the normal regression surface in a curve having the shape of a normal distribution.

3. If the exponent of e in (13.2.1) is

$$\frac{-1}{102}[(x+2)^2 - 2.8(x+2)(y-1) + 4(y-1)^2]$$

find μ_1, μ_2, σ_1, σ_2, and ρ.

4. With reference to Exercise 3, what is the conditional density of $\mathbf{y}$ given x?

5. If the exponent of e in (13.2.1) is

$$-\tfrac{1}{54}(x^2 + 4y^2 + 2xy + 2x + 8y + 4)$$

find σ_1, σ_2, and ρ, given that $\mu_1 = 0$ and $\mu_2 = -1$.

6. If $\mathbf{x}$ and $\mathbf{y}$ have the bivariate normal distribution (13.2.1), $\mathbf{u} = \mathbf{x} + \mathbf{y}$ and $\mathbf{v} = \mathbf{x} - \mathbf{y}$, find an expression for the correlation coefficient of $\mathbf{u}$ and $\mathbf{v}$.

7. It can be shown that the *joint moment generating function* of two random variables $\mathbf{x}$ and $\mathbf{y}$ having the bivariate normal density (13.2.1) is given by

$$\begin{aligned} M_{\mathbf{x},\mathbf{y}}(t_1, t_2) &= E(e^{t_1\mathbf{x}+t_2\mathbf{y}}) \\ &= e^{t_1\mu_1+t_2\mu_2+\frac{1}{2}(\sigma_1{}^2t_1{}^2+2\rho\sigma_1\sigma_2t_1t_2+\sigma_2{}^2t_2{}^2)} \end{aligned}$$

Verify that

(a) the first partial derivative of this function with respect to t_1 at $t_1 = 0$ and $t_2 = 0$ is μ_1;

(b) the second partial derivative with respect to t_1 at $t_1 = 0$ and $t_2 = 0$ is $\sigma_1^2 + \mu_1^2$;

(c) the second partial derivative with respect to t_1 and t_2 at $t_1 = 0$ and $t_2 = 0$ is $\rho\sigma_1\sigma_2 + \mu_1\mu_2$.

8. Show that if $\mathbf{x}$ and $\mathbf{y}$ have the bivariate normal distribution (13.2.1), then $\mathbf{u} = a\mathbf{x} + b\mathbf{y}$, where a and b are real constants, is normally distributed. (*Hint:* Find the moment generating function of $\mathbf{u}$ making use of the joint moment generating function of $\mathbf{x}$ and $\mathbf{y}$ given in Exercise 7.)

9. Given a pair of random variables $\mathbf{x}$ and $\mathbf{y}$ having the bivariate normal distribution, find

(a) the joint density of $\mathbf{x}$ and $\mathbf{v}_2$ where

$$\mathbf{v}_2 = \frac{\mathbf{y} - \left\{\mu_2 + \rho\dfrac{\sigma_2}{\sigma_1}(\mathbf{x} - \mu_1)\right\}}{\sigma_2\sqrt{1-\rho^2}}$$

(b) the joint density of $\mathbf{v}_1$ and $\mathbf{v}_2$ where $\mathbf{v}_2$ is as defined in part (a) and

$$\mathbf{v}_1 = \frac{\mathbf{x} - \mu_1}{\sigma_1}$$

[*Hint:* In part (a) write the joint density of $\mathbf{x}$ and $\mathbf{y}$ as the product of $g(x)$ and $\varphi(y|x)$ as given by (13.2.2) and (13.2.3).]

10. Given a pair of random variables having the bivariate normal distribution (13.2.1), show that the probability of obtaining a point (x, y) inside the ellipse

$$\frac{1}{2(1-\rho^2)}\left[\left(\frac{x-\mu_1}{\sigma_1}\right)^2 - 2\rho\left(\frac{x-\mu_1}{\sigma_1}\right)\left(\frac{y-\mu_2}{\sigma_2}\right) + \left(\frac{y-\mu_2}{\sigma_2}\right)^2\right] = k^2$$

is $1 - e^{-k^2}$. (*Hint:* Use the result obtained in Exercise 9 and change to polar coordinates.)

APPLIED EXERCISES

11. The center of a target at which a missile is aimed is taken as the origin of a rectangular system of coordinates, with reference to which the point of impact has the coordinates $\mathbf{x}$ and $\mathbf{y}$. Given that $\mathbf{x}$ and $\mathbf{y}$ have a bivariate normal density with $\mu_1 = 0$, $\mu_2 = 0$, $\sigma_1 = 100$ feet, $\sigma_2 = 100$ feet, and $\rho = 0$, find the probabilities that the point of impact will be

(a) inside a square with sides of 150 feet, whose center is at the origin and whose sides are parallel to the coordinate axes;

(b) inside a circle with a radius of 60 feet whose center is at the origin.

12. If $\mathbf{x}$ and $\mathbf{y}$ have the circular normal distribution with $\sigma_1 = \sigma_2 = 10$, find

(a) the probability of getting a point (x, y) inside the circle $x^2 + y^2 = 25$;

(b) the value of c for which the probability of getting a point (x, y) inside the circle $x^2 + y^2 = c^2$ is 0.90.

13. Suppose that the height $\mathbf{x}$ and the weight $\mathbf{y}$ of certain animals have the bivariate normal distribution with $\mu_1 = 18$ inches, $\mu_2 = 15$ pounds, $\sigma_1 = 3$ inches, $\sigma_2 = 2$ pounds, and $\rho = \frac{3}{4}$. Find

(a) the expected weight of one of these animals which is 16 inches tall;

(b) the probability that the weight of one of these animals which is 16 inches tall is between 12 and 16 pounds.

13.2.1 Normal Correlation Analysis

When analyzing sets of paired data (x_1, y_1), (x_2, y_2), . . ., and (x_n, y_n), it is important to make the following distinction: *If x_i and y_i (for $i = 1, 2, \ldots, n$) are values of corresponding random variables $\mathbf{x}_i$ and $\mathbf{y}_i$ having the joint* *density* $f(x_i, y_i)$*, the analysis of the data is called correlation analysis; if, on the other hand, the* x_i *are constants while the* y_i *are values of corresponding random variables* $\mathbf{y}_i$ *having the conditional densities* $\varphi(y_i|x_i)$*, the analysis of the data is called regression analysis.* Thus, if we want to analyze data on the height and weight of certain animals, and height and weight are both looked upon as random variables, this is a problem of *correlation analysis.* However, if we want to analyze data on the ages and prices of used cars, treating the ages as known constants and the prices as values of random variables, this is a problem of *regression analysis.* Leaving regression analysis to Section 13.2.2, we shall devote this section to some of the basic problems of *normal correlation analysis*, that is, correlation analysis where the joint density $f(x_i, y_i)$ of each pair of random variables $\mathbf{x}_i$ and $\mathbf{y}_i$ is the same bivariate normal density (13.2.1).

Thus, let us consider a set of paired data (x_1, y_1), (x_2, y_2), . . ., and (x_n, y_n), which constitute a random sample from a bivariate normal population, and indicate how to *estimate* its parameters μ_1, μ_2, σ_1, σ_2, and ρ. Using the method of maximum likelihood, we shall have to maximize the likelihood

$$L = \prod_{i=1}^{n} f(x_i, y_i) \tag{13.2.7}$$

where $f(x_i, y_i)$ is given by (13.2.1), and to this end we shall have to differentiate L, or $\ln L$, partially with respect to μ_1, μ_2, σ_1, σ_2, and ρ, equate the resulting expressions to zero, and then solve the resulting system of equations for the five parameters. Leaving the details to the reader, let us merely state the result that when $\dfrac{\partial \ln L}{\partial \mu_1}$ and $\dfrac{\partial \ln L}{\partial \mu_2}$ are equated to zero, we get

$$-\frac{\sum_{i=1}^{n}(x_i - \mu_1)}{\sigma_1^2} + \frac{\rho \sum_{i=1}^{n}(y_i - \mu_2)}{\sigma_1\sigma_2} = 0 \tag{13.2.8}$$

and

$$-\frac{\rho \sum_{i=1}^{n}(x_i - \mu_1)}{\sigma_1\sigma_2} + \frac{\sum_{i=1}^{n}(y_i - \mu_2)}{\sigma_2^2} = 0 \tag{13.2.9}$$

Solving these two equations for μ_1 and μ_2, we find that the maximum likelihood estimates of these two parameters are $\hat{\mu}_1 = \bar{x}$ and $\hat{\mu}_2 = \bar{y}$, namely, the respective sample means. Subsequently, equating $\dfrac{\partial \ln L}{\partial \sigma_1}$,

$\dfrac{\partial \ln L}{\partial \sigma_2}$, and $\dfrac{\partial \ln L}{\partial \rho}$ to zero and substituting $\bar{x}$ and $\bar{y}$ for μ_1 and μ_2, we obtain a system of equations whose solution is

$$\hat{\sigma}_1 = \sqrt{\frac{\sum_{i=1}^{n} (x_i - \bar{x})^2}{n}}, \qquad \hat{\sigma}_2 = \sqrt{\frac{\sum_{i=1}^{n} (y_i - \bar{y})^2}{n}} \tag{13.2.10}$$

$$\hat{\rho} = \frac{\sum_{i=1}^{n} (x_i - \bar{x})(y_i - \bar{y})}{\sqrt{\sum_{i=1}^{n} (x_i - \bar{x})^2}\sqrt{\sum_{i=1}^{n} (y_i - \bar{y})^2}} \tag{13.2.11}$$

(A detailed derivation of these maximum likelihood estimates is referred to at the end of this chapter.) It is of interest to note that the maximum likelihood estimates of σ_1 and σ_2 are identical with the one obtained on page 269 for the standard deviation of the univariate normal distribution; they differ from the respective sample standard deviations s_1 and s_2 only by the factor $\sqrt{\dfrac{n-1}{n}}$.

The estimate $\hat{\rho}$, called the *sample correlation coefficient*, is usually denoted by the letter r and its calculation is facilitated by use of the formula

$$r = \frac{n \cdot \sum_{i=1}^{n} x_i y_i - \left(\sum_{i=1}^{n} x_i\right)\left(\sum_{i=1}^{n} y_i\right)}{\sqrt{n \cdot \sum_{i=1}^{n} x_i^2 - \left(\sum_{i=1}^{n} x_i\right)^2}\sqrt{n \cdot \sum_{i=1}^{n} y_i^2 - \left(\sum_{i=1}^{n} y_i\right)^2}} \tag{13.2.12}$$

It will be left to the reader in Exercise 1 on page 387 to verify that (13.2.11) and (13.2.12) are, indeed, equivalent.

There are many problems in which the estimation of ρ and tests concerning ρ are of special interest, for ρ measures the strength of the *linear relationship* between $\mathbf{x}$ and $\mathbf{y}$. When $\rho = 0$, the two random variables are uncorrelated and, as we have already seen, in the case of the bivariate normal distribution this means that they are also independent. When $\rho \to 1$, the probability of obtaining a pair of values of $\mathbf{x}$ and $\mathbf{y}$ yielding a point outside any narrow strip containing the line $y = \mu_2 + \dfrac{\sigma_2}{\sigma_1}(x - \mu_1)$ approaches 0, and this means that there is a very strong linear relationship between $\mathbf{x}$ and $\mathbf{y}$. [If a set of data points (x_i, y_i) actually do fall on a straight line, then r will equal $+1$ or -1, depending on whether this line has an upward or a downward slope. The significance of intermediate values of r or ρ, say, -0.60 or $+0.50$, will be discussed on page 384.

Since the sampling distribution of **r** for random samples from bivariate normal populations is rather complicated, it is common practice to base tests concerning ρ on the statistic $\frac{1}{2}\cdot\ln\frac{1+\mathbf{r}}{1-\mathbf{r}}$, whose distribution is approximately normal with the mean $\frac{1}{2}\cdot\ln\frac{1+\rho}{1-\rho}$ and the variance $\frac{1}{n-3}$. Thus,

$$z = \frac{\frac{1}{2}\cdot\ln\frac{1+r}{1-r} - \frac{1}{2}\cdot\ln\frac{1+\rho}{1-\rho}}{1/\sqrt{n-3}}$$

$$= \frac{\sqrt{n-3}}{2}\cdot\ln\frac{(1+r)(1-\rho)}{(1-r)(1+\rho)} \tag{13.2.13}$$

can be looked upon as a value of a random variable having approximately the standard normal distribution. Using this approximation, we can test hypotheses concerning ρ as is illustrated below, or calculate confidence intervals for ρ as is illustrated in Exercise 3 on page 387.

To give an example of a test concerning ρ, suppose we want to determine on the basis of the following data whether there is a relationship between the time (in minutes) it takes a secretary to complete a certain form in the morning and in the late afternoon:

Morning x	*Afternoon* y
8.2	8.7
9.6	9.6
7.0	6.9
9.4	8.5
10.9	11.3
7.1	7.6
9.0	9.2
6.6	6.3
8.4	8.4
10.5	12.3

Since $n = 10$, $\Sigma\, x = 86.7$, $\Sigma\, x^2 = 771.35$, $\Sigma\, y = 88.8$, $\Sigma\, y^2 = 819.34$, and $\Sigma\, xy = 792.92$, substitution into (13.2.12) yields $r = 0.936$, which seems to indicate a *positive association* between the time it takes a secretary to perform the given task in the morning and in the late afternoon. This association is also apparent from the *scattergram* of Figure 13.6.

To justify any definite statement to the effect that there *is* a relation-

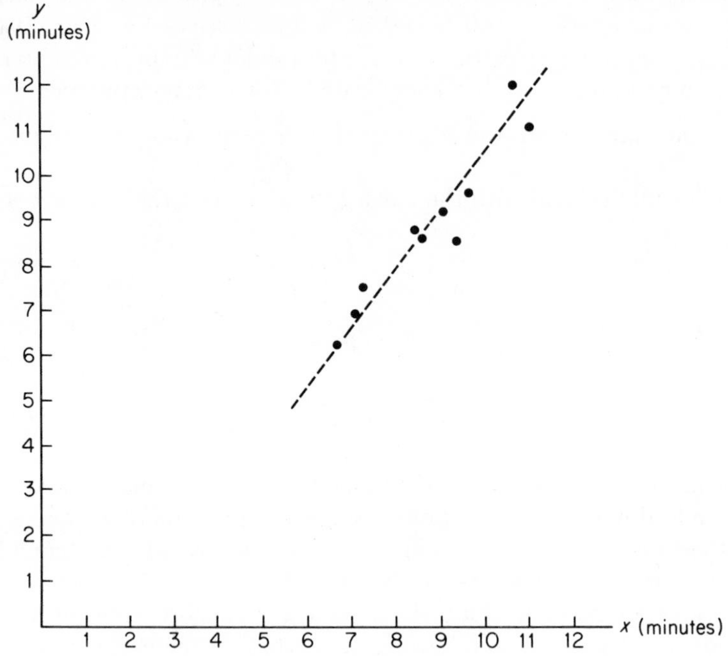

FIGURE 13.6 Data on time it takes to complete form.

ship, we shall have to test the null hypothesis $\rho = 0$ against the alternative hypothesis $\rho \neq 0$, say, at the level of significance $\alpha = 0.01$. Substituting $r = 0.936$, $n = 10$, and $\rho = 0$ into (13.2.13), we get

$$z = \frac{\sqrt{7}}{2} \cdot \ln \frac{1.936}{0.064} = 4.5$$

and since this exceeds $z_{.005} = 2.58$, the null hypothesis of no correlation must be rejected. We conclude that there *is* a relationship between the time it takes a secretary to complete the given form in the morning and in the late afternoon. Of course, this test is approximate and we also had to assume that the given data can be looked upon as a random sample from a bivariate normal population.

13.2.2 Normal Regression Analysis

In *normal regression analysis* we consider paired data (x_1, y_1), (x_2, y_2), ..., and (x_n, y_n), where the x's are constants and the y's are values of corresponding *independent* random variables having the conditional densities

$$\varphi(y_i|x_i) = \frac{1}{\sigma\sqrt{2\pi}} \cdot e^{-\frac{1}{2}\left(\frac{y_i-(\alpha+\beta x_i)}{\sigma}\right)^2} \tag{13.2.14}$$

for $-\infty < y_i < \infty$, where α, β, and σ are the same for each i. Given such a random sample of paired data, normal regression analysis concerns itself mainly with the estimation of the regression coefficients α and β, with tests of hypotheses concerning the three parameters in (13.1.14), and with predictions based on the estimated regression equation $y = \hat{\alpha} + \hat{\beta}x$, where $\hat{\alpha}$ and $\hat{\beta}$ are estimates of α and β.

To obtain maximum likelihood estimates of the parameters α, β, and σ in (13.1.14), we partially differentiate the likelihood function (or its logarithm, which is easier) with respect to α, β, and σ, equate the expressions to zero, and then solve the resulting system of equations. Thus, differentiating

$$\ln L = -n \cdot \ln \sigma - \frac{n}{2} \cdot \ln 2\pi - \frac{1}{2\sigma^2} \cdot \sum_{i=1}^{n} [y_i - (\alpha + \beta x_i)]^2$$

partially with respect to α, β, and σ, and equating the expressions which we obtain to zero, we get

$$\frac{\partial \ln L}{\partial \alpha} = \frac{1}{\sigma^2} \cdot \sum_{i=1}^{n} [y_i - (\alpha + \beta x_i)] = 0 \tag{13.2.15}$$

$$\frac{\partial \ln L}{\partial \beta} = \frac{1}{\sigma^2} \cdot \sum_{i=1}^{n} x_i[y_i - (\alpha + \beta x_i)] = 0 \tag{13.2.16}$$

$$\frac{\partial \ln L}{\partial \sigma} = -\frac{n}{\sigma} + \frac{1}{\sigma^3} \cdot \sum_{i=1}^{n} [y_i - (\alpha + \beta x_i)]^2 = 0 \tag{13.2.17}$$

As the reader will be asked to verify in Exercise 6 on page 388, the first two of these equations yield the maximum likelihood estimates

$$\hat{\alpha} = \frac{\left(\sum_{i=1}^{n} x_i^2\right)\left(\sum_{i=1}^{n} y_i\right) - \left(\sum_{i=1}^{n} x_i\right)\left(\sum_{i=1}^{n} x_i y_i\right)}{n\left(\sum_{i=1}^{n} x_i^2\right) - \left(\sum_{i=1}^{n} x_i\right)^2} \tag{13.2.18}$$

$$\hat{\beta} = \frac{n\left(\sum_{i=1}^{n} x_i y_i\right) - \left(\sum_{i=1}^{n} x_i\right)\left(\sum_{i=1}^{n} y_i\right)}{n\left(\sum_{i=1}^{n} x_i^2\right) - \left(\sum_{i=1}^{n} x_i\right)^2} \tag{13.2.19}$$

which are identical with (13.1.16) and (13.1.17). To simplify the calculating of these estimated regression coefficients, it is common practice to determine $\hat{\beta}$ first with the use of (13.2.19), and then $\hat{\alpha}$ by means of the formula

$$\hat{\alpha} = \frac{\sum_{i=1}^{n} y_i - \hat{\beta}\cdot\sum_{i=1}^{n} x_i}{n} \tag{13.2.20}$$

In Exercise 7 on page 388 the reader will be asked to verify that these two expressions for $\hat{\alpha}$ are, indeed, equivalent.

If we substitute $\hat{\alpha}$ and $\hat{\beta}$ for α and β into (13.2.17), it also follows immediately that the maximum likelihood estimate of σ is given by

$$\hat{\sigma} = \sqrt{\frac{1}{n}\cdot\sum_{i=1}^{n} [y_i - (\hat{\alpha} + \hat{\beta}x_i)]^2} \tag{13.2.21}$$

although, in actual practice, it is generally easier to find $\hat{\sigma}$ by means of the formula

$$\hat{\sigma}^2 = (1 - r^2)\hat{\sigma}_2^2 \tag{13.2.22}$$

where r is the sample correlation coefficient (13.2.12) and $\hat{\sigma}_2$ is the maximum likelihood estimate (13.2.10). It will be left to the reader to show in Exercise 8 on page 388 that (13.2.21) and (13.2.22) are, in fact, equivalent.

Formula (13.2.22) not only simplifies the calculation of $\hat{\sigma}^2$, but it also serves to tie together the concepts of regression and correlation. Solving for r^2 and multiplying by 100, we get

$$100r^2 = \frac{\hat{\sigma}_2^2 - \hat{\sigma}^2}{\hat{\sigma}_2^2}\cdot 100$$

where $\hat{\sigma}_2^2$ measures the *total variation* of the y's, $\hat{\sigma}^2$ measures the *conditional variation* of the y's for fixed values of x, and, hence, $\hat{\sigma}_2^2 - \hat{\sigma}^2$ measures *that part of the total variation of the y's which is accounted for by the relationship with x*. Thus, $100r^2$ *is the percentage of the total variation of the y's that is accounted for by the relationship with x*, and it is in this sense that we interpret values of r between 0 and $+1$ or -1. For instance, when $r = 0.5$

then 25 percent of the variation of the y's is accounted for by the relationship with x, when $r = 0.7$ then 49 percent of the variation of the y's is accounted for by the relationship with x, and we might thus say that a correlation of $r = 0.7$ is almost "twice as strong" as a correlation of $r = 0.5$; similarly, we might say that a correlation of $r = 0.6$ is "nine times as strong" as a correlation of $r = 0.20$.

Having obtained maximum likelihood estimators of the regression coefficients, let us now investigate their use in testing hypothesis concerning α and β, and in constructing confidence intervals for these two parameters. Since problems concerning β are usually of more immediate interest than problems concerning α (the null hypothesis $\beta = 0$ is equivalent to the null hypothesis $\rho = 0$), we shall discuss some of the sampling theory connected with β in the text, while leaving corresponding theory concerning α to Exercises 9 and 10 on page 388.

To study the sampling distribution of $\hat{\boldsymbol{\beta}}$, let us rewrite (13.2.19) as

$$\hat{\boldsymbol{\beta}} = \frac{n \cdot \sum_{i=1}^{n} (x_i - \bar{x})\mathbf{y}_i}{n \cdot \sum_{i=1}^{n} x_i^2 - \left(\sum_{i=1}^{n} x_i\right)^2} = \sum_{i=1}^{n} \left(\frac{x_i - \bar{x}}{n \cdot \hat{\sigma}_1^2}\right) \cdot \mathbf{y}_i \tag{13.2.23}$$

where $\hat{\sigma}_1^2$ is given by (13.2.10). Thus, it can be seen by inspection that $\hat{\boldsymbol{\beta}}$ is a *linear combination* of the independent random variables $\mathbf{y}_i$ having the normal distributions (13.2.14), and it follows from Exercise 6 on page 194 and Theorem 6.3 that $\hat{\boldsymbol{\beta}}$, itself, has a normal distribution with

$$E(\hat{\boldsymbol{\beta}}) = \sum_{i=1}^{n} \left[\frac{x_i - \bar{x}}{n \cdot \hat{\sigma}_1^2}\right] \cdot E(\mathbf{y}_i | x_i)$$

$$= \sum_{i=1}^{n} \left[\frac{x_i - \bar{x}}{n \cdot \hat{\sigma}_1^2}\right] (\alpha + \beta x_i) = \beta \tag{13.2.24}$$

and

$$\text{var}(\hat{\boldsymbol{\beta}}) = \sum_{i=1}^{n} \left[\frac{x_i - \bar{x}}{n \cdot \hat{\sigma}_1^2}\right]^2 \cdot \text{var}(\mathbf{y}_i | x_i)$$

$$= \sum_{i=1}^{n} \left[\frac{x_i - \bar{x}}{n \cdot \hat{\sigma}_1^2}\right]^2 \cdot \sigma^2 = \frac{\sigma^2}{n \cdot \hat{\sigma}_1^2} \tag{13.2.25}$$

In order to apply this theory to construct confidence intervals for β or test hypotheses concerning β, we shall have to make use of the following theorem, a proof of which is referred to on page 392:

THEOREM 13.1

Under the assumptions of normal regression analysis, $\frac{n \cdot \hat{\sigma}^2}{\sigma^2}$, *with* $\hat{\sigma}^2$ *given by* (13.2.21), *is a value of a random variable having the chi-square distribution with* $n - 2$ *degrees of freedom. Furthermore, the two random variables* $\frac{n \cdot \hat{\boldsymbol{\sigma}}^2}{\sigma^2}$ *and* $\hat{\boldsymbol{\beta}}$ *are independent.*

Making use of this theorem as well as the previously proved result that $\hat{\boldsymbol{\beta}}$ has a normal distribution whose mean and variance are given by (13.2.24) and (13.2.25), the definition of the t distribution in Section 7.2.4 leads to the fact that

$$t = \frac{\dfrac{\hat{\beta} - \beta}{\dfrac{\sigma}{\sqrt{n} \cdot \hat{\sigma}_1}}}{\sqrt{\dfrac{n \cdot \hat{\sigma}^2}{\sigma^2} \Big/ (n-2)}} = \frac{(\hat{\beta} - \beta)\hat{\sigma}_1 \sqrt{n-2}}{\hat{\sigma}} \tag{13.2.26}$$

is a value of a random variable having the t distribution with $n - 2$ degrees of freedom.

Using the t statistic whose values are given by (13.2.26), we can now construct a $1 - \alpha$ confidence interval for β by manipulating the double inequality

$$-t_{\alpha/2,n-2} \leq t \leq t_{\alpha/2,n-2}$$

so that the middle term is β while the two limits to not involve this parameter; the result is

$$\hat{\beta} - t_{\alpha/2,n-2} \cdot \frac{\hat{\sigma}}{\hat{\sigma}_1 \sqrt{n-2}} \leq \beta \leq \hat{\beta} + t_{\alpha/2,n-2} \cdot \frac{\hat{\sigma}}{\hat{\sigma}_1 \sqrt{n-2}} \tag{13.2.27}$$

Similarly, to test the null hypothesis $\beta = \beta_0$ against one of the alternatives $\beta > \beta_0$, $\beta < \beta_0$, or $\beta \neq \beta_0$, the appropriate critical regions of size α are $t \geq t_{\alpha,n-2}$, $t \leq -t_{\alpha,n-2}$, and $|t| \geq t_{\alpha/2,n-2}$, where t is given by (13.2.26) with β_0 substituted for β.

To illustrate the methods discussed in the preceding paragraph, let us refer again to the length of diet and weight loss data on page 364, and let us determine a 0.95 confidence interval for the regression coefficient β, for which we had previously obtained the point estimate $\beta = 3.47$. Adding $\Sigma y^2 = 18{,}002$ to the sums given on page 366, we find that $\hat{\sigma}_1 = 6.13$, $\hat{\sigma}_2^2 = 475.24$, $r = 0.976$, so that $\hat{\sigma}^2 = (1 - 0.9526)(475.24) = 22.53$, and, hence, $\hat{\sigma} = 4.75$ and

$$3.47 - 2.306 \cdot \frac{4.75}{6.13\sqrt{8}} \leq \beta \leq 3.47 + 2.306 \cdot \frac{4.75}{6.13\sqrt{8}}$$

$$2.84 \leq \beta \leq 4.10$$

since $t_{.025,8} = 2.306$ according to Table IV.

THEORETICAL EXERCISES

1. Verify equations (13.2.8) and (13.2.9), and also show that (13.2.11) and (13.2.12) are equivalent.

2. The calculation of r can often be simplified by adding the same constant to each x, adding the same constant to each y, or by multiplying each x and/or each y by the same positive constant.

(a) Show that if r_{xy} is the correlation coefficient calculated for a set of paired data (x_1, y_1), (x_2, y_2), $\ldots$, and (x_n, y_n), then r_{uv}, the correlation coefficient for $u_i = ax_i + b$ and $v_i = cy_i + d$ (with $a \neq 0$ and $c \neq 0$), is given by $r_{uv} = r_{xy}$.

(b) Recalculate the correlation coefficient for the illustration on page 381 by first multiplying each x and each y by 10, and then subtracting 70 from each x and 60 from each y.

3. Construct a $1 - \alpha$ confidence interval for ρ by solving the double inequality $-z_{\alpha/2} \leq z \leq z_{\alpha/2}$ with z given by (13.2.13) so that the middle term is ρ and the two limits do not depend on ρ. Use the result to construct a 0.95 confidence interval for ρ for the example on page 381 where we had $n = 10$ and $r = 0.936$.

4. If the ranks 1, 2, $\ldots$, and n are assigned to the x's arranged according to size, the y's are ranked in a similar fashion, and in either case there are no ties, show that

$$r = 1 - \frac{6 \cdot \sum_{i=1}^{n} d_i^2}{n(n^2 - 1)}$$

where d_i is the difference between the ranks assigned to x_i and y_i. When the x's and y's are thus replaced by their ranks, the resulting correlation coefficient is called the *coefficient of rank correlation* (or *Spearman's rank correlation coefficient*). In actual practice, the above formula is also used when there are ties, in which case tied observations are assigned the ranks which they jointly occupy.

5. Calculate the coefficient of rank correlation (see Exercise 4) for the data on page 381.
6. Verify that the solution of (13.2.15) and (13.2.16) is given by (13.2.18) and (13.2.19).
7. Verify that (13.2.18) and (13.2.20) are equivalent.
8. Verify that (13.2.21) and (13.2.22) are equivalent.
9. Show that
   (a) (13.2.20) can be written in the form

$$\hat{\alpha} = \sum_{i=1}^{n} \left[\frac{\hat{\sigma}_1^2 + \bar{x}^2 - x_i\bar{x}}{n\hat{\sigma}_1^2} \right] y_i$$

   (b) $\hat{\boldsymbol{\alpha}}$ has a normal distribution with

$$E(\hat{\boldsymbol{\alpha}}) = \alpha \quad \text{and} \quad \text{var}(\hat{\boldsymbol{\alpha}}) = \frac{\sigma^2(\hat{\sigma}_1^2 + \bar{x}^2)}{n\hat{\sigma}_1^2}$$

   Also, use Theorem 6.4 to show that

$$\text{cov}(\hat{\boldsymbol{\alpha}}, \hat{\boldsymbol{\beta}}) = \frac{-\sigma^2\bar{x}}{n\hat{\sigma}_1^2}$$

10. Using the result of part (b) of Exercise 9, show that

$$z = \frac{(\hat{\alpha} - \alpha)\hat{\sigma}_1\sqrt{n}}{\sigma\sqrt{\hat{\sigma}_1^2 + \bar{x}^2}}$$

   is a value of a random variable having the standard normal distribution. Also, using the first part of Theorem 13.1 and the fact that $\hat{\boldsymbol{\alpha}}$ and $\frac{n\hat{\boldsymbol{\sigma}}^2}{\sigma^2}$ are independent, show that

$$t = \frac{(\hat{\alpha} - \alpha)\hat{\sigma}_1\sqrt{n-2}}{\hat{\sigma}\sqrt{\hat{\sigma}_1^2 + \bar{x}^2}} \tag{13.2.28}$$

   is a value of a random variable having the t distribution with $n - 2$ degrees of freedom.
11. Verify that the t statistic (13.2.26) can be written as

$$t = \left(1 - \frac{\beta}{\hat{\beta}}\right) \frac{r\sqrt{n-2}}{\sqrt{1-r^2}}$$

where r is the sample correlation coefficient (13.2.12). Also use this result to derive the following $1 - \alpha$ confidence limits for β:

$$\hat{\beta}\left[1 \pm t_{\alpha/2,n-2}\cdot\frac{\sqrt{1-r^2}}{r\sqrt{n-2}}\right]$$

12. Use the first part of Exercise 11 to show that if the assumptions underlying normal regression analysis are met and $\beta = 0$, then $\mathbf{r}^2$ has a *beta distribution* (4.3.7) with the mean $\frac{1}{n-1}$.

13. Using (13.2.24), (13.2.25), and the results of Exercise 9, show that $\mathbf{y} - \hat{\boldsymbol{\alpha}} - \hat{\boldsymbol{\beta}}x_0$ is a random variable having a normal distribution with the mean 0 and the variance

$$\sigma^2\left[1 + \frac{1}{n} + \frac{(x_0 - \bar{x})^2}{n\hat{\sigma}_1^2}\right]$$

Also, using the first part of Theorem 13.1 as well as the fact that $\mathbf{y} - \hat{\boldsymbol{\alpha}} - \hat{\boldsymbol{\beta}}x_0$ and $\frac{n\hat{\boldsymbol{\sigma}}^2}{\sigma^2}$ are independent, show that

$$t = \frac{(y - \hat{\alpha} - \hat{\beta}x_0)\sqrt{n-2}}{\hat{\sigma}\sqrt{1 + n + \frac{(x_0 - \bar{x})^2}{\hat{\sigma}_1^2}}} \qquad (13.2.29)$$

is a value of a random variable having the t distribution with $n - 2$ degrees of freedom.

14. Solve the double inequality $-t_{\alpha/2,n-2} \leq t \leq t_{\alpha/2,n-2}$, with t given by (13.2.29), so that the middle term is y and the two limits can be calculated without knowledge of y. Note that although the resulting double inequality should be interpreted like a confidence interval, it is not designed to estimate a parameter; instead, it provides *limits of prediction* for the value of y which corresponds to (the given or observed value) x_0.

APPLIED EXERCISES

15. An objective achievement test is said to be reliable if a student who takes the test several times would consistently get high (or low) scores. One way of checking the reliability of a test is to divide it into two parts, usually the even-numbered problems and the odd-numbered problems, and to check the correlation between the scores which students get in both halves of the test. Thus, the following data represent the grades, x and y, which 20 students obtained for the even-numbered problems and the odd-numbered problems of a new objective test designed to test eighth grade achievement in general science:

x	y	x	y
27	29	33	42
36	44	39	31
44	49	38	38
32	27	24	22
27	35	33	34
41	33	32	37
38	29	37	38
44	40	33	35
30	27	34	32
27	38	39	43

Calculate r for these data and test its significance, namely, the null hypothesis $\rho = 0$, at $\alpha = 0.05$.

16. Use the result of Exercise 3 to calculate a 0.95 confidence interval for ρ on the basis of the data of Exercise 15.

17. The following data pertain to x, the amount of fertilizer (in pounds) which a farmer applies to his soil, and y, his yield of wheat (in bushels per acre):

x	y	x	y	x	y
112	33	88	24	37	27
92	28	44	17	23	9
72	38	132	36	77	32
66	17	23	14	142	38
112	35	57	25	37	13
88	31	111	40	127	23
42	8	69	29	88	31
126	37	19	12	48	37
72	32	103	27	61	25
52	20	141	40	71	14
28	17	77	26	113	26

Assuming that the data can be looked upon as a sample from a bivariate normal population, calculate r and test its significance at $\alpha = 0.01$. Judging from a scattergram of these paired data, does the assumption seem reasonable?

18. Use the result of Exercise 3 to calculate a 0.95 confidence interval for ρ on the basis of the data of Exercise 17.

19. Calculate the coefficient of rank correlation (see Exercise 4) for the data of Exercise 15 and compare it with the value of r obtained in that exercise. Break ties in rank among the x's or the y's by assigning tied observations the mean of the ranks which they jointly occupy.

20. The following are the rankings given by two critics, who were asked to indicate their preference among 12 television programs which appeared on the NBC network in 1969:

	Critic A	*Critic B*
Bonanza	4	2
Walt Disney's World	6	9
Mothers-in-Law	12	10
Get Smart	9	3
Laugh-In	2	5
Julia	5	6
Music Hall	10	11
Outsider	7	7
Ironside	1	4
Dragnet	8	12
Dean Martin	3	1
High Chapparal	11	8

Calculate the coefficient of rank correlation (see Exercise 4) as a measure of the *consistency* (or *inconsistency*) of the two critics.

21. Referring to the length of diet and weight loss data on page 364 (as well as the calculations already performed in the text), test the null hypothesis $\beta = 3.20$ at the level of significance $\alpha = 0.05$.

22. Referring to the illustration on page 381 pertaining to the time it takes a secretary to perform a certain task in the morning and in the late afternoon, test the null hypothesis $\beta = 1.1$ at the level of significance 0.05.

23. Referring to Exercise 17 on page 369, construct a 0.95 confidence interval for the regression coefficient β.

24. Referring to Exercise 18 on page 370, construct a 0.99 confidence interval for the regression coefficient β.

25. Use the result of Exercise 11 to calculate a 0.95 confidence interval for β for the length of diet and weight loss data on page 364, and compare this interval with the one obtained in the text on page 387.

26. Referring to the example on page 364, test the null hypothesis $\alpha = 1.50$ against the alternative $\alpha \neq 1.50$ with the use of the theory of Exercise 10. Use a level of significance of 0.01.

27. Use the result of Exercise 10 to calculate a 0.95 confidence interval for α for the data of Exercise 17 on page 369.

28. Use the result of Exercise 10 to calculate a 0.99 confidence interval for α with reference to Exercise 18 on page 370.

29. Use the result of Exercise 14 (as well as the calculations already performed in the text) to construct 0.95 limits of prediction for the weight loss of a person who has been on the diet for 14 months.

30. Use the result of Exercise 14 and the data of Exercise 17 on page 369 to find 0.99 limits of prediction of the number of deaths (in a group of 25 mice) when the dosage is 9 milligrams.

REFERENCES

A detailed treatment of many of the mathematical and statistical properties of the bivariate normal surface may be found in

Yule, G. U., and Kendall, M. G., *An Introduction to the Theory of Statistics*, 14th ed. New York: Hafner Publishing Co., Inc., 1950.

A proof of Theorem 13.1 and other mathematical details left out in the text may be found in the book by S. S. Wilks referred to on page 230; information about the distribution of the statistic whose values are given by $\frac{1}{2}\cdot \ln \frac{1+r}{1-r}$ may be found in the book by M. G. Kendall and A. Stuart referred to on page 181.

A derivation of the maximum likelihood estimates given by (13.2.10) and (13.2.11) is given in

Hoel, P., *Introduction to Mathematical Statistics*, 3rd ed., New York: John Wiley & Sons, Inc., 1962.

14

ANALYSIS OF VARIANCE

14.1 EXPERIMENTAL DESIGN

In the performance of statistical tests we may well decide that one kind of tire is better than another, that the students in one school get better instruction than those in another, or that one rocket is more accurate than another, completely overlooking the fact that the second kind of tire may have been tested over much rougher roads, that the students in the second school did poorly in the test (on which the judgment was based) because they happened to be thinking about a big football game scheduled for that evening, and that the person who fired the second kind of rocket was a much poorer marksman than the one who fired the first. All this serves to point out that experiments purported to test one thing often test another, or that they cannot serve any useful purpose at all.

There are essentially two ways of avoiding situations like those described. One is to perform a *rigorously controlled* experiment in which all variables are held fixed except for the one with which we are concerned. For instance, if we wanted to compare the performance of two kinds of gasoline, A and B, all test runs could be made with the same car (which is carefully inspected after each run), with the same driver, and over identical routes. In that case, if there *is* a significant difference in the average mileage yield of the two gasolines, we know that it is *not* due to differences in cars, drivers, and routes. On the positive side, we know that one of the gasolines performs better than the other *provided it is used in a certain kind of car, by a given driver, and over a certain kind of route.* However, this does not tell us very much, as it would be dangerous and risky to infer that similar results would be obtained with different kinds of cars, different drivers, and over different routes (say, in freeway driving instead of crawling in heavy city traffic).

The other way of handling problems of this kind is to *design* the experiment in such a way that we can not only compare the merits of the two gasolines under more general conditions, but that we can also test whether the other variables really affect their performance. To illustrate how this might be done, suppose that the test runs are performed in two cars, a low-priced car L and a high-priced car H, by two drivers, a good driver Mr. G and a poor driver Mr. P, and over two routes, a city route C and a freeway route F. Suppose, furthermore, that each test run is performed with a gallon of the respective gasoline, and that the experiment consists of the following 16 test runs:

Test Run	*Gasoline*	*Car*	*Driver*	*Route*
1	B	L	Mr. G	C
2	B	H	Mr. P	C
3	A	H	Mr. P	C
4	B	L	Mr. P	F
5	A	H	Mr. G	C
6	B	H	Mr. G	C
7	B	L	Mr. G	F
8	A	L	Mr. P	F
9	A	L	Mr. G	F
10	A	H	Mr. P	F
11	B	L	Mr. P	C
12	A	H	Mr. G	F
13	A	L	Mr. G	C
14	B	H	Mr. G	F
15	A	L	Mr. P	C
16	B	H	Mr. P	F

This means that the first test run is performed with gasoline B in the low-priced car, by the good driver, over the city route; the second test run is performed with gasoline B in the high-priced car, by the poor driver, over the city route; the third test run is performed with gasoline A in the high-priced car, by the poor driver, over the city route; . . . ; and the sixteenth test run is performed with gasoline B in the high-priced car, by the poor driver, over the freeway route. It is customary to refer to this kind of scheme as a *completely balanced design*—each gasoline is used once with each possible combination of cars, drivers, and routes.

Another important feature of the above scheme (which may not be apparent) is that we protected ourselves by *randomization*. First we wrote down the 16 possible ways in which we can select one of the two gasolines, one of the two cars, one of the two drivers, and one of the two routes, and then we *randomly* selected the order in which the test runs are to be performed. If we did not randomize the experiment in this fashion, extraneous factors might conceivably upset the results; for instance, if we used gasoline A in the first 8 test runs and gasoline B in the others, the results might be affected by deterioration of the equipment, driver fatigue, or differences in traffic conditions along the chosen routes. Similarly, we might have asked for trouble if we had performed the first 8 test runs with car L and the others with car H, or if we had performed the first 8 test runs with Mr. G and the others with Mr. P.

Another factor which is of importance in the design of an experiment is that of *replication* or *repetition*. In order to be able to decide whether observed differences between sample means are significant, we must have an estimate of *chance variation*, or as it is usually called, the *experimental error*. Such an estimate is usually obtained by repeating all or part of the experimental scheme, and in our example we might thus perform 32 test runs, 2 of each of the 16 possible combinations listed above. Whatever differences there are between the corresponding pairs would then be attributed to chance.

The purpose of this example has been to introduce some of the basic ideas of experimental design, and it is important to realize that the 16 test runs would not only enable us to decide whether there really is a difference between the gasolines, but also whether their performance is affected by differences in cars, differences in drivers, and differences in driving conditions. The actual analysis of a *four-factor experiment* like the one just described is fairly complicated and it will not be discussed in this text; conceptually, though, its analysis is a straightforward generalization of that of the *one-factor* and *two-factor experiments* which will be described in Sections 14.2 and 14.3.

The subject of experimental design and, correspondingly, the analysis of the data which are obtained with these designs is so vast that we can at best touch upon some of the basic ideas. In Sections 14.2 and 14.3 we shall introduce a method of analyzing experimental data called the *analysis of variance*, ANOVA for short, which is undoubtedly the most important technique of experimental statistics. Some further considerations will be touched upon briefly in Section 14.4.

14.2 ONE-WAY ANALYSIS OF VARIANCE

Let us now generalize the work of Section 11.2.2 and consider the problem of deciding whether observed differences among more than two sample means can be attributed to chance, or whether they are indicative of actual differences among the means of the corresponding populations. For instance, we may want to decide on the basis of sample data whether there really is a difference in the effectiveness of three methods of teaching computer programming, we may want to compare the average yield per acre of six varieties of wheat, or we may want to decide on the basis of samples whether there really is a difference in the average mechanical aptitude of the students attending four large schools. To give a concrete example, suppose that Mr. Martin can drive to work along four different routes, and that the following are the number of minutes in which he timed himself on five different occasions for each route:

Route 1	*Route 2*	*Route 3*	*Route 4*
22	25	26	26
26	27	29	28
25	28	33	27
25	26	30	30
31	29	33	30

The means of these four samples are 25.8, 27.0, 30.2, 28.2, and what we would like to know is whether the differences among these means are significant or whether they can be attributed to chance—after all, the size of the samples is *very small.*

To treat this kind of problem in a general way, suppose that we have k independent random samples of size n from k populations, and that the jth observation of the sample from the ith population is denoted x_{ij}. In the analysis which follows, we shall assume that the corresponding random variables $\mathbf{x}_{ij}$ are *independent* and that they have normal distributions with the respective means μ_i and the common variance σ^2. Stating this assumption somewhat differently, we could specify the *underlying model* by expressing the observations as

$$x_{ij} = \mu_i + e_{ij} \qquad \text{for } i = 1, 2, \ldots, k \quad j = 1, 2, \ldots, n \tag{14.2.1}$$

where the e_{ij} are values of $n \cdot k$ independent random variables having normal distributions with zero means and the common variance σ^2. To permit the generalization of (14.2.1) to more complicated kinds of models, it will prove to be useful to write it as

$$x_{ij} = \mu + \alpha_i + e_{ij} \qquad \text{for } i = 1, 2, \ldots, k \quad j = 1, 2, \ldots, n \tag{14.2.2}$$

where μ is referred to as the *grand mean*, and the *treatment effects* α_i are such that $\sum_{i=1}^{k} \alpha_i = 0$. Note that although (14.2.2) may look more complicated than (14.2.1), we have merely written the mean of the ith population as $\mu_i = \mu + \alpha_i$ and imposed the condition $\sum_{i=1}^{k} \alpha_i = 0$ so that the *mean* of the μ_i equals the grand mean μ. [The practice of referring to the different populations as different *treatments* is accounted for by the fact that the method which we are discussing in this section was used first in the analysis of agricultural experiments where different treatments (that is, different kinds of fertilizer, different kinds of seeds, different amounts of irrigation, etc.) were applied to the soil.]

The null hypothesis we shall want to test is that the means of the k populations are all equal, namely, that $\mu_1 = \mu_2 = \cdots = \mu_k$ or that

$$\alpha_i = 0 \qquad \text{for } i = 1, 2, \ldots, k$$

Correspondingly, the alternative hypothesis is that the population means are *not all equal*, namely, that $\alpha_i \neq 0$ for at least one value of i. The method which we shall use to test this hypothesis is to analyse a measure of the *total variability of the combined data*, namely, $nk - 1$ times their variance, which is given by

$$\sum_{i=1}^{k} \sum_{j=1}^{n} (x_{ij} - \bar{x})^2 \qquad \text{where} \qquad \bar{x} = \frac{1}{n \cdot k} \cdot \sum_{i=1}^{k} \sum_{j=1}^{n} x_{ij} \tag{14.2.3}$$

If the null hypothesis is true, all this variability is due to chance, but if it is *not true*, then part of the *sum of squares* (14.2.3) is due to the differences among the means of the k populations. To isolate, or separate, these two contributions to the total variability of the data, let us prove the following theorem:

THEOREM 14.1

$$\sum_{i=1}^{k} \sum_{j=1}^{n} (x_{ij} - \bar{x})^2 = \sum_{i=1}^{k} \sum_{j=1}^{n} (x_{ij} - \bar{x}_{i.})^2 + n \cdot \sum_{i=1}^{k} (\bar{x}_{i.} - \bar{x})^2$$

where $\bar{x}_{i.}$ is the mean of the sample from the ith population, namely, $\bar{x}_{i.} = \frac{1}{n} \cdot \sum_{j=1}^{n} x_{ij}$.

Writing $x_{ij} - \bar{x}$ as $(x_{ij} - \bar{x}_{i.}) + (\bar{x}_{i.} - \bar{x})$ and applying the binomial theorem, we can write

$$\begin{aligned}
\sum_{i=1}^{k}\sum_{j=1}^{n}(x_{ij} - \bar{x})^2 &= \sum_{i=1}^{k}\sum_{j=1}^{n}[(x_{ij} - \bar{x}_{i.}) + (\bar{x}_{i.} - \bar{x})]^2 \\
&= \sum_{i=1}^{k}\sum_{j=1}^{n}(x_{ij} - \bar{x}_{i.})^2 + 2\cdot\sum_{i=1}^{k}\sum_{j=1}^{n}(x_{ij} - \bar{x}_{i.})(\bar{x}_{i.} - \bar{x}) \\
&\quad + \sum_{i=1}^{k}\sum_{j=1}^{n}(\bar{x}_{i.} - \bar{x})^2 \\
&= \sum_{i=1}^{k}\sum_{j=1}^{n}(x_{ij} - \bar{x}_{i.})^2 + 2\cdot\sum_{i=1}^{k}\left[(\bar{x}_{i.} - \bar{x})\cdot\sum_{j=1}^{n}(x_{ij} - \bar{x}_{i.})\right] \\
&\quad + n\cdot\sum_{i=1}^{k}(\bar{x}_{i.} - \bar{x})^2 \\
&\ \sum_{i=1}^{k}\sum_{j=1}^{n}(x_{ij} - \bar{x}_{i.})^2 + n\cdot\sum_{i=1}^{k}(\bar{x}_{i.} - \bar{x})^2
\end{aligned}$$

since $\sum_{j=1}^{n}(x_{ij} - \bar{x}_{i.}) = 0$ for each value of i, and this completes the proof of Theorem 14.1.

It is customary to refer to the expression on the left-hand side of the equation of Theorem 14.1 as the *total sum of squares* SST, to the first term of the expression on the right-hand side as the *error sum of squares* SSE, and to the second term as the *treatment sum of squares* SS(Tr), and we can thus write

$$\text{SST} = \text{SSE} + \text{SS(Tr)} \tag{14.2.4}$$

Note that we have now accomplished what we originally set out to do: *We have partitioned SST, a measure of the total variability of the combined data, into two components: The first component, SSE, measures chance variation (namely, the variation within the samples) regardless of whether the null hypothesis is true or false; the second component, SS(Tr), also measures chance variation when the null hypothesis is true, but it is affected by the variation among the population means when the null hypothesis is false.*

If the null hypothesis is true, the nk observations x_{ij} can be looked upon as values of *one* random sample of size nk from a normal population with the mean μ and the variance σ^2, and it follows from Theorem 7.6 that

$$\frac{1}{\sigma^2}\cdot\sum_{i=1}^{k}\sum_{j=1}^{n}(\mathbf{x}_{ij} - \bar{\mathbf{x}})^2$$

is a random variable having the chi-square distribution with $nk - 1$ degrees of freedom, and that for each value of i

$$\frac{1}{\sigma^2} \cdot \sum_{j=1}^{n} (\mathbf{x}_{ij} - \bar{\mathbf{x}}_{i.})^2$$

is a random variable having the chi-square distribution with $n - 1$ degrees of freedom. Hence, according to Theorem 7.4

$$\frac{1}{\sigma^2} \cdot \sum_{i=1}^{k} \sum_{j=1}^{n} (\mathbf{x}_{ij} - \bar{\mathbf{x}}_{i.})^2 \tag{14.2.5}$$

is a random variable having the chi-square distribution with $k(n - 1)$ degrees of freedom, and since the mean of a chi-square distribution equals its degrees of freedom (see page 212), we find that $\frac{1}{\sigma^2} \cdot \text{SSE}$ is a value of a random variable whose distribution has the mean $k(n - 1)$. *Thus,* $\frac{\text{SSE}}{k(n-1)}$ *provides an estimate of* σ^2.

Also, since under the null hypothesis the $\bar{x}_{i.}$ are values of independent random variables having the normal distribution with the mean μ and the variance $\frac{\sigma^2}{n}$, it follows from Theorem 7.6 that

$$\frac{n}{\sigma^2} \cdot \sum_{i=1}^{k} (\bar{\mathbf{x}}_{i.} - \bar{\mathbf{x}})^2 \tag{14.2.6}$$

is a random variable having the chi-square distribution with $k - 1$ degrees of freedom. Since the mean of this distribution is $k - 1$, *it follows that* $\frac{\text{SS(Tr)}}{k-1}$ *also provides an estimate of* σ^2. Of course, *if the null hypothesis is not true, then* $\frac{\text{SS(Tr)}}{k-1}$ *provides an estimate of* σ^2 *plus whatever variability there may be among the population means.*

This suggests that we reject the null hypothesis $\mu_1 = \mu_2 = \cdots = \mu_k$ when $\frac{\text{SS(Tr)}}{k-1}$ is *considerably greater* than $\frac{\text{SSE}}{k(n-1)}$, and to put this decision on a precise basis, we shall have to assume without proof that the two random variables (14.2.5) and (14.2.6) are independent.* With this

*A proof of this independence may be found in the book by H. Scheffé listed on page 418.

assumption we can utilize Theorem 7.7, according to which

$$\frac{\frac{SS(Tr)}{(k-1)\sigma^2}}{\frac{SSE}{k(n-1)\sigma^2}} = \frac{k(n-1)SS(Tr)}{(k-1)SSE}$$

is a value of a random variable having the F distribution with $k-1$ and $k(n-1)$ degrees of freedom. Thus, we reject the null hypothesis when the value which we obtain for

$$F = \frac{k(n-1)SS(Tr)}{(k-1)SSE} \tag{14.2.7}$$

exceeds $F_{\alpha, k-1, k(n-1)}$. [As we pointed out in the preceding paragraph, differences among the population means tend to make $\frac{SS(Tr)}{k-1}$ greater than $\frac{SSE}{k(n-1)}$.]

The kind of analysis we have been discussing in this section, called a *one-way analysis of variance* because we are studying the effect of only *one source of variation* (other than chance), is usually presented in the following kind of *analysis of variance table:*

Source of Variation	*Degrees of Freedom*	*Sum of Squares*	*Mean Square*	*F*
Treatments	$k-1$	SS(Tr)	$MS(Tr) = \frac{SS(Tr)}{k-1}$	$\frac{MS(Tr)}{MSE}$
Error	$k(n-1)$	SSE	$MSE = \frac{SSE}{k(n-1)}$	
Total	$kn-1$	SST		

Note that the *mean squares* are simply the sums of squares divided by the corresponding degrees of freedom.

Let us now return to the numerical example on page 396, where we wanted to judge whether the differences among the average times it took a certain person to get to work along four different routes are significant. To determine F, we could calculate SS(Tr) and SSE according to their definition, but to simplify the calculations we shall instead use the following short-cut formulas (which the reader will be asked to verify in Exercise 1 on page 402):

Theorem 14.2

If $T_{i.}$ is the total of the observations for the ith treatment and $T_{..}$ is the grand total of all observations,

$$\text{SST} = \sum_{i=1}^{k}\sum_{j=1}^{n} x_{ij}^2 - \frac{1}{kn}\cdot T_{..}^2$$

$$\text{SS(Tr)} = \frac{1}{n}\cdot\sum_{i=1}^{k} T_{i.}^2 - \frac{1}{kn}\cdot T_{..}^2$$

Then, SSE can be found by subtraction; that is, by means of the formula SSE = SST − SS(Tr).

Calculating the necessary sums and sum of squares for the data on page 396, we get $T_{1.} = 129$, $T_{2.} = 135$, $T_{3.} = 151$, $T_{4.} = 141$, $T_{..} = 556$, $\sum_{i=1}^{k} T_{i.}^2 = 77{,}548$, and $\sum_{i=1}^{k}\sum_{j=1}^{n} x_{ij}^2 = 15{,}610$, and, hence,

$$\text{SST} = 15{,}610 - \tfrac{1}{20}(556)^2 = 153.2$$

$$\text{SS(Tr)} = \tfrac{1}{5}(77{,}548) - \tfrac{1}{20}(556)^2 = 52.8$$

and

$$\text{SSE} = 153.2 - 52.8 = 100.4$$

The remaining calculations are shown in the following analysis of variance table:

Source of Variation	*Degrees of Freedom*	*Sum of Squares*	*Mean Square*	*F*
Treatments	3	52.8	$\frac{52.8}{3} = 17.6$	$\frac{17.6}{6.28} = 2.80$
Error	16	100.4	$\frac{100.4}{16} = 6.28$	
Total	19	153.2		

and since $F = 2.80$ does *not* exceed $F_{.05,3,16} = 3.24$, *the null hypothesis cannot be rejected.* Even though Route 1 seems to be the fastest and Route 3 seems to be the slowest, all this may very well be due to chance.

The parameters of the model (14.2.2), namely, μ and the α_i, are usually estimated by the method of least squares. In other words, their estimates are the values which minimize

$$\sum_{i=1}^{k} \sum_{j=1}^{n} [x_{ij} - (\mu + \alpha_i)]^2$$

subject to the restriction that $\sum_{i=1}^{k} \alpha_i = 0$, and as the reader will be asked to verify in Exercise 5 on page 403, these *least squares estimates* are $\hat{\mu} = \bar{x}$ and $\hat{\alpha}_i = \bar{x}_{i.} - \bar{x}$.

THEORETICAL EXERCISES

1. Verify the short-cut formulas of Theorem 14.2.
2. If the sample sizes are unequal in a one-way analysis of variance and there are n_i observations for the ith treatment, show that analogous to Theorem 14.1 we have

$$\sum_{i=1}^{k} \sum_{j=1}^{n_i} (x_{ij} - \bar{x})^2 = \sum_{i=1}^{k} \sum_{j=1}^{n_i} (x_{ij} - \bar{x}_{i.})^2 + \sum_{i=1}^{k} n_i(\bar{x}_{i.} - \bar{x})^2$$

   and that the number of degrees of freedom for SST, SSE, and SS(Tr) are, respectively, $N - 1$, $N - k$, and $k - 1$, where $N = \sum_{i=1}^{k} n_i$.
3. With reference to Exercise 2 where the sample sizes are unequal, show that the short-cut computing formulas for the sums of squares become

$$\text{SST} = \sum_{i=1}^{k} \sum_{j=1}^{n_i} x_{ij}^2 - \frac{1}{N} \cdot T_{..}^2$$

$$\text{SS(Tr)} = \sum_{i=1}^{k} \frac{T_{i.}^2}{n_i} - \frac{1}{N} \cdot T_{..}^2$$

   and

$$\text{SSE} = \text{SST} - \text{SS(Tr)}$$

   where the notation is the same as in Theorem 14.2.

4. Show that when $k = 2$ the F test based on (14.2.7) is *equivalent* to the t test of Section 11.2.2 based on (11.2.5) with $\delta = 0$.

5. Use Lagrange multipliers (see reference on page 417) to verify that the least squares estimates of the parameters of (14.2.2) are $\hat{\mu} = \bar{x}$ and $\hat{\alpha}_i = \bar{x}_{i.} - \bar{x}$.

APPLIED EXERCISES

6. To compare the effectiveness of three different types of phosphorescent coatings of airplane instrument dials, eight dials each are coated with the three types. Then the dials are illuminated by an ultraviolet light, and the following are the number of minutes each glowed after the light source was shut off:

Type 1	*Type 2*	*Type 3*
52.9	58.4	71.3
62.1	55.0	66.6
57.4	59.8	63.4
50.0	62.5	64.7
59.3	64.7	75.8
61.2	59.9	65.6
60.8	54.7	72.9
53.1	58.4	67.3

Test the null hypothesis that there is no difference in the effectiveness of the three coatings at the level of significance $\alpha = 0.01$.

7. The following are the yields of wheat (in pounds per plot) which an agronomist obtained for three test plots each of four different varieties:

Variety A	*Variety B*	*Variety C*	*Variety D*
51	38	44	47
49	43	46	48
56	50	52	53

Test at the level of significance $\alpha = 0.05$ whether the differences among the average yields of the four varieties are significant.

8. Three groups of six guinea pigs each were injected, respectively, with 0.5 mg, 1.0 mg, and 1.5 mg of a new tranquilizer, and the following are

the number of seconds it took them to fall asleep:

The 0.5 mg group: 11, 13, 9, 14, 15, 13
The 1.0 mg group: 9, 11, 10, 8, 12, 10
The 1.5 mg group: 10, 5, 8, 9, 6, 10

Test at $\alpha = 0.05$ whether we can reject the null hypothesis that differences in dosage have no effect. Also estimate the parameters μ, α_1, α_2, and α_3 of the model used in this analysis.

9. The following are the scores obtained in an achievement test by random samples of students from four different schools:

School 1	*School 2*	*School 3*	*School 4*
88	78	80	71
99	62	61	65
96	98	74	90
68	83	92	46
85	61	78	53
90	88	54	67
99	74	77	
76	59	83	
	73		

Use the formulas of Exercise 3 to calculate the sums of squares which are required to test the null hypothesis that the differences among the sample means can be attributed to chance. Use $\alpha = 0.01$.

10. A consumer testing service, wishing to test the accuracy of the thermostats of three different kinds of electric irons, set them at 480°F and obtained the following actual temperature readings by means of a thermocouple:

Iron X: 474, 496, 467, 471
Iron Y: 492, 498
Iron Z: 460, 495, 490

Use $\alpha = 0.05$ to test the null hypothesis that there is no difference in the true average setting at 480°F of the thermostats of the three kinds of irons. (*Hint:* Use the formulas of Exercises 2 and 3.)

14.3 TWO-WAY ANALYSIS OF VARIANCE

In the example on page 396 we were unable to prove that there is a difference in the average time it takes the given person to get to work even though the sample means varied all the way from 25.8 to 30.2 minutes.

Of course, there may be no difference or our samples may have been too small, but it is also possible that SSE, the variability which we ascribed to *chance* (or *experimental error*), may actually have been "inflated" by identifiable sources of variation. For instance, if we look at the figures on page 396, careful scrutiny shows that *in each case the first sample value was the smallest while the last sample value was the largest.* This seems rather strange, and upon inquiring into the way the data were obtained, we might discover that in each case the first observation was taken on a Monday, the second on a Tuesday, the third on a Wednesday, the fourth on a Thursday, and the fifth on a Friday. Thus, the variability which we have been ascribing to chance (or at least part of it) may well be due to difference in traffic conditions on different days of the week.

This suggests that we should have performed a *two-way analysis of variance,* in which the *total variability of the data* is partitioned into one component which we ascribe to possible differences due to one variable (the *treatments* of Section 14.2), a second component which we ascribe to possible differences due to a second variable (referred to as *blocks,* again due to the origin of this method in agricultural research), while the remainder of the variability is ascribed to chance. With reference to the illustration on page 396, we would thus refer to the different routes as treatments and to different days as blocks.

Before we go into any details, let us point out that there are essentially two different ways of analyzing such two-variable experiments—they depend on whether the variables are *independent* or whether they *interact.* To illustrate what we mean here by "interact," suppose that a tire manufacturer is experimenting with different kinds of treads, and that he finds that one kind is especially good for use on dirt roads while another kind is specially good for use on icy roads. If this is the case, we say that there is an *interaction* between road conditions and the design of the treads. On the other hand, if all of the treads behaved equally well (or equally poorly) under all kinds of road conditions, we would say that there is *no interaction,* namely, that the two variables (tread design and road conditions) are independent. The *no interaction* case will be taken up first in Section 14.3.1, while a method which is suitable for testing for *interactions* will be described in Section 14.3.2.

14.3.1 Two-way Analysis without Interaction

If x_{ij} for $i = 1, 2, \ldots, k$ and $j = 1, 2, \ldots, n$ are values of independent random variables having normal distributions with the respective means μ_{ij} and the common variance σ^2, we write the model for a two-way analysis of variance without interaction as

$$x_{ij} = \mu + \alpha_i + \beta_j + e_{ij} \qquad \text{for } i = 1, 2, \ldots, k \quad j = 1, 2, \ldots, n \tag{14.3.1}$$

where μ is the *grand mean,* α_i is the ith *treatment effect,* β_j is the jth *block effect,* $\sum_{i=1}^{k} \alpha_i = 0$, $\sum_{j=1}^{n} \beta_j = 0$, and the e_{ij} are values of independent random variables having normal distributions with zero means and the common variance σ^2. Note that in this model $\mu_{ij} = \mu + \alpha_i + \beta_j$, and the *mean* of the μ_{ij} equals the grand mean μ by virtue of the restrictions imposed on the α_i and β_j.

The two null hypotheses we shall want to test are that *the treatment effects are all equal to zero* and that *the block effects are all equal to zero,* namely, that

$$\alpha_i = 0 \qquad \text{for } i = 1, 2, \ldots, k$$

and

$$\beta_j = 0 \qquad \text{for } j = 1, 2, \ldots, n$$

The alternative to the first null hypothesis is that the treatment effects are *not all equal to zero,* namely, that $\alpha_i \neq 0$ for at least one value of i; correspondingly, the alternative to the second null hypothesis is that the block effects are *not all equal to zero,* namely, that $\beta_j \neq 0$ for at least one value of j. The analysis on which we shall base these tests requires the following generalization of Theorem 14.1, which the reader will be asked to prove in Exercise 1 on page 412:

THEOREM 14.3

$$\sum_{i=1}^{k} \sum_{j=1}^{n} (x_{ij} - \bar{x})^2 = \sum_{i=1}^{k} \sum_{j=1}^{n} (x_{ij} - \bar{x}_{i.} - \bar{x}_{.j} + \bar{x})^2 + n \cdot \sum_{i=1}^{k} (\bar{x}_{i.} - \bar{x})^2 + k \cdot \sum_{j=1}^{n} (\bar{x}_{.j} - \bar{x})^2$$

where $\bar{x}_{i.}$ is the mean of the observations for the ith treatment, $\bar{x}_{.j}$ is the mean of the observations for the jth block, and $\bar{x}$ is the mean of all the nk observations.

Note that the expression on the left-hand side of the equation of Theorem 14.3 is the *total sum of squares* SST as defined on page 398, and that the second term on the right is the *treatment sum of squares* SS(Tr). Analogously, the third term on the right is the *block sum of squares* SSB—it measures the variation of the $\bar{x}_{.j}$ — and the only term which remains, the first term on the right, is the *new error sum of squares* SSE.

We thus have SST = SSE + SS(Tr) + SSB analogous to (14.2.4), and it can be shown that if the null hypothesis concerning the treatment effects α_i is true, then $\frac{\text{SS(Tr)}}{\sigma^2}$ and $\frac{\text{SSE}}{\sigma^2}$ are values of independent random variables having chi-square distributions with $k - 1$ and $(n - 1)(k - 1)$ degrees of freedom. If this null hypothesis is *not true*, then SS(Tr) will also reflect the variation among the α_i, and according to Theorem 7.7 the null hypothesis can be rejected if $F_{\text{Tr}} \geq F_{\alpha,k-1,(n-1)(k-1)}$, where

$$F_{\text{Tr}} = \frac{\dfrac{\text{SS(Tr)}}{k-1}}{\dfrac{\text{SEE}}{(n-1)(k-1)}} = \frac{(n-1)\cdot\text{SS(Tr)}}{\text{SSE}} \tag{14.3.2}$$

Similarly, if the null hypothesis concerning the block effects β_j is true, then $\frac{\text{SSB}}{\sigma^2}$ and $\frac{\text{SSE}}{\sigma^2}$ are values of independent random variables having chi-square distributions with $n - 1$ and $(n - 1)(k - 1)$ degrees of freedom. If this null hypothesis is *not true*, then SSB will also reflect the variation among the β_j, and according to Theorem 7.7 the null hypothesis can be rejected if $F_B \geq F_{\alpha,n-1,(n-1)(k-1)}$, where

$$F_B = \frac{\dfrac{\text{SSB}}{n-1}}{\dfrac{\text{SSE}}{(n-1)(k-1)}} = \frac{(k-1)\text{SSB}}{\text{SSE}} \tag{14.3.3}$$

The *analysis of variance table* for this kind of two-way analysis is usually presented in the following fashion:

Source of Variation	*Degrees of Freedom*	*Sum of Squares*	*Mean Square*	*F*
Treatments	$k - 1$	SS(Tr)	$\text{MS(Tr)} = \frac{\text{SS(Tr)}}{k-1}$	$F_{\text{Tr}} = \frac{\text{MS(Tr)}}{\text{MSE}}$
Blocks	$n - 1$	SSB	$\text{MSB} = \frac{\text{SSB}}{n-1}$	$F_B = \frac{\text{MSB}}{\text{MSE}}$
Error	$(n-1)(k-1)$	SSE	$\text{MSE} = \frac{\text{SSE}}{(n-1)(k-1)}$	
Total	$nk - 1$	SST		

where the mean squares are again the sums of squares divided by the corresponding degrees of freedom. To simplify the calculations, SST and SS(Tr) can be computed by means of the short-cut formulas of Theorem 14.2, and SSB can be determined by means of the following formula (which the reader will be asked to verify in Exercise 2 on page 412):

THEOREM 14.4

If $T_{.j}$ is the total of all the observations for the jth block and $T_{..}$ is the grand total of all the observations,

$$\text{SSB} = \frac{1}{k}\cdot\sum_{j=1}^{n} T_{.j}^2 - \frac{1}{kn}\cdot T_{..}^2$$

Then, SSE can be found by subtraction; that is, by means of the formula SSE = SST − SS(Tr) − SSB.

Returning now to the numerical example on page 396, we find that $\sum_{j=1}^{n} T_{.j}^2 = 62{,}120$ and, hence, that

$$\text{SSB} = \tfrac{1}{4}(62{,}120) - \tfrac{1}{20}(556)^2 = 73.2$$

Since we have already shown on page 401 that SST = 153.2 and SS(Tr) = 52.8, it follows that now SSE = 153.2 − 52.8 − 73.2 = 27.2, and the remaining calculations are given in the following table:

Source of Variation	*Degrees of Freedom*	*Sum of Squares*	*Mean Square*	*F*
Treatments	3	52.8	$\frac{52.8}{3} = 17.6$	$\frac{17.6}{2.27} = 7.75$
Blocks	4	73.2	$\frac{73.2}{4} = 18.3$	$\frac{18.3}{2.27} = 8.06$
Error	12	27.2	$\frac{27.2}{12} = 2.27$	
Total	19	153.2		

Since $F_{\text{Tr}} = 7.75$ *exceeds* $F_{.05,3,12} = 3.49$ and $F_B = 8.06$ *exceeds* $F_{.05,4,12} = 3.26$, we find that both null hypotheses must be rejected. In other words, the difference between the means obtained for the four routes are significant, and so are the differences between the means obtained for the different days of the week. Note, however, that we cannot conclude that Route 1 is necessarily fastest and that on Fridays traffic conditions are necessarily always the worst. All we have shown by means of the analysis is that *differences exist*, and if we want to go one step further and pinpoint the nature of the differences, we will have to use one of the so-called *multiple comparisons tests* referred to on page 418.

14.3.2 Two-way Analysis with Interaction

Looking again at the data on page 396, we cannot rule out the possibility that Route 1 is always good on Mondays (perhaps because a large restaurant along the way is always closed on that day of the week), and we cannot rule out the possibility that Route 3 is always bad during the latter part of the week (perhaps due to the work schedule of crews working on road repairs). To detect such interactions, we should originally have written the mean which corresponds to the ith treatment (the ith route) and the jth block (the jth day) as

$$\mu_{ij} = \mu + \alpha_i + \beta_j + \gamma_{ij} \tag{14.3.4}$$

where the parameters γ_{ij} account for the interactions. Comparison with (14.3.1) shows that we will have to find some way of distinguishing between the *interaction effects* γ_{ij} and the e_{ij} (which we attributed to chance), and to accomplish this we shall *replicate*, namely take r observations for each combination of treatments and blocks. Thus, the model equation becomes

$$x_{ijh} = \mu + \alpha_i + \beta_j + \gamma_{ij} + e_{ijh} \qquad \text{for } \begin{aligned} i &= 1, 2, \ldots, k \\ j &= 1, 2, \ldots, n \\ h &= 1, 2, \ldots, r \end{aligned} \tag{14.3.5}$$

where $\sum_{i=1}^{k} \alpha_i = 0$, $\sum_{j=1}^{n} \beta_j = 0$, $\sum_{i=1}^{k} \gamma_{ij} = 0$ for each j, and $\sum_{j=1}^{n} \gamma_{ij} = 0$ for each i; also, the e_{ijh} are the values of knr independent random variables having normal distributions with zero means and the common variance σ^2.

The three null hypotheses we shall want to test are that *the treatment effects are all equal to zero*, that *the block effects are all equal to zero*, and that *the interaction effects are all equal to zero,* namely, that

$$\alpha_i = 0 \qquad \text{for } i = 1, 2, \ldots, k$$
$$\beta_j = 0 \qquad \text{for } j = 1, 2, \ldots, n$$

and

$$\gamma_{ij} = 0 \qquad \text{for } i = 1, 2, \ldots, k \text{ and } j = 1, 2, \ldots, n$$

The alternative to the first hypothesis is that the treatment effects are *not all equal to zero,* namely, that $\alpha_i \neq 0$ for at least one value of i; the alternative to the second hypothesis is that the block effects are *not all equal to zero,* namely, that $\beta_j \neq 0$ for at least one value of j; and the alternative to the third null hypothesis is that the interaction effects are *not all equal to zero,* namely, that $\gamma_{ij} \neq 0$ for at least one pair of values of i and j.

The analysis on which we base these tests requires the following generalization of Theorems 14.1 and 14.3, which we shall merely state without proof:

THEOREM 14.5

$$\sum_{i=1}^{k}\sum_{j=1}^{n}\sum_{h=1}^{r}(x_{ijh}-\bar{x})^2 = \sum_{i=1}^{k}\sum_{j=1}^{n}\sum_{h=1}^{r}(x_{ijh}-\bar{x}_{ij.})^2 + nr\cdot\sum_{i=1}^{k}(\bar{x}_{i..}-\bar{x})^2 + kr\cdot\sum_{j=1}^{n}(x_{.j.}-\bar{x})^2 + r\cdot\sum_{i=1}^{k}\sum_{j=1}^{n}(\bar{x}_{ij.}-\bar{x}_{i..}-\bar{x}_{.j.}+\bar{x})^2$$

where $\bar{x}$ is again the mean of all the data, $\bar{x}_{i..}$ is the mean of all the observations for the ith treatment, $\bar{x}_{.j.}$ is the mean of all the observations for the jth block, and $\bar{x}_{ij.}$ is the mean of the r observations where the ith treatment is used in combination with the jth block.

The expression on the left-hand side of the equation of this theorem measures the total variation of all the data and it is denoted SST; the first term on the right measures the variation *within* the sets of data which correspond to each of the nk combinations of treatments and blocks, that is, it measures the variation which is due to chance, and it is denoted SSE; the second and third terms on the right are, respectively, the treatment and block sums of squares SS(Tr) and SSB; and the fourth term on the right accounts for the remaining source of variation, interactions, and it is denoted SSI.

In terms of these quantities, the analysis is then performed as indicated in the following *analysis of variance table:*

Source of Variation	*Degrees of Freedom*	*Sum of Squares*	*Mean Square*	*F*
Treatments	$k - 1$	SS(Tr)	$\text{MS(Tr)} = \dfrac{\text{SS(Tr)}}{k-1}$	$F_{\text{Tr}} = \dfrac{\text{MS(Tr)}}{\text{MSE}}$
Blocks	$n - 1$	SSB	$\text{MSB} = \dfrac{\text{SSB}}{n-1}$	$F_B = \dfrac{\text{MSB}}{\text{MSE}}$
Interaction	$(n-1)(k-1)$	SSI	$\text{MSI} = \dfrac{\text{SSI}}{(n-1)(k-1)}$	$F_I = \dfrac{\text{MSI}}{\text{MSE}}$
Error	$kn(r-1)$	SSE	$\text{MSE} = \dfrac{\text{SSE}}{kn(r-1)}$	
Total	$rkn - 1$	SST		

and we *reject* the null hypothesis about the treatments if $F_{\text{Tr}} \geq F_{\alpha,k-1,kn(r-1)}$, the null hypothesis about the blocks if $F_B \geq F_{\alpha,n-1,kn(r-1,)}$ and the null hypothesis about interactions if $F_I \geq F_{\alpha,(n-1)(k-1),kn(r-1)}$.

To illustrate this kind of analysis, suppose that the following are the number of defective pieces produced by four machine operators, each working on three different occasions four different machines:

	Machine Operator			
	Mr. A	*Mr. B*	*Mr. C*	*Mr. D*
Machine 1	16, 13, 19	18, 17, 21	14, 16, 13	13, 14, 16
Machine 2	9, 15, 11	15, 13, 12	7, 12, 9	3, 1, 9
Machine 3	22, 25, 17	14, 16, 12	11, 14, 12	13, 17, 14
Machine 4	14, 16, 12	20, 16, 17	8, 11, 10	6, 8, 12

Calculating the necessary sums of squares by means of the short-cut formulas the reader will be asked to derive in Exercise 3 on page 413, we find that SST = 977.5, SS(Tr) = 290.4 where the different "treatments" are Mr. A, Mr. B, Mr. C, and Mr. D, SSB = 305.3 where the different "blocks" are the four machines, SSI = 181.1, and SSE = 200.7. Thus

Source of Variation	*Degrees of Freedom*	*Sum of Squares*	*Mean Square*	F
Treatments	3	290.4	$\frac{290.4}{3} = 96.8$	$\frac{96.8}{6.27} = 15.4$
Blocks	3	305.3	$\frac{305.3}{3} = 101.8$	$\frac{101.8}{6.27} = 16.2$
Interaction	9	181.1	$\frac{181.1}{9} = 20.1$	$\frac{20.1}{6.27} = 3.20$
Error	32	200.7	$\frac{200.7}{32} = 6.27$	
Total	47	977.5		

The required values of $F_{.05,3,32}$ and $F_{.05,9,32}$ are not given in the table at the end of this book, but since $F_{.05,3,32}$ is *less than* 2.92 and $F_{.05,9,32}$ is *less than* 2.21, we find that all three of the null hypotheses will have to be *rejected* at the level of significance 0.05 (and in Exercise 9 on page 415 the reader will be asked to verify that the same is true also for the level of significance 0.01).

The fact that there are significant interactions is most important, as it requires that we be very careful before we judge that one machine operator is better than another, or that one machine performs better (or worse) than the rest. However, it is apparent from the data that Mr. A should be kept away from Machine 3 and that Machine 2 should be reserved for Mr. D. (These things do not follow from the F tests, but they are observations which seem to be on fairly safe grounds.)

THEORETICAL EXERCISES

1. Prove Theorem 14.3, making use of the identity

$$x_{ij} - \bar{x} = (x_{ij} - \bar{x}_{i.} - \bar{x}_{.j} + \bar{x}) + (\bar{x}_{i.} - \bar{x}) + (\bar{x}_{.j} - \bar{x})$$

2. Verify the computing formula for SSB given in Theorem 14.4.

3. Show that the sums of squares required for the analysis of Section 14.3.2 can be computed by means of the following formulas:

$$\text{SST} = \sum_{i=1}^{k}\sum_{j=1}^{n}\sum_{h=1}^{r} x_{ijh}^2 - \frac{1}{knr}\cdot T_{\cdots}^2$$

$$\text{SS(Tr)} = \frac{1}{nr}\cdot\sum_{i=1}^{k} T_{i\cdot\cdot}^2 - \frac{1}{knr}\cdot T_{\cdots}^2$$

$$\text{SSB} = \frac{1}{kr}\cdot\sum_{j=1}^{n} T_{\cdot j\cdot}^2 - \frac{1}{knr}\cdot T_{\cdots}^2$$

$$\text{SSI} = \frac{1}{r}\cdot\sum_{i=1}^{k}\sum_{j=1}^{n} T_{ij\cdot}^2 - \frac{1}{knr}\cdot T_{\cdots}^2 - \text{SS(Tr)} - \text{SSB}$$

and

$$\text{SSE} = \text{SST} - \text{SS(Tr)} - \text{SSB} - \text{SSI}$$

where $T_{\cdots}$ is the grand total of all the observations, $T_{i\cdot\cdot}$ is the total of all the observations for the ith treatment, $T_{\cdot j\cdot}$ is the total of all the observations for the jth block, and $T_{ij\cdot}$ is the total of the r observations where the ith treatment is used in combination with the jth block.

4. A *Latin square* is a square array in which each letter (or other symbol) appears exactly once in each row and once in each column. For instance,

A	*B*	*C*	*D*
B	*C*	*D*	*A*
C	*D*	*A*	*B*
D	*A*	*B*	*C*

is a Latin square of side 4. If we look upon the m *rows* of a Latin square as the levels of one variable, the m columns as the levels of a second variable, and A, B, C, . . . , as m "treatments," namely, as the levels of a third variable, it is possible to test hypotheses concerning all three of these variables on the basis of as few as m^2 observations (provided there are no interactions). Letting $x_{ij(k)}$ denote the observation in the ith row and the jth column of a Latin square (so that k, the treatment, is determined when we give i and j), we write the model equation as

$$x_{ij(k)} = \mu + \alpha_i + \beta_j + \tau_k + e_{ij} \qquad \text{for} \qquad \begin{aligned} i &= 1, 2, \ldots, m \\ j &= 1, 2, \ldots, m \\ k &= 1, 2, \ldots, m \end{aligned}$$

where μ is the *grand mean*, α_i is the ith *row effect*, β_j is the jth *column effect*, τ_k is the kth *treatment effect*, $\sum_{i=1}^{m} \alpha_i = 0$, $\sum_{j=1}^{m} \beta_j = 0$, $\sum_{k=1}^{m} \tau_k = 0$, and the e_{ij} are values of independent random variables having normal distributions with zero means and the common variance σ^2. The null hypotheses we shall want to test (against the appropriate alternatives) are that the row effects are all zero, that the column effects are all zero, and that the treatment effects are all zero.

(a) Show that

$$\sum_{i=1}^{m}\sum_{j=1}^{m}(x_{ij(k)} - \bar{x})^2 = \sum_{i=1}^{m}\sum_{j=1}^{m}(x_{ij(k)} - \bar{x}_{i.} - \bar{x}_{.j} - \bar{x}_{(k)} + 2\bar{x})^2$$
$$+ m\cdot\sum_{i=1}^{m}(\bar{x}_{i.} - \bar{x})^2 + m\cdot\sum_{j=1}^{m}(\bar{x}_{.j} - \bar{x})^2$$
$$+ m\cdot\sum_{k=1}^{m}(\bar{x}_{(k)} - \bar{x})^2$$

where $\bar{x}_{(k)}$ is the mean of all the observations for the kth treatment and the other means are as defined in Theorem 14.3. The expression on the left-hand side of the equation is the *total sum of squares* SST, while those on the right-hand side are, respectively, the *error sum of squares* SSE, the *row sum of squares* SSR, the *column sum of squares* SSC, and the *treatment sum of squares* SS(Tr).

(b) Construct an analysis of variance table for this kind of experiment, determining the degrees of freedom for SSE by subtracting those for SSR, SSC, and SS(Tr) from $m^2 - 1$ (the degrees of freedom for SST).

APPLIED EXERCISES

5. An experiment was performed to judge the effect of four different fuels and three different types of launchers on the range of a certain rocket. Test, on the basis of the following data, whether there is a significant effect due to differences in fuels and whether there is a significant effect due to differences in launchers (use $\alpha = 0.01$):

	Fuel 1	Fuel 2	Fuel 3	Fuel 4
Launcher X	45.9	57.6	52.2	41.7
Launcher Y	46.0	51.0	50.1	38.8
Launcher Z	45.7	56.9	55.3	48.1

6. To study the performance of three different detergents at three different water temperatures, the following "whiteness" readings were obtained with specially designed equipment:

	Detergent A	Detergent B	Detergent C
Cold Water	45	43	55
Warm Water	37	40	56
Hot Water	42	44	46

Perform a two-way analysis of variance, using the level of significance $\alpha = 0.05$.

7. Suppose that in the experiment of Exercise 7 on page 403 three different kinds of fertilizers were applied to the soil; in fact, the first value for each variety of wheat was obtained with Fertilizer I, the second with Fertilizer II, and the third with Fertilizer III. Perform a two-way analysis of variance, using the level of significance $\alpha = 0.05$.

8. Suppose that in the experiment of Exercise 6 they tested three loads of washing at each combination of detergents and water temperature, and that they obtained the following results:

	Detergent A	Detergent B	Detergent C
Cold Water	45, 39, 46	43, 46, 41	55, 48, 53
Warm Water	37, 32, 43	40, 37, 46	56, 51, 53
Hot Water	42, 42, 46	44, 45, 38	46, 49, 42

Use the level of significance $\alpha = 0.01$ to test for differences between the detergents, differences due to water temperature, and interactions.

9. Use the formulas of Exercise 3 to verify the values given for the various sums of squares in the analysis of variance table on page 412, and check also whether the three null hypotheses could have been rejected at the level of significance $\alpha = 0.01$.

10. The following data are the lifetimes (in hours) of three different designs of an airplane wing subjected twice each to two different kinds of vibration (at constant frequency and sweeping back and forth over a given band width):

	Design 1	*Design 2*	*Design 3*
Vibration 1	1676, 1510	883, 961	1212, 1090
Vibration 2	1915, 1763	948, 1062	1091, 1212

Perform a two-way analysis of variance, testing all three null hypotheses at the level of significance $\alpha = 0.05$.

11. To study the market for a new breakfast food, a research organization tests it in four different cities, with four different kinds of packaging, and with four different advertising campaigns (denoted A, B, C, and D in the following *Latin square*). The figures are one week's sales in $10,000.

		Cities 1	2	3	4
Packaging	I	*A* 52	*B* 51	*C* 55	*D* 56
	II	*B* 50	*C* 45	*D* 49	*A* 51
	III	*C* 39	*D* 41	*A* 37	*B* 39
	IV	*D* 43	*A* 41	*B* 42	*C* 42

Analyze this experiment by the method of Exercise 4, using the level of significance $\alpha = 0.05$.

14.4 SOME FURTHER CONSIDERATIONS

The topics presented in this chapter gave us a brief introduction to some of the basic ideas of experimental design and the analysis of variance. The scope of these subjects, which is closely interrelated, is vast, and new methods are constantly being developed as their need arises in experiments where "standard" assumptions do not apply.

The designs which we have discussed all had the special feature that there were observations corresponding to all "treatments" and "blocks," namely, for all possible combinations of the values (levels) of the variables under consideration. To show that this can be very impractical or even impossible, we have only to consider an experiment in which we want to compare the yield of 25 varieties of corn and the effect of 12 different fertilizers. To use a model like that of Section 14.3.1 we would have to plant 300 plots, and it does not require much imagination to see how difficult it would be to find that many plots for which soil composition, irrigation, slope, ..., are constant or otherwise controllable. Consequently, there is a need for designs which make it possible to test hypotheses concerning the most relevant (though not all) parameters of the model on the basis of experiments which are feasible from a practical point of view. This leads to so-called *incomplete block designs*, which are discussed in some of the general references on experimental design listed on page 418.

Further complications arise when there are extraneous variables which can be measured but not controlled. For example, in a comparison of various kinds of "teaching machines" it may be impossible to use individuals who all have the same I.Q., but at least their I.Q.'s can be known. In a situation like this we might use an *analysis of covariance* model such as

$$x_{ij} = \mu + \alpha_i + \beta y_i + e_{ij}$$

which differs from (14.2.2) in that we added the term βy_i, where y_i are the known I.Q.'s. Note that in this model the estimation of β is essentially a problem of regression analysis.

Other difficulties arise when the α_i and β_j in (14.3.1) are not constants, but values of random variables. This kind of situation would arise, for example, when there are 30 varieties of corn and 12 kinds of fertilizer and we randomly select, say, six of the varieties of corn and four of the fertilizers to be included in the experiment. These are just some of the generalizations of the methods which we have presented in this chapter; they are treated in great detail in the special texts on experimental statistics which are listed below.

REFERENCES

The method of Lagrange multipliers (required for Exercise 5 on page 403) is treated in most textbooks on advanced calculus; for example, in

Widder, D. V., *Advanced Calculus*, 2nd ed. Englewood Cliffs, N.J.: Prentice-Hall, Inc., 1961.

A proof of the independence of the random variables whose values constitute the various sums of squares in an analysis of variance [for instance, SS(Tr) and SSE on page 399] may be found in

Scheffé, H., *The Analysis of Variance.* New York: John Wiley & Sons, Inc., 1959,

and a treatment of various multiple comparisons tests (see page 409) is given in

Federer, W. T., *Experimental Design, Theory and Application.* New York: The Macmillan Company, 1955.

The following are some general texts on experimental design and analysis of variance:

Cochran, W. G., and Cox, G. M., *Experimental Design,* 2nd ed. New York: John Wiley & Sons, Inc., 1957.

Finney, D. J., *An Introduction to Experimental Design.* Chicago: University of Chicago Press, 1960.

Fisher, R. A., *Statistical Methods for Research Workers,* 10th ed. New York: Hafner Publishing Co., 1948.

Graybill, F. A., *An Introduction to Linear Statistical Models,* Vol. I. New York: McGraw-Hill Book Company, 1961.

Hicks, C. R., *Fundamental Concepts in the Design of Experiments.* New York: Holt, Rinehart & Winston, Inc., 1964.

Kempthorne, O., *The Design and Analysis of Experiments.* New York: John Wiley & Sons, Inc., 1952.

Snedecor, G. W., and Cochran, W. G., *Statistical Methods,* 6th ed. Ames, Iowa: Iowa State University Press, 1967.

APPENDIX: SUMS AND PRODUCTS

A.1 RULES FOR SUMS AND PRODUCTS

To simplify expressions involving sums and products, the Σ and Π notations are widely used in statistics. In case the reader is not familiar with the use of these symbols,

$$\sum_{i=a}^{b} x_i = x_a + x_{a+1} + x_{a+2} + \cdots + x_b$$

and

$$\prod_{i=a}^{b} x_i = x_a \cdot x_{a+1} \cdot x_{a+2} \cdot \ldots \cdot x_b$$

for any non-negative integers a and b with $b \geq a$.

When working with sums or products, it is often helpful to apply the following rules, which can all be verified directly by writing the respective expressions in full, that is, without the Σ and Π notations:

$$\sum_{i=1}^{n} kx_i = k \cdot \sum_{i=1}^{n} x_i \tag{A.1.1}$$

$$\sum_{i=1}^{n} k = nk \tag{A.1.2}$$

$$\sum_{i=1}^{n} (x_i + y_i) = \sum_{i=1}^{n} x_i + \sum_{i=1}^{n} y_i \tag{A.1.3}$$

$$\prod_{i=1}^{n} kx_i = k^n \cdot \prod_{i=1}^{n} x_i \tag{A.1.4}$$

$$\prod_{i=1}^{n} k = k^n \tag{A.1.5}$$

$$\prod_{i=1}^{n} x_i y_i = \left(\prod_{i=1}^{n} x_i\right)\left(\prod_{i=1}^{n} y_i\right) \tag{A.1.6}$$

$$\ln \prod_{i=1}^{n} x_i = \sum_{i=1}^{n} \ln x_i \tag{A.1.7}$$

Double sums, triple sums, ..., are also widely used in statistics, and repeatedly applying the definition of Σ given above, we have, for example,

$$\begin{aligned}\sum_{i=1}^{m}\sum_{j=1}^{n} x_{ij} &= \sum_{i=1}^{m} (x_{i1} + x_{i2} + \cdots + x_{in}) \\ &= (x_{11} + x_{12} + \cdots + x_{1n}) + (x_{21} + x_{22} + \cdots + x_{2n}) \\ &\quad + \cdots + (x_{m1} + x_{m2} + \cdots + x_{mn})\end{aligned}$$

When working with double sums, the following theorem, which is an immediate consequence of the multinomial expansion of

$$(x_1 + x_2 + \cdots + x_n)^2$$

is of special interest:

THEOREM A.1

$$\sum_{i<j}\sum x_i x_j = \frac{1}{2}\left[\left(\sum_{i=1}^{n} x_i\right)^2 - \sum_{i=1}^{n} x_i^2\right]$$

where

$$\sum_{i<j}\sum x_i x_j = \sum_{i=1}^{n-1}\sum_{j=i+1}^{n} x_i x_j$$

A.2 SOME SPECIAL SUMS

In the theory of nonparametric statistics, particularly when dealing with rank sums, we often need expressions for sums of powers of the first n positive integers, namely, expressions for

$$S(n, r) = 1^r + 2^r + 3^r + \cdots + n^r$$

for $r = 0, 1, 2, 3, \ldots$. The following theorem, which the reader will be asked to prove in Exercise 1 below, provides an easy way of obtaining these sums:

Theorem A.2

For any positive integers n and k,

$$\sum_{r=0}^{k-1} \binom{k}{r} S(n, r) = (n+1)^k - 1$$

The only disadvantage of this theorem is that we have to find the sums $S(n, r)$ one at a time, first for $r = 0$, then for $r = 1$, then for $r = 2$, and so forth. For instance, for $k = 1$ we get

$$\binom{1}{0} S(n, 0) = (n+1) - 1 = n$$

and, hence, $S(n, 0) = 1^0 + 2^0 + \cdots + n^0 = n$. Similarly, for $k = 2$ we get

$$\binom{2}{0} S(n, 0) + \binom{2}{1} S(n, 1) = (n+1)^2 - 1$$
$$n + 2S(n, 1) = n^2 + 2n$$

and, hence, $S(n, 1) = 1^1 + 2^1 + \cdots + n^1 = \frac{1}{2}n(n+1)$. Using the same technique, the reader will be asked to show in Exercise 2 below that

$$S(n, 2) = \tfrac{1}{6}n(n+1)(2n+1) \quad \text{and} \quad S(n, 3) = \tfrac{1}{4}n^2(n+1)^2$$

THEORETICAL EXERCISES

1. Prove Theorem A.2 by making use of the fact that

$$(m+1)^k - m^k = \sum_{r=0}^{k-1} \binom{k}{r} m^r$$

   which follows from the binomial expansion of $(m+1)^k$, and then summing on m from 1 to n.

2. Verify the formulas for $S(n, 2)$ and $S(n, 3)$ given in the text, and find one for $S(n, 4)$.

STATISTICAL TABLES

TABLE I

BINOMIAL PROBABILITIES*

		θ									
n	x	.05	.10	.15	.20	.25	.30	.35	.40	.45	.50
1	0	.9500	.9000	.8500	.8000	.7500	.7000	.6500	.6000	.5500	.5000
	1	.0500	.1000	.1500	.2000	.2500	.3000	.3500	.4000	.4500	.5000
2	0	.9025	.8100	.7225	.6400	.5625	.4900	.4225	.3600	.3025	.2500
	1	.0950	.1800	.2550	.3200	.3750	.4200	.4550	.4800	.4950	.5000
	2	.0025	.0100	.0225	.0400	.0625	.0900	.1225	.1600	.2025	.2500
3	0	.8574	.7290	.6141	.5120	.4219	.3430	.2746	.2160	.1664	.1250
	1	.1354	.2430	.3251	.3840	.4219	.4410	.4436	.4320	.4084	.3750
	2	.0071	.0270	.0574	.0960	.1406	.1890	.2389	.2880	.3341	.3750
	3	.0001	.0010	.0034	.0080	.0156	.0270	.0429	.0640	.0911	.1250
4	0	.8145	.6561	.5220	.4096	.3164	.2401	.1785	.1296	.0915	.0625
	1	.1715	.2916	.3685	.4096	.4219	.4116	.3845	.3456	.2995	.2500
	2	.0135	.0486	.0975	.1536	.2109	.2646	.3105	.3456	.3675	.3750
	3	.0005	.0036	.0115	.0256	.0469	.0756	.1115	.1536	.2005	.2500
	4	.0000	.0001	.0005	.0016	.0039	.0081	.0150	.0256	.0410	.0625
5	0	.7738	.5905	.4437	.3277	.2373	.1681	.1160	.0778	.0503	.0312
	1	.2036	.3280	.3915	.4096	.3955	.3602	.3124	.2592	.2059	.1562
	2	.0214	.0729	.1382	.2048	.2637	.3087	.3364	.3456	.3369	.3125
	3	.0011	.0081	.0244	.0512	.0879	.1323	.1811	.2304	.2757	.3125
	4	.0000	.0004	.0022	.0064	.0146	.0284	.0488	.0768	.1128	.1562
	5	.0000	.0000	.0001	.0003	.0010	.0024	.0053	.0102	.0185	.0312
6	0	.7351	.5314	.3771	.2621	.1780	.1176	.0754	.0467	.0277	.0156
	1	.2321	.3543	.3993	.3932	.3560	.3025	.2437	.1866	.1359	.0938
	2	.0305	.0984	.1762	.2458	.2966	.3241	.3280	.3110	.2780	.2344
	3	.0021	.0146	.0415	.0819	.1318	.1852	.2355	.2765	.3032	.3125
	4	.0001	.0012	.0055	.0154	.0330	.0595	.0951	.1382	.1861	.2344
	5	.0000	.0001	.0004	.0015	.0044	.0102	.0205	.0369	.0609	.0938
	6	.0000	.0000	.0000	.0001	.0002	.0007	.0018	.0041	.0083	.0156
7	0	.6983	.4783	.3206	.2097	.1335	.0824	.0490	.0280	.0152	.0078
	1	.2573	.3720	.3960	.3670	.3115	.2471	.1848	.1306	.0872	.0547
	2	.0406	.1240	.2097	.2753	.3115	.3177	.2985	.2613	.2140	.1641
	3	.0036	.0230	.0617	.1147	.1730	.2269	.2679	.2903	.2918	.2734
	4	.0002	.0026	.0109	.0287	.0577	.0972	.1442	.1935	.2388	.2734
	5	.0000	.0002	.0012	.0043	.0115	.0250	.0466	.0774	.1172	.1641
	6	.0000	.0000	.0001	.0004	.0013	.0036	.0084	.0172	.0320	.0547
	7	.0000	.0000	.0000	.0000	.0001	.0002	.0006	.0016	.0037	.0078
8	0	.6634	.4305	.2725	.1678	.1001	.0576	.0319	.0168	.0084	.0039
	1	.2793	.3826	.3847	.3355	.2670	.1977	.1373	.0896	.0548	.0312
	2	.0515	.1488	.2376	.2936	.3115	.2965	.2587	.2090	.1569	.1094
	3	.0054	.0331	.0839	.1468	.2076	.2541	.2786	.2787	.2568	.2188
	4	.0004	.0046	.0185	.0459	.0865	.1361	.1875	.2322	.2627	.2734

* Entries in this table are values of $\binom{n}{x}\theta^x(1-\theta)^{n-x}$ for the indicated values of n, x, and θ. Reproduced by permission from *Handbook of Probability and Statistics with Tables*, by Burington and May, 1953, McGraw-Hill Book Co., Inc.

TABLE I (continued)

		θ									
n	x	.05	.10	.15	.20	.25	.30	.35	.40	.45	.50
8	5	.0000	.0004	.0026	.0092	.0231	.0467	.0808	.1239	.1719	.2188
	6	.0000	.0000	.0002	.0011	.0038	.0100	.0217	.0413	.0703	.1094
	7	.0000	.0000	.0000	.0001	.0004	.0012	.0033	.0079	.0164	.0312
	8	.0000	.0000	.0000	.0000	.0000	.0001	.0002	.0007	.0017	.0039
9	0	.6302	.3874	.2316	.1342	.0751	.0404	.0207	.0101	.0046	.0020
	1	.2985	.3874	.3679	.3020	.2253	.1556	.1004	.0605	.0339	.0176
	2	.0629	.1722	.2597	.3020	.3003	.2668	.2162	.1612	.1110	.0703
	3	.0077	.0446	.1069	.1762	.2336	.2668	.2716	.2508	.2119	.1641
	4	.0006	.0074	.0283	.0661	.1168	.1715	.2194	.2508	.2600	.2461
	5	.0000	.0008	.0050	.0165	.0389	.0735	.1181	.1672	.2128	.2461
	6	.0000	.0001	.0006	.0028	.0087	.0210	.0424	.0743	.1160	.1641
	7	.0000	.0000	.0000	.0003	.0012	.0039	.0098	.0212	.0407	.0703
	8	.0000	.0000	.0000	.0000	.0001	.0004	.0013	.0035	.0083	.0176
	9	.0000	.0000	.0000	.0000	.0000	.0000	.0001	.0003	.0008	.0020
10	0	.5987	.3487	.1969	.1074	.0563	.0282	.0135	.0060	.0025	.0010
	1	.3151	.3874	.3474	.2684	.1877	.1211	.0725	.0403	.0207	.0098
	2	.0746	.1937	.2759	.3020	.2816	.2335	.1757	.1209	.0763	.0439
	3	.0105	.0574	.1298	.2013	.2503	.2668	.2522	.2150	.1665	.1172
	4	.0010	.0112	.0401	.0881	.1460	.2001	.2377	.2508	.2384	.2051
	5	.0001	.0015	.0085	.0264	.0584	.1029	.1536	.2007	.2340	.2461
	6	.0000	.0001	.0012	.0055	.0162	.0368	.0689	.1115	.1596	.2051
	7	.0000	.0000	.0001	.0008	.0031	.0090	.0212	.0425	.0746	.1172
	8	.0000	.0000	.0000	.0001	.0004	.0014	.0043	.0106	.0229	.0439
	9	.0000	.0000	.0000	.0000	.0000	.0001	.0005	.0016	.0042	.0098
	10	.0000	.0000	.0000	.0000	.0000	.0000	.0000	.0001	.0003	.0010
11	0	.5688	.3138	.1673	.0859	.0422	.0198	.0088	.0036	.0014	.0005
	1	.3293	.3835	.3248	.2362	.1549	.0932	.0518	.0266	.0125	.0054
	2	.0867	.2131	.2866	.2953	.2581	.1998	.1395	.0887	.0513	.0269
	3	.0137	.0710	.1517	.2215	.2581	.2568	.2254	.1774	.1259	.0806
	4	.0014	.0158	.0536	.1107	.1721	.2201	.2428	.2365	.2060	.1611
	5	.0001	.0025	.0132	.0388	.0803	.1321	.1830	.2207	.2360	.2256
	6	.0000	.0003	.0023	.0097	.0268	.0566	.0985	.1471	.1931	.2256
	7	.0000	.0000	.0003	.0017	.0064	.0173	.0379	.0701	.1128	.1611
	8	.0000	.0000	.0000	.0002	.0011	.0037	.0102	.0234	.0462	.0806
	9	.0000	.0000	.0000	.0000	.0001	.0005	.0018	.0052	.0126	.0269
	10	.0000	.0000	.0000	.0000	.0000	.0000	.0002	.0007	.0021	.0054
	11	.0000	.0000	.0000	.0000	.0000	.0000	.0000	.0000	.0002	.0005
12	0	.5404	.2824	.1422	.0687	.0317	.0138	.0057	.0022	.0008	.0002
	1	.3413	.3766	.3012	.2062	.1267	.0712	.0368	.0174	.0075	.0029
	2	.0988	.2301	.2924	.2835	.2323	.1678	.1088	.0639	.0339	.0161
	3	.0173	.0852	.1720	.2362	.2581	.2397	.1954	.1419	.0923	.0537
	4	.0021	.0213	.0683	.1329	.1936	.2311	.2367	.2128	.1700	.1208
	5	.0002	.0038	.0193	.0532	.1032	.1585	.2039	.2270	.2225	.1934
	6	.0000	.0005	.0040	.0155	.0401	.0792	.1281	.1766	.2124	.2256
	7	.0000	.0000	.0006	.0033	.0115	.0291	.0591	.1009	.1489	.1934
	8	.0000	.0000	.0001	.0005	.0024	.0078	.0199	.0420	.0762	.1208
	9	.0000	.0000	.0000	.0001	.0004	.0015	.0048	.0125	.0277	.0537

TABLE I (continued)

n	x	θ .05	.10	.15	.20	.25	.30	.35	.40	.45	.50
12	10	.0000	.0000	.0000	.0000	.0000	.0002	.0008	.0025	.0068	.0161
	11	.0000	.0000	.0000	.0000	.0000	.0000	.0001	.0003	.0010	.0029
	12	.0000	.0000	.0000	.0000	.0000	.0000	.0000	.0000	.0001	.0002
13	0	.5133	.2542	.1209	.0550	.0238	.0097	.0037	.0013	.0004	.0001
	1	.3512	.3672	.2774	.1787	.1029	.0540	.0259	.0113	.0045	.0016
	2	.1109	.2448	.2937	.2680	.2059	.1388	.0836	.0453	.0220	.0095
	3	.0214	.0997	.1900	.2457	.2517	.2181	.1651	.1107	.0660	.0349
	4	.0028	.0277	.0838	.1535	.2097	.2337	.2222	.1845	.1350	.0873
	5	.0003	.0055	.0266	.0691	.1258	.1803	.2154	.2214	.1989	.1571
	6	.0000	.0008	.0063	.0230	.0559	.1030	.1546	.1968	.2169	.2095
	7	.0000	.0001	.0011	.0058	.0186	.0442	.0833	.1312	.1775	.2095
	8	.0000	.0000	.0001	.0011	.0047	.0142	.0336	.0656	.1089	.1571
	9	.0000	.0000	.0000	.0001	.0009	.0034	.0101	.0243	.0495	.0873
	10	.0000	.0000	.0000	.0000	.0001	.0006	.0022	.0065	.0162	.0349
	11	.0000	.0000	.0000	.0000	.0000	.0001	.0003	.0012	.0036	.0095
	12	.0000	.0000	.0000	.0000	.0000	.0000	.0000	.0001	.0005	.0016
	13	.0000	.0000	.0000	.0000	.0000	.0000	.0000	.0000	.0000	.0001
14	0	.4877	.2288	.1028	.0440	.0178	.0068	.0024	.0008	.0002	.0001
	1	.3593	.3559	.2539	.1539	.0832	.0407	.0181	.0073	.0027	.0009
	2	.1229	.2570	.2912	.2501	.1802	.1134	.0634	.0317	.0141	.0056
	3	.0259	.1142	.2056	.2501	.2402	.1943	.1366	.0845	.0462	.0222
	4	.0037	.0349	.0998	.1720	.2202	.2290	.2022	.1549	.1040	.0611
	5	.0004	.0078	.0352	.0860	.1468	.1963	.2178	.2066	.1701	.1222
	6	.0000	.0013	.0093	.0322	.0734	.1262	.1759	.2066	.2088	.1833
	7	.0000	.0002	.0019	.0092	.0280	.0618	.1082	.1574	.1952	.2095
	8	.0000	.0000	.0003	.0020	.0082	.0232	.0510	.0918	.1398	.1833
	9	.0000	.0000	.0000	.0003	.0018	.0066	.0183	.0408	.0762	.1222
	10	.0000	.0000	.0000	.0000	.0003	.0014	.0049	.0136	.0312	.0611
	11	.0000	.0000	.0000	.0000	.0000	.0002	.0010	.0033	.0093	.0222
	12	.0000	.0000	.0000	.0000	.0000	.0000	.0001	.0005	.0019	.0056
	13	.0000	.0000	.0000	.0000	.0000	.0000	.0000	.0001	.0002	.0009
	14	.0000	.0000	.0000	.0000	.0000	.0000	.0000	.0000	.0000	.0001
15	0	.4633	.2059	.0874	.0352	.0134	.0047	.0016	.0005	.0001	.0000
	1	.3658	.3432	.2312	.1319	.0668	.0305	.0126	.0047	.0016	.0005
	2	.1348	.2669	.2856	.2309	.1559	.0916	.0476	.0219	.0090	.0032
	3	.0307	.1285	.2184	.2501	.2252	.1700	.1110	.0634	.0318	.0139
	4	.0049	.0428	.1156	.1876	.2252	.2186	.1792	.1268	.0780	.0417
	5	.0006	.0105	.0449	.1032	.1651	.2061	.2123	.1859	.1404	.0916
	6	.0000	.0019	.0132	.0430	.0917	.1472	.1906	.2066	.1914	.1527
	7	.0000	.0003	.0030	.0138	.0393	.0811	.1319	.1771	.2013	.1964
	8	.0000	.0000	.0005	.0035	.0131	.0348	.0710	.1181	.1647	.1964
	9	.0000	.0000	.0001	.0007	.0034	.0116	.0298	.0612	.1048	.1527
	10	.0000	.0000	.0000	.0001	.0007	.0030	.0096	.0245	.0515	.0916
	11	.0000	.0000	.0000	.0000	.0001	.0006	.0024	.0074	.0191	.0417
	12	.0000	.0000	.0000	.0000	.0000	.0001	.0004	.0016	.0052	.0139
	13	.0000	.0000	.0000	.0000	.0000	.0000	.0001	.0003	.0010	.0032
	14	.0000	.0000	.0000	.0000	.0000	.0000	.0000	.0000	.0001	.0005
	15	.0000	.0000	.0000	.0000	.0000	.0000	.0000	.0000	.0000	.0000

TABLE I (continued)

						θ					
n	x	.05	.10	.15	.20	.25	.30	.35	.40	.45	.50
16	0	.4401	.1853	.0743	.0281	.0100	.0033	.0010	.0003	.0001	.0000
	1	.3706	.3294	.2097	.1126	.0535	.0228	.0087	.0030	.0009	.0002
	2	.1463	.2745	.2775	.2111	.1336	.0732	.0353	.0150	.0056	.0018
	3	.0359	.1423	.2285	.2463	.2079	.1465	.0888	.0468	.0215	.0085
	4	.0061	.0514	.1311	.2001	.2252	.2040	.1553	.1014	.0572	.0278
	5	.0008	.0137	.0555	.1201	.1802	.2099	.2008	.1623	.1123	.0667
	6	.0001	.0028	.0180	.0550	.1101	.1649	.1982	.1983	.1684	.1222
	7	.0000	.0004	.0045	.0197	.0524	.1010	.1524	.1889	.1969	.1746
	8	.0000	.0001	.0009	.0055	.0197	.0487	.0923	.1417	.1812	.1964
	9	.0000	.0000	.0001	.0012	.0058	.0185	.0442	.0840	.1318	.1746
	10	.0000	.0000	.0000	.0002	.0014	.0056	.0167	.0392	.0755	.1222
	11	.0000	.0000	.0000	.0000	.0002	.0013	.0049	.0142	.0337	.0667
	12	.0000	.0000	.0000	.0000	.0000	.0002	.0011	.0040	.0115	.0278
	13	.0000	.0000	.0000	.0000	.0000	.0000	.0002	.0008	.0029	.0085
	14	.0000	.0000	.0000	.0000	.0000	.0000	.0000	.0001	.0005	.0018
	15	.0000	.0000	.0000	.0000	.0000	.0000	.0000	.0000	.0001	.0002
	16	.0000	.0000	.0000	.0000	.0000	.0000	.0000	.0000	.0000	.0000
17	0	.4181	.1668	.0631	.0225	.0075	.0023	.0007	.0002	.0000	.0000
	1	.3741	.3150	.1893	.0957	.0426	.0169	.0060	.0019	.0005	.0001
	2	.1575	.2800	.2673	.1914	.1136	.0581	.0260	.0102	.0035	.0010
	3	.0415	.1556	.2359	.2393	.1893	.1245	.0701	.0341	.0144	.0052
	4	.0076	.0605	.1457	.2093	.2209	.1868	.1320	.0796	.0411	.0182
	5	.0010	.0175	.0668	.1361	.1914	.2081	.1849	.1379	.0875	.0472
	6	.0001	.0039	.0236	.0680	.1276	.1784	.1991	.1839	.1432	.0944
	7	.0000	.0007	.0065	.0267	.0668	.1201	.1685	.1927	.1841	.1484
	8	.0000	.0001	.0014	.0084	.0279	.0644	.1134	.1606	.1883	.1855
	9	.0000	.0000	.0003	.0021	.0093	.0276	.0611	.1070	.1540	.1855
	10	.0000	.0000	.0000	.0004	.0025	.0095	.0263	.0571	.1008	.1484
	11	.0000	.0000	.0000	.0001	.0005	.0026	.0090	.0242	.0525	.0944
	12	.0000	.0000	.0000	.0000	.0001	.0006	.0024	.0081	.0215	.0472
	13	.0000	.0000	.0000	.0000	.0000	.0001	.0005	.0021	.0068	.0182
	14	.0000	.0000	.0000	.0000	.0000	.0000	.0001	.0004	.0016	.0052
	15	.0000	.0000	.0000	.0000	.0000	.0000	.0000	.0001	.0003	.0010
	16	.0000	.0000	.0000	.0000	.0000	.0000	.0000	.0000	.0000	.0001
	17	.0000	.0000	.0000	.0000	.0000	.0000	.0000	.0000	.0000	.0000
18	0	.3972	.1501	.0536	.0180	.0056	.0016	.0004	.0001	.0000	.0000
	1	.3763	.3002	.1704	.0811	.0338	.0126	.0042	.0012	.0003	.0001
	2	.1683	.2835	.2556	.1723	.0958	.0458	.0190	.0069	.0022	.0006
	3	.0473	.1680	.2406	.2297	.1704	.1046	.0547	.0246	.0095	.0031
	4	.0093	.0700	.1592	.2153	.2130	.1681	.1104	.0614	.0291	.0117
	5	.0014	.0218	.0787	.1507	.1988	.2017	.1664	.1146	.0666	.0327
	6	.0002	.0052	.0301	.0816	.1436	.1873	.1941	.1655	.1181	.0708
	7	.0000	.0010	.0091	.0350	.0820	.1376	.1792	.1892	.1657	.1214
	8	.0000	.0002	.0022	.0120	.0376	.0811	.1327	.1734	.1864	.1669
	9	.0000	.0000	.0004	.0033	.0139	.0386	.0794	.1284	.1694	.1855

TABLE I (continued)

		θ									
n	*x*	.05	.10	.15	.20	.25	.30	.35	.40	.45	.50
18	10	.0000	.0000	.0001	.0008	.0042	.0149	.0385	.0771	.1248	.1669
	11	.0000	.0000	.0000	.0001	.0010	.0046	.0151	.0374	.0742	.1214
	12	.0000	.0000	.0000	.0000	.0002	.0012	.0047	.0145	.0354	.0708
	13	.0000	.0000	.0000	.0000	.0000	.0002	.0012	.0045	.0134	.0327
	14	.0000	.0000	.0000	.0000	.0000	.0000	.0002	.0011	.0039	.0117
	15	.0000	.0000	.0000	.0000	.0000	.0000	.0000	.0002	.0009	.0031
	16	.0000	.0000	.0000	.0000	.0000	.0000	.0000	.0000	.0001	.0006
	17	.0000	.0000	.0000	.0000	.0000	.0000	.0000	.0000	.0000	.0001
	18	.0000	.0000	.0000	.0000	.0000	.0000	.0000	.0000	.0000	.0000
19	0	.3774	.1351	.0456	.0144	.0042	.0011	.0003	.0001	.0000	.0000
	1	.3774	.2852	.1529	.0685	.0268	.0093	.0029	.0008	.0002	.0000
	2	.1787	.2852	.2428	.1540	.0803	.0358	.0138	.0046	.0013	.0003
	3	.0533	.1796	.2428	.2182	.1517	.0869	.0422	.0175	.0062	.0018
	4	.0112	.0798	.1714	.2182	.2023	.1491	.0909	.0467	.0203	.0074
	5	.0018	.0266	.0907	.1636	.2023	.1916	.1468	.0933	.0497	.0222
	6	.0002	.0069	.0374	.0955	.1574	.1916	.1844	.1451	.0949	.0518
	7	.0000	.0014	.0122	.0443	.0974	.1525	.1844	.1797	.1443	.0961
	8	.0000	.0002	.0032	.0166	.0487	.0981	.1489	.1797	.1771	.1442
	9	.0000	.0000	.0007	.0051	.0198	.0514	.0980	.1464	.1771	.1762
	10	.0000	.0000	.0001	.0013	.0066	.0220	.0528	.0976	.1449	.1762
	11	.0000	.0000	.0000	.0003	.0018	.0077	.0233	.0532	.0970	.1442
	12	.0000	.0000	.0000	.0000	.0004	.0022	.0083	.0237	.0529	.0961
	13	.0000	.0000	.0000	.0000	.0001	.0005	.0024	.0085	.0233	.0518
	14	.0000	.0000	.0000	.0000	.0000	.0001	.0006	.0024	.0082	.0222
	15	.0000	.0000	.0000	.0000	.0000	.0000	.0001	.0005	.0022	.0074
	16	.0000	.0000	.0000	.0000	.0000	.0000	.0000	.0001	.0005	.0018
	17	.0000	.0000	.0000	.0000	.0000	.0000	.0000	.0000	.0001	.0003
	18	.0000	.0000	.0000	.0000	.0000	.0000	.0000	.0000	.0000	.0000
	19	.0000	.0000	.0000	.0000	.0000	.0000	.0000	.0000	.0000	.0000
20	0	.3585	.1216	.0388	.0115	.0032	.0008	.0002	.0000	.0000	.0000
	1	.3774	.2702	.1368	.0576	.0211	.0068	.0020	.0005	.0001	.0000
	2	.1887	.2852	.2293	.1369	.0669	.0278	.0100	.0031	.0008	.0002
	3	.0596	.1901	.2428	.2054	.1339	.0716	.0323	.0123	.0040	.0011
	4	.0133	.0898	.1821	.2182	.1897	.1304	.0738	.0350	.0139	.0046
	5	.0022	.0319	.1028	.1746	.2023	.1789	.1272	.0746	.0365	.0148
	6	.0003	.0089	.0454	.1091	.1686	.1916	.1712	.1244	.0746	.0370
	7	.0000	.0020	.0160	.0545	.1124	.1643	.1844	.1659	.1221	.0739
	8	.0000	.0004	.0046	.0222	.0609	.1144	.1614	.1797	.1623	.1201
	9	.0000	.0001	.0011	.0074	.0271	.0654	.1158	.1597	.1771	.1602
	10	.0000	.0000	.0002	.0020	.0099	.0308	.0686	.1171	.1593	.1762
	11	.0000	.0000	.0000	.0005	.0030	.0120	.0336	.0710	.1185	.1602
	12	.0000	.0000	.0000	.0001	.0008	.0039	.0136	.0355	.0727	.1201
	13	.0000	.0000	.0000	.0000	.0002	.0010	.0045	.0146	.0366	.0739
	14	.0000	.0000	.0000	.0000	.0000	.0002	.0012	.0049	.0150	.0370
	15	.0000	.0000	.0000	.0000	.0000	.0000	.0003	.0013	.0049	.0148
	16	.0000	.0000	.0000	.0000	.0000	.0000	.0000	.0003	.0013	.0046
	17	.0000	.0000	.0000	.0000	.0000	.0000	.0000	.0000	.0002	.0011
	18	.0000	.0000	.0000	.0000	.0000	.0000	.0000	.0000	.0000	.0002
	19	.0000	.0000	.0000	.0000	.0000	.0000	.0000	.0000	.0000	.0000
	20	.0000	.0000	.0000	.0000	.0000	.0000	.0000	.0000	.0000	.0000

TABLE II

POISSON PROBABILITIES*

	λ									
x	0.1	0.2	0.3	0.4	0.5	0.6	0.7	0.8	0.9	1.0
0	.9048	.8187	.7408	.6703	.6065	.5488	.4966	.4493	.4066	.3679
1	.0905	.1637	.2222	.2681	.3033	.3293	.3476	.3595	.3659	.3679
2	.0045	.0164	.0333	.0536	.0758	.0988	.1217	.1438	.1647	.1839
3	.0002	.0011	.0033	.0072	.0126	.0198	.0284	.0383	.0494	.0613
4	.0000	.0001	.0002	.0007	.0016	.0030	.0050	.0077	.0111	.0153
5	.0000	.0000	.0000	.0001	.0002	.0004	.0007	.0012	.0020	.0031
6	.0000	.0000	.0000	.0000	.0000	.0000	.0001	.0002	.0003	.0005
7	.0000	.0000	.0000	.0000	.0000	.0000	.0000	.0000	.0000	.0001

	λ									
x	1.1	1.2	1.3	1.4	1.5	1.6	1.7	1.8	1.9	2.0
0	.3329	.3012	.2725	.2466	.2231	.2019	.1827	.1653	.1496	.1353
1	.3662	.3614	.3543	.3452	.3347	.3230	.3106	.2975	.2842	.2707
2	.2014	.2169	.2303	.2417	.2510	.2584	.2640	.2678	.2700	.2707
3	.0738	.0867	.0998	.1128	.1255	.1378	.1496	.1607	.1710	.1804
4	.0203	.0260	.0324	.0395	.0471	.0551	.0636	.0723	.0812	.0902
5	.0045	.0062	.0084	.0111	.0141	.0176	.0216	.0260	.0309	.0361
6	.0008	.0012	.0018	.0026	.0035	.0047	.0061	.0078	.0098	.0120
7	.0001	.0002	.0003	.0005	.0008	.0011	.0015	.0020	.0027	.0034
8	.0000	.0000	.0001	.0001	.0001	.0002	.0003	.0005	.0006	.0009
9	.0000	.0000	.0000	.0000	.0000	.0000	.0001	.0001	.0001	.0002

	λ									
x	2.1	2.2	2.3	2.4	2.5	2.6	2.7	2.8	2.9	3.0
0	.1225	.1108	.1003	.0907	.0821	.0743	.0672	.0608	.0550	.0498
1	.2572	.2438	.2306	.2177	.2052	.1931	.1815	.1703	.1596	.1494
2	.2700	.2681	.2652	.2613	.2565	.2510	.2450	.2384	.2314	.2240
3	.1890	.1966	.2033	.2090	.2138	.2176	.2205	.2225	.2237	.2240
4	.0992	.1082	.1169	.1254	.1336	.1414	.1488	.1557	.1622	.1680
5	.0417	.0476	.0538	.0602	.0668	.0735	.0804	.0872	.0940	.1008
6	.0146	.0174	.0206	.0241	.0278	.0319	.0362	.0407	.0455	.0504
7	.0044	.0055	.0068	.0083	.0099	.0118	.0139	.0163	.0188	.0216
8	.0011	.0015	.0019	.0025	.0031	.0038	.0047	.0057	.0068	.0081
9	.0003	.0004	.0005	.0007	.0009	.0011	.0014	.0018	.0022	.0027
10	.0001	.0001	.0001	.0002	.0002	.0003	.0004	.0005	.0006	.0008
11	.0000	.0000	.0000	.0000	.0000	.0001	.0001	.0001	.0002	.0002
12	.0000	.0000	.0000	.0000	.0000	.0000	.0000	.0000	.0000	.0001

* Entries in this table are values of $(e^{-\lambda}\lambda^x/x!)$ for the indicated values of x and λ. Reproduced by permission from *Handbook of Probability and Statistics with Tables*, by Burington and May, 1953, McGraw-Hill Book Co., Inc.

TABLE II (continued)

	λ									
x	3.1	3.2	3.3	3.4	3.5	3.6	3.7	3.8	3.9	4.0
0	.0450	.0408	.0369	.0334	.0302	.0273	.0247	.0224	.0202	.0183
1	.1397	.1304	.1217	.1135	.1057	.0984	.0915	.0850	.0789	.0733
2	.2165	.2087	.2008	.1929	.1850	.1771	.1692	.1615	.1539	.1465
3	.2237	.2226	.2209	.2186	.2158	.2125	.2087	.2046	.2001	.1954
4	.1734	.1781	.1823	.1858	.1888	.1912	.1931	.1944	.1951	.1954
5	.1075	.1140	.1203	.1264	.1322	.1377	.1429	.1477	.1522	.1563
6	.0555	.0608	.0662	.0716	.0771	.0826	.0881	.0936	.0989	.1042
7	.0246	.0278	.0312	.0348	.0385	.0425	.0466	.0508	.0551	.0595
8	.0095	.0111	.0129	.0148	.0169	.0191	.0215	.0241	.0269	.0298
9	.0033	.0040	.0047	.0056	.0066	.0076	.0089	.0102	.0116	.0132
10	.0010	.0013	.0016	.0019	.0023	.0028	.0033	.0039	.0045	.0053
11	.0003	.0004	.0005	.0006	.0007	.0009	.0011	.0013	.0016	.0019
12	.0001	.0001	.0001	.0002	.0002	.0003	.0003	.0004	.0005	.0006
13	.0000	.0000	.0000	.0000	.0001	.0001	.0001	.0001	.0002	.0002
14	.0000	.0000	.0000	.0000	.0000	.0000	.0000	.0000	.0000	.0001

	λ									
x	4.1	4.2	4.3	4.4	4.5	4.6	4.7	4.8	4.9	5.0
0	.0166	.0150	.0136	.0123	.0111	.0101	.0091	.0082	.0074	.0067
1	.0679	.0630	.0583	.0540	.0500	.0462	.0427	.0395	.0365	.0337
2	.1393	.1323	.1254	.1188	.1125	.1063	.1005	.0948	.0894	.0842
3	.1904	.1852	.1798	.1743	.1687	.1631	.1574	.1517	.1460	.1404
4	.1951	.1944	.1933	.1917	.1898	.1875	.1849	.1820	.1789	.1755
5	.1600	.1633	.1662	.1687	.1708	.1725	.1738	.1747	.1753	.1755
6	.1093	.1143	.1191	.1237	.1281	.1323	.1362	.1398	.1432	.1462
7	.0640	.0686	.0732	.0778	.0824	.0869	.0914	.0959	.1002	.1044
8	.0328	.0360	.0393	.0428	.0463	.0500	.0537	.0575	.0614	.0653
9	.0150	.0168	.0188	.0209	.0232	.0255	.0280	.0307	.0334	.0363
10	.0061	.0071	.0081	.0092	.0104	.0118	.0132	.0147	.0164	.0181
11	.0023	.0027	.0032	.0037	.0043	.0049	.0056	.0064	.0073	.0082
12	.0008	.0009	.0011	.0014	.0016	.0019	.0022	.0026	.0030	.0034
13	.0002	.0003	.0004	.0005	.0006	.0007	.0008	.0009	.0011	.0013
14	.0001	.0001	.0001	.0001	.0002	.0002	.0003	.0003	.0004	.0005
15	.0000	.0000	.0000	.0000	.0001	.0001	.0001	.0001	.0001	.0002

	λ									
x	5.1	5.2	5.3	5.4	5.5	5.6	5.7	5.8	5.9	6.0
0	.0061	.0055	.0050	.0045	.0041	.0037	.0033	.0030	.0027	.0025
1	.0311	.0287	.0265	.0244	.0225	.0207	.0191	.0176	.0162	.0149
2	.0793	.0746	.0701	.0659	.0618	.0580	.0544	.0509	.0477	.0446
3	.1348	.1293	.1239	.1185	.1133	.1082	.1033	.0985	.0938	.0892
4	.1719	.1681	.1641	.1600	.1558	.1515	.1472	.1428	.1383	.1339

TABLE II (continued)

x	λ 5.1	5.2	5.3	5.4	5.5	5.6	5.7	5.8	5.9	6.0
5	.1753	.1748	.1740	.1728	.1714	.1697	.1678	.1656	.1632	.1606
6	.1490	.1515	.1537	.1555	.1571	.1584	.1594	.1601	.1605	.1606
7	.1086	.1125	.1163	.1200	.1234	.1267	.1298	.1326	.1353	.1377
8	.0692	.0731	.0771	.0810	.0849	.0887	.0925	.0962	.0998	.1033
9	.0392	.0423	.0454	.0486	.0519	.0552	.0586	.0620	.0654	.0688
10	.0200	.0220	.0241	.0262	.0285	.0309	.0334	.0359	.0386	.0413
11	.0093	.0104	.0116	.0129	.0143	.0157	.0173	.0190	.0207	.0225
12	.0039	.0045	.0051	.0058	.0065	.0073	.0082	.0092	.0102	.0113
13	.0015	.0018	.0021	.0024	.0028	.0032	.0036	.0041	.0046	.0052
14	.0006	.0007	.0008	.0009	.0011	.0013	.0015	.0017	.0019	.0022
15	.0002	.0002	.0003	.0003	.0004	.0005	.0006	.0007	.0008	.0009
16	.0001	.0001	.0001	.0001	.0001	.0002	.0002	.0002	.0003	.0003
17	.0000	.0000	.0000	.0000	.0000	.0001	.0001	.0001	.0001	.0001

x	λ 6.1	6.2	6.3	6.4	6.5	6.6	6.7	6.8	6.9	7.0
0	.0022	.0020	.0018	.0017	.0015	.0014	.0012	.0011	.0010	.0009
1	.0137	.0126	.0116	.0106	.0098	.0090	.0082	.0076	.0070	.0064
2	.0417	.0390	.0364	.0340	.0318	.0296	.0276	.0258	.0240	.0223
3	.0848	.0806	.0765	.0726	.0688	.0652	.0617	.0584	.0552	.0521
4	.1294	.1249	.1205	.1162	.1118	.1076	.1034	.0992	.0952	.0912
5	.1579	.1549	.1519	.1487	.1454	.1420	.1385	.1349	.1314	.1277
6	.1605	.1601	.1595	.1586	.1575	.1562	.1546	.1529	.1511	.1490
7	.1399	.1418	.1435	.1450	.1462	.1472	.1480	.1486	.1489	.1490
8	.1066	.1099	.1130	.1160	.1188	.1215	.1240	.1263	.1284	.1304
9	.0723	.0757	.0791	.0825	.0858	.0891	.0923	.0954	.0985	.1014
10	.0441	.0469	.0498	.0528	.0558	.0588	.0618	.0649	.0679	.0710
11	.0245	.0265	.0285	.0307	.0330	.0353	.0377	.0401	.0426	.0452
12	.0124	.0137	.0150	.0164	.0179	.0194	.0210	.0227	.0245	.0264
13	.0058	.0065	.0073	.0081	.0089	.0098	.0108	.0119	.0130	.0142
14	.0025	.0029	.0033	.0037	.0041	.0046	.0052	.0058	.0064	.0071
15	.0010	.0012	.0014	.0016	.0018	.0020	.0023	.0026	.0029	.0033
16	.0004	.0005	.0005	.0006	.0007	.0008	.0010	.0011	.0013	.0014
17	.0001	.0002	.0002	.0002	.0003	.0003	.0004	.0004	.0005	.0006
18	.0000	.0001	.0001	.0001	.0001	.0001	.0001	.0002	.0002	.0002
19	.0000	.0000	.0000	.0000	.0000	.0000	.0000	.0001	.0001	.0001

x	λ 7.1	7.2	7.3	7.4	7.5	7.6	7.7	7.8	7.9	8.0
0	.0008	.0007	.0007	.0006	.0006	.0005	.0005	.0004	.0004	.0003
1	.0059	.0054	.0049	.0045	.0041	.0038	.0035	.0032	.0029	.0027
2	.0208	.0194	.0180	.0167	.0156	.0145	.0134	.0125	.0116	.0107
3	.0492	.0464	.0438	.0413	.0389	.0366	.0345	.0324	.0305	.0286
4	.0874	.0836	.0799	.0764	.0729	.0696	.0663	.0632	.0602	.0573
5	.1241	.1204	.1167	.1130	.1094	.1057	.1021	.0986	.0951	.0916
6	.1468	.1445	.1420	.1394	.1367	.1339	.1311	.1282	.1252	.1221
7	.1489	.1486	.1481	.1474	.1465	.1454	.1442	.1428	.1413	.1396
8	.1321	.1337	.1351	.1363	.1373	.1382	.1388	.1392	.1395	.1396
9	.1042	.1070	.1096	.1121	.1144	.1167	.1187	.1207	.1224	.1241

TABLE II (continued)

x	λ 7.1	7.2	7.3	7.4	7.5	7.6	7.7	7.8	7.9	8.0
10	.0740	.0770	.0800	.0829	.0858	.0887	.0914	.0941	.0967	.0993
11	.0478	.0504	.0531	.0558	.0585	.0613	.0640	.0667	.0695	.0722
12	.0283	.0303	.0323	.0344	.0366	.0388	.0411	.0434	.0457	.0481
13	.0154	.0168	.0181	.0196	.0211	.0227	.0243	.0260	.0278	.0296
14	.0078	.0086	.0095	.0104	.0113	.0123	.0134	.0145	.0157	.0169
15	.0037	.0041	.0046	.0051	.0057	.0062	.0069	.0075	.0083	.0090
16	.0016	.0019	.0021	.0024	.0026	.0030	.0033	.0037	.0041	.0045
17	.0007	.0008	.0009	.0010	.0012	.0013	.0015	.0017	.0019	.0021
18	.0003	.0003	.0004	.0004	.0005	.0006	.0006	.0007	.0008	.0009
19	.0001	.0001	.0001	.0002	.0002	.0002	.0003	.0003	.0003	.0004
20	.0000	.0000	.0001	.0001	.0001	.0001	.0001	.0001	.0001	.0002
21	.0000	.0000	.0000	.0000	.0000	.0000	.0000	.0000	.0001	.0001

x	λ 8.1	8.2	8.3	8.4	8.5	8.6	8.7	8.8	8.9	9.0
0	.0003	.0003	.0002	.0002	.0002	.0002	.0002	.0002	.0001	.0001
1	.0025	.0023	.0021	.0019	.0017	.0016	.0014	.0013	.0012	.0011
2	.0100	.0092	.0086	.0079	.0074	.0068	.0063	.0058	.0054	.0050
3	.0269	.0252	.0237	.0222	.0208	.0195	.0183	.0171	.0160	.0150
4	.0544	.0517	.0491	.0466	.0443	.0420	.0398	.0377	.0357	.0337
5	.0882	.0849	.0816	.0784	.0752	.0722	.0692	.0663	.0635	.0607
6	.1191	.1160	.1128	.1097	.1066	.1034	.1003	.0972	.0941	.0911
7	.1378	.1358	.1338	.1317	.1294	.1271	.1247	.1222	.1197	.1171
8	.1395	.1392	.1388	.1382	.1375	.1366	.1356	.1344	.1332	.1318
9	.1256	.1269	.1280	.1290	.1299	.1306	.1311	.1315	.1317	.1318
10	.1017	.1040	.1063	.1084	.1104	.1123	.1140	.1157	.1172	.1186
11	.0749	.0776	.0802	.0828	.0853	.0878	.0902	.0925	.0948	.0970
12	.0505	.0530	.0555	.0579	.0604	.0629	.0654	.0679	.0703	.0728
13	.0315	.0334	.0354	.0374	.0395	.0416	.0438	.0459	.0481	.0504
14	.0182	.0196	.0210	.0225	.0240	.0256	.0272	.0289	.0306	.0324
15	.0098	.0107	.0116	.0126	.0136	.0147	.0158	.0169	.0182	.0194
16	.0050	.0055	.0060	.0066	.0072	.0079	.0086	.0093	.0101	.0109
17	.0024	.0026	.0029	.0033	.0036	.0040	.0044	.0048	.0053	.0058
18	.0011	.0012	.0014	.0015	.0017	.0019	.0021	.0024	.0026	.0029
19	.0005	.0005	.0006	.0007	.0008	.0009	.0010	.0011	.0012	.0014
20	.0002	.0002	.0002	.0003	.0003	.0004	.0004	.0005	.0005	.0006
21	.0001	.0001	.0001	.0001	.0001	.0002	.0002	.0002	.0002	.0003
22	.0000	.0000	.0000	.0000	.0001	.0001	.0001	.0001	.0001	.0001

x	λ 9.1	9.2	9.3	9.4	9.5	9.6	9.7	9.8	9.9	10
0	.0001	.0001	.0001	.0001	.0001	.0001	.0001	.0001	.0001	.0000
1	.0010	.0009	.0009	.0008	.0007	.0007	.0006	.0005	.0005	.0005
2	.0046	.0043	.0040	.0037	.0034	.0031	.0029	.0027	.0025	.0023
3	.0140	.0131	.0123	.0115	.0107	.0100	.0093	.0087	.0081	.0076
4	.0319	.0302	.0285	.0269	.0254	.0240	.0226	.0213	.0201	.0189

TABLE II (continued)

	λ									
x	9.1	9.2	9.3	9.4	9.5	9.6	9.7	9.8	9.9	10
5	.0581	.0555	.0530	.0506	.0483	.0460	.0439	.0418	.0398	.0378
6	.0881	.0851	.0822	.0793	.0764	.0736	.0709	.0682	.0656	.0631
7	.1145	.1118	.1091	.1064	.1037	.1010	.0982	.0955	.0928	.0901
8	.1302	.1286	.1269	.1251	.1232	.1212	.1191	.1170	.1148	.1126
9	.1317	.1315	.1311	.1306	.1300	.1293	.1284	.1274	.1263	.1251
10	.1198	.1210	.1219	.1228	.1235	.1241	.1245	.1249	.1250	.1251
11	.0991	.1012	.1031	.1049	.1067	.1083	.1098	.1112	.1125	.1137
12	.0752	.0776	.0799	.0822	.0844	.0866	.0888	.0908	.0928	.0948
13	.0526	.0549	.0572	.0594	.0617	.0640	.0662	.0685	.0707	.0729
14	.0342	.0361	.0380	.0399	.0419	.0439	.0459	.0479	.0500	.0521
15	.0208	.0221	.0235	.0250	.0265	.0281	.0297	.0313	.0330	.0347
16	.0118	.0127	.0137	.0147	.0157	.0168	.0180	.0192	.0204	.0217
17	.0063	.0069	.0075	.0081	.0088	.0095	.0103	.0111	.0119	.0128
18	.0032	.0035	.0039	.0042	.0046	.0051	.0055	.0060	.0065	.0071
19	.0015	.0017	.0019	.0021	.0023	.0026	.0028	.0031	.0034	.0037
20	.0007	.0008	.0009	.0010	.0011	.0012	.0014	.0015	.0017	.0019
21	.0003	.0003	.0004	.0004	.0005	.0006	.0006	.0007	.0008	.0009
22	.0001	.0001	.0002	.0002	.0002	.0002	.0003	.0003	.0004	.0004
23	.0000	.0001	.0001	.0001	.0001	.0001	.0001	.0001	.0002	.0002
24	.0000	.0000	.0000	.0000	.0000	.0000	.0000	.0001	.0001	.0001

	λ									
x	11	12	13	14	15	16	17	18	19	20
0	.0000	.0000	.0000	.0000	.0000	.0000	.0000	.0000	.0000	.0000
1	.0002	.0001	.0000	.0000	.0000	.0000	.0000	.0000	.0000	.0000
2	.0010	.0004	.0002	.0001	.0000	.0000	.0000	.0000	.0000	.0000
3	.0037	.0018	.0008	.0004	.0002	.0001	.0000	.0000	.0000	.0000
4	.0102	.0053	.0027	.0013	.0006	.0003	.0001	.0001	.0000	.0000
5	.0224	.0127	.0070	.0037	.0019	.0010	.0005	.0002	.0001	.0001
6	.0411	.0255	.0152	.0087	.0048	.0026	.0014	.0007	.0004	.0002
7	.0646	.0437	.0281	.0174	.0104	.0060	.0034	.0018	.0010	.0005
8	.0888	.0655	.0457	.0304	.0194	.0120	.0072	.0042	.0024	.0013
9	.1085	.0874	.0661	.0473	.0324	.0213	.0135	.0083	.0050	.0029
10	.1194	.1048	.0859	.0663	.0486	.0341	.0230	.0150	.0095	.0058
11	.1194	.1144	.1015	.0844	.0663	.0496	.0355	.0245	.0164	.0106
12	.1094	.1144	.1099	.0984	.0829	.0661	.0504	.0368	.0259	.0176
13	.0926	.1056	.1099	.1060	.0956	.0814	.0658	.0509	.0378	.0271
14	.0728	.0905	.1021	.1060	.1024	.0930	.0800	.0655	.0514	.0387
15	.0534	.0724	.0885	.0989	.1024	.0992	.0906	.0786	.0650	.0516
16	.0367	.0543	.0719	.0866	.0960	.0992	.0963	.0884	.0772	.0646
17	.0237	.0383	.0550	.0713	.0847	.0934	.0963	.0936	.0863	.0760
18	.0145	.0256	.0397	.0554	.0706	.0830	.0909	.0936	.0911	.0844
19	.0084	.0161	.0272	.0409	.0557	.0699	.0814	.0887	.0911	.0888
20	.0046	.0097	.0177	.0286	.0418	.0559	.0692	.0798	.0866	.0888
21	.0024	.0055	.0109	.0191	.0299	.0426	.0560	.0684	.0783	.0846
22	.0012	.0030	.0065	.0121	.0204	.0310	.0433	.0560	.0676	.0769
23	.0006	.0016	.0037	.0074	.0133	.0216	.0320	.0438	.0559	.0669
24	.0003	.0008	.0020	.0043	.0083	.0144	.0226	.0328	.0442	.0557

TABLE II (continued)

	λ									
x	11	12	13	14	15	16	17	18	19	20
25	.0001	.0004	.0010	.0024	.0050	.0092	.0154	.0237	.0336	.0446
26	.0000	.0002	.0005	.0013	.0029	.0057	.0101	.0164	.0246	.0343
27	.0000	.0001	.0002	.0007	.0016	.0034	.0063	.0109	.0173	.0254
28	.0000	.0000	.0001	.0003	.0009	.0019	.0038	.0070	.0117	.0181
29	.0000	.0000	.0001	.0002	.0004	.0011	.0023	.0044	.0077	.0125
30	.0000	.0000	.0000	.0001	.0002	.0006	.0013	.0026	.0049	.0083
31	.0000	.0000	.0000	.0000	.0001	.0003	.0007	.0015	.0030	.0054
32	.0000	.0000	.0000	.0000	.0001	.0001	.0004	.0009	.0018	.0034
33	.0000	.0000	.0000	.0000	.0000	.0001	.0002	.0005	.0010	.0020
34	.0000	.0000	.0000	.0000	.0000	.0000	.0001	.0002	.0006	.0012
35	.0000	.0000	.0000	.0000	.0000	.0000	.0000	.0001	.0003	.0007
36	.0000	.0000	.0000	.0000	.0000	.0000	.0000	.0001	.0002	.0004
37	.0000	.0000	.0000	.0000	.0000	.0000	.0000	.0000	.0001	.0002
38	.0000	.0000	.0000	.0000	.0000	.0000	.0000	.0000	.0000	.0001
39	.0000	.0000	.0000	.0000	.0000	.0000	.0000	.0000	.0000	.0001

TABLE III

THE STANDARD NORMAL DISTRIBUTION

z	.00	.01	.02	.03	.04	.05	.06	.07	.08	.09
0.0	.0000	.0040	.0080	.0120	.0160	.0199	.0239	.0279	.0319	.0359
0.1	.0398	.0438	.0478	.0517	.0557	.0596	.0636	.0675	.0714	.0753
0.2	.0793	.0832	.0871	.0910	.0948	.0987	.1026	.1064	.1103	.1141
0.3	.1179	.1217	.1255	.1293	.1331	.1368	.1406	.1443	.1480	.1517
0.4	.1554	.1591	.1628	.1664	.1700	.1736	.1772	.1808	.1844	.1879
0.5	.1915	.1950	.1985	.2019	.2054	.2088	.2123	.2157	.2190	.2224
0.6	.2257	.2291	.2324	.2357	.2389	.2422	.2454	.2486	.2517	.2549
0.7	.2580	.2611	.2642	.2673	.2704	.2734	.2764	.2794	.2823	.2852
0.8	.2881	.2910	.2939	.2967	.2995	.3023	.3051	.3078	.3106	.3133
0.9	.3159	.3186	.3212	.3238	.3264	.3289	.3315	.3340	.3365	.3389
1.0	.3413	.3438	.3461	.3485	.3508	.3531	.3554	.3577	.3599	.3621
1.1	.3643	.3665	.3686	.3708	.3729	.3749	.3770	.3790	.3810	.3830
1.2	.3849	.3869	.3888	.3907	.3925	.3944	.3962	.3980	.3997	.4015
1.3	.4032	.4049	.4066	.4082	.4099	.4115	.4131	.4147	.4162	.4177
1.4	.4192	.4207	.4222	.4236	.4251	.4265	.4279	.4292	.4306	.4319
1.5	.4332	.4345	.4357	.4370	.4382	.4394	.4406	.4418	.4429	.4441
1.6	.4452	.4463	.4474	.4484	.4495	.4505	.4515	.4525	.4535	.4545
1.7	.4554	.4564	.4573	.4582	.4591	.4599	.4608	.4616	.4625	.4633
1.8	.4641	.4649	.4656	.4664	.4671	.4678	.4686	.4693	.4699	.4706
1.9	.4713	.4719	.4726	.4732	.4738	.4744	.4750	.4756	.4761	.4767
2.0	.4772	.4778	.4783	.4788	.4793	.4798	.4803	.4808	.4812	.4817
2.1	.4821	.4826	.4830	.4834	.4838	.4842	.4846	.4850	.4854	.4857
2.2	.4861	.4864	.4868	.4871	.4875	.4878	.4881	.4884	.4887	.4890
2.3	.4893	.4896	.4898	.4901	.4904	.4906	.4909	.4911	.4913	.4916
2.4	.4918	.4920	.4922	.4925	.4927	.4929	.4931	.4932	.4934	.4936
2.5	.4938	.4940	.4941	.4943	.4945	.4946	.4948	.4949	.4951	.4952
2.6	.4953	.4955	.4956	.4957	.4959	.4960	.4961	.4962	.4963	.4964
2.7	.4965	.4966	.4967	.4968	.4969	.4970	.4971	.4972	.4973	.4974
2.8	.4974	.4975	.4976	.4977	.4977	.4978	.4979	.4979	.4980	.4981
2.9	.4981	.4982	.4982	.4983	.4984	.4984	.4985	.4985	.4986	.4986
3.0	.4987	.4987	.4987	.4988	.4988	.4989	.4989	.4989	.4990	.4990

TABLE IV

VALUES OF $t_{\alpha,\nu}$*

ν	$\alpha = .10$	$\alpha = .05$	$\alpha = .025$	$\alpha = .01$	$\alpha = .005$	ν
1	3.078	6.314	12.706	31.821	63.657	1
2	1.886	2.920	4.303	6.965	9.925	2
3	1.638	2.353	3.182	4.541	5.841	3
4	1.533	2.132	2.776	3.747	4.604	4
5	1.476	2.015	2.571	3.365	4.032	5
6	1.440	1.943	2.447	3.143	3.707	6
7	1.415	1.895	2.365	2.998	3.499	7
8	1.397	1.860	2.306	2.896	3.355	8
9	1.383	1.833	2.262	2.821	3.250	9
10	1.372	1.812	2.228	2.764	3.169	10
11	1.363	1.796	2.201	2.718	3.106	11
12	1.356	1.782	2.179	2.681	3.055	12
13	1.350	1.771	2.160	2.650	3.012	13
14	1.345	1.761	2.145	2.624	2.977	14
15	1.341	1.753	2.131	2.602	2.947	15
16	1.337	1.746	2.120	2.583	2.921	16
17	1.333	1.740	2.110	2.567	2.898	17
18	1.330	1.734	2.101	2.552	2.878	18
19	1.328	1.729	2.093	2.539	2.861	19
20	1.325	1.725	2.086	2.528	2.845	20
21	1.323	1.721	2.080	2.518	2.831	21
22	1.321	1.717	2.074	2.508	2.819	22
23	1.319	1.714	2.069	2.500	2.807	23
24	1.318	1.711	2.064	2.492	2.797	24
25	1.316	1.708	2.060	2.485	2.787	25
26	1.315	1.706	2.056	2.479	2.779	26
27	1.314	1.703	2.052	2.473	2.771	27
28	1.313	1.701	2.048	2.467	2.763	28
29	1.311	1.699	2.045	2.462	2.756	29
inf.	1.282	1.645	1.960	2.326	2.576	inf.

* This table is abridged from Table IV of R. A. Fisher, *Statistical Methods for Research Workers*, published by Oliver and Boyd, Ltd., Edinburgh, by permission of the author and publishers.

TABLE V

VALUES OF $\chi^2_{\alpha,\nu}$*

ν	$\alpha = .995$	$\alpha = .99$	$\alpha = .975$	$\alpha = .95$	$\alpha = .05$	$\alpha = .025$	$\alpha = .01$	$\alpha = .005$	ν
1	.0000393	.000157	.000982	.00393	3.841	5.024	6.635	7.879	1
2	.0100	.0201	.0506	.103	5.991	7.378	9.210	10.597	2
3	.0717	.115	.216	.352	7.815	9.348	11.345	12.838	3
4	.207	.297	.484	.711	9.488	11.143	13.277	14.860	4
5	.412	.554	.831	1.145	11.070	12.832	15.086	16.750	5
6	.676	.872	1.237	1.635	12.592	14.449	16.812	18.548	6
7	.989	1.239	1.690	2.167	14.067	16.013	18.475	20.278	7
8	1.344	1.646	2.180	2.733	15.507	17.535	20.090	21.955	8
9	1.735	2.088	2.700	3.325	16.919	19.023	21.666	23.589	9
10	2.156	2.558	3.247	3.940	18.307	20.483	23.209	25.188	10
11	2.603	3.053	3.816	4.575	19.675	21.920	24.725	26.757	11
12	3.074	3.571	4.404	5.226	21.026	23.337	26.217	28.300	12
13	3.565	4.107	5.009	5.892	22.362	24.736	27.688	29.819	13
14	4.075	4.660	5.629	6.571	23.685	26.119	29.141	31.319	14
15	4.601	5.229	6.262	7.261	24.996	27.488	30.578	32.801	15
16	5.142	5.812	6.908	7.962	26.296	28.845	32.000	34.267	16
17	5.697	6.408	7.564	8.672	27.587	30.191	33.409	35.718	17
18	6.265	7.015	8.231	9.390	28.869	31.526	34.805	37.156	18
19	6.844	7.633	8.907	10.117	30.144	32.852	36.191	38.582	19
20	7.434	8.260	9.591	10.851	31.410	34.170	37.566	39.997	20
21	8.034	8.897	10.283	11.591	32.671	35.479	38.932	41.401	21
22	8.643	9.542	10.982	12.338	33.924	36.781	40.289	42.796	22
23	9.260	10.196	11.689	13.091	35.172	38.076	41.638	44.181	23
24	9.886	10.856	12.401	13.848	36.415	39.364	42.980	45.558	24
25	10.520	11.524	13.120	14.611	37.652	40.646	44.314	46.928	25
26	11.160	12.198	13.844	15.379	38.885	41.923	45.642	48.290	26
27	11.808	12.879	14.573	16.151	40.113	43.194	46.963	49.645	27
28	12.461	13.565	15.308	16.928	41.337	44.461	48.278	50.993	28
29	13.121	14.256	16.047	17.708	42.557	45.722	49.588	52.336	29
30	13.787	14.953	16.791	18.493	43.773	46.979	50.892	53.672	30

* This table is based on Table 8 of *Biometrika Tables for Statisticians, Volume I,* by permission of the *Biometrika* trustees.

TABLE VIa

VALUES OF $F_{.05,\nu_1,\nu_2}$*

ν_1 = Degrees of freedom for numerator

ν_2 = Degrees of freedom for denominator	1	2	3	4	5	6	7	8	9	10	12	15	20	24	30	40	60	120	∞
1	161	200	216	225	230	234	237	239	241	242	244	246	248	249	250	251	252	253	254
2	18.5	19.0	19.2	19.2	19.3	19.3	19.4	19.4	19.4	19.4	19.4	19.4	19.4	19.5	19.5	19.5	19.5	19.5	19.5
3	10.1	9.55	9.28	9.12	9.01	8.94	8.89	8.85	8.81	8.79	8.74	8.70	8.66	8.64	8.62	8.59	8.57	8.55	8.53
4	7.71	6.94	6.59	6.39	6.26	6.16	6.09	6.04	6.00	5.96	5.91	5.86	5.80	5.77	5.75	5.72	5.69	5.66	5.63
5	6.61	5.79	5.41	5.19	5.05	4.95	4.88	4.82	4.77	4.74	4.68	4.62	4.56	4.53	4.50	4.46	4.43	4.40	4.37
6	5.99	5.14	4.76	4.53	4.39	4.28	4.21	4.15	4.10	4.06	4.00	3.94	3.87	3.84	3.81	3.77	3.74	3.70	3.67
7	5.59	4.74	4.35	4.12	3.97	3.87	3.79	3.73	3.68	3.64	3.57	3.51	3.44	3.41	3.38	3.34	3.30	3.27	3.23
8	5.32	4.46	4.07	3.84	3.69	3.58	3.50	3.44	3.39	3.35	3.28	3.22	3.15	3.12	3.08	3.04	3.01	2.97	2.93
9	5.12	4.26	3.86	3.63	3.48	3.37	3.29	3.23	3.18	3.14	3.07	3.01	2.94	2.90	2.86	2.83	2.79	2.75	2.71
10	4.96	4.10	3.71	3.48	3.33	3.22	3.14	3.07	3.02	2.98	2.91	2.85	2.77	2.74	2.70	2.66	2.62	2.58	2.54
11	4.84	3.98	3.59	3.36	3.20	3.09	3.01	2.95	2.90	2.85	2.79	2.72	2.65	2.61	2.57	2.53	2.49	2.45	2.40
12	4.75	3.89	3.49	3.26	3.11	3.00	2.91	2.85	2.80	2.75	2.69	2.62	2.54	2.51	2.47	2.43	2.38	2.34	2.30
13	4.67	3.81	3.41	3.18	3.03	2.92	2.83	2.77	2.71	2.67	2.60	2.53	2.46	2.42	2.38	2.34	2.30	2.25	2.21
14	4.60	3.74	3.34	3.11	2.96	2.85	2.76	2.70	2.65	2.60	2.53	2.46	2.39	2.35	2.31	2.27	2.22	2.18	2.13
15	4.54	3.68	3.29	3.06	2.90	2.79	2.71	2.64	2.59	2.54	2.48	2.40	2.33	2.29	2.25	2.20	2.16	2.11	2.07
16	4.49	3.63	3.24	3.01	2.85	2.74	2.66	2.59	2.54	2.49	2.42	2.35	2.28	2.24	2.19	2.15	2.11	2.06	2.01
17	4.45	3.59	3.20	2.96	2.81	2.70	2.61	2.55	2.49	2.45	2.38	2.31	2.23	2.19	2.15	2.10	2.06	2.01	1.96
18	4.41	3.55	3.16	2.93	2.77	2.66	2.58	2.51	2.46	2.41	2.34	2.27	2.19	2.15	2.11	2.06	2.02	1.97	1.92
19	4.38	3.52	3.13	2.90	2.74	2.63	2.54	2.48	2.42	2.38	2.31	2.23	2.16	2.11	2.07	2.03	1.98	1.93	1.88
20	4.35	3.49	3.10	2.87	2.71	2.60	2.51	2.45	2.39	2.35	2.28	2.20	2.12	2.08	2.04	1.99	1.95	1.90	1.84
21	4.32	3.47	3.07	2.84	2.68	2.57	2.49	2.42	2.37	2.32	2.25	2.18	2.10	2.05	2.01	1.96	1.92	1.87	1.81
22	4.30	3.44	3.05	2.82	2.66	2.55	2.46	2.40	2.34	2.30	2.23	2.15	2.07	2.03	1.98	1.94	1.89	1.84	1.78
23	4.28	3.42	3.03	2.80	2.64	2.53	2.44	2.37	2.32	2.27	2.20	2.13	2.05	2.01	1.96	1.91	1.86	1.81	1.76
24	4.26	3.40	3.01	2.78	2.62	2.51	2.42	2.36	2.30	2.25	2.18	2.11	2.03	1.98	1.94	1.89	1.84	1.79	1.73
25	4.24	3.39	2.99	2.76	2.60	2.49	2.40	2.34	2.28	2.24	2.16	2.09	2.01	1.96	1.92	1.87	1.82	1.77	1.71
30	4.17	3.32	2.92	2.69	2.53	2.42	2.33	2.27	2.21	2.16	2.09	2.01	1.93	1.89	1.84	1.79	1.74	1.68	1.62
40	4.08	3.23	2.84	2.61	2.45	2.34	2.25	2.18	2.12	2.08	2.00	1.92	1.84	1.79	1.74	1.69	1.64	1.58	1.51
60	4.00	3.15	2.76	2.53	2.37	2.25	2.17	2.10	2.04	1.99	1.92	1.84	1.75	1.70	1.65	1.59	1.53	1.47	1.39
120	3.92	3.07	2.68	2.45	2.29	2.18	2.09	2.02	1.96	1.91	1.83	1.75	1.66	1.61	1.55	1.50	1.43	1.35	1.25
∞	3.84	3.00	2.60	2.37	2.21	2.10	2.01	1.94	1.88	1.83	1.75	1.67	1.57	1.52	1.46	1.39	1.32	1.22	1.00

* This table is reproduced from M. Merrington and C. M. Thompson, "Tables of percentage points of the inverted beta (*F*) distribution," *Biometrika*, Vol. 33 (1943), by permission of the *Biometrika* trustees.

TABLE VIb

VALUES OF $F_{.01,\nu_1,\nu_2}$*

ν_1 = Degrees of freedom for numerator

ν_2 = Degrees of freedom for denominator	1	2	3	4	5	6	7	8	9	10	12	15	20	24	30	40	60	120	∞
1	4,052	5,000	5,403	5,625	5,764	5,859	5,928	5,982	6,023	6,056	6,106	6,157	6,209	6,235	6,261	6,287	6,313	6,339	6,366
2	98.5	99.0	99.2	99.2	99.3	99.3	99.4	99.4	99.4	99.4	99.4	99.4	99.4	99.5	99.5	99.5	99.5	99.5	99.5
3	34.1	30.8	29.5	28.7	28.2	27.9	27.7	27.5	27.3	27.2	27.1	26.9	26.7	26.6	26.5	26.4	26.3	26.2	26.1
4	21.2	18.0	16.7	16.0	15.5	15.2	15.0	14.8	14.7	14.5	14.4	14.2	14.0	13.9	13.8	13.7	13.7	13.6	13.5
5	16.3	13.3	12.1	11.4	11.0	10.7	10.5	10.3	10.2	10.1	9.89	9.72	9.55	9.47	9.38	9.29	9.20	9.11	9.02
6	13.7	10.9	9.78	9.15	8.75	8.47	8.26	8.10	7.98	7.87	7.72	7.56	7.40	7.31	7.23	7.14	7.06	6.97	6.88
7	12.2	9.55	8.45	7.85	7.46	7.19	6.99	6.84	6.72	6.62	6.47	6.31	6.16	6.07	5.99	5.91	5.82	5.74	5.65
8	11.3	8.65	7.59	7.01	6.63	6.37	6.18	6.03	5.91	5.81	5.67	5.52	5.36	5.28	5.20	5.12	5.03	4.95	4.86
9	10.6	8.02	6.99	6.42	6.06	5.80	5.61	5.47	5.35	5.26	5.11	4.96	4.81	4.73	4.65	4.57	4.48	4.40	4.31
10	10.0	7.56	6.55	5.99	5.64	5.39	5.20	5.06	4.94	4.85	4.71	4.56	4.41	4.33	4.25	4.17	4.08	4.00	3.91
11	9.65	7.21	6.22	5.67	5.32	5.07	4.89	4.74	4.63	4.54	4.40	4.25	4.10	4.02	3.94	3.86	3.78	3.69	3.60
12	9.33	6.93	5.95	5.41	5.06	4.82	4.64	4.50	4.39	4.30	4.16	4.01	3.86	3.78	3.70	3.62	3.54	3.45	3.36
13	9.07	6.70	5.74	5.21	4.86	4.62	4.44	4.30	4.19	4.10	3.96	3.82	3.66	3.59	3.51	3.43	3.34	3.25	3.17
14	8.86	6.51	5.56	5.04	4.70	4.46	4.28	4.14	4.03	3.94	3.80	3.66	3.51	3.43	3.35	3.27	3.18	3.09	3.00
15	8.68	6.36	5.42	4.89	4.56	4.32	4.14	4.00	3.89	3.80	3.67	3.52	3.37	3.29	3.21	3.13	3.05	2.96	2.87
16	8.53	6.23	5.29	4.77	4.44	4.20	4.03	3.89	3.78	3.69	3.55	3.41	3.26	3.18	3.10	3.02	2.93	2.84	2.75
17	8.40	6.11	5.19	4.67	4.34	4.10	3.93	3.79	3.68	3.59	3.46	3.31	3.16	3.08	3.00	2.92	2.83	2.75	2.65
18	8.29	6.01	5.09	4.58	4.25	4.01	3.84	3.71	3.60	3.51	3.37	3.23	3.08	3.00	2.92	2.84	2.75	2.66	2.57
19	8.19	5.93	5.01	4.50	4.17	3.94	3.77	3.63	3.52	3.43	3.30	3.15	3.00	2.92	2.84	2.76	2.67	2.58	2.49
20	8.10	5.85	4.94	4.43	4.10	3.87	3.70	3.56	3.46	3.37	3.23	3.09	2.94	2.86	2.78	2.69	2.61	2.52	2.42
21	8.02	5.78	4.87	4.37	4.04	3.81	3.64	3.51	3.40	3.31	3.17	3.03	2.88	2.80	2.72	2.64	2.55	2.46	2.36
22	7.95	5.72	4.82	4.31	3.99	3.76	3.59	3.45	3.35	3.26	3.12	2.98	2.83	2.75	2.67	2.58	2.50	2.40	2.31
23	7.88	5.66	4.76	4.26	3.94	3.71	3.54	3.41	3.30	3.21	3.07	2.93	2.78	2.70	2.62	2.54	2.45	2.35	2.26
24	7.82	5.61	4.72	4.22	3.90	3.67	3.50	3.36	3.26	3.17	3.03	2.89	2.74	2.66	2.58	2.49	2.40	2.31	2.21
25	7.77	5.57	4.68	4.18	3.86	3.63	3.46	3.32	3.22	3.13	2.99	2.85	2.70	2.62	2.53	2.45	2.36	2.27	2.17
30	7.56	5.39	4.51	4.02	3.70	3.47	3.30	3.17	3.07	2.98	2.84	2.70	2.55	2.47	2.39	2.30	2.21	2.11	2.01
40	7.31	5.18	4.31	3.83	3.51	3.29	3.12	2.99	2.89	2.80	2.66	2.52	2.37	2.29	2.20	2.11	2.02	1.92	1.80
60	7.08	4.98	4.13	3.65	3.34	3.12	2.95	2.82	2.72	2.63	2.50	2.35	2.20	2.12	2.03	1.94	1.84	1.73	1.60
120	6.85	4.79	3.95	3.48	3.17	2.96	2.79	2.66	2.56	2.47	2.34	2.19	2.03	1.95	1.86	1.76	1.66	1.53	1.38
∞	6.63	4.61	3.78	3.32	3.02	2.80	2.64	2.51	2.41	2.32	2.18	2.04	1.88	1.79	1.70	1.59	1.47	1.32	1.00

* This table is reproduced from M. Merrington and C. M. Thompson, "Tables of percentage points of the inverted beta (*F*) distribution," *Biometrika*, Vol. 33 (1943), by permission of the *Biometrika* trustees.

TABLE VII

FACTORIALS

n	$n!$	$\log n!$
0	1	0.0000
1	1	0.0000
2	2	0.3010
3	6	0.7782
4	24	1.3802
5	120	2.0792
6	720	2.8573
7	5,040	3.7024
8	40,320	4.6055
9	362,880	5.5598
10	3,628,800	6.5598
11	39,916,800	7.6012
12	479,001,600	8.6803
13	6,227,020,800	9.7943
14	87,178,291,200	10.9404
15	1,307,674,368,000	12.1165

BINOMIAL COEFFICIENTS

n	$\binom{n}{0}$	$\binom{n}{1}$	$\binom{n}{2}$	$\binom{n}{3}$	$\binom{n}{4}$	$\binom{n}{5}$	$\binom{n}{6}$	$\binom{n}{7}$	$\binom{n}{8}$	$\binom{n}{9}$	$\binom{n}{10}$
0	1										
1	1	1									
2	1	2	1								
3	1	3	3	1							
4	1	4	6	4	1						
5	1	5	10	10	5	1					
6	1	6	15	20	15	6	1				
7	1	7	21	35	35	21	7	1			
8	1	8	28	56	70	56	28	8	1		
9	1	9	36	84	126	126	84	36	9	1	
10	1	10	45	120	210	252	210	120	45	10	1
11	1	11	55	165	330	462	462	330	165	55	11
12	1	12	66	220	495	792	924	792	495	220	66
13	1	13	78	286	715	1287	1716	1716	1287	715	286
14	1	14	91	364	1001	2002	3003	3432	3003	2002	1001
15	1	15	105	455	1365	3003	5005	6435	6435	5005	3003
16	1	16	120	560	1820	4368	8008	11440	12870	11440	8008
17	1	17	136	680	2380	6188	12376	19448	24310	24310	19448
18	1	18	153	816	3060	8568	18564	31824	43758	48620	43758
19	1	19	171	969	3876	11628	27132	50388	75582	92378	92378
20	1	20	190	1140	4845	15504	38760	77520	125970	167960	184756

TABLE VIII

VALUES OF e^x AND e^{-x}

x	e^x	e^{-x}	x	e^x	e^{-x}
0.0	1.000	1.000	**2.5**	12.18	0.082
0.1	1.105	0.905	**2.6**	13.46	0.074
0.2	1.221	0.819	**2.7**	14.88	0.067
0.3	1.350	0.741	**2.8**	16.44	0.061
0.4	1.492	0.670	**2.9**	18.17	0.055
0.5	1.649	0.607	**3.0**	20.09	0.050
0.6	1.822	0.549	**3.1**	22.20	0.045
0.7	2.014	0.497	**3.2**	24.53	0.041
0.8	2.226	0.449	**3.3**	27.11	0.037
0.9	2.460	0.407	**3.4**	29.96	0.033
1.0	2.718	0.368	**3.5**	33.12	0.030
1.1	3.004	0.333	**3.6**	36.60	0.027
1.2	3.320	0.301	**3.7**	40.45	0.025
1.3	3.669	0.273	**3.8**	44.70	0.022
1.4	4.055	0.247	**3.9**	49.40	0.020
1.5	4.482	0.223	**4.0**	54.60	0.018
1.6	4.953	0.202	**4.1**	60.34	0.017
1.7	5.474	0.183	**4.2**	66.69	0.015
1.8	6.050	0.165	**4.3**	73.70	0.014
1.9	6.686	0.150	**4.4**	81.45	0.012
2.0	7.389	0.135	**4.5**	90.02	0.011
2.1	8.166	0.122	**4.6**	99.48	0.010
2.2	9.025	0.111	**4.7**	109.95	0.009
2.3	9.974	0.100	**4.8**	121.51	0.008
2.4	11.023	0.091	**4.9**	134.29	0.007

TABLE VIII (continued)

x	e^x	e^{-x}	x	e^x	e^{-x}
5.0	148.4	0.0067	**7.5**	1,808.0	0.00055
5.1	164.0	0.0061	**7.6**	1,998.2	0.00050
5.2	181.3	0.0055	**7.7**	2,208.3	0.00045
5.3	200.3	0.0050	**7.8**	2,440.6	0.00041
5.4	221.4	0.0045	**7.9**	2,697.3	0.00037
5.5	244.7	0.0041	**8.0**	2,981.0	0.00034
5.6	270.4	0.0037	**8.1**	3,294.5	0.00030
5.7	298.9	0.0033	**8.2**	3,641.0	0.00027
5.8	330.3	0.0030	**8.3**	4,023.9	0.00025
5.9	365.0	0.0027	**8.4**	4,447.1	0.00022
6.0	403.4	0.0025	**8.5**	4,914.8	0.00020
6.1	445.9	0.0022	**8.6**	5,431.7	0.00018
6.2	492.8	0.0020	**8.7**	6,002.9	0.00017
6.3	544.6	0.0018	**8.8**	6,634.2	0.00015
6.4	601.8	0.0017	**8.9**	7,332.0	0.00014
6.5	665.1	0.0015	**9.0**	8,103.1	0.00012
6.6	735.1	0.0014	**9.1**	8,955.3	0.00011
6.7	812.4	0.0012	**9.2**	9,897.1	0.00010
6.8	897.8	0.0011	**9.3**	10,938	0.00009
6.9	992.3	0.0010	**9.4**	12,088	0.00008
7.0	1,096.6	0.0009	**9.5**	13,360	0.00007
7.1	1,212.0	0.0008	**9.6**	14,765	0.00007
7.2	1,339.4	0.0007	**9.7**	16,318	0.00006
7.3	1,480.3	0.0007	**9.8**	18,034	0.00006
7.4	1,636.0	0.0006	**9.9**	19,930	0.00005

TABLE IX

SAMPLE PAGES OF RANDOM NUMBERS

17623	47441	27821	91845	01654	50375	23941	44848
45054	58410	92081	97624	73750	68343	40727	81203
73700	58730	06111	64486	64163	22132	22896	14305
58374	03905	06865	95353	88445	85544	23627	79176
04981	17531	97372	39558	94180	71108	19121	11958
45639	02485	43905	01823	11433	12220	36719	35435
98832	38188	24080	24519	61838	68801	49856	21739
66638	03619	90906	95370	59908	68103	36855	19127
77580	87772	86877	57085	41018	69556	06402	03436
67125	98175	00912	11246	85095	01581	92299	06166
83808	98997	71829	99430	59705	78103	66740	41743
56462	58166	97302	86828	75094	55208	77905	20705
52728	15101	72070	33706	78425	68672	79455	94334
44873	35302	04511	38088	89088	86918	20787	05691
67352	41526	23497	75440	83345	95889	39333	86027
35370	14915	16569	54945	62896	00342	66647	57096
06081	74957	87787	68849	96498	38270	80532	54307
58478	99297	43519	62410	30974	47335	04918	42974
21211	77299	74967	99038	57901	06163	99162	53285
98964	64425	33536	15079	88494	80633	47785	53996
81496	23996	56872	71401	34883	00045	89682	86664
71361	41989	92589	69788	24373	46438	28935	63903
36341	20326	37489	34626	16828	79262	23678	05509
34183	22856	18724	60422	33723	27646	92335	87136
98272	13969	12429	03093	01542	75066	73921	97188
75689	70722	88553	83300	00100	12787	74100	95536
64204	95212	31320	03783	82697	03389	19303	21646
16574	42305	56300	84227	28137	17549	22698	72955
93552	74363	30951	41367	02248	21570	33796	83789
48907	79840	34607	62668	56175	82515	23348	42207
42569	82391	20435	79306	80020	21622	67659	07878
81618	15125	48087	01250	20271	23094	48372	77621
31413	33756	15218	81976	38734	98044	02658	90698
77600	15175	67415	88801	48183	24263	49297	32923
42009	78616	45210	73186	48163	34158	03177	51696
02955	84348	46436	77911	45658	15024	66664	18730
93611	93346	71212	24405	71128	15524	55666	14763
11509	95853	02747	61889	19041	42899	49464	93965
63899	30932	90572	98971	32672	67506	93040	94527
73581	76780	03842	64009	15823	48310	04391	15521
49604	14809	12317	78062	82810	18981	62581	31642
81825	76822	87170	77235	74772	80840	05816	29023
60534	44842	16954	99466	52931	38199	85632	23761
41788	74409	76177	55519	95395	87644	09722	99251
73818	74454	02371	94693	76695	33451	57139	90612
81585	55439	98095	55578	83560	50374	04410	57272
21858	39489	39251	70450	28355	62002	85994	35807
99438	68184	62119	20229	84684	54861	41330	66808
03249	74135	43003	63132	21135	92001	43896	55887
59638	31226	89860	45191	24236	01536	43897	41294

TABLE IX (continued)

79154	30193	15271	93296	02946	96520	81881	56247
91727	06463	12248	57567	85697	62000	87957	07258
45642	71580	21558	66457	26734	68426	52067	23123
32502	34568	78777	35179	47829	32353	95941	72169
33308	75930	24865	17426	76603	99339	40571	41186
77727	78493	94580	70091	47526	26522	11045	83565
95701	11180	73936	17628	70100	85732	19741	92951
56637	06383	83182	76927	86819	50200	50889	06493
89803	95604	55735	66978	41614	30074	23403	03656
71078	31823	64316	95567	17930	26194	53836	53692
69177	59277	47629	63874	24649	31845	25736	75231
32408	16630	01242	63119	79899	34061	54308	59358
23292	90422	75540	15843	76801	49594	81002	30397
97309	96673	16389	51437	62567	08480	61873	63162
24680	51909	97230	58136	49723	15275	09399	11211
84913	67895	18804	17691	42658	70183	89417	57676
07885	38892	50990	96766	65080	35569	79392	14937
19294	72581	77377	04652	02906	38119	72407	71427
27341	02507	41858	08436	75153	86376	63852	60557
57058	04222	54488	12019	14192	49525	78844	13664
92195	42593	56488	35402	32059	11548	86264	74406
14722	10715	58795	42800	81716	80301	96704	57214
23733	95318	77730	87614	43315	50483	02950	09611
88062	21506	01750	71326	27510	10769	09921	46721
31250	41996	31680	41783	81782	04769	36716	82519
42359	01761	28842	71562	19975	48346	91029	78902
22532	81701	03425	28914	98356	76855	18769	52843
59849	02370	02784	13711	29708	17814	31556	68610
72981	96423	68791	91684	88014	27583	78167	25057
87644	57353	90349	16448	94491	19238	17396	10592
17586	65524	20162	04712	56957	05072	53948	07850
32889	19595	66500	28064	50915	31924	80621	17495
72563	15076	23780	52815	49631	93771	80200	84622
94406	63865	44336	27224	99683	58162	45516	39761
57795	31725	14403	29856	86017	20264	94618	85979
40671	92727	68626	81631	77339	64605	82583	85011
13729	51708	54104	81331	61714	57933	37342	26000
14960	88896	72784	82054	15232	48027	15832	62924
31556	80163	80203	90928	41447	34275	10779	83515
79255	69253	60254	01653	23244	43524	16382	36340
42693	78972	60322	90462	53460	83542	25224	70378
67410	12916	87933	78840	53442	16897	61578	05032
99084	48028	07184	41635	55548	19096	04130	23104
97129	70847	91864	08549	18185	69329	02340	63111
11918	90871	60965	23555	02372	45690	38595	23121
36705	51302	93147	29479	51715	35492	61371	87132
84810	14186	51153	78998	24717	16786	42786	86985
65231	14168	45193	27156	78022	32604	87259	93702
35319	03793	60344	95970	35995	08275	62405	43313
45551	46877	58631	82654	29192	86922	31908	42703

ANSWERS TO ODD-NUMBERED EXERCISES

In numerical exercises involving extensive calculations, the reader may get answers differing somewhat from those given here due to rounding at various intermediate stages.

Page 14

5. Postulates 5, 6, 4, 2, 5, and 6.
7. Write $A \cap (B_1 \cup B_2 \cup B_3)$ as $A \cap [(B_1 \cup B_2) \cup B_3]$ and then use the first distributive law twice.
9. (a) (0, 0), (1, 0), (2, 0), (3, 0), (4, 0), (5, 0), (0, 1), (0, 2), (0, 3), (0, 4); (b) (0, 2), (1, 1), (2, 0), (0, 4), (1, 3), (2, 2), (3, 1), (4, 0), (2, 4), (3, 3), (4, 2) (5, 1); (c) (0, 0), (1, 1), (2, 2), (3, 3), (4, 4); (d) (0, 0), (1, 0), (2, 0), (3, 0), (4, 0), (5, 0), (0, 1), (0, 2), (0, 3), (0, 4), (1, 1), (1, 3), (2, 2), (3, 1), (2, 4), (3, 3), (4, 2), (5, 1); (e) (0, 4), (0, 2), (2, 0), (4, 0); (f) (0, 2), (1, 1), (2, 0), (0, 4), (1, 3), (2, 2), (3, 1), (4, 0), (2, 4), (3, 3), (4, 2), (5, 1), (0, 0), (4, 4); (g) (0, 4), (0, 3), (0, 2), (0, 1), (0, 0), (1, 0), (2, 0), (3, 0), (4, 0), (5, 0), (1, 4), (1, 2), (2, 3), (2, 1), (3, 4), (3, 2), (4, 4), (4, 3), (4, 1), (5, 4), (5, 3), (5, 2); (h) (0, 4), (0, 3), (0, 2), (0, 1), (1, 0), (2, 0), (3, 0), (4, 0), (5, 0); (i) (1, 4), (1, 2), (2, 3), (2, 1), (3, 4), (3, 2), (4, 4), (4, 3), (4, 1), (5, 4), (5, 3), (5, 2).
11. (a) {Car 5, Car 6, Car 7, Car 8}; (b) {Car 2, Car 4, Car 5, Car 7}; (c) {Car 1, Car 8}; (d) {Car 3, Car 4, Car 7, Car 8}; (e) he chooses a car with air-conditioning; (f) he chooses a car which has either no power steering or it does have bucket seats; (g) he chooses a 2 or 3 year-old car with bucket seats; (h) same as part (g).
13. 45.
15. (a) 12; (b) 6; (c) 20.

Page 29

9. (b) 1, 6, 15, 20, 15, 6, 1; 1, 7, 21, 35, 35, 21, 7, 1.
11. (a) 28; (b) $\binom{n+r-1}{n-1}$.
13. (a) $-15/384$ and -10.
15. (b) 4; (c) 4.
17. (b) 240; (c) 160; (d) 120.
19. 60 and 120.
21. $4^{12} = 16{,}777{,}216$.
23. 492,960.
25. 40,320.
27. (a) 1,140; (b) 184,756; (c) 211.
29. 47,520.
31. $\frac{1145}{512} = 2.236$.

Page 46

5. (a) 0.67; (b) 0.58; (c) 0.75; (d) 0; (e) 0.33; (f) 0.25.
9. Mr. Jones.
11. The probability is at least $\frac{5}{8}$ but less than $\frac{3}{4}$.
13. (a) The sum of the probabilities exceeds 1; (b) the sum of the probabilities is less than 1; (c) the first probability cannot exceed the second; (d) the second probability cannot exceed the first; (e) the given probabilities yield $0.72 + 0.85 - 0.51 = 1.06$ for the probability that the person will see the giraffes or the bears, and, hence, they are inconsistent.
15. (a) 0.56; (b) 0.54; (c) 0.66.
17. (a) 0.41; (b) 0.44; (c) 0.37.
19. (a) $\frac{1}{26}$; (b) $\frac{1}{2}$; (c) $\frac{4}{13}$; (d) $\frac{1}{13}$.
21. 0.34.
23. (a) $\frac{167}{180}$; (b) $\frac{13}{180}$.
25. 0.98.

Page 63

9. $\frac{11}{23}$.
11. (a) $\frac{52}{74}$; (b) $\frac{34}{52}$; (c) $\frac{24}{88}$; (d) $\frac{12}{30}$.
13. $\frac{91}{323}$.
17. 0.113.
19. $\frac{148}{600}$.
21. 0.84
23. (a) $\frac{23}{148}$; (b) $\frac{147}{452}$; (c) $\frac{5}{9}$; (d) $\frac{8}{33}$.

Page 83

13. $p(x+1) = \dfrac{\lambda}{x+1}\cdot p(x)$.
17. 0.5905, 0.3280, 0.0729, 0.0081, 0.0004.
19. (a) 0.2254; (b) 0.2254.
21. (a) 0.0008; (b) 0.7454; (c) 0.0479; (d) 0.0354.
23. (a) $\frac{6}{9}$, $\frac{2}{9}$, and $\frac{1}{9}$; (b) 0.645, 0.300, and 0.054.
25. (a) $\frac{57}{92}$; (b) $\frac{91}{276}$; (c) $\frac{3}{23}$.
27. 0.299; binomial approximation yields 0.279, namely, an error of 0.020.
29. (a) 0.059; (b) 0.0645; (c) 0.048.
31. 0.30.
33. (a) 0.1221; (b) 0.7148; (c) 0.0425.
35. (a) 0.1653; (b) 0.1087; (c) 0.7306.
37. (a) 0.1420; (b) 0.0674; (c) 0.0358; (d) 0.7315.

Page 96

1. (a) $g(1) = \frac{8}{30}$, $g(2) = \frac{10}{30}$, $g(3) = \frac{12}{30}$; (b) $h(1) = \frac{12}{30}$, $h(2) = \frac{10}{30}$, $h(3) = \frac{8}{30}$; (c) $\varphi(1|2) = 0.1$, $\varphi(2|2) = 0.5$, $\varphi(3|2) = 0.4$; (d) $\pi(1|1) = \frac{3}{4}$, $\pi(2|1) = \frac{1}{8}$, $\pi(3|1) = \frac{1}{8}$; the random variables are not independent.
3. (a) $m(x, y) = \dfrac{xy}{36}$ for $x = 1, 2, 3$ and $y = 1, 2, 3$; (b) $n(x, z) = \dfrac{xz}{18}$ for $x = 1, 2, 3$ and $z = 1, 2$; (c) $g(x) = \dfrac{x}{6}$ for $x = 1, 2, 3$; (d) $\varphi(z|x, y) = \dfrac{z}{3}$ for $z = 1, 2$; (e) $\psi(y, z|x) = \dfrac{yz}{18}$ for $y = 1, 2, 3$ and $z = 1, 2$.
5. (a) $\frac{9}{16}$; (b) $\frac{9}{25}$; (c) $\frac{1}{4}$; (d) $\frac{3}{5}$.
11. 0.026.
13. 0.006.
15. (a) $\frac{6720}{125970}$; (b) $\frac{840}{125970}$.

Page 108

1. (a) $\frac{3}{4}$, $\frac{21}{25}$, and $\frac{1}{16}$; (b) $f(x) = 8/x^3$ for $x > 2$ and $f(x) = 0$ elsewhere.
3. (a) $k = 2$; (b)
$$F(x) = \begin{cases} 0 & \text{for } x \le 0 \\ 2x - x^2 & \text{for } 0 < x < 1 \\ 1 & \text{for } x \ge 1 \end{cases}$$
(c) $\frac{3}{4}$.
5.
$$F(x) = \begin{cases} 0 & \text{for } x \le 0 \\ \frac{1}{2}x & \text{for } 0 < x < \frac{1}{2} \\ \frac{3}{4} & \text{for } x = \frac{1}{2} \\ \frac{1}{2}(1+x) & \text{for } \frac{1}{2} < x < 1 \\ 1 & \text{for } x \ge 1 \end{cases}$$
7. (a) $\frac{325}{864}$; (b) $\frac{45}{288}$; (c) $\frac{160}{864}$; (d) 0.

Page 116

5. (a) 0.405; (b) 0.92.
11. $k = \frac{a}{\pi}$.
15. $\frac{1}{2}$.
17. 0.054.
19. 0.06.
21. 0.18.

Page 126

1. $f(y) = \theta(1-\theta)^{\frac{1-y}{2}}$ for $y = 1, -1, -3, -5, -7, \ldots$.
3. (a) $f(0) = \frac{1}{2}$ and $f(1) = \frac{1}{2}$; (b) $g(-1) = \frac{1}{9}$, $g(0) = \frac{6}{9}$, and $g(1) = \frac{2}{9}$.
5. $f(y) = \frac{1}{y\sqrt{2\pi}} \cdot e^{-\frac{1}{2}(\ln y)^2}$ for $y > 0$ and $f(y) = 0$ elsewhere.
7. Beta distribution with $\alpha = 4$ and $\beta = 2$; $k = 320$.
9. -0.77, -0.44, -0.74, 0.44.

Page 136

1. (a) $\frac{1}{2}$; (b) $\frac{5}{16}$; (c) $\frac{5}{9}$.
3. (a) 0.849; (b) 0.982; (c) 0.964; (d) 0.865; (e) 0.002.
5. (a) $\frac{1}{128}$; (b) $\frac{133}{3888}$; (c) $\frac{427}{65536}$.
7. (a) $f(x_2, x_3) = (x_2 + \frac{1}{2})e^{-x_3}$ for $0 < x_2 < 1$ and $x_3 > 0$, and $f(x_2, x_3) = 0$ elsewhere; (b) $n(x_1) = \frac{1}{2} + x_1$ for $0 < x_1 < 1$, and $n(x_1) = 0$ elsewhere; (c) $r(x_3) = e^{-x_3}$ for $x_3 > 0$, and $r(x_3) = 0$ elsewhere.
9. $\psi(x_2|x_1) = \frac{x_1 + 2x_2}{x_1 + 1}$ for $0 < x_2 < 1$, and $\psi(x_2|x_1) = 0$ elsewhere.
11. (a) Independent; (b) not independent.
13. (a) $g(p) = 5$ for $0.2 < p < 0.4$, and $g(p) = 0$ elsewhere; (b) $\varphi(s|p) = pe^{-ps}$ for $s > 0$, and $\varphi(s|p) = 0$ elsewhere; (c) 0.607.

Page 144

3. (a) 0.548; (b) 0.548.
7. \$1.20.
9. \$600.
11. (a) \$1.50; (b) \$1.67; (c) \$1.60.
13. 20,000 miles.
15. 0.60.

Page 153

1. $\mu = 0$, $\mu_2' = 4$, and $\sigma^2 = 4$.
3. $\sigma^2 = \frac{35}{12}$.
9. (a) $\alpha_3 = 0$; (b) $\alpha_3 = -0.17$.
13. The probability is less than or equal to σ^2/c^2.
17. The probability of getting less than 1.36 or more than 5.64 is less than or equal to 0.64; the actual probability is $\frac{1}{3}$.

Page 168

5. (c) $F_{\mathbf{x}}(t) = [1 + \theta(t - 1)]^n$; (d) $F_{\mathbf{x}}(t) = e^{\lambda(t-1)}$.
9. $\mu_2 = \lambda$, $\mu_3 = \lambda$, $\mu_4 = \lambda + 3\lambda^2$
25. The probability is less than or equal to 0.16.
27. (a) 2.26; (b) -1.34; (c) -0.62; (d) 1.25.
29. (a) 0.1056; (b) 0.0122; (c) 0.3830.
31. (a) 0.0401; (b) 0.7486; (c) 0.4649.

Page 177

3. (a) 0.12; (b) 0.17.
5. 0.53.
7. 0.014.

Page 180

3. (a) $-\frac{1}{144}$; (b) 0.
5. (a) not independent; (b) 0.

Page 189

1. (a) $f(y) = \dfrac{1}{\theta_1 - \theta_2}[e^{-y/\theta_1} - e^{-y/\theta_2}]$ for $y > 0$, and $f(y) = 0$ elsewhere;

   (b) if $\theta_1 = \theta_2 = \theta$, $f(y) = \dfrac{1}{\theta^2}\cdot ye^{-y/\theta}$ for $y > 0$, and $f(y) = 0$ elsewhere.
3. (a) $F(y) = 0$; (b) $F(y) = \frac{1}{2}y^2$; (c) $F(y) = 2y - \frac{1}{2}y^2 - 1$; (d) $F(y) = 1$.

Page 193

3. $M_{\mathbf{y}}(t) = (1 - \beta t)^{-\alpha n}$; a gamma distribution with the parameters αn and β.
7. (a) 0.2125; (b) 0.6265; (c) 0.0050.
9. (a) 0.08; (b) 0.29; (c) 0.54.

Page 201

1. (a) $\mu = 27$ and $\sigma^2 = 272$; (b) $\mu = 2$ and $\sigma^2 = 102$.
3. $\text{var}(\mathbf{x} + \mathbf{y}) = \text{var}(\mathbf{x}) + \text{var}(\mathbf{y}) + 2\text{cov}(\mathbf{x}, \mathbf{y})$; $\text{var}(\mathbf{x} - \mathbf{y}) = \text{var}(\mathbf{x}) + \text{var}(\mathbf{y}) - 2\text{cov}(\mathbf{x}, \mathbf{y})$; $\text{cov}(\mathbf{x} + \mathbf{y}, \mathbf{x} - \mathbf{y}) = \text{var}(\mathbf{x}) - \text{var}(\mathbf{y})$.
9. $\mu = 424.5$ inches and $\sigma = 0.74$ inches.
11. $\mu = 4.75$ and $\sigma = 1.58$.
13. The probability is less than or equal to $\frac{1}{4}$.
15. It will fall between -0.51 and 0.61.

Page 208

5. Table III does not give a value for $z = 4$, but the probability exceeds 0.9999.

Page 216

11. (a) 0.036; (b) 0.0485.
13. (a) 0.0228; (b) 0.76; (c) 0.24.
15. 0.8414.
17. 0.2158.
19. 0.025.

Page 223

11. $F = 4$; the data do not support the claim.
13. $t = -1.33$; the information seems to support the claim.

Page 227

3. $\mu = \dfrac{1}{n+1}$ and $\sigma^2 = \dfrac{n}{(n+1)^2(n+2)}$.
5. Without replacement: $g_1(1) = \frac{4}{10}$, $g_1(2) = \frac{3}{10}$, $g_1(3) = \frac{2}{10}$, $g_1(4) = \frac{1}{10}$; with replacement: $g_1(1) = \frac{9}{25}$, $g_1(2) = \frac{7}{25}$, $g_1(3) = \frac{5}{25}$, $g_1(4) = \frac{3}{25}$, $g_1(5) = \frac{1}{25}$.
7. $h(y_1, R) = n(n-1)f(y_1)f(y_1 + R)\left[\int_{y_1}^{y_1+R} f(x)\,dx\right]^{n-2}$.

9. $g(R) = n(n-1)(1-R)R^{n-2}$ for $0 < R < 1$ and $g(R) = 0$ elsewhere;

$$\mu = \frac{n-1}{n+1} \text{ and } \sigma^2 = \frac{2(n-1)}{(n+1)^2(n+2)}.$$

13. $(2.197)(0.9)^9 = 0.085$.
15. 0.625.

Page 239

3. Continue drilling.
5. (a) Candy; (b) wine; (c) candy.
7. (a) Choose Hotel B; (b) send candy.
9. Saddle points are (a) at I and 2, (b) at II and 1, (c) at I and 1; (d) at I and 2.
15. $\frac{1}{6}$ and $\frac{5}{6}$; the value is $10,333,333.

Page 250

1. (a)

	$a_1(\theta = 0)$	$a_2(\theta = \frac{1}{2})$	$a_3(\theta = 1)$
$\theta_1(\theta = 0)$	0	50	100
$\theta_2(\theta = \frac{1}{2})$	50	0	50
$\theta_3(\theta = 1)$	100	50	0

(b) $d_1(0) = 0$ and $d_1(1) = 0$, $d_2(0) = 0$ and $d_2(1) = \frac{1}{2}$, $d_3(0) = 0$ and $d_3(1) = 1$, $d_4(0) = \frac{1}{2}$ and $d_4(1) = 0$, $d_5(0) = \frac{1}{2}$ and $d_5(1) = \frac{1}{2}$, $d_6(0) = \frac{1}{2}$ and $d_6(1) = 1$, $d_7(0) = 1$ and $d_7(1) = 0$, $d_8(0) = 1$ and $d_8(1) = \frac{1}{2}$, $d_9(0) = 1$ and $d_9(1) = 1$; (c) only $\mathbf{d}_2$, $\mathbf{d}_3$, $\mathbf{d}_5$, and $\mathbf{d}_6$ are admissible, and in each case the maximum value of the risk function equals 50; (d) $\mathbf{d}_3$.

5. $k = \dfrac{\theta_1\theta_2}{\sqrt{\theta_1^2 + \theta_2^2}}$.

Page 264

11. From $\frac{1}{2} - \frac{1}{2}\sqrt{\frac{n+1}{2n+1}}$ to $\frac{1}{2} + \frac{1}{2}\sqrt{\frac{n+1}{2n+1}}$
15. (a) Biased; (b) consistent.

Page 270

1. $2m_1'$.
3. (a) $\dfrac{m_1'}{1 - m_1'}$; (b) $\dfrac{-n}{\sum_{i=1}^{n} \ln x_i}$.
5. $\sqrt{\dfrac{\sum_{i=1}^{n} (x_i - \mu)^2}{n}}$
7. The smallest sample value.
9. $\dfrac{1}{n\alpha} \cdot \sum_{i=1}^{n} x_i$.

Page 277

1. $k = \dfrac{-1}{\ln(1 - \alpha)}$.
3. $c = \dfrac{1 + \sqrt{1 - \alpha}}{\alpha}$
5. $\bar{x}_1 - \bar{x}_2 \pm z_{\alpha/2} \sqrt{\dfrac{\sigma_1^2}{n_1} + \dfrac{\sigma_2^2}{n_2}}$

11. $31.44 < \mu < 33.56$.
13. $152{,}579 < \mu < 184{,}379$.
15. (a) $0.39 < \theta < 0.45$; (b) $0.39 < \theta < 0.45$.
17. (a) $0.16 < \theta < 0.22$; (b) $0.15 < \theta < 0.23$.
19. $n = 4{,}161$.
21. $7{,}544 < \sigma < 34{,}184$.
23. $7.24 < \sigma^2 < 18.23$.

Page 285

5. 0.29.
7. (a) 0.48; (b) 696.4 − 733.6.
9. (a) 112; (b) 100; (c) 108.

Page 297

1. (a) Simple; (b) composite; (c) composite; (d) composite; (e) simple; (f) composite; (g) composite; (h) composite.
5. 0.223 and 0.451.
7. The critical region is $\sum_{i=1}^{n} x_i \geq K$, where K can be evaluated by making use of the fact that $\sum_{i=1}^{n} \mathbf{x}_i$ has a gamma distribution with $\alpha = n$ and $\beta = \theta_0$.
9. 0.45.
11. $\sum_{i=1}^{n} x_i^2 \geq \sigma_0^2 \cdot \chi^2_{\alpha,n}$.
13. He would be committing a Type I error if he erroneously concluded that the executive cannot take on additional responsibilities; he would be committing a Type II error if he erroneously concluded that the executive can take on additional responsibilities.

Page 309

1. (a) 0, 0, and $\frac{1}{21}$; (b) $\frac{18}{21}$, $\frac{15}{21}$, $\frac{11}{21}$, $\frac{6}{21}$, and 0.
3. (a) 0.852; (b) 0.016, 0.085, 0.145, 0.134, 0.121.
5. (a) 0.0375, 0.0203, 0.0107, 0.0055, 0.0027; (b) 0.9329, 0.7585, 0.3840, 0.0419.

Page 321

3. $n = 891$.
5. $z = -1.90$; the figure 4,200 is too high.
7. $t = -1.17$; the null hypothesis cannot be rejected.
9. $z = 1.44$; the null hypothesis cannot be rejected.
11. $t = 3.25$; the difference is significant.
13. $t = -0.50$; the null hypothesis cannot be rejected.

Page 326

3. $\chi^2 = 33.5$; reject the null hypothesis.
5. (a) $\chi^2 = 23$; the null hypothesis cannot be rejected; (b) $z = -0.8$; the null hypothesis cannot be rejected.
7. $F = 2.47$; assumption was reasonable.
9. $F = 3.86$; reject the null hypothesis.

Page 332

7. $k_{.01} = 11$; 0.95, 0.87, and 0.75.
9. $k_{.005} = 12$ and $k'_{.005} = 1$; 0.97, 0.99, 0.98, and 0.96.
11. $z = -1.52$; the null hypothesis $\theta = 0.70$ cannot be rejected.
13. $z = 3$; reject the null hypothesis $\theta_1 = \theta_2$.
15. $\chi^2 = 0.7$; the difference is not significant.
17. $\chi^2 = 0.75$; null hypothesis cannot be rejected (in fact, the data strongly support it).

Page 338

3. $\chi^2 = 38$; reject the null hypothesis.
5. $\chi^2 = 1.6$; no significant difference in quality.
7. $\chi^2 = 29.2$; reject the null hypothesis.
9. $\chi^2 = 5.8$; null hypothesis cannot be rejected (good fit).

Page 349

3. Reject the null hypothesis.
5. Null hypothesis cannot be rejected.
7. $z = 3.4$; reject the null hypothesis.
9. Null hypothesis cannot be rejected.
11. (a) Null hypothesis cannot be rejected; (b) null hypothesis cannot be rejected.
13. Null hypothesis cannot be rejected.

Page 355

3. $\frac{11}{42}$.
7. $z = -3.2$; not random.
9. $z = -0.96$; cannot reject hypothesis of randomness.

Page 367

3. $\frac{1+x}{2}$, $\frac{(1-x)^2}{12}$, and $\frac{2}{3}y$.
5. 2.3 and $\frac{11}{8}$.
13. $$\begin{aligned} \Sigma\, y &= n\cdot\beta_0 + \beta_1(\Sigma\, x) + \beta_2(\Sigma\, x^2) \\ \Sigma\, xy &= \beta_0(\Sigma\, x) + \beta_1(\Sigma\, x^2) + \beta_2(\Sigma\, x^3) \\ \Sigma\, x^2y &= \beta_0(\Sigma\, x^2) + \beta_1(\Sigma\, x^3) + \beta_2(\Sigma\, x^4) \end{aligned}$$

15. $\Sigma z = n\alpha + \beta_1(\Sigma x) + \beta_2(\Sigma y)$
$\Sigma xz = \alpha(\Sigma x) + \beta_1(\Sigma x^2) + \beta_2(\Sigma xy)$
$\Sigma yz = \alpha(\Sigma y) + \beta_1(\Sigma xy) + \beta_2(\Sigma y^2)$
17. (a) $y = -5.96 + 1.55x$; (b) $y = -5.96 + 1.55x$.
19. $y = 1.37(1.38)^x$.
21. $y = 10.5 - 2.0x + 0.2x^2$.

Page 376

3. $\mu_1 = -2$, $\mu_2 = 1$, $\sigma_1 = 10$, $\sigma_2 = 5$, and $\rho = 0.7$.
5. $\sigma_1 = 6$, $\sigma_2 = 3$, and $\rho = -0.50$.
9. (b) Bivariate normal distribution with zero means, unit variances, and zero covariance.
11. (a) 0.30; (b) 0.165.
13. (a) 14; (b) 0.87.

Page 387

3. $\dfrac{1 + r - (1 - r)e^{\pm 2z_{\alpha/2}/\sqrt{n-3}}}{1 + r + (1 - r)e^{\pm 2z_{\alpha/2}/\sqrt{n-3}}}$.
5. 0.93.
15. $r = 0.55$; significant.
17. $r = 0.73$; significant.
19. 0.48.
21. Null hypothesis cannot be rejected.
23. $1.33 < \beta < 1.77$.
25. $2.84 < \beta < 4.10$; results are the same.
27. $-8.58 < \alpha < -3.34$.
29. 37.3 − 63.3.

Page 402

7. $F = 1.92$; not significant.
9. $F = 3.59$; not significant.

Page 412

5. For the launchers F_{Tr}, $= 4.43$, which is not significant; for the fuels $F_B = 17.06$, which is significant.
7. For the varieties of wheat $F_{Tr} = 11.7$, which is significant; for the fertilizers $F_B = 21.3$, which is significant.
11. Only differences due to packaging are significant, $F = 42.9$.

INDEX